Evolving Connectionist Systems

Nikola Kasabov

Evolving Connectionist Systems

The Knowledge Engineering Approach

Second edition

Professor Nikola Kasabov, PhD, FRSNZ
Director and Chief Scientist
Knowledge Engineering and Discovery Research Institute
Auckland University of Technology
Auckland, New Zealand

British Library Cataloguing in Publication Data
A catalogue record for this book is available from the British Library

Library of Congress Control Number: 2006940182

ISBN 978-1-84628-345-1 e-ISBN 978-1-84628-347-5

Printed on acid-free paper

9 8 7 6 5 4 3 2 1

Springer Science+Business Media
springer.com

To my daughters Assia and Kapka,
for all their love, understanding, and support throughout my academic career.

Foreword I

This second edition provides fully integrated, up-to-date support for knowledge-based computing in a broad range of applications by students and professionals. Part I retains well-organized introductory chapters on modeling dynamics in evolving connectionist systems adapted through supervised, unsupervised, and reinforcement learning; it includes chapters on spiking neural networks, neuro-fuzzy inference systems, and evolutionary computation. Part II develops promising new and expanded applications in gene regulation, DNA-protein interactions, adaptive speech recognition, multimodal signal processing, and adaptive robotics. Emphasis is placed on multilayered adaptive systems in which the rules that govern parameter optimization are themselves subject to evolutionary pressures and modification in accord with strategic reformulation of pathways to problem solving. The human brain is treated both as a source of new concepts to be incorporated into knowledge engineering and as a prime target for application of novel techniques for analyzing and modeling brain data. Brains are material systems that operate simultaneously at all levels of scientific study, from quantum fields through cellular networks and neural populations to the dynamics of social, economic, and ecological systems. All these levels find place and orientation in this succinct presentation of universal tools, backed by an extended glossary, selected appendices, and indices referenced to Internet resources.

Professor Walter J. Freeman
University of California at Berkeley

Foreword II

This book is an important update on the first edition, taking account of exciting new developments in adaptive evolving systems. Evolving processes, through both individual development and population/generation evolution, inexorably led the human race to our supreme intelligence and our superior position in the animal kingdom. Nikola Kasabov has captured the essence of this powerful natural tool by adding various forms of adaptivity implemented by neural networks. The new edition of the book brings the treatment of the first edition to the cutting edge of modern research. At the same time Kasabov has kept the treatment to the two-part format of generic level and applications, with demonstrations showing how important problems can be handled by his techniques such as in gene and protein interactions and in brain imaging and modelling, as well as in other exciting areas. In all, this new edition is a very important book, and Nik should be congratulated on letting his enthusiasm shine through, but at the same time keeping his expertise as the ultimate guide. A must for all in the field!

Professor John G. Taylor
King's College London

Preface

This second edition of the book reflects on the new development in the area of computational intelligence and especially the adaptive evolving systems. Even though the structure of the book is preserved and lots of the material from the first edition is also included here, there are new topics that make the new edition a contemporary and an advanced reading in the field of computational intelligence. In terms of generic methods, these are: spiking neural networks, transductive neuro-fuzzy inference methods, personalised modelling, methods for integrating data and models, quantum inspired neural networks, neuro-genetic models, and others. In terms of applications, these are: gene-regulatory network modelling, computational neuro-genetic modelling, adaptive robots, and modelling adaptive socioeconomic and ecological systems.

The emphasis in the second edition is more on the following aspects.

1. Evolving intelligent systems: systems that apply evolving rules to evolve their structure, functionality, and knowledge through incremental adaptive learning and interaction, where the evolving rules may change as well
2. The knowledge discovery aspect of computational modelling across application areas such as bioinformatics, brain study, engineering, environment, and social sciences, i.e. the discovery of the evolving rules that drive the processes and the resulting patterns in the collected data
3. The interaction between different levels of information processing in one system, e.g. parameters (genes), internal clusters, and output data (behaviour)
4. Challenges for the future development in the field of computational intelligence (e.g. personalised modelling, quantum inspired neuro-computing, gene regulatory network discovery)

The book covers contemporary computational modelling and machine learning techniques and their applications, where in the core of the models are artificial neural networks and hybrid models (e.g. neuro-fuzzy) inspired by the evolving nature of processes in the brain, in proteins and genes in a cell, and by some quantum principles. The book also covers population-generation-based methods for optimisation of model parameters and features (variables), but the emphasis is on the learning and the development of the structure and the functionality of an individual model. In this respect, the book has a much wider scope than some earlier work on evolutionary (population-generation) based training of artificial neural networks, also called there 'evolving neural networks'.

The second edition of the book includes new applications to gene and protein interaction modelling, brain data analysis and brain model creation, computational neuro-genetic modelling, adaptive speech, image and multimodal recognition, language modelling, adaptive robotics, and modelling dynamic financial, socioeconomic, and ecological processes.

Overall, the book is more about problem solving and intelligent systems rather than about mathematical proofs of theoretical models. Additional resources for practical model creation, model validation, and problem solving, related to topics presented in some parts of the book, are available from: http://www.kedri.info/ and http://www.theneucom.com.

Evolving Connectionist Systems is aimed at students and practitioners interested in developing and using intelligent computational models and systems to solve challenging real-world problems in computer science, engineering, bioinformatics, and neuro-informatics. The book challenges scientists and practitioners with open questions about future directions for information sciences inspired by nature.

The book argues for a further development of humanlike and human-oriented information-processing methods and systems. In this context, 'humanlike' means that some principles from the brain and genetics are used for the creation of new computational methods, whereas 'human-oriented' means that these methods can be used to discover and understand more about the functioning of the brain and the genes, about speech and language, about image and vision, and about our society and our environment.

It is likely that future progress in many important areas of science (e.g. bioinformatics, brain science, information science, physics, communication engineering, and social sciences) can be achieved only if the areas of artificial intelligence, brain science, bioinformatics, and quantum physics share their methods and their knowledge. This book offers some steps in this direction. This book introduces and applies similar or identical fundamental information-processing methods to different domain areas. In this respect the conception of this work was inspired by the wonderful book by Douglas Hofstadter, *Godel, Escher, Bach: An Eternal Golden Braid* (1979), and by the remarkable *Handbook of Brain Science and Neural Networks,* edited by Michael Arbib (1995, 2002).

The book consists of two parts. The first part presents generic methods and techniques. The second part presents specific techniques and applications in bioinformatics, neuro-informatics, speech and image recognition, robotics, finance, economics, and ecology. The last chapter presents a new promising direction: quantum inspired evolving intelligent systems.

Each chapter of the book stands on its own. In order to understand the details of the methods and the applications, one may need to refer to some relevant entries in the extended glossary, or to a textbook on neural networks, fuzzy systems, and knowledge engineering (see, for example Kasabov (1996)). The glossary contains brief descriptions and explanations of many of the basic notions of information science, statistical analysis, artificial intelligence, biological neurons, brain organization, artificial neural networks, molecular biology, bioinformatics, evolutionary computation, etc.

This work was partially supported by the research grant AUTX02001 'Connectionist-based intelligent information systems', funded by the New Zealand

Foundation for Research, Science, and Technology and the New Economy Research Fund, and also by the Auckland University of Technology.

I am grateful for the support and encouragement I received from the editorial team of Springer-Verlag, London, especially from Professor John G. Taylor and the assistant editor Helen Desmond.

There are a number of people whom I would like to thank for their participation in some sections of the book. These are several colleagues, research associates, and postgraduate students I have worked with at the Knowledge Engineering and Discovery Research Institute in Auckland, New Zealand, in the period from 2002 till 2007: Dr. Qun Song, Mrs. Joyce D'Mello, Dr. Zeke S. Chan, Dr. Lubica Benuskova, Dr. Paul S. Pang, Dr. Liang Goh, Dr. Mark Laws, Dr. Richard Kilgour, Akbar Ghobakhlou, Simei Wysosky, Vishal Jain, Tian-Min Ma (Maggie), Dr. Mark Marshall, Dougal Greer, Peter Hwang, David Zhang, Dr. Matthias Futschik, Dr. Mike Watts, Nisha Mohan, Dr. Ilkka Havukkala, Dr. Sue Worner, Snjezana Soltic, Dr. DaDeng, Dr. Brendon Woodford, Dr. John R. Taylor, Prof. R. Kozma. I would like to thank again Mrs. Kirsty Richards who helped me with the first edition of the book, as most of the figures from the first edition are included in this second one.

The second edition became possible due to the time I had during my sabbatical leave in 2005/06 as a Guest Professor funded by the German DAAD (Deutscher Akademisher Austausch Dienst) organisation for exchange of academics, and hosted by Professor Andreas Koenig and his group at the TU Kaiserslautern.

I have presented parts of the book at conferences and I appreciate the discussions I had with a number of colleagues. Among them are Walter Freeman and Lotfi Zadeh, both from the University of California at Berkeley; Takeshi Yamakawa, Kyushu Institute of Technology; John G. Taylor, Kings College, London; Ceese van Leuwen and his team, RIKEN, Japan; Michael Arbib, University of Southern California; Dimiter Dimitrov, National Cancer Institute in Frederick, Maryland; Jaap van den Herik and Eric Postma, University of Maastricht; Wlodeck Duch, Copernicus University; Germano Resconi, Catholic University in Brescia; Alessandro Villa, University of Lausanne; UK; Peter Erdi, Budapest; Max Bremer, University of Plymouth; Bill Howell, National Research Council of Canada; Mario Fedrizzi, University of Trento in Italy; Plamen Angelov, the University of Lancaster, UK, Dimitar Filev, FORD; Bogdan Gabrich, University of Bournemouth Dr. G. Coghill, University of Auckland; Dr V. Brusic, Harvard; Prof. Jim Wright, Auckland and many more.

I remember a comment by Walter Freeman, when I first presented the concept of evolving connectionist systems (ECOS) at the Iizuka'98 conference in Japan: 'Throw the "chemicals" and let the system grow, is that what you are talking about, Nik?' After the same presentation at Iizuka'98, Robert Hecht-Nielsen made the following comment, 'This is a powerful method! Why don't you apply it to challenging real world problems?' Later on, in November, 2001, Walter Freeman made another comment at the ICONIP conference in Shanghai: 'Integrating genetic level and neuronal level in brain modelling and intelligent machines is a very important and a promising approach, but how to do that is the big question.' Michael Arbib said in 2004 'If you include genetic information in your models, you may need to include atomic information as well....'

Max Bremer commented after my talk in Cambridge, at the 25th anniversary of the AI SIG of the BCS in December 2005: 'A good keynote speech is the one that makes at least half of the audience abandon their previous research topics and start researching on the problems and topics presented by the speaker.'

All those comments encouraged me and at the same time challenged me in my research. I hope that some readers would follow on some of the techniques, applications, and future directions presented in the book, and later develop their own methods and systems, as the book offers many open questions and directions for further research in the area of evolving intelligent systems (EIS).

Nikola Kasabov
23 May 2007
Auckland

Contents

Abstract

This book covers contemporary computational modelling and machine-learning techniques and their applications, where in the core of the models are artificial neural networks and hybrid models (e.g. neuro-fuzzy) that evolve to develop their structure and functionality through incremental adaptive learning. This is inspired by the evolving nature of processes in the brain, the proteins, and the genes in a cell, and by some quantum principles. The book also covers population/generation-based optimisation of model parameters and features (variables), but the emphasis is on the learning and the development of the structure and the functionality of an individual model. In this respect, the book has a much wider scope than some earlier work on evolutionary (population/generation)-based training of artificial neural networks, called 'evolving neural networks'.

This second edition of the book includes new methods, such as online incremental feature selection, spiking neural networks, transductive neuro-fuzzy inference, adaptive data and model integration, cellular automata and artificial life systems, particle swarm optimisation, ensembles of evolving systems, and quantum inspired neural networks.

In this book new applications are included to gene and protein interaction modelling, brain data analysis and brain model creation, computational neuro-genetic modelling, adaptive speech, image and multimodal recognition, language modelling, adaptive robotics, modelling dynamic financial and socioeconomic structures, and ecological and environmental event prediction. The main emphasis here is on adaptive modelling and knowledge discovery from complex data.

A new feature of the book is the attempt to connect different structural and functional elements in a single computational model. It looks for inspiration at some functional relationships in natural systems, such as genetic and brain activity.

Overall, this book is more about problem solving and intelligent systems than about mathematical proofs of theoretical models. Additional resources for practical model creation, model validation, and problem solving, related to topics presented in some parts of the book, are available from http://www.kedri.info/ ->books, and from http://www.theneucom.com.

Evolving Connectionist Systems is aimed at students and practitioners interested in developing and using intelligent computational models and systems to solve challenging real-world problems in computer science, engineering, bioinformatics, and neuro-informatics. The book challenges scientists with open questions about future directions of information sciences.

PART I
Evolving Connectionist Methods

This part presents some existing connectionist and hybrid techniques for adaptive learning and knowledge discovery and also introduces some new evolving connectionist techniques. Three types of evolving adaptive methods are presented, namely unsupervised, supervised, and reinforcement learning. They include: evolving clustering, evolving self-organising maps, evolving fuzzy neural networks, spiking neural networks, knowledge manipulation, and structure optimisation with the use of evolutionary computation. The last chapter of this part, Chapter 7, suggests methods for data, information, and knowledge integration into multimodel adaptive systems and also methods for evolving ensembles of ECOS. The extended glossary at the end of the book can be used for a clarification of some of the used concepts.

Introduction

Modelling and Knowledge Discovery from Evolving Information Processes

This introductory chapter presents the main concepts used in the book and gives a justification for the development of this field. The emphasis is on a process/system evolvability based on evolving rules (laws). To model such processes, to extract the rules that drive the evolving processes, and to trace how they change over time are among the main objectives of the knowledge engineering approach that we take in this book. The introductory chapter consists of the following sections.

- Everything is evolving, but what are the evolving rules?
- Evolving intelligent systems (EIS) and evolving connectionist systems (ECOS)
- Biological inspirations for EIS and ECOS
- About the book
- Further reading

I.1 Everything Is Evolving, but What Are the Evolving Rules?

According to the *Concise Oxford English Dictionary* (1983), 'evolving' means 'revealing', 'developing'. It also means 'unfolding, changing'. We define an evolving process as a process that is developing, changing over time in a continuous manner. Such a process may also interact with other processes in the environment. It may not be possible to determine in advance the course of interaction, though. For example, there may be more or fewer variables related to a process at a future time than at the time when the process started.

Evolving processes are difficult to model because some of their evolving rules (laws) may not be known a priori; they may dynamically change due to unexpected perturbations, and therefore they are not strictly predictable in a longer term. Thus, modelling of such processes is a challenging task with a lot of practical applications in life sciences and engineering.

When a real process is evolving, a modelling system needs to be able to trace the dynamics of the process and to adapt to changes in the process. For example, a speech recognition system has to be able to adapt to various new accents, and to learn new languages incrementally. A system that models cognitive tasks of the human brain needs to be adaptive, as all cognitive processes are evolving by nature. (We never stop learning!) In bioinformatics, a gene expression modelling system has to be able to adapt to new information that would define how a gene could become inhibited by another gene, the latter being triggered by a third gene, etc. There are an enormous number of tasks from life sciences where the processes evolve over time.

It would not be an overstatement to say that everything in nature evolves. But what are the rules, the laws that drive these processes, the evolving rules? And how do they change over time? If we know these rules, we can make a model that can evolve in a similar manner as the real evolving process, and use this model to make predictions and to understand the real processes. But if we do not know these rules, we can try to discover them from data collected from this process using the knowledge engineering approach presented in this book.

The term 'evolving' is used here in a broader sense than the term 'evolutionary'. The latter is related to a population of individual systems traced over generations (Charles Darwin; Holland, 1992), whereas the former, as it is used in this book, is mainly concerned with the development of the structure and functionality of an individual system during its lifetime (Kasabov, 1998a; Weng *et al.*, 2001). An evolutionary (population/generation) optimisation of the system can be applied as well.

The most obvious example of an evolving process is life. Life is defined in the *Concise Oxford English Dictionary* (1983) as 'a state of functional activity and continual change peculiar to organized matter, and especially to the portion of it constituting an animal or plant before death, animate existence, being alive.' Continual change, along with certain stability, is what characterizes life. Modelling living systems requires that the continuous changes are represented in the model; i.e. the model adapts in a lifelong mode and at the same time preserves some features and principles that are characteristic to the process. The 'stability–plasticity' dilemma is a well-known principle of life that is also widely used in connectionist computational models (Grossberg, 1969, 1982).

In a living system, evolving processes are observed at different levels (Fig. I.1).

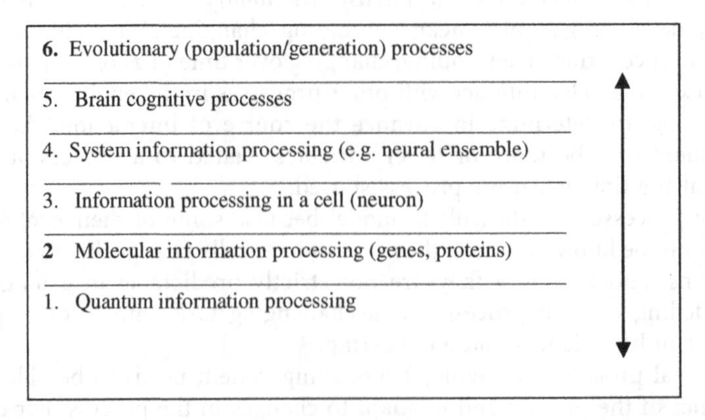

Fig. I.1 Six levels of evolving processes in a higher-order living organism: evolution, cognitive brain processes, brain functions in neural networks, single neuron functions, molecular processes, and quantum processes.

At the quantum level, particles are in a complex evolving state all the time, being in a superposion of several locations at the same time, which is defined by probabilities. General evolving rules are defined by several principles, such as entanglement, superposition, etc. (see Chapter 15).

At a molecular level, RNA and protein molecules, for example, evolve and interact in a continuous way based on the DNA information and on the environment. The central dogma of molecular biology constitutes a general evolving rule, but what are the specific rules for different species and individuals? The area of science that deals with the information processing and data manipulation at this level is bioinformatics. Modelling evolving processes at the molecular level is discussed in Chapter 8.

At the cellular level (e.g. a neuronal cell) all the metabolic processes, the cell growing, cell division, etc., are evolving processes. Modelling evolving processes in cells and neurons is discussed in Chapter 8.

At the level of cell ensembles, or at a neural network level, an ensemble of cells (neurons) operates in concert, defining the function of the ensemble or the network through learning, for instance, perception of sound, perception of an image, or learning languages. An example of a general evolving rule is the Hebbian learning rule (Hebb, 1949); see Chapter 9.

In the human brain, complex dynamic interactions between groups of neurons can be observed when certain cognitive functions are performed, e.g. speech and language learning, visual pattern recognition, reasoning, and decision making. Modelling such processes is presented in Chapters 9 and 10.

At the level of population of individuals, species evolve through evolution. A biological system evolves its structure and functionality through both lifelong learning of an individual and the evolution of populations of many such individuals (Charles Darwin; Holland, 1992). In other words, an individual is a result of the evolution of many generations of populations, as well as a result of its own developmental lifelong learning processes. The Mendelian and Darwinian rules of evolution have inspired the creation of computational modelling techniques called evolutionary computation, EC (Holland, 1992; Goldberg, 1989). EC is discussed in Chapter 6, mainly from the point of view of optimisation of some parameters of an evolving system.

All processes in Fig. I.1 are evolving. Everything is evolving, the living organisms being more evolving than the other, but what are the evolving rules, the laws that govern these processes? Are there any common evolving rules for every material item and for every living organism, along with their specific evolving rules? And what are the specific rules? Do these rules change over time; i.e. do they evolve as well?

An evolving process, characterised by its evolving governing rules, manifests itself in a certain way and produces data that in many cases can be measured. Through analysis of these data, one can extract relationship rules that describe the data, but do they describe the evolving rules of the process as well?

Processes at different levels from Fig. I.1 are characterised by general characteristics, such as frequency spectrum, energy, information, and interaction as explained below.

1. Frequency spectrum

Frequency, denoted F, is defined as the number of a signal/event repetition over a period of time T (seconds, minutes, centuries). Some processes have stable frequencies, but others change their frequencies over time. Different processes from Fig. I.1 are characterised by different frequencies, defined by their physical parameters. Usually, a process is characterised by a spectrum of frequencies. Different frequency spectrums characterise brain oscillations (e.g. delta waves, Chapter 9), speech signals (Chapter 10), image signals (Chapter 11), or quantum processes (Chapter 15).

2. Energy

Energy is a major characteristic of any object and organism. Albert Einstein's most celebrated energy formula defines energy E as depending on the mass of the object m and the speed of light c:

$$E = mc^2 \tag{I.1}$$

The energy of a protein, for example, depends not only on the DNA sequence that is translated into this protein, but on the 3D shape of the protein and on external factors.

3. Information

Generally speaking, *information* is a report, a communication, a measure, a representation of news, events, facts, knowledge not known earlier. This is a characteristic that can be defined in different ways. One of them is entropy; see Chapter 1.

4. Interaction, connection with other elements of the system (e.g. objects, particles)

There are many interactions within each of the six levels from Fig. I.1 and across these levels. Interactions are what make a living organism a complex one, and that is also a challenge for computational modelling. For example, there are complex interactions between genes in a genome, and between proteins and DNA. There are complex interactions between the genes and the functioning of each neuron, a neural network, and the whole brain. Abnormalities in some of these interactions are known to have caused brain diseases and many of them are unknown at present (see the section on computational neuro-genetic modelling in Chapter 9).

An example of interactions between genes and neuronal functions is the observed dependence between long-term potentiation (learning) in the synapses and the expression of the immediate early genes and their corresponding proteins such as Zif/268 (Abraham *et al.*, 1993). Genetic reasons for several brain diseases have been already discovered (see Chapter 9).

Generally speaking, neurons from different parts of the brain, associated with different functions, such as memory, learning, control, hearing, and vision, function in a similar way. Their functioning is defined by evolving rules and factors, one of them being the level of neuro-transmitters. These factors are controlled at a genetic level. There are genes that are known to regulate the level of neuro-transmitters for different types of neurons from different areas of the brain (RIKEN, 2001). The functioning of these genes and the proteins produced can be controlled through nutrition and drugs. This is a general principle that can be exploited for different models of the processes from Fig. I.1 and for different systems performing different tasks, e.g. memory and learning; see Benuskova and Kasabov (2007). We refer to the above in the book as neuro-genetic interactions (Chapter 9).

Based on the evolving rules, an evolving process would manifest different behaviour:

- *Random:* There is no rule that governs the process in time and the process is not predictable.
- *Chaotic:* The process is predictable but only in a short time ahead, as the process at a time moment depends on the process at previous time moments via a nonlinear function.
- *Quasi-periodic:* The process is predictable subject to an error. The same rules apply over time, but slightly modified each time.
- *Periodic:* The process repeats the same patterns of behaviour over time and is fully predictable (there are fixed rules that govern the process and the rules do not change over time).

Many complex processes in engineering, social sciences, physics, mathematics, economics, and other sciences are evolving by nature. Some dynamic time series in nature manifest chaotic behaviour; i.e. there are some vague patterns of repetition over time, and the time series are approximately predictable in the near future, but not in the long run (Gleick, 1987; Barndorff-Nielsen *et al.*, 1993; Hoppensteadt, 1989; McCauley, 1994). Chaotic processes are usually described by mathematical equations that use some parameters to evaluate the next state of the process from its previous states. Simple formulas may describe a very complicated behaviour over time: e.g. a formula that describes fish population growth $F(t+1)$ is based on the current fish population $F(t)$ and a parameter g (Gleick, 1987):

$$F(t+1) = 4gF(t)(1 - F(t)) \tag{I.2}$$

When $g > 0.89$, the function becomes chaotic.

A chaotic process is defined by evolving rules, so that the process lies on the continuum of 'orderness' somewhere between random processes (not predictable at all) and quasi-periodic processes (predictable in a longer timeframe, but only to a certain degree). Modelling a chaotic process in reality, especially if the process changes its rules over time, is a task for an adaptive system that captures the changes in the process in time, e.g. the value for the parameter g from the formula above.

All problems from engineering, economics, and social sciences that are characterised by evolving processes require continuously adapting models to model them. A speech recognition system, an image recognition system, a multimodal information processing system, a stock prediction system, or an intelligent robot, for example, a system that predicts the emergence of insects based on climate, etc. should always adjust its structure and functionality for a better performance over time, which is the topic of Part II of the book, evolving intelligent systems.

I.2 Evolving Intelligent Systems (EIS) and Evolving Connectionist Systems (ECOS)

Despite the successfully developed and used methods of computational intelligence (CI), such as artificial neural networks (ANN), fuzzy systems (FS), evolutionary computation, hybrid systems, and other methods and techniques for adaptive machine learning, there are a number of problems while applying these techniques to complex evolving processes:

1. *Difficulty in preselecting the system's architecture:* Usually a CI model has a fixed architecture (e.g. a fixed number of neurons and connections). This makes it difficult for the system to adapt to new data of unknown distribution. A fixed architecture would definitely prevent the ANN from learning in a lifelong learning mode.
2. *Catastrophic forgetting:* The system would forget a significant amount of old knowledge while learning from new data.
3. *Excessive training time required:* Training an ANN in a batch mode usually requires many iterations of data propagation through the ANN structure. This may not be acceptable for an adaptive online system, which would require fast adaptation.
4. *Lack of knowledge representation facilities:* Many of the existing CI architectures capture statistical parameters during training, but do not facilitate extracting the evolving rules in terms of linguistically meaningful information. This problem is called the 'black box' problem. It occurs when only limited information is learned from the data and essential aspects, that may be more appropriate and more useful for the future work of the system, are missed forever.

To overcome the above problems, improved and new connectionist and hybrid methods and techniques are required both in terms of learning algorithms and system development.

Intelligence is seen by some authors as a set of features or fixed properties of the mind that are stable and static. According to this approach, intelligence is genetically defined – given – rather than developed. Contrary to this view, intelligence is viewed by other authors as a constant and continuous adaptation. Darwin's contemporary H. Spencer proposed in 1855 the law of intelligence, stating that 'the fundamental condition of vitality is that the internal state shall be continually adjusted to the external order' (Richardson, 1999, p. 14). Intelligence is 'the faculty of adapting oneself to circumstances,' according to Henri Simon and Francis Binet, the authors of the first IQ test (see Newell and Simon (1972)). In Plotkyn (1994), intelligence is defined as 'the human capacity to acquire knowledge, to acquire a set of adaptations and to achieve adaptation.'

Knowledge representation, concept formation, reasoning, and adaptation are obviously the main characteristics of intelligence upon which all authors agree (Rosch and Lloyd, 1978; Smith and Medin, 1981). How these features can be implemented and achieved in a computer model is the main objective of the area of artificial intelligence (AI).

AI develops methods, tools, techniques, and systems that make possible the implementation of intelligence in computer models. This is a 'soft' definition of

AI, which is in contrast to the first definition of AI (the 'hard' one) given by Alan Turing in 1950. According to the Turing test for AI, if a person communicates in natural language with another person or an artificial system behind a barrier without being able to distinguish between the two, and also without being able to identify whether this is a male or a female, as the system should be able to fool the human in this respect, then if it is a system behind the barrier, it can be considered an AI system. The Turing test points to an ultimate goal of AI, which is the understanding of concepts and language, but on the other hand it points to no direction or criteria to develop useful AI systems.

In a general sense, information systems should help trace and understand the dynamics of the modelled processes, automatically evolve rules, 'knowledge' that captures the essence of these processes, 'take a short cut' while solving a problem in a complex problem space, and improve their performance all the time. These requirements define a subset of AI which is called here evolving intelligent systems (EIS). The emphasis here is not on achieving the ultimate goal of AI, as defined by Turing, but rather on creating systems that learn all the time, improve their performance, develop a knowledge representation for the problem in hand, and become more intelligent.

A constructivist working definition of EIS is given below. It emphasises the dynamic and the knowledge-based structural and functional self-development of a system.

EIS is an information system that develops its structure, functionality, and knowledge in a continuous, self-organised, adaptive, and interactive way from incoming information, possibly from many sources, and performs intelligent tasks typical for humans (e.g. adaptive pattern recognition, concept formation, language learning, intelligent control) thus improving its performance.

David Fogel (2002), in his highly entertaining and highly sophisticated book, *Blondie 24 – Playing at the Edge of AI*, describes a case of EIS as a system that learns to play checkers online without using any instructions, and improves after every game. The system uses connectionist structure and evolutionary algorithms along with statistical analysis methods.

EIS are presented here in this book in the form of methods of evolving connectionist systems (ECOS) and their applications. An ECOS is an adaptive, incremental learning and knowledge representation system that evolves its structure and functionality, where in the core of the system is a connectionist architecture that consists of neurons (information processing units) and connections between them. An ECOS is a CI system based on neural networks, but using other techniques of CI that operate continuously in time and adapt their structure and functionality through a continuous interaction with the environment and with other systems (Fig. I.2). The adaptation is defined through:

1. A set of evolving rules
2. A set of parameters ("genes") that are subject to change during the system operation
3. An incoming continuous flow of information, possibly with unknown distribution
4. Goal (rationale) criteria (also subject to modification) that are applied to optimise the performance of the system over time

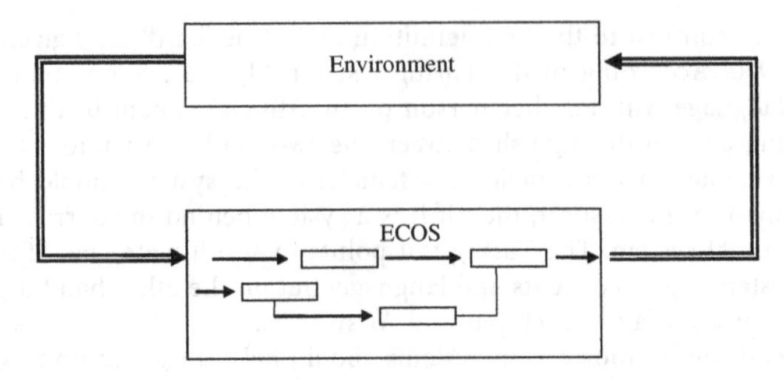

Fig. I.2 EIS, and ECOS in particular, evolve their structure and functionality through incremental (possibly online) learning in time and interaction with the environment.

The methods of ECOS presented in the book can be used as traditional CI techniques, but they also have some specific characteristics that make them applicable to more complex problems:

1. They may evolve in an open space, where the dimensions of the space can change.
2. They learn via incremental learning, possibly in an online mode.
3. They may learn continuously in a lifelong learning mode.
4. They learn both as individual systems and as an evolutionary population of such systems.

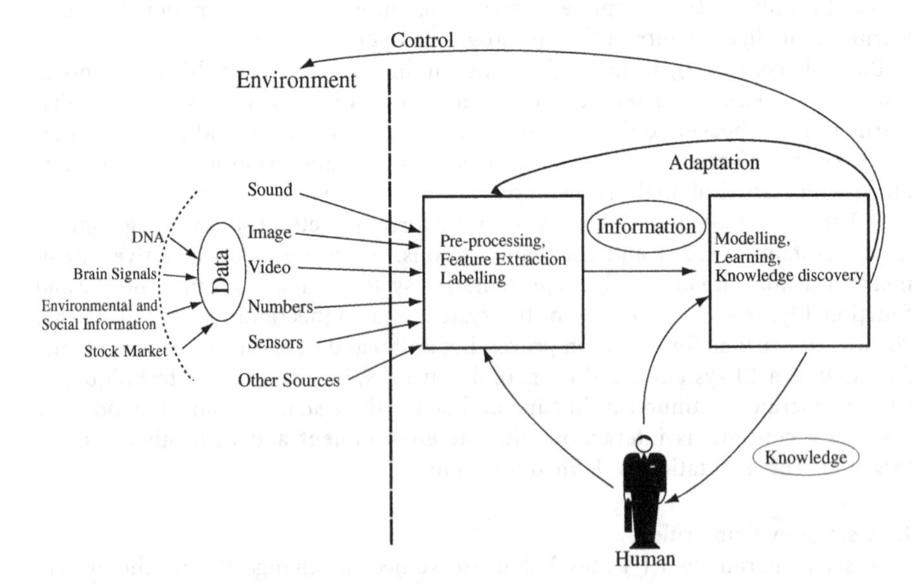

Fig. I.3 An EIS, and ECOS in particular, consists of four parts: data acquisition, feature extraction, modelling, and knowledge acquisition. They process different types of information in a continuous adaptive way, and communicate with the user in an intelligent way providing knowledge (rules). Data can come from different sources: DNA (Chapter 8), brain signals (Chapter 9), socioeconomic and ecological data (Chapter 14), and from many others.

5. They use constructive learning and have evolving structures.
6. They learn and partition the problem space locally, thus allowing for a fast adaptation and tracing the evolving processes over time.
7. They evolve different types of knowledge representation from data, mostly a combination of memory-based, statistical, and symbolic knowledge.

Each EIS system, and an ECOS in particular, consists of four main parts:

1. Data acquisition
2. Preprocessing and feature evaluation
3. Modelling
4. Knowledge acquisition

Figure I.3 illustrates the different parts of an EIS that processes different types of information in a continuous adaptive way. The online processing of all this information makes it possible for the ECOS to interact with users in an intelligent way. If human–system interaction can be achieved in this way, this can be used to extend system–system interactions as well.

I.3 Biological Inspirations for EIS and ECOS

Some of the methods for EIS and ECOS presented in Chapters 2 through 6 use principles from the human brain, as discussed here and in many publications (e.g. Arbib (1995, 2002) and Kitamura (2001)).

It is known that the human brain develops before the child is born. During learning the brain allocates neurons to respond to certain stimuli and develops their connections. Some parts of the brain develop connections and also retain their ability to create neurons during the person's lifetime. Such an area is the hippocampus (Erikson *et al.*, 1998; McClelland *et al.*, 1995). According to McClelland *et al.* (1995) the sequential acquisition of knowledge is due to the continuous interaction between the hippocampus and the neocortex. New nerve cells have a crucial role in memory formation.

The process of brain development and learning is based on several principles (van Owen, 1994; Wong, 1995; Amit, 1989; Arbib, 1972, 1987, 1998, 1995, 2002; Churchland and Sejnowski, 1992; J. G. Taylor, 1998, 1999; Deacon, 1988, 1998; Freeman, 2001; Grossberg, 1982; Weng *et al.*, 2001), some of them used as inspirations for the development of ECOS:

1. Evolution is achieved through both genetically defined information and learning.
2. The evolved neurons in the brain have a spatial–temporal representation where similar stimuli activate close neurons.
3. Redundancy is the evolving process in the brain leading to the creation of a large number of neurons involved in each learned task, where many neurons are allocated to respond to a single stimulus or to perform a single task; e.g. when a word is heard, there are hundreds of thousands of neurons that are immediately activated.

4. Memory-based learning, i.e. the brain stores exemplars of facts that can be recalled at a later stage. Bernie Widrow (2006) argues that learning is a process of memorising and everything we do is based on our memory.
5. Evolving through interaction with the environment.
6. Inner processes take place, e.g. sleep-learning and information consolidation.
7. The evolving process is continuous and lifelong.
8. Through learning, higher-level concepts emerge that are embodied in the evolved brain structure, and can be represented as a level of abstraction (e.g. acquisition and the development of speech and language, especially in multilingual subjects).

Learning and structural evolution coexist in the brain. The neuronal structures eventually implement a long-term memory. Biological facts about growing neural network structures through learning and adaptation are presented in Joseph (1998).

The observation that humans (and animals) learn through memorising sensory information, and then interpreting it in a context-driven way, has been known for a long time. This is demonstrated in the consolidation principle that is widely accepted in physiology. It states that in the first five or so hours after presenting input stimuli to a subject, the brain is learning to 'cement' what it has perceived. This has been used to explain retrograde amnesia (a trauma of the brain that results in loss of memory about events that occurred several hours before the event of the trauma). The above biological principle is used in some methods of ECOS in the form of sleep, eco-training mode.

During the ECOS learning process, exemplars (or patterns) are stored in a long-term memory. Using stored patterns in the eco-training mode is similar to the task rehearsal mechanism (TRM). The TRM assumes that there are long-term and short-term centres for learning (McClelland *et al.*, 1995). According to the authors, the TRM relies on long-term memory for the production of virtual examples of previously learned task knowledge (background knowledge). A functional transfer method is then used to selectively bias the learning of a new task that is developed in short-term memory. The representation of this short-term memory is then transferred to long-term memory, where it can be used for learning yet another new task in the future. Note that explicit examples of a new task need not be stored in long-term memory, only the representation of the task, which can be used later to generate virtual examples. These virtual examples can be used to rehearse previously learned tasks in concert with a new 'related' task.

But if a system is working in a real-time mode, it may not be able to adapt to new data due to lack of sufficient processing speed. This phenomenon is known in psychology as 'loss of skills'. The brain has a limited amount of working short-term memory. When encountering important new information, the brain stores it simply by erasing some old information from the working memory. The prior information gets erased from the working memory before the brain has time to transfer it to a more permanent or semi-permanent location for actual learning. ECOS sleep-training is based on similar principles.

In Freeman (2000), intelligence is described as related to an active search for information, goal-driven information processing, and constant adaptation. In this respect an intelligent system has to be actively selecting data from the environment. This feature can be modelled and is present in ECOS through data and feature selection for the training process. The filtering part of the ECOS architecture from Fig. I.3 serves as an active filter to select only 'appropriate' data and features from the data streams. Freeman (2000) describes learning as a reinforcement process which is also goal-driven.

Part of the human brain works as associative memory (Freeman, 2000). The ECOS models can be used as associative memories, where the first part is trained in an unsupervised mode and the second part in a reinforcement or supervised learning mode.

Humans are always seeking information. Is it because of the instinct for survival? Or is there another instinct, an instinct for information? If that is true, how is this instinct defined, and what are its rules? Although Perlovski (2006) talks about cognitive aspects of this instinct; here we refer to the genetic aspects of the instinct. In Chapter 9 we refer to genes that are associated with long-term potentiation in synapses, which is a basic neuronal operation of learning and memory (Abraham *et al.*, 1993; Benuskova and Kasabov, 2006). We also refer to genes associated with loss of memory and other brain diseases that affect information processing in the brain, mainly the learning and the memory functions. It is now accepted that learning and memory are both defined genetically and developed during the life of an individual through interaction with the environment.

Principles of brain and gene information processing have been used as an inspiration to develop the methods of ECOS and to apply them in different chapters of the book.

The challenge for the scientific area of computational modelling, and for the ECOS paradigm in particular, is how to create structures and algorithms that solve complex problems to enable progress in many scientific areas.

I.4 About the Book

Figure I.4 represents a diagram that links the inspirations/principles, the ECOS methods, and their applications covered in different chapters of the book.

I.5 Further Reading

- *The Nature of Knowledge* (Plotkyn, 1994)
- *Cognition and Categorization* (Rosch and Lloyd, 1978)
- *Categories and Concepts* (Smith and Medin, 1981).
- *Chaotic Processes* (Barndorff-Nielsen *et al.*, 1993; Gleick, 1987; Hoppensteadt, 1989; McCauley, 1994; Erdi, 2007)
- *Emergence and Evolutionary Processes* (Holland, 1998)
- *Different Aspects of Artificial Intelligence* (Dean *et al.*, 1995; Feigenbaum, 1989; Hofstadter, 1979; Newell and Simon, 1972)
- *Alan Turing's Test for AI* (Fogel, 2002; Hofstadter, 1979)
- *Emerging Intelligence* (Fogel, 2002)
- *Evolving Connectionist Systems as Evolving Intelligence* (Kasabov, 1998–2006)
- *Evolving Processes in the Brain* (Freeman, 2000, 2001)

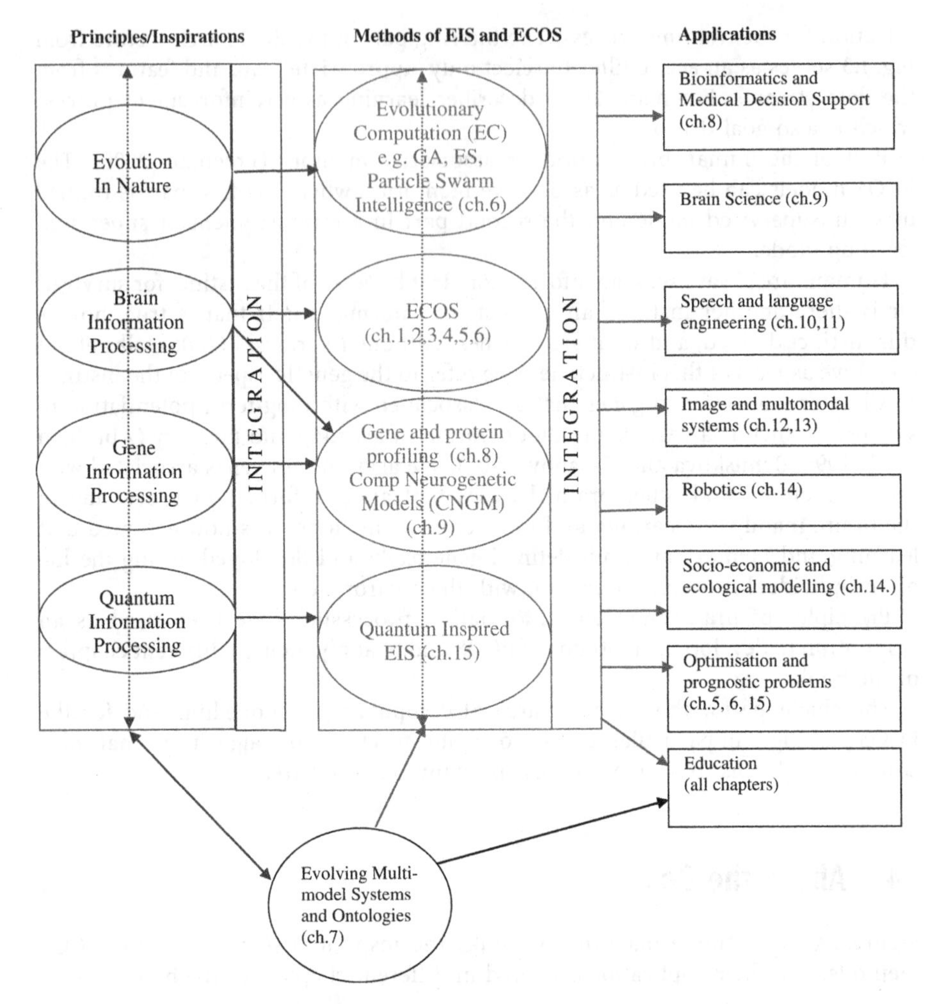

Fig. I.4 A block diagram schematically showing principles, methods, and applications covered in the book in their relationship.

- *Evolving Consciousness* (Taylor, 1999, 2005)
- *Principles of the Development of the Human Brain* (Amit, 1989; Arbib, 1972, 1987, 1998, 1995, 2002; Churchland and Sejnowski, 1992; Deacon, 1988, 1998; Freeman, 2001; Grossberg, 1982; Joseph, 1998; J. G. Taylor, 1998, 1999; van Owen, 1994; Wong, 1995)
- *Learning in the Hippocampus Brain* (Durand *et al.*, 1996; Eriksson *et al.*, 1998; Grossberg and Merrill, 1996; McClelland *et al.*, 1995)
- *Biological Motivations Behind ECOS* (Kasabov, 1998; Kitamura, 2001)
- *Autonomous Mental Development* (J.Weng *et al.*, 2001)

1. Feature Selection, Model Creation, and Model Validation

This chapter presents background information, methods, and techniques of computational modelling that are used in the other chapters. They include methods for feature selection, statistical learning, and model validation. Special attention is paid to several contemporary issues such as incremental feature selection and feature evaluation, inductive versus transductive learning and reasoning, and a comprehensive model validation. The chapter is presented in the following sections.

- Feature selection and feature evaluation
- Incremental feature selection
- Machine learning methods – a classification scheme
- Probability and information measure. Bayesian classifiers, hidden Markov models, and multiple linear regressions
- Support vector machines
- Inductive versus transductive learning and reasoning. Global versus local models
- Model validation
- Exercise
- Summary and open problems
- Further reading

1.1 Feature Selection and Feature Evaluation

Feature selection is the process of choosing the most appropriate features (variables) when creating a computational model (Pal, 1999).

Feature evaluation is the process of establishing how relevant to the problem in hand (e.g. the classification of gene expression microarray data) are the features (e.g. the genes) used in the model.

Features can be:

- Original variables, used in the first instance to specify the problem (e.g. raw pixels of an image, an amplitude of a signal, etc.)
- Transformed variables, obtained through mapping the original variable space into a new one (e.g. principle component analysis (PCA); linear discriminant analysis (LDA); fast Fourier transformation (FFT), SVM, etc.)

There are different groups of methods for feature selection:

- Filtering methods: The features are 'filtered', selected, and ranked in advance, before a model is created (e.g. a classification model).
- Wrapping methods: Features are selected on the basis of how well the created model performs using these features.

Traditional filtering methods are: correlation, t-test, and signal-to-noise ratio (SNR).

Correlation coefficients represent the relationship between the variables, including a class variable if such is available. For every variable x_i $(i = 1, 2, \ldots, d_1)$ its correlation coefficients $Corr(x_i, y_j)$ with all other variables, including output variables y_j $(j = 1, 2, \ldots, d_2)$, are calculated. The following is the formula to calculate the Pearson correlation between two variables x and y based on n values for each of them:

$$Corr = \sum_{i=1}^{n}((x_i - Mx)(y_i - My))/[(n-1)\ Stdx\ Stdy] \qquad (1.1)$$

where Mx and My are the mean values of the two variables x and y, and $Stdx$ and $Stdy$ are their respective standard deviations.

The t-test and the SNR methods evaluate how important a variable is to discriminate samples belonging to different classes. For the case of a two-class problem, a SNR ranking coefficient for a variable x is calculated as an absolute difference between the mean value $M1x$ of the variable for class 1 and the mean $M2x$ of this variable for class 2, divided to the sum of the respective standard deviations:

$$\text{SNR\_}x = abs\ (M1x - M2x)/(Std1x + Std2x). \qquad (1.2)$$

A similar formula is used for the t-test:

$$t\text{-test\_}x = abs(M1x - M2x)/(Std1x^2/N1 + Std2x^2/N2) \qquad (1.3)$$

where $N1$ and $N2$ are the numbers of samples in class 1 and class 2 respectively.

Figure 1.1a shows a graphical representation of the correlation coefficients of all four inputs and the class variables of the Iris benchmark data, and Fig. 1.1b gives the SNR ranking of the variables. The Iris benchmark data consist of 150 samples defined by four variables: sepal length, sepal width, petal length, petal width (in cm) Each of these samples belongs to one of three classes: Setosa, Versicolour, or Virginica (Fisher, 1936). There are 50 samples of each class.

Principal component analysis aims at finding a representation of a problem space X defined by its variables $X = \{x1, x2, \ldots, xn\}$ into another orthogonal space having a smaller number of dimensions defined by another set of variables $Z = \{z1, z2, \ldots, zm\}$, such that every data vector $\boldsymbol{x}$ from the original space is projected into a vector $\boldsymbol{z}$ of the new space, so that the distance between different vectors in the original space X is maximally preserved after their projection into the new space Z. A PCA projection of the Iris data is shown in Fig. 1.2a.

Linear discriminant analysis is a transformation of classification data from the original space into a new space of LDA coefficients that has an objective function

(a)

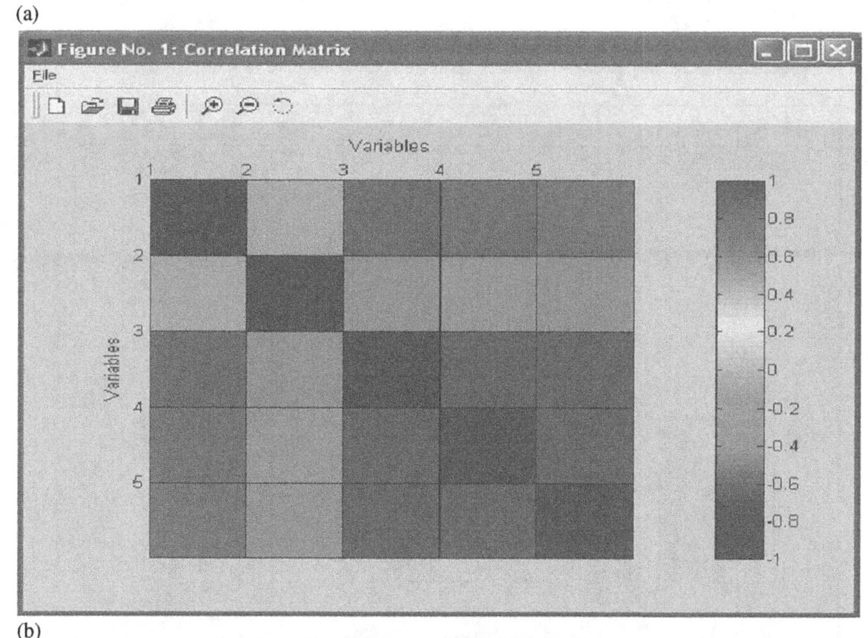

(b)

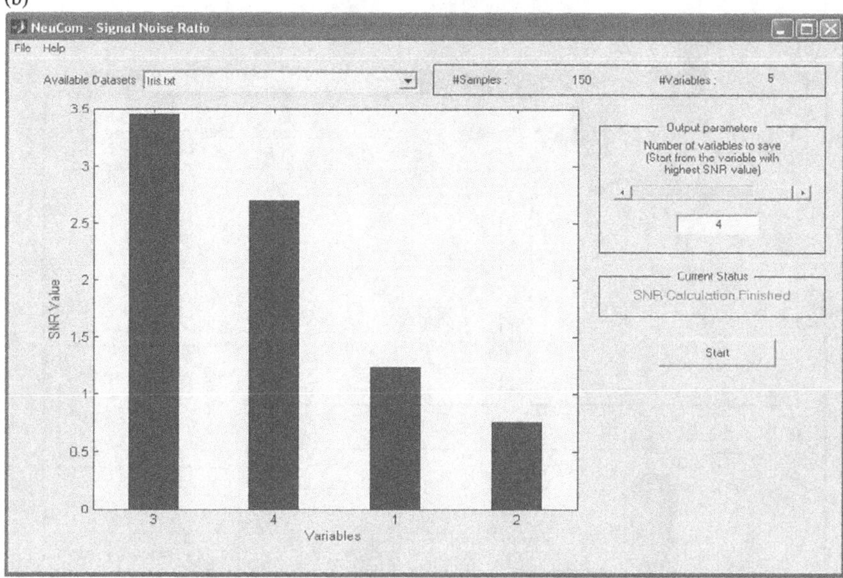

Fig. 1.1 (a) Correlation coefficients between the five variables in the Iris data set (four input variables and one class variable encoding class Setosa as 1, Versicolour as 2, and Virginica as 3); (b) SNR ranking of the four variables of the Iris case study data (variable 3, petal length is ranked the highest). A colour version of this figure is available from www.kedri.info.

to preserve the distance between the samples using also the class label to make them more distinguishable between the classes. An LDA projection of the Iris data is shown in Fig. 1.2a.

(a)

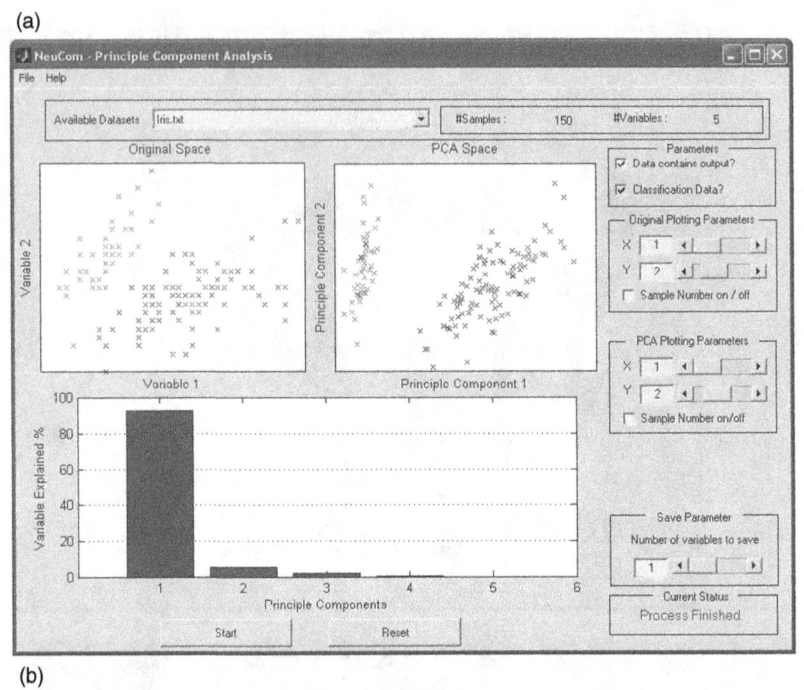

(b)

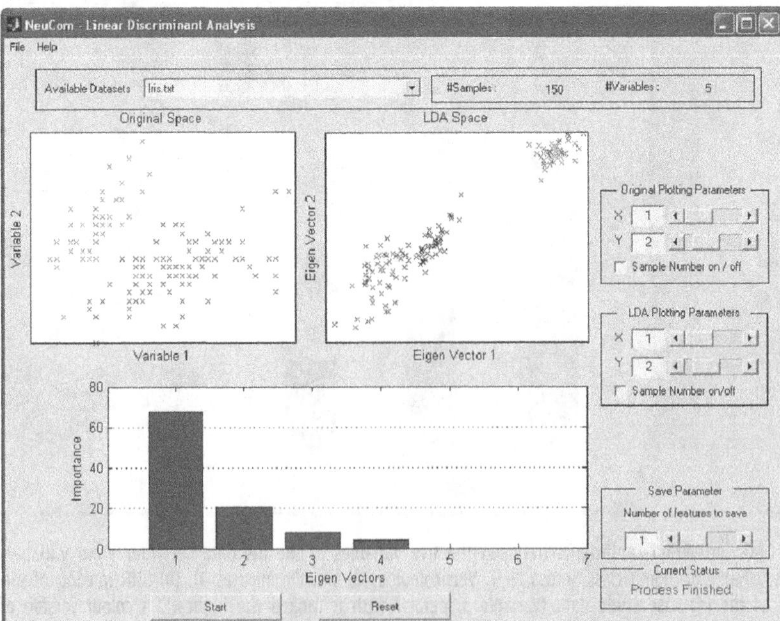

Fig. 1.2 (a) PCA transformation (unsupervised, uses only input variables) of the Iris case study data: the first principal component alone accounts for more than 90% of the variation among the data samples; (b) LDA transformation of the Iris case study data gives a better discrimination than the PCA as it uses class labels to achieve the transformation (it is supervised). (See a colour version at www.kedri.info)

Another benchmark dataset used in the book is the gas-furnace time-series data (Box and Jenkins, 1970). A quantity of methane gas (representing the first independent variable) is fed continuously into a furnace and the CO_2 gas produced is measured every minute (a second independent variable). This process can theoretically run forever, supposing that there is a constant supply of methane and the burner remains mechanically intact. The process of CO_2 emission is an evolving process. In this case it depends on the quantity of the methane supplied and on the parameters of the environment. For simplicity, only 292 values of CO_2 are taken in the well-known gas-furnace benchmark problem. Given the values of methane at a particular moment $(t-4)$ and the value of CO_2 at the moment (t) the task is to predict the value for CO_2 at the moment $(t+1)$ (output variable). The CO_2 data from Box and Jenkins (1970), along with some of their statistical characteristics, are plotted in Fig. 1.3. It shows the 292 points from the time series, the 3D phase space, the histogram, and the power spectrum of the frequency characteristics of the process. The program used for this analysis as well as for some other time-series analysis and visualisation in this book is given in Appendix A.

Several dynamic benchmark time series have been used in the literature and also in this book. We develop and test evolving models to model the well-known Mackey–Glass chaotic time series $x(t)$, defined by the Mackey–Glass time delay differential equation (see Farmer and Sidorovich (1987)):

$$\frac{dx}{d(t)} = \frac{ax(t-\tau)}{1+x^{10}\,(t-\tau)} - bx(t) \tag{1.4}$$

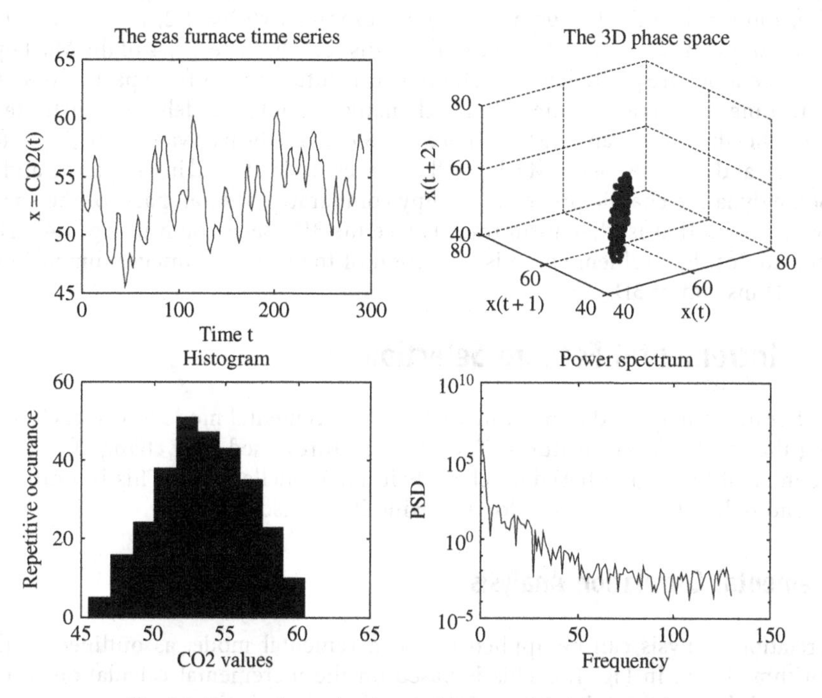

Fig. 1.3 The gas-furnace benchmark dataset, statistical characteristics.

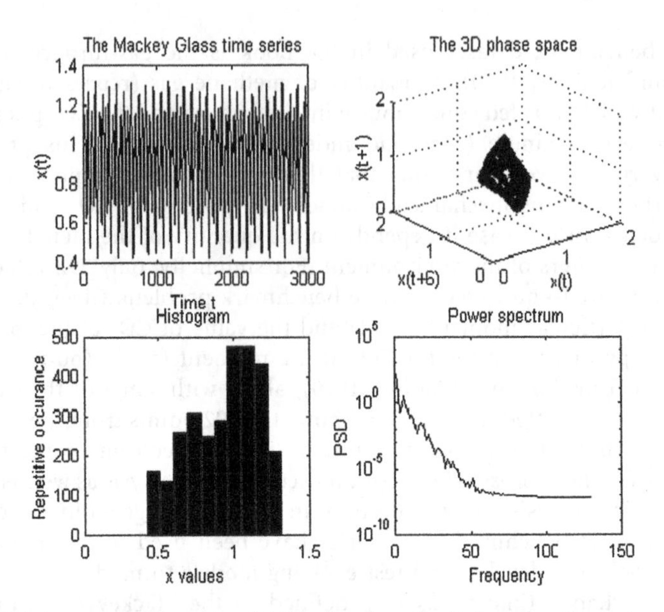

Fig. 1.4 Statistical parameters of a series of 3000 values from the Mackey–Glass time series. Top left: 3000 data points from the time series; top right: a 3D phase space; bottom left: the histogram of the time-series values; bottom right: the power spectrum, showing on the x-axis frequencies of repetition, and on the y-axis: the power of this repetition.

This series behaves as a chaotic time series for some values of the parameters τ, a, and b, and for the initial value of x, $x(0)$; for example, $x(0) = 1.2$, $\tau = 17$, $a = 0.2$, $b = 0.1$, and $x(t) = 0$ for $t < 0$. Some of the statistical characteristics of the Mackey–Glass time series are plotted in Fig. 1.4. Predicting future values from past values of a chaotic time series is a problem with which many computer models deal. Such a task is to predict the future values, $x(t+6)$ or $x(t+85)$, from the past values $x(t)$, $x(t-6)$, $x(t-12)$, and $x(t-18)$ of the Mackey–Glass series, as illustrated in Chapters 2 and 3.

Some dynamic chaotic processes occupy comparatively small parts of the space they evolve in; that is, they form fractals (see the 3D space graph in Fig. 1.4). The dimension of the problem space is a fraction of the standard integer dimensions, e.g. 2.1D instead of 3D.

1.2 Incremental Feature Selection

In EIS features may need to be evaluated in an incremental mode too, at each time using the most relevant features. The set of features used may change from one time interval to another based on changes in the modelled data. This is a difficult task and only a few techniques for achieving it are discussed here.

Incremental Correlation Analysis

Correlation analysis can be applied in an incremental mode, as outlined in the algorithm shown in Fig. 1.5. This is based on the incremental calculation of the mean and the standard deviation of the variable.

```
Calculating the online correlation coefficient CorrXY
between two variables: an input variable X, and an output variable
Y

SumX=0;
SumY=0;
SumXY=0;
SumX2=0;
SumY2=0;
CorrXY=[];

WHILE there are data pairs (x,y) from the input stream, DO
{
 INPUT the current data pair (x(i), y(i));

 SumX=SumX+x(i);
 SumY=SumY+y(i);
 AvX=SumX/i;
 AvY=SumY/i;
 SumXY=SumXY + (x(i) – AvX)*(y(i) – AvY);
 SumX2=SumX2 + (x(i) – AvX)^2;
 SumY2=SumY2 + (y(i) – AvY)^2;
 the current value for the correlation coefficient is:
 CorrXY(i)=SumXY / sqrt(SumX2 * SumY2);
}
```

Fig. 1.5 An illustrative algorithm for online correlation analysis. The operations in the algorithm are whole-vector operations expressed in the notation of MATLAB.

An example is given in Fig. 1.6. The figure shows the graphs of the euro/US$ exchange rate and the Dow Jones index over time, along with the online calculated correlation between the two time series using the algorithm from Fig. 1.5 (the bottom line). It can be seen that the correlation coefficient changes over time.

In many classification problems there is a data stream that contains different chunks of data, each having a different number of samples of each class. New class samples can emerge as well; see Fig. 1.7a (see Ozawa *et al.* (2005, 2006)).

Incremental selection of features is a complex procedure. S. Ozawa *et al.*, (2005, 2006) have introduced a method for incremental PCA feature selection where after the presentation of a new sample from the input data stream (or a chunk of data) a new PCA axis may be created (Fig.1.7b) or an axis can be rotated (Fig.1.7c) based on the position of the new sample in the PCA space. An algorithm for incremental LDA feature selection is proposed in S. Pang *et al.* (2005).

1.3 Machine Learning Methods – A Classification Scheme

Machine learning is an area of information science concerned with the creation of information models from data, with the representation of knowledge, and with the elucidation of information and knowledge from processes and objects. Machine learning includes methods for feature selection, model creation, model validation, and knowledge extraction (see Fig. I.3).

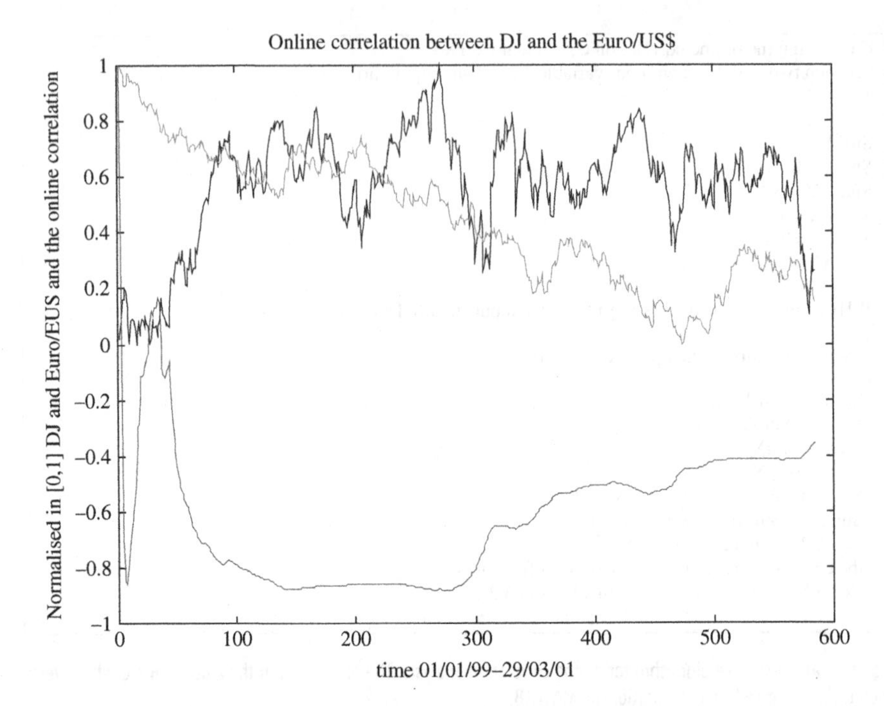

Fig. 1.6 The values of the euro/US$ exchange rate normalized in the interval [0,1] and the Dow Jones stock index as evolving time series, and the online correlation (the bottom line) between them for the period 1 January 1999 until 29 March 2001. The correlation coefficient changes over time and it is important to be able to trace this process. The first 100 or so values of the calculated correlation coefficient should be ignored, as they do not represent a meaningful statistical dependence.

Here we talk mainly about learning in connectionist systems (neural networks, ANN) even though the principles of these methods and the classification scheme presented below are valid for other machine learning methods as well.

Most of the known ANN learning algorithms are influenced by a concept introduced by Donald O. Hebb (1949). He proposed a model for unsupervised learning

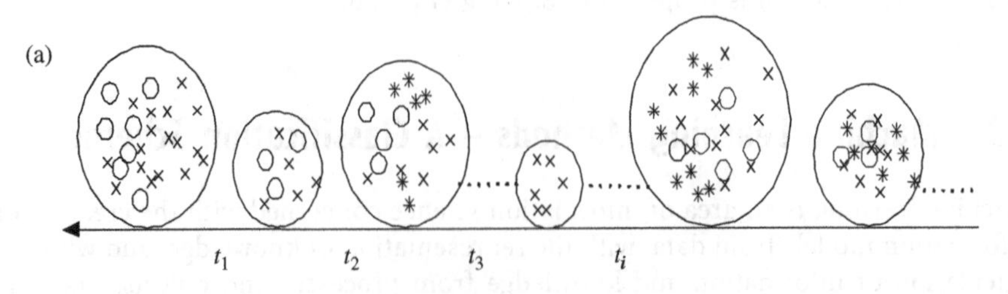

Fig. 1.7 (a) A data stream that contains chunks of data characterised by different numbers of samples (vectors, examples) from different classes; (*Continued overleaf*)

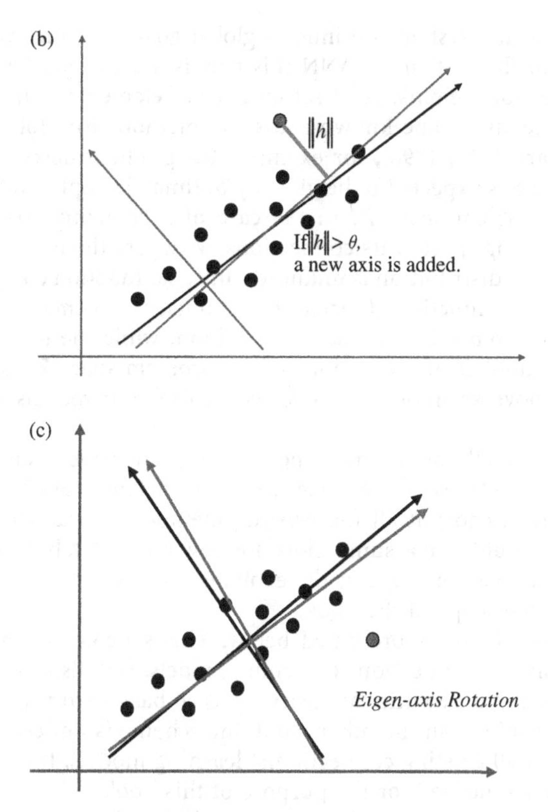

Fig. 1.7 (*continued*) (b) incremental PCA through new PCA axis creation: when a new data vector is entered and the distance between this vector and the existing eigenvector (PCA axis) is larger than a threshold, a new eigenaxis is created; (c) incremental PCA with axis rotation: when a new vector is added, the eigenvectors may need to be rotated (from Ozawa *et al.* (2005, 2006)).

in which the synaptic strength (weight) is increased if both the source and the destination neurons become simultaneously activated. It is expressed as

$$w_{ij}\ (t+1) = w_{ij}\ (t) + c.\ o_{i.}\ o_j, \tag{1.5}$$

where $w_{ij}(t)$ is the weight of the connection between the ith and jth neurons at the moment t, and o_i and o_j are the output signals of neurons i and j at the same moment t. The weight $w_{ij}(t+1)$ is the adjusted weight at the next time moment $(t+1)$.

In general terms, a connectionist system $\{S, W, P, F, L, J\}$ that is defined by its structure S, its parameter set P, its connection weights W, its function F, its goal function J, and a learning procedure L, learns if the system optimises its structure and its function F when observing events $z1, z2, z3, \ldots$ from a problem space Z. Through a learning process, the system improves its reaction to the observed events and captures useful information that may be later represented as knowledge. In Tsypkin (1973) the goal of a learning system is defined as finding the minimum of an objective function $J(S)$ named 'the expected risk function'. The function $J(S)$ can be represented by a loss function $Q(Z, S)$ and an unknown probability distribution $\mathrm{Prob}(Z)$.

Most of the learning systems optimise a global goal function over a fixed part of the structure of the system. In ANN this part is a set of predefined and fixed number of connection weights, i.e. a set number of elements in the set W. As an optimisation procedure, some known statistical methods for global optimisation are applied (Amari, 1967, 1990), for example, the gradient descent method. The obtained structure S is expected to be globally optimal, i.e. optimal for data drawn from the whole problem space Z. In the case of a changing structure S and a changing (e.g. growing) part of its connections W, where the input stream of data is continuous and its distribution is unknown, the goal function could be expressed as a sum of local goal functions J, each one optimised in a small subspace $Z' \subset Z$ as data are drawn from this subspace. In addition, while the learning process is taking place, the number of dimensions of the problem space Z may also change over time. The above scenarios are reflected in different models of learning, as explained next.

There are many methods for machine learning that have been developed for connectionist architectures (for a review, see Arbib (1995, 2002)). It is difficult and quite risky to try to put all the existing methods into a clear classification structure (which should also assume 'slots' for new methods), but this is necessary here in order to define the scope of the evolving connectionist system paradigm. This also defines the scope of the book.

A classification scheme is presented below. This scheme is a general one, as it is valid not only for connectionist learning models, but also for other learning paradigms, for example, evolutionary learning, case-based learning, analogy-based learning, and reasoning. On the other hand, the scheme is not comprehensive, as it does not present all existing connectionist learning models. It is only a working classification scheme needed for the purpose of this book.

A (connectionist) system that learns from observations $z1$, $z2$, $z3$, ... from a problem space Z can be designed to perform learning in different ways. The following classification scheme outlines the main questions and issues and their alternative solutions when constructing a connectionist learning system.

1. *In what space is the learning system developing?*
   (a) The learning system is developing in the original data space Z.

   The structural elements (nodes) of the connectionist learning system are points in the d-dimensional original data space Z (Fig. 1.8a). This is the case in some clustering and prototype learning systems. One of the problems here is that if the original space is high-dimensional (e.g. 30,000 gene expression space) it is difficult to visualise the structure of the system and observe some important patterns. For this purpose, special visualisation techniques, such as principal component analysis, or Sammon mapping, are used to project the system structure S into a visualisation space V.

   (b) The learning system is developing in its own machine learning space M.

   The structural elements (nodes) of the connectionist learning system are created in a system (machine) space M, different from the d-dimensional original data space Z (Fig. 1.8b). An example is the self-organising map (SOM) NN (Kohonen, 1977, 1982, 1990, 1993, 1997). SOMs develop in two-, three-, or more-dimensional topological spaces (maps) from the original data.

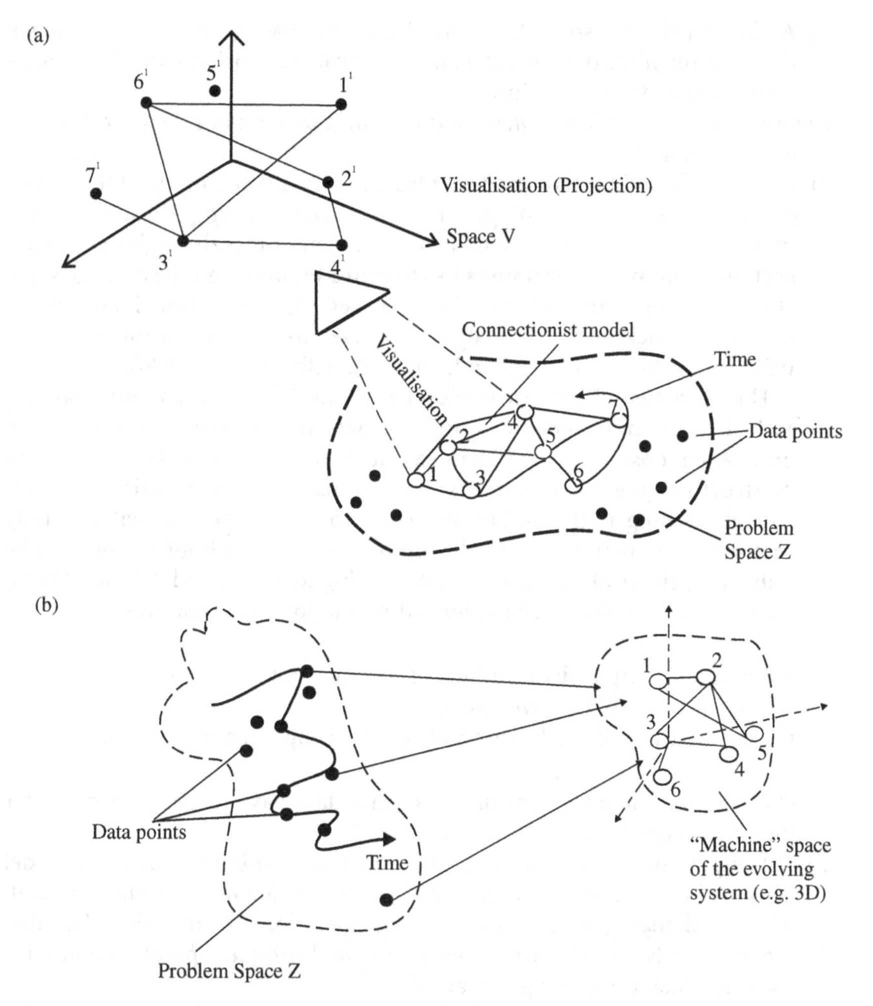

Fig. 1.8 (a) A computational model is built in the original data space; i.e. the original problem variables are used and a network of connections is built to model their interaction; a special visualisation procedure may be used to visualise the model in a different space; (b) a computational model is built in a new ('machine') space, where the original variables are transformed into a new set of variables.

2. *Is the problem space open?*
   (a) An open problem space is characterised by an unknown probability distribution $P(Z)$ of the incoming data and a possible change in its dimensionality. Sometimes the dimensionality of the data space may change over time, involving more or fewer dimensions, for example, adding new modalities to a person identification system. In this case the methods discussed in the previous section for incremental feature selection would be appropriate.

(b) A closed problem space has a fixed dimensionality, and either a known distribution of the data or the distribution can be approximated in advance through statistical procedures.

3. *Is learning performed in an incremental or in a batch mode, in an off-line or in an online mode?*
   (a) Batch-mode and pattern modes of learning: In a batch mode of learning a predefined learning (training) set of data $\{z1, z2, \ldots, zp\}$ is learned by the system through propagating this dataset several times through the system. Each time the system optimises its structure W, based on the average value of the goal function over the whole dataset. Many traditional algorithms, such as the backpropagation algorithm, use this type of learning (Werbos, 1990; Rumelhart and McLelland, 1986; Rumelhart *et al.*, 1986).

   The incremental pattern mode of learning is concerned with learning each data example separately and the data might exist only for a short time. After observing each data example, the system makes changes in its structure (the W parameters) to optimise the goal function J. Incremental learning is the ability of an NN to learn new data without fully destroying the patterns learned from old data and without the need to be trained on either old or new data. According to Schaal and Atkeson (1998) incremental learning is characterized by the following features.

   - Input and output distributions of data are not known and these distributions may change over time.
   - The structure of the learning system W is updated incrementally.

   Only limited memory is available so that data have to be discarded after they have been used.

   (b) Off-line versus online learning: In an off-line learning mode, a NN model is trained on data and then implemented to operate in a real environment, without changing its structure during operation. In an online learning mode, the NN model learns from new data during its operation and once used the data are no longer available.

   A typical simulation scenario for online learning is when data examples are drawn randomly from a problem space and fed into the system one by one for training. Although there are chances of drawing the same examples twice or several times, this is considered as a special case in contrast to off-line learning when one example is presented to the system many times as part of the training procedure. Methods for online learning in NN are studied in Albus (1975), Fritzke (1995), and Saad (1999). In Saad (1999), a review of some statistical methods for online learning, mainly gradient descent methods applied to fixed-size connectionist structures, is presented. Some other types of learning, such as incremental learning and lifelong learning, are closely related to online learning.

   Online learning, incremental learning, and lifelong learning are typical adaptive learning methods. Adaptive learning aims at solving the well-known stability/plasticity dilemma, which means that the system is stable enough to retain patterns learned from previously observed data, while being flexible enough to learn new patterns from new incoming data.

Adaptive learning is typical for many biological systems and is also useful in engineering applications such as robotic systems and process control. Significant progress in adaptive learning has been achieved due to the adaptive resonance theory (ART; Carpenter and Grossberg (1987, 1990, 1991) and Carpenter *et al.* (1991)) and its various models, which include unsupervised models (ART1, ART2, FuzzyART) and supervised versions (ARTMAP, FuzzyARTMAP – FAM).

(c) Combined online and off-line learning: In this mode the system may work for some of the time in an online mode, after which it switches to off-line mode, etc. This is often used for optimisation purposes, where a small 'window' of data from the continuous input stream can be kept aside, and the learning system, which works in an online mode, can be locally or globally optimised through off-line learning on this window of data through 'window-based' optimisation of the goal function $J(W)$.

4. *Is the learning process lifelong?*

(a) Single session learning: The learning process happens only once over the whole set of available data (even though it may take many iterations during training). After that the system is set in operation and never trained again. This is the most common learning mode in many existing connectionist methods and relates to the off-line, batch mode of training. But how can we expect that once a system is trained on certain (limited) data, it will always operate perfectly well in a future time, on any new data, regardless of where they are located in the problem space?

(b) Lifelong learning is concerned with the ability of a system to learn from continuously incoming data in a changing environment during its entire existence. Growing, as well as pruning, may be involved in the lifelong learning process, as the system needs to restrict its growth while always maintaining a good learning and generalisation ability. Lifelong learning relates to incremental, online learning modes, but requires more sophisticated methods.

5. *Are there desired output data and in what form are they available?*

The availability of examples with desired output data (labels) that can be used for comparison with what the learning system produces on its outputs defines four types of learning.

(a) Unsupervised learning: There are no desired output data attached to the examples $z1, z2, z3, \ldots$. The data are considered as coming from an input space Z only.

(b) Supervised learning: There are desired output data attached to the examples $z1, z2, z3, \ldots$. The data are considered as coming in (x, y) pairs from both an input space X and an output space Y that collectively define the problem space Z. The connectionist learning system associates data from the input space X to data from the output space Y (see Fig. 1.9).

(c) Reinforcement learning: In this case there are no exact desired output data, but some hints about the 'goodness' of the system reaction are available. The system learns and adjusts its structural parameters from these hints. In many robotic systems a robot learns from the feedback from the environment, which may be used as, for example, a qualitative indication of the correct movement of the robot.

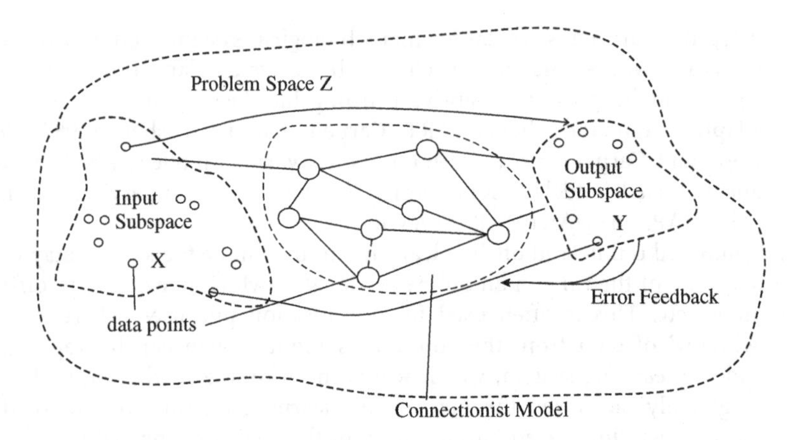

Fig. 1.9 A supervised learning model maps the input subspace into the output subspace of the problem space Z.

(d) Combined learning: This is the case when a connectionist system can operate in more than one of the above learning modes.

6. *Does learning include populations of individuals over generations?*
   (a) Individual development-based learning: A system is developing independently and is not part of a population of individual systems over generations.
   (b) Evolutionary (population/generation based) learning: Here, learning is concerned with the performance not only of an individual system, but of a population of systems that improve their performance through generations (see Chapter 6). The best individual system is expected to emerge and evolve from such populations. Evolutionary computation (EC) methods, such as genetic algorithms (GA), have been widely used for optimising ANN structures (Yao, 1993; Fogel *et al.*, 1990; Watts and Kasabov, 1998). They utilise ideas from Darwinism. Most of the evolutionary EC methods developed thus far assume that the problem space is fixed, i.e. that the evolution takes place within a predefined problem space and this space does not change dynamically. Therefore these methods do not allow for modelling real online adaptation. In addition they are very time-consuming, which also prevents them from being used in real-world applications.
7. *Is the structure of the learning system of a fixed size, or it is evolving?*
   Here we refer again to the bias/variance dilemma (see, e.g. Carpenter and Grossberg (1991) and Grossberg (1969, 1982)). For an NN structure the dilemma states that if the structure is too small, the NN is biased to certain patterns, and if the NN structure is too large there are too many variances, which may result in overtraining, poor generalization, etc. In order to avoid this problem, an NN structure should change dynamically during the learning process, thus better representing the patterns in the data and the changes in the environment.
   (a) Fixed-size structure: This type of learning assumes that the size of the structure S is fixed (e.g. number of neurons, number of connections), and

through learning the system changes some structural parameters (e.g. W, the values of connection weights). This is the case in many multilayer perceptron ANNs trained with the backpropagation algorithm (Rosenblatt, 1962; Amari, 1967, 1990; Arbib, 1972, 1987, 1995, 2002; Werbos, 1990; Hertz *et al.*, 1991; Rumelhart *et al.*, 1986).

(b) Dynamically changing structure: According to Heskes and Kappen (1993) there are three different approaches to dynamically changing structures: constructivism, selectivism, and a hybrid approach. Connectionist constructivism is about developing ANNs that have a simple initial structure and grow during their operation through inserting new nodes using evolving rules. This theory is supported by biological facts (see Saad (1999)). The insertion can be controlled by either a similarity measure of input vectors, by the output error measure, or by both, depending on whether the system performs an unsupervised or supervised mode of learning. A measure of difference between an input pattern and already stored ones is used for deciding whether to insert new nodes in the adaptive resonance theory models ART1 and ART2 (Carpenter and Grossberg, 1987) for unsupervised learning. There are other methods that insert nodes based on the evaluation of the local error. Such methods are the growing cell structure and growing neural gas (Fritzke, 1995). Other methods insert nodes based on a global error to evaluate the performance of the whole NN. One such method is the cascade correlation method (Fahlman and Lebiere, 1990). Methods that use both similarity and output error for node insertion are used in Fuzzy ARTMAP (Carpenter *et al.*, 1991) and also in EFuNN (Chapter 3). Connectionist selectivism is concerned with pruning unnecessary connections in an NN that starts its learning with many, in most cases redundant, connections (Rummery and Niranjan, 1994; Sankar and Mammone, 1993). Pruning connections that do not contribute to the performance of the system can be done by using several methods: optimal brain damage (Le Cun *et al.*, 1990), optimal brain surgeon (Hassibi and Stork, 1992), and structural learning with forgetting (Ishikawa, 1996).

8. *How does structural modification in the learning system partition the problem space?*

   When a machine learning (e.g. connectionist) model is created, in either a supervised or an unsupervised mode, the nodes and the connections partition the problem space Z into segments. Each segment of the input subspace is mapped onto a segment from the output subspace in the case of supervised learning. The partitioning in the input subspace imposed by the model can be one of the following types.

(a) Global partitioning (global learning): Learning causes global partitioning of the space. Usually the space is partitioned by hyperplanes that are modified either after every example is presented (in the case of incremental learning), or after all of them being presented together.

   Through the gradient descent learning algorithm, for example, the problem space is partitioned globally. This is one of the reasons why global learning in multilayer perceptrons suffers from the catastrophic forgetting phenomenon (Robins, 1996; Miller *et al.*, 1996). Catastrophic forgetting (also called unlearning) is the inability of the system to learn new patterns without forgetting previously learned patterns. Methods to deal with this problem include rehearsing the NN on a selection of past data, or on new data points generated from the problem space (Robins, 1996). Other techniques that use global partitioning are support vector machines (SVM; Vapnik (1998)). SVM optimise the positioning of the hyperplanes to achieve maximum distance from all data items on both sides of the plane (Kecman, 2001).

(b) Local partitioning (local learning): In the case of local learning, structural modifications of the system affect the partitioning of only a small part of the space from where the current data example is drawn. Examples are given in Figs. 1.10a and b, where the space is partitioned by circles and squares in a two-dimensional space. Each circle or square is the subspace defined by a neuron. The activation of each neuron is defined by local functions imposed on its subspace. Kernels, as shown in Fig. 1.10a, are examples of such local functions. Other examples of local partitioning are shown in Figs. 1.11a and b, where the space is partitioned by hypercubes and fractals in a 3D space.

Before creating a model it is important to choose which type of partitioning would be more suitable for the task in hand. In the ECOS presented later in this book, the partitioning is local. Local partitioning is easier to adapt in an online mode, faster to calculate, and does not cause catastrophic forgetting.

9. *What knowledge representation is facilitated in the learning system?*

It is a well-known fact that one of the most important characteristics of the brain is that it can retain and build knowledge. However, it is not yet known exactly how the activities of the neurons in the brain are transferred into knowledge. For the purpose of the discussion in this book, knowledge can be defined as the information learned by a system such that the system and humans can interpret it to obtain new facts and new knowledge. Traditional neural networks and connectionist systems have been known as poor facilitators of representing and processing knowledge, despite some early investigations (Hinton, 1987, 1990).

However, some of the issues of knowledge representation in connectionist systems have already been addressed in the so-called knowledge-based neural net- works (KBNN) (Towell and Shavlik, 1993, 1994; Cloete and Zurada, 2000). KBNN are neural networks that are prestructured in a way that allows for data and knowledge manipulation, which includes learning, knowledge insertion, knowledge extraction, adaptation, and reasoning. KBNN have been developed either as a combination of symbolic AI systems and NN (Towell *et al.*, 1990), or as a combina- tion of fuzzy logic systems and NN (Yamakawa and Tomoda, 1989; Yamakawa *et al.*, 1992, 1993; Furuhashi *et al.*, 1993, 1994; Hauptmann and Heesche, 1995; Jang, 1993; Kasabov, 1996). Rule insertion and rule extraction operations are examples of how a KBNN can accommodate existing knowledge along with data, and how it can 'explain' what it has learned. There are different methods for rule extraction that are applied to practical problems (Hayashi, 1991; Mitra and Hayashi, 2000; Duch *et al.*, 1998; Kasabov, 1996, 1998c, 2001c).

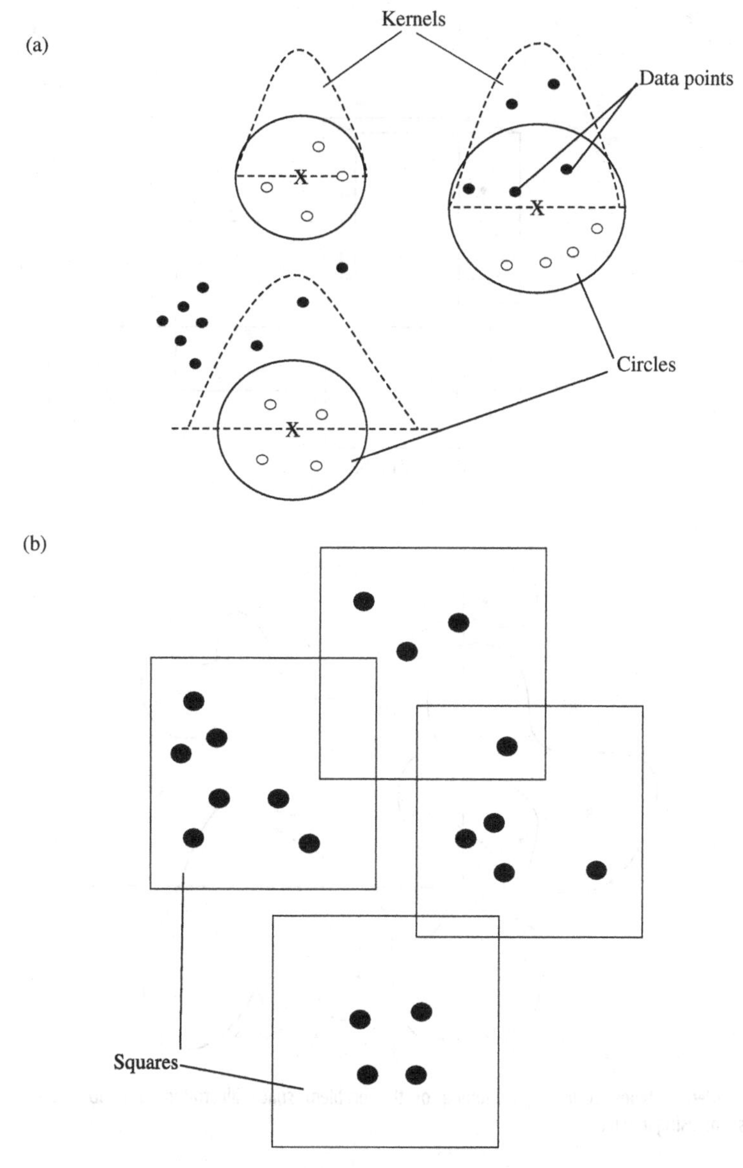

Fig. 1.10 Local partitioning of the problem space using different types of kernels: (a) hyperspheres and Gaussian functions defined on them; (b) squares, where a simple function can be defined on them (e.g.: yes, the data vector belongs to the square, and no, it does not belong). Local partitioning can be used for local learning.

Generally speaking, learning systems can be distinguished based on the type of knowledge they represent.

(a) No explicit knowledge representation is facilitated in the system: An example for such a connectionist system is the traditional multilayer perceptron network trained with the backpropagation algorithm

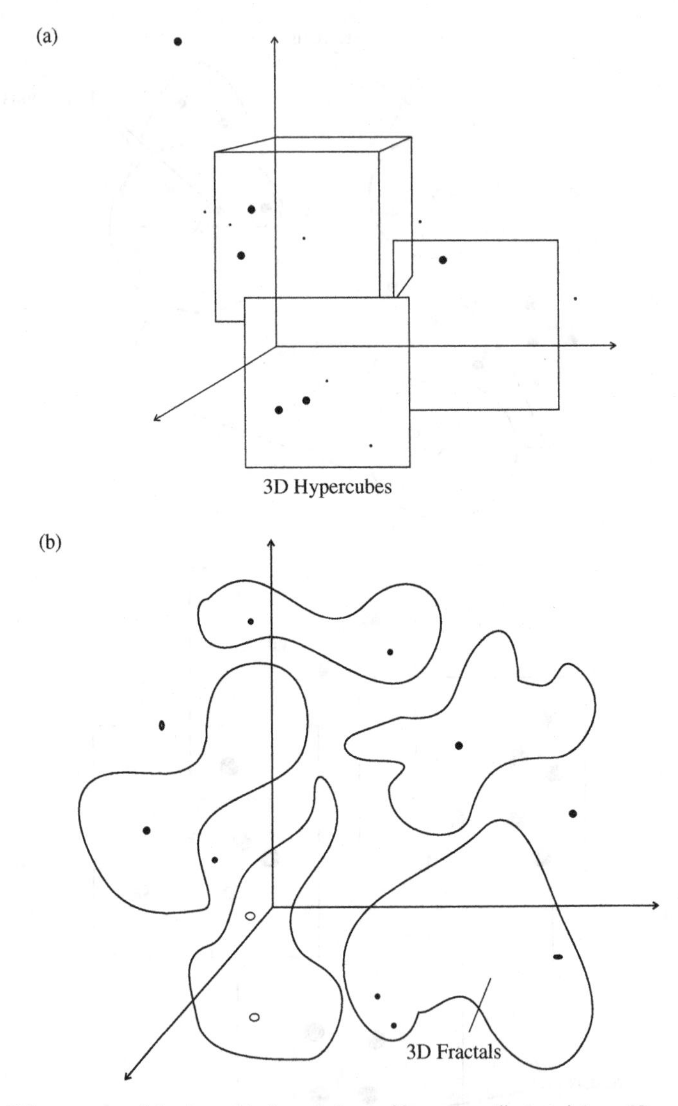

Fig. 1.11 Different types of local partitioning of the problem space illustrated in a 3D space: (a) using hypercubes; (b) using fractals.

(Rosenblatt, 1962; Amari, 1967; Arbib, 1972, 1987, 1995, 2002; Werbos, 1990; Hertz *et al.*, 1991; Rumelhart *et al.*, 1986).

(b) Memory-based knowledge: The system retains examples, patterns, prototypes, and cases, for example, instance-based learning (Aha *et al.*, 1991), case-based reasoning systems (Mitchell, 1997), and exemplar-based reasoning systems (Salzberg, 1990).

(c) Statistical knowledge: The system captures conditional probabilities, probability distribution, clusters, correlation, principal components, and other statistical parameters.

(d) Analytical knowledge: The system learns an analytical function $f: X -> Y$, that represents the mapping of the input space X into the output space Y. Regression techniques and kernel regressions in particular, are well established.
(e) Symbolic knowledge: Through learning, the system associates information with predefined symbols. Different types of symbolic knowledge can be facilitated in a learning system as further discussed below.
(f) Combined knowledge: The system facilitates learning of several types of knowledge.
(g) Meta-knowledge: The system learns a hierarchical level of knowledge representation where meta-knowledge is also learned, for example, which piece of knowledge is applicable and when.
(h) 'Consciousness' (the system knows about itself): The system becomes 'aware' of what it is, what it can do, and where its position is among the rest of the systems in the problem space.
(i) 'Creativity' (e.g. generating new knowledge): An ultimate type of knowledge would be such knowledge that allows the system to act creatively, to create scenarios, and possibly to reproduce itself, for example, a system that generates other systems (programs) improves in time based on its performance in the past.

Without the ability of a system to represent knowledge, it cannot capture knowledge from data, it cannot capture the evolving rules of the process that the system is modelling or controlling, and it cannot help much to better understand the processes. In this book we indeed take a knowledge-engineering approach to modelling and building evolving intelligent systems (EIS), where knowledge representation, knowledge extraction, and knowledge refinement in an evolving structure are the focus of our study along with the issues of adaptive learning.

10. *If symbolic knowledge is represented in the system, of what type is it?*

If we can represent the knowledge learned in a learning system as symbols, different types of symbolic knowledge can be distinguished.
(a) Propositional rules
(b) First-order logic rules
(c) Fuzzy rules
(d) Semantic maps
(e) Schemata
(f) Meta-rules
(g) Finite automata
(h) Higher-order logic

11. *If the system's knowledge can be represented as fuzzy rules, what type of fuzzy rules are they?*

Different types of fuzzy rules can be used, for example:
(a) Zadeh–Mamdani fuzzy rules (Zadeh, 1965; Mamdani, 1977).
(b) Takagi–Sugeno fuzzy rules (Takagi and Sugeno, 1985).
(c) Other types of fuzzy rules, for example, type-2 fuzzy rules (for a comprehensive reading, see Mendel (2001)).

The above types of rules are explained in Chapter 5. Generally speaking, different types of knowledge can be learned from a process or from an object in different ways, all of them involving human participation. Some of these ways are shown in Fig. 1.12. They include direct learning by humans, simple problem representation as graphs, analytical formulas, using NN for learning and rule extraction, and so on. All these forms can be viewed as alternative and possibly equivalent forms in terms of final results obtained after a reasoning mechanism is applied on them. Elaborating analytical knowledge in a changing environment is a very difficult process involving changing parameters and formulas with the change of the data. If evolving processes are to be learned in a system and also understood by humans, neural networks that are trained in an incremental mode and their structure interpreted as knowledge are the most promising models at present. This is the approach that is taken and developed in this book.

12. *Is learning active?*

    Humans and animals are selective in terms of processing only important information. They are searching actively for new information (Freeman, 2000; J.G. Taylor, 1999). Similarly, we can have two types of learning in an intelligent system:

    (a) Active learning: In terms of data selection, filtering, and searching for relevant data.
    (b) Passive learning: The system accepts all incoming data.

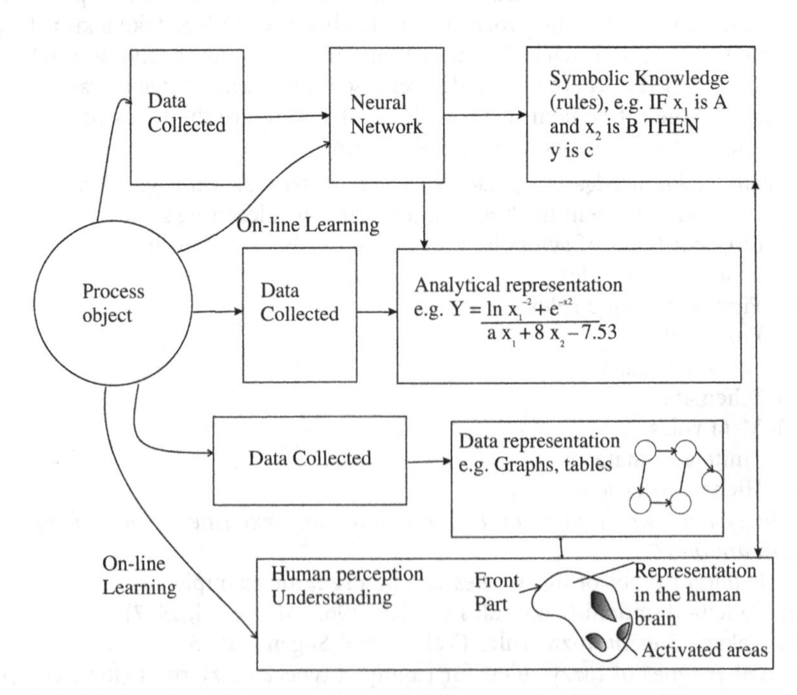

Fig. 1.12 Different learning and knowledge representation techniques applied to modelling a process or an object, including: off-line versus online learning; connectionist versus analytical (e.g. regression) learning; learning in a model versus learning in the brain.

Both approaches are applied in the different methods and techniques of evolving connections systems presented in Chapters 2 to 7.

1.4 Probability and Information Measure. Bayesian Classifiers, Hidden Markov Models. Multiple Linear Regression

Probability Characteristics of Events

Many learning methods are based on probability of events that are learned from data, and then used to predict new events. The formal theory of probability relies on the following three axioms, where $p(E)$ is the probability of an event E to happen and $p(\neg E)$ is the probability of an event not happening. $E1, E2, \ldots, Ek$ is a set of mutually exclusive events that form an universe U:

Box 1.1. Probability axioms

Axiom 1. $0 <= p(E) <= 1$.
Axiom 2. $\Sigma p(Ei) = 1$, $E1 \cup E2 \cup \ldots \cup Ek = U$, U- problem space.
Corollary: $p(E) + p(\neg E) = 1$.
Axiom 3. $p(E1 \text{ OR } E2) = p(E1) + p(E2)$, where $E1$ and $E2$ are mutually exclusive events.

Probabilities are defined as:

- Theoretical – some rules are used to evaluate a probability of an event.
- Experimental – probabilities are learned from data and experiments: throw dice 1000 times and measure how many times the event 'getting 6' has happened.
- Subjective – probabilities are based on common-sense human knowledge, such as defining that the probability of getting '6' after throwing dice is 1/6th , without really throwing it many times.

Information and Entropy Characteristics of Events

A random variable x is characterised at any moment of time by its uncertainty in terms of what value this variable will take in the next moment, its entropy. A measure of uncertainty $h(x_i)$ can be associated with each random value x_i of a random variable x, and the total uncertainty $H(x)$, called entropy, measures our lack of knowledge, the seeming disorder in the space of the variable x:

$$H(X) = \Sigma_{i=1,\ldots,n} p_i \cdot h(x_i) \tag{1.6}$$

where p_i is the probability of the variable x taking the value of x_i.
The following axioms for the entropy $H(x)$ apply.

- Monotonicity: If $n > n'$ are the number of events (values) that a variable x can take, then $Hn(x) > Hn'(x)$, so the more values x can take, the greater the entropy.
- Additivity: If x and y are independent random variables, then the joint entropy $H(x, y)$, meaning $H(x$ AND $y)$, is equal to the sum of $H(x)$ and $H(y)$.

The following log function satisfies the above two axioms.

$$h(x_i) = \log(1/p_i) \tag{1.7}$$

If the log has a basis of 2, the uncertainty is measured in [bits], and if it is the natural logarithm ln, then the uncertainty is measured in [nats].

$$H(X) = \Sigma_{i=1,\ldots,n} (p_i \,.\, h(x_i)) = -c.\, \Sigma_{i=1,\ldots,n} (p_i \,.\, \log p_i) \tag{1.8}$$

where c is a constant.
Based on the Shannon measure of uncertainty, entropy, we can calculate an overall probability for a successful prediction for all states of a random variable x, or the predictability of the variable as a whole:

$$P(x) = 2^{-H(x)} \tag{1.9}$$

The max entropy is calculated when all the n values of the random variable x are equiprobable; i.e. they have the same probability $1/n$, a uniform probability distribution:

$$H(X) = -\Sigma_{i=1,\ldots,n} p_i \,.\, \log p_i <= \log n \tag{1.10}$$

Example
Let us assume that it is known that a stock market crashes (goes to an extremely low value that causes many people and companies to lose their shares and assets and to go bankrupt) every six years. What are the uncertainty and the predictability of the crash in terms of determining the year of crash, if: (a) the last crash happened two years ago? (b) The same as in (a), plus we know for sure that there will be no crash in the current year, nor in the last year of the six-year period.
The solution will be:

(a) The possible years for a crash are the current one (year 3) and also years 4, 5, and 6. The random variable x which has the meaning of 'annual crash of the stock market' can take any of the values 3, 4, 5, and 6, therefore $n = 4$ and the maximum entropy is $H(x) = \log_2 4 = 2$. The predictability is $P(x) = 2^{-2} = 1/4$.

(b) The possible values for the variable x are reduced to 2 (years 4 and 5) as we have some extra knowledge of the stock market. In this case the maximum entropy will be 1 and the predictability will be 1/2.

Joint entropy between two random variables x and y (e.g. an input and an output variable in a system) is defined by the formulas:

$$H(x,y) = -\Sigma_{i=1,\ldots,n} p(x_i \text{ AND } y_j).\log p(x_i \text{ AND } y_j) \tag{1.11}$$

$$H(x,y) <= H(x) + H(y) \tag{1.12}$$

Conditional entropy, that is, measuring the uncertainty of a variable y (output variable) after observing the value of a variable x (input variable), is defined as follows.

$$H(y|x) = -\Sigma_{i=1,\ldots,n} p(x_i, y_j) \ . \ \log p(y_j|x_i) \tag{1.13}$$

$$H(y|x) <= H(y) \tag{1.14}$$

Entropy can be used as a measure of the information associated with a random variable x, its uncertainty, and its predictability.The mutual information between two random variables, also simply called information, can be measured as follows.

$$I(y,x) = H(y) - H(y|x) \tag{1.15}$$

The process of online information entropy evaluation is important as in a time series of events, after each event has happened, the entropy changes and its value needs to be re-evaluated.

Models based on probability are:

Bayesian classifiers
Hidden Markov models

A Bayesian classifier uses a conditional probability estimate to predict a class for new data. The following formula, which represents the conditional probability between two events C and A, is known as Bayes' formula (Thomas Bayes, 18th century):

$$p(A|C) = p(C|A).p(A)/p(C) \tag{1.16}$$

It follows from the equations,

$$p(A \text{ AND } C) = p(C \text{ AND } A) = p(A|C)p(C) = p(C|A)p(A) \tag{1.17}$$

Example

Evaluating the probability $p(A|C)$ of a patient having a flu (event A) based on the evidence that the patient has a high temperature (fact C). In order to accomplish this, we need the prior probability $p(C)$ of all ill patients having a high temperature, the prior probability of all people suffering of a flu at the time $p(A)$, and the conditional probability $p(C|A)$ of patients who, if having a flu A, have a high temperature C.

Problems with the Bayesian learning models relate to unknown prior probabilities and the requirement of a large amount of data for more accurate probability calculation. This is especially true for a chain of events $A, B, C, \ldots$, where probabilities $p(C|A, B), \ldots$, etc. need to be evaluated. The latter problem is addressed in techniques called hidden Markov models (HMM).

HMM is a technique for modelling the temporal structure of a time series signal, or of a sequence of events (Rabiner, 1989). It is a probabilistic pattern-matching approach that models a sequence of patterns as the output of a random process. The HMM consists of an underlying Markov chain.

$$P(q(t+1)|q(t), q(t-1), q(t-2), \ldots, q(t-n)) \approx P(q(t+1)|q(t)) \quad (1.18)$$

where $q(t)$ is state q sampled at a time t.

Example

Weather forecast problem as a Markov chain of events. Given today is sunny (S), what is the probability that the next following five days are S, Cloudy (C), or Rainy (R)? The answer can be derived using Table 1.1a.

HMM can be used not only to model time series of events, but a sequence of events in space. An example is modelling DNA sequences of four basic molecules: A, C, T, G (see Chapter 8) based on a probability matrix of having all 16 pairs of these molecules derived from a large enough segment of DNA (see Table 1.1b). Building a HMM from a DNA sequence and using this HMM to predict segments

Table 1.1a Representation of conditional probabilities for a HMM for weather forecast of tomorrow's weather from the weather today. Using this probability matrix, we can build a HMM for prediction of the weather several days ahead, starting from any day named 'Today'.

		P (Tomorrow | Today)			*Table*
		Tomorrow			
		(S)	(C)	(R)	
	Sunny(S)	.7	.2	.1	
Today	Cloudy(C)	.05	.8	.15	
	Rainy(R)	.15	.25	.6	

Table 1.1b Probability for a Pair of Neighbouring Molecular Nucleotides to Appear in a DNA Sequence of a Species.

	A	*C*	*T*	*G*
A (Adenine)	**0.3**	**0.5**	**0.1**	**0.1**
C (Cytosine)	**0.1**	**0.1**	**0.2**	**0.6**
T (Thymine)	**0.3**	**0.1**	**0.2**	**0.4**
G (Guanine)	**0.25**	**0.15**	**0.3**	**0.3**

of a DNA that will be translated into proteins is the main purpose of the software system GeneMark (Lukashin and Borodovski, 1998).

Multiple Linear Regression Methods (MLR)

The purpose of multiple linear regression is to establish a quantitative relationship between a group of p predictor variables (X) and a response y.

This relationship is useful for:

- Understanding which predictors have the greatest effect
- Knowing the direction of the effect (i.e. increasing x increases/decreases y).
- Using the model to predict future values of the response when only the predictors are currently known

A linear model takes its common form of:

$$Y = X\,A + b \tag{1.19}$$

where p is the number of the predictor variables; y is an n-by-1 vector of observations, X is an n-by-p matrix of regressors, A is a p-by-1 vector of parameters, and b is an n-by-1 vector of random disturbances. The solution to the problem is a vector A' which estimates the unknown vector of parameters. The least squares solution is used, so that the linear regression formula approximates the data with the least root mean square error (RMSE) as follows,

$$\mathrm{RMSE} = \mathrm{SQRT}(\mathrm{SUM}_{i=1,2,\ldots,n}((y_i - y_i')^2)/n) \tag{1.20}$$

where y_i is the desired value from the dataset corresponding to an input vector x_i, y_i' is the value obtained through the regression formula for the same input vector x_i, and n is the number of the samples (vectors) in the dataset.

Another error measure is also used to evaluate the performance of the regression model – a nondimensional error index (NDEI) – the RMSE divided to the standard deviation of the dataset:

$$\mathrm{NDEI} = \mathrm{RMSE}/\mathrm{Std} \tag{1.21}$$

Example 1

Linear regression modelling of the gas furnace benchmark data. Fig. 1.13 shows the regression formula that approximates the data, the desired versus the approximated by the formula values of the time series, and the two error measures, root mean square and the nondimensional error index.

Example 2

The following linear regression approximates the Mackey–Glass benchmark data (data are normalised).

$$Y = 0.93 - 0.3\,x1 - 0.01\,x2 - 0.56\,x3 + 0.86\,x4 \tag{1.22}$$

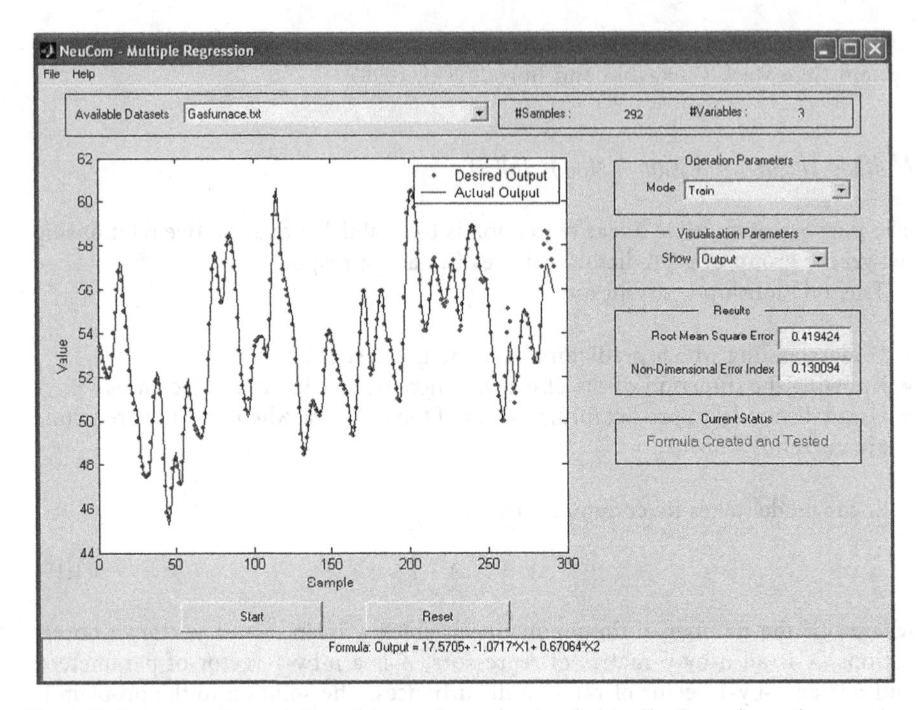

Fig. 1.13 Linear regression modelling of the gas-furnace benchmark data. The figure shows: the regression formula that approximates the data; the desired versus the approximated by the formula values of the time series; and the two error measures: root mean square error and the nondimensional error index.

Example 3

The following multiple linear regression model (three linear regressions) discriminates the samples among the three classes of the Iris data (data are normalised).

Class 1 (Setosa) $= 0.12 + 0.06\ x1 + 0.24\ x2 - 0.22\ x3 - 0.06\ x4$
Class 2 (Versicolor) $= 1.5 - 0.02\ x1 - 0.44\ x2 + 0.22\ x3 - 0.48\ x4$
Class 3 (Virginica) $= -0.68 - 0.04\ x1 + 0.2\ x2 + 0.004x3 + 0.55\ x4$

1.5 Support Vector Machines (SVM)

The support vector machine was first proposed by Vapnik and his group at AT&T Bell laboratories (Vapnik, 1998). For a typical learning task defined as probability estimation of output values y depending on input vectors $\vec{x}$:

$$P(\vec{x}, y) = P(y|\vec{x})P(\vec{x}) \tag{1.23}$$

an *inductive SVM* classifier aims to build a decision function

$$f_L : \vec{x} \rightarrow \{-1, +1\} \tag{1.24}$$

based on a training set,

$$f_L = L(S_{train})$$
$$where : S_{train} = (\vec{x}_1, y_1), (\vec{x}_2, y_2), \ldots, (\vec{x}_n, y_n) \quad (1.25)$$

In the SVM theory, the computation of f_L can be traced back to the classical structural risk minimization approach, which determines the classification decision function by minimizing the empirical risk, as

$$R = \frac{1}{l} \sum_{i=1}^{N} \left| f(\vec{x}_i) - y_i \right| \quad (1.26)$$

where N and f represent the size of the set of examples and the classification decision function, respectively; l is a constant for normalization. For SVM, the primary concern is determining an optimal separating hyperplane that gives a low generalization error. Usually, the classification decision function in the linearly separable problem is represented by

$$f_{\vec{w},b} = sign(\vec{w} \cdot \vec{x} + b) \quad (1.27)$$

In SVM, this optimal separating hyperplane is determined by giving the largest margin of separation between vectors that belong to different classes. It bisects the shortest line between the convex hulls of the two classes, which is required to satisfy the following constrained minimization conditions.

$$Minimize : \frac{1}{2} \vec{w}^T \vec{w}$$
$$Subject\ to : y_i(\vec{w} \cdot \vec{x} + b) \geq 1. \quad (1.28)$$

For the linearly nonseparable case, the minimization problem needs to be modified to allow misclassified data points. This modification results in a soft margin classifier that allows but penalizes errors, by introducing a new set of variables $\xi_{i=1}^l$ as the measurement of violation of the constraints (Fig. 1.14a):

$$Minimize : \frac{1}{2} \vec{w}^T \vec{w} + C(\textstyle\sum_{i=1}^{L} \xi_i)^k$$
$$Subject\ to : y_i(\vec{w} \cdot \varphi(\vec{x}_i) + b) \geq 1 - \xi_i. \quad (1.29)$$

where C and k are used to weight the penalizing variables $\xi_{i=1}^l$, and $\varphi(\cdot)$ is a nonlinear function that maps the input space into a higher-dimensional space. In order to solve the above equation, we need to construct a set of functions and implement the classical risk minimization on this set. Here, a Lagrangian method is used to solve the above problem. Then, the above equation can be written as

$$Minimize : F(\Lambda) = \Lambda \cdot 1 - \frac{1}{2} \Lambda \cdot D \cdot \Lambda$$
$$Subject\ to : \Lambda \cdot y = 0; \ \Lambda \leq C; \ \Lambda > 0, \quad (1.30)$$

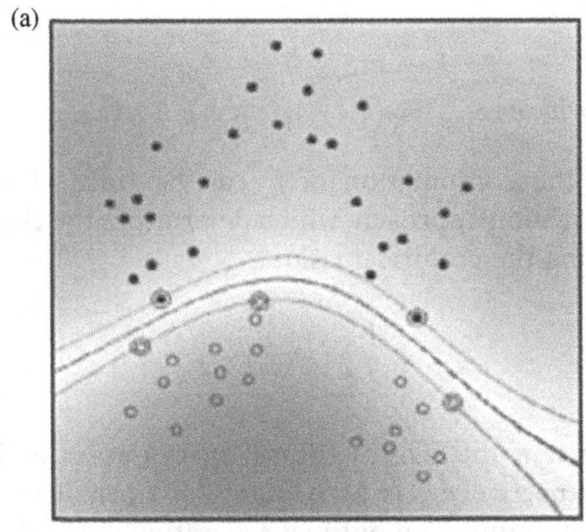

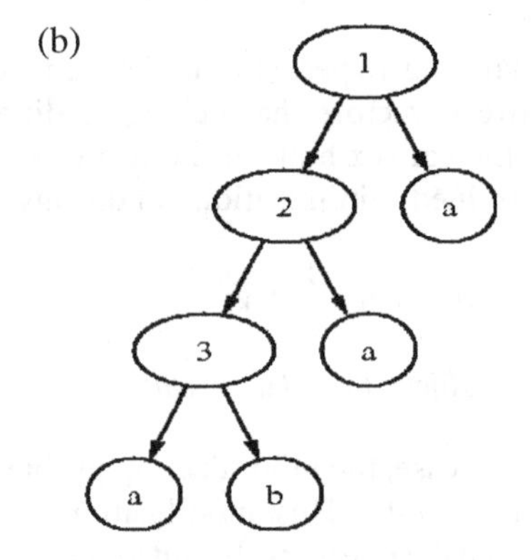

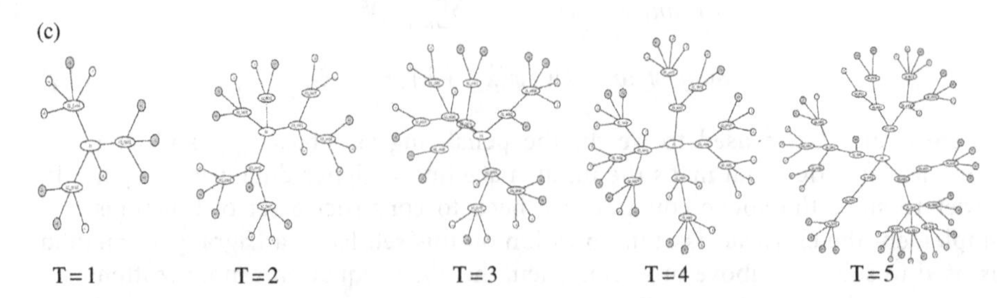

Fig. 1.14 (a) A SVM classifier builds a hyperplane to separate samples from different classes in a higher-dimensional space; the new vectors on the border are called support vectors; (b) a SVM tree, where each node is a SVM; (c) an evolving SVM tree evolves new nodes incrementally (from S. Pang *et al.* (2005)).

where $\Lambda = (\lambda_1, \cdots, \lambda_l)$, $D = y_i y_j \vec{x}_i \cdot \vec{x}_j$ for binary classification and the decision function can be rewritten as

$$f(x) = sign(\sum_{i=1}^{l} y_i \lambda_i^* (\vec{x} \cdot \varphi(\vec{x}_i) + b^*) \tag{1.31}$$

For more details, the reader is referred to Vapniak (1998) and Cherkassky and Mulier (1998).

Transductive SVM (TSVM)

In contrast to the inductive SVM learning method described above, transductive SVM (TSVM) learning includes knowledge of test set S_{test} in the training procedure, thus the above learning function of inductive SVM can be reformulated as (Kasabov and Pang, 2004)

$$\begin{aligned} f_L &= L(S_{train}, S_{test}). \\ where: S_{train} &= (\vec{x}_1^*, y_1^*), (\vec{x}_2^*, y_2^*), \ldots, (\vec{x}_n^*, y_n^*) \end{aligned} \tag{1.32}$$

Therefore, in a linearly separable data case, to find a labelling $y_1^*, y_2^*, \cdots, y_n^*$ of the test data, the hyper plane $< \vec{w}, b >$ should separate both training and test data with maximum margin:

$$\begin{aligned} &\textit{Minimize} \ \mathrm{Over}(\mathrm{y}_1^*, \mathrm{y}_2^*, \cdots, \mathrm{y}_3^*, \vec{\mathrm{w}}, \mathrm{b}): \\ &\qquad \frac{1}{2}\vec{w}^T\vec{w} \\ &\textit{Subject to}: \quad y_i(\vec{w} \cdot \vec{x}_i + b) \geq 1 \\ &\qquad\qquad y_j^*(\vec{w} \cdot \vec{x}_j^* + b) \geq 1. \end{aligned} \tag{1.33}$$

To be able to handle nonseparable data, similar to the way in the above inductive SVM, the learning process of transductive SVM can be formulated as the following optimization problem.

$$\begin{aligned} &\textit{Minimize} \ \mathrm{Over} \\ &(\mathrm{y}_1^*, \mathrm{y}_2^*, \cdots, \mathrm{y}_3^*, \vec{\mathrm{w}}, \mathrm{b}, \xi_1, \cdots, \xi_n, \xi_1^*, \cdots \xi_k^*): \\ &\qquad \frac{1}{2}\vec{w}^T\vec{w} + C(\sum_{i=1}^{L} \xi_i)^k + C(\sum_{j=}^{K} \xi_j^*)^k \\ &\textit{Subject to}: \quad y_i(\vec{w} \cdot \varphi(\vec{x}_i) + b) \geq 1 - \xi_i \\ &\qquad\qquad y_j^*(\vec{w} \cdot \varphi(\vec{x}_j^*) + b) \geq 1 - \xi_j^* \end{aligned} \tag{1.34}$$

where C^* is the effect factor of the query examples, and $C^* \xi_i^*$ is the effect term of the *i*th query example in the above objective function.

SVM Tree (SVMT)

The SVM tree is constructed by a divide-and-conquer approach using a binary class-specific clustering and SVM classification technique; see, for example, Fig. 1.14b (Pang and Kasabov, 2004; Pang *et al.*, 2006).

Basically, we perform two procedures at each node in the above tree generation. First, the class-specific clustering performs a rough classification because it splits the data into two disjoint subsets based on the global features. Next, the SVM classifier performs a 'fine' classification based on training supported by the previous separation result.

Figure 1.14b is an example of the SVM tree which is derived from the above SVM tree construction. As mentioned, the SVM test starts at the root node 1. If the test $T1\ (x) = +1$ is observed, the test $T2\ (x)$ is performed. If the condition $T1\ (x) = +1$ and $T2\ (x) = -1$ is observed, then the input data x are assigned to class a, and so forth.

SVM trees can evolve new nodes, new local SVM to accommodate new data from an input data stream. An example of an evolving SVM tree is shown in Fig.1.14c (Pang *et al.*, 2006).

1.6 Inductive Versus Transductive Learning and Reasoning. Global, Local, and 'Personalised' Modelling

1.6.1 Inductive Global and Local Modelling

Most learning models and systems in artificial intelligence developed and implemented thus far are based on inductive inference methods, where a model (a function) is derived from data representing the problem space and this model is further applied to new data (Fig. 1.15a). The model is usually created without taking into account any information about a particular new data vector (test data). An error is measured to estimate how well the new data fit into the model.

The models are in most cases global models, covering the whole problem space. Such models are, for example, regression functions, some NN models, and also some SVM models, depending on the kernel function they use. These models are difficult to update on new data without using old data previously used to derive the models. Creating a global model (function) that would be valid for the whole problem space is a difficult task, and in most cases it is not necessary to solve.

Some global models may consist of many local models that collectively cover the whole space and can be adjusted incrementally on new data. The output for a new vector is calculated based on the activation of one or several neighbouring local models. Such systems are the evolving connectionist systems (ECOS) – for example, EFuNN and DENFIS – presented in Chapters 3 and 5, respectively.

The inductive learning and inference approach is useful when a global model ('the big picture') of the problem is needed even in its very approximate form. In some models (e.g. ECOS) it is possible to apply incremental online learning to adjust this model on new data and trace its evolution.

1.6.2 Transductive Modelling. WKNN

In contrast to the inductive learning and inference methods, transductive inference methods estimate the value of a potential model (function) only in a single point of the space (the new data vector) utilising additional information related to this point (Vapnik, 1998). This approach seems to be more appropriate for clinical and medical applications of learning systems, where the focus is not on the model, but on the individual patient. Each individual data vector (e.g. a patient in the medical area, a future time moment for predicting a time series, or a target day for predicting a stock index) may need an individual local model that best fits the new data, rather than a global model. In the latter case the new data are matched into a model without taking into account any specific information about these data.

Transductive inference is concerned with the estimation of a function in a single point of the space only. For every new input vector $\boldsymbol{xi}$ that needs to be processed for a prognostic task, the Ni nearest neighbours, which form a subdata set Di, are derived from an existing dataset D and, if necessary, generated from an existing model M. A new model Mi is dynamically created from these samples to approximate the function in the point $\boldsymbol{xi}$. The system is then used to calculate the output value y_i for this input vector $\boldsymbol{x}_i$ (Fig. 1.15b,c).

A very simple transductive inference method, the k-nearest neighbour method (K-NN) is briefly introduced here. In the K-NN method, the output value y_i for a new vector $\boldsymbol{xi}$ is calculated as the average of the output values of the k nearest samples from the dataset Di. In the weighted K-NN method (WKNN) the output y_i is calculated based on the distance of the Ni nearest neighbour samples to $\boldsymbol{x}_i$:

$$yi = \frac{\sum_{j=1}^{Ni} w_j y_j}{\sum_{j=1}^{Ni} w_j} \tag{1.35}$$

where y_j is the output value for the sample $\boldsymbol{x}_j$ from Di and w_j are their weights measured as

$$wj = \frac{\max(\boldsymbol{d}) - [dj - \min(\boldsymbol{d})]}{\max(\boldsymbol{d})} \tag{1.36}$$

The vector $\boldsymbol{d} = [d1, d2, \ldots, d_{Ni}]$ is defined as the distances between the new input vector $\boldsymbol{x}_i$ and Ni nearest neighbours $(\boldsymbol{x}_j, y_j)$ for $j = 1$ to Ni; max(d) and min(d) are the maximum and minimum values in d, respectively. The weights wj have the values between min(d)/max(d) and 1; the sample with the minimum distance to the new input vector has the weight value of 1, and it has the value min(d)/max(d) in case of maximum distance.

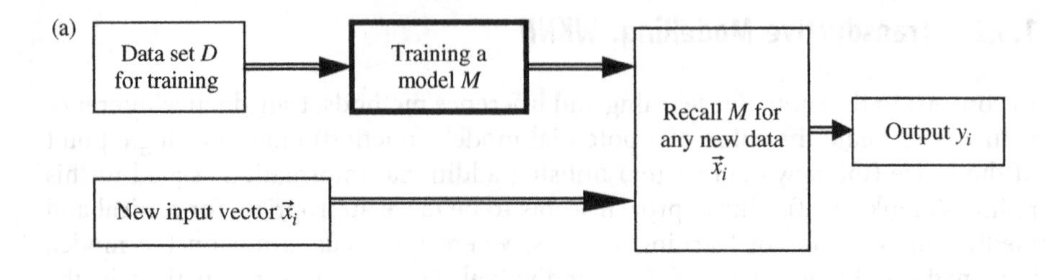

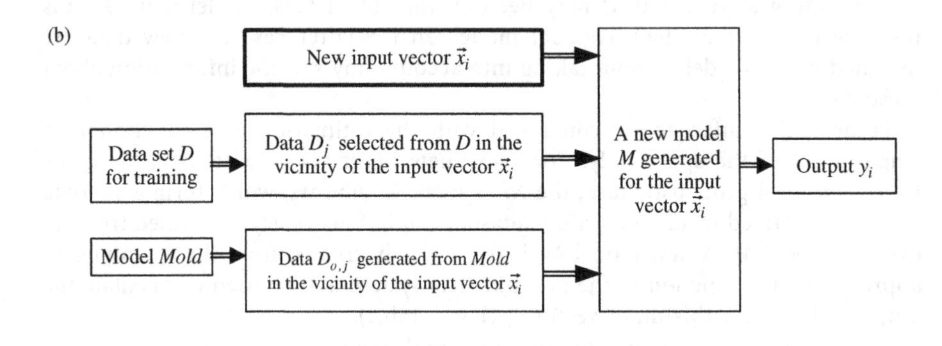

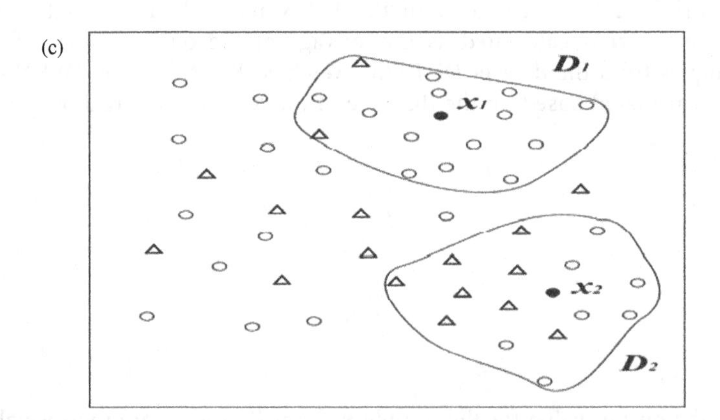

Fig. 1.15 (a) Inductive learning: given a training set, construct a model M that will accurately represent the examples in the set; recall the model M on a new example *xi* to evaluate the output *yi*. (b) Transductive learning: for every new input vector *xi*, a new model *Mi* is dynamically created from the available samples to approximate the function in the locality of the point *xi*. (c) A transductive model is created with a subtraining dataset of neighbouring samples for each new input vector. This is shown here as two vectors ***x*1** and ***x*2**.

Distance is usually measured as Euclidean distance:

$$\|\mathbf{x}-\mathbf{y}\| = \left[\frac{1}{P}\sum_{j=1}^{P}\left|x_j - y_j\right|^2\right]^{\frac{1}{2}} \tag{1.37}$$

Distance can be also measured as Pearson correlation distance, Hamming distance, cosine distance, etc. (Cherkassky and Mulier, 1998).

1.6.3 Weighted Examples – Weighted Variables K-NN: WWKNN

In the WKNN the calculated output for a new input vector depends not only on the number of its neighbouring vectors and their output values (class labels), as in the KNN method, but on the distance between these vectors and the new vector which is represented as a weight vector (W). It is assumed that all v input variables are used and the distance is measured in a v-dimensional Euclidean space with all variables having the same impact on the output variable.

But when the variables are ranked in terms of their discriminative power of class samples over the whole v-dimensional space, we can see that different variables have different importance to separate samples from different classes, therefore a different impact on the performance of a classification model. If we measure the discriminative power of the same variables for a subspace (local space) of the problem space, the variables may have a different ranking.

Using the ranking of the variables in terms of a discriminative power within the neighborhood of K vectors, when calculating the output for the new input vector, is the main idea behind the WWKNN algorithm (Kasabov, 2007b), which includes one more weight vector to weigh the importance of the variables. The distance d_j between a new vector $\boldsymbol{x}_i$ and a neighboring one $\boldsymbol{x}_j$ in 1.36 is calculated now as:

$$d_j = \text{sqr}[\text{sum}_{l=1 \text{ to } v}(c_{i,l}(x_{i,l} - x_{j,l}))^2] \tag{1.38}$$

where $c_{i,l}$ is the coefficient weighing variable x_l in a neighbourhood of $\boldsymbol{x}_i$. It can be calculated using a signal-to-noise ratio procedure that ranks each variable across all vectors in the neighbourhood set Di of $\boldsymbol{Ni}$ vectors:

$$\boldsymbol{Ci} = (c_{i,1}, c_{i,2}, \ldots, c_{i,v}) \tag{1.39}$$

$$c_{i,l} = S_l/\text{sum}(S_l) \text{ for } l = 1, 2, \ldots, v$$

where

$$S_l = abs(M_l^{(\text{class } 1)} - M_l^{(\text{class } 2)})/(\text{Std}_l^{(\text{class } 1)} + \text{Std}_l^{(\text{class2})}) \tag{1.40}$$

Here $M_l^{(\text{class } 1)}$ and $\text{Std}_l^{(\text{class } 1)}$ are, respectively, the mean value and the standard deviation of variable x_l for all vectors in Di that belong to class 1.

The new distance measure, that weighs all variables according to their importance as discriminating factors in the neighbourhood area Di, is the new element in the WWKNN algorithm when compared to the WKNN.

Using the WWKNN algorithm, a 'personalised' profile of the variable importance can be derived for any new input vector that represents a new piece of 'personalised' knowledge.

Weighting variables in personalised models is used in the TWNFI models (transductive weighted neuro-fuzzy inference) in Song and Kasabov (2005, 2006).

There are several open problems related to transductive learning and reasoning, e.g. how to choose the optimal number of vectors in a neighbourhood and the optimal number of variables, which for different new vectors may be different (Mohan and Kasabov, 2005).

1.7 Model Validation

When a machine learning model is built based on a dataset *S*, it needs to be validated in terms of its generalisation ability to produce good results on new, unseen data samples. There are several ways to validate a model:

1) Train-test split of data: Splitting the dataset *S* into two sets: *Str* for training, and *Sts* for testing the model.
2) *K*-fold cross validation (e.g. 3, 5, 10): in this case the dataset *S* is split randomly into *k* subsets S1,S2,..., *Sk* and $i = 1, 2, \ldots, k$ times a model *Mi* is created on the dataset *S–Si* and tested on the set *Si*; the mean accuracy across all *k* experiments is calculated.
3) Leave-one-out cross-validation (a partial case of the above method when the dataset *S* is split *N* times; in each subset there is only one sample).

What concerns the whole task of feature selection, model creation, and model validation, the above methods can be applied in two different ways:

1) A 'biased' way – features are selected from the whole set *S* using a filtering-based method, and then a model is created and validated on the selected features.
2) An 'unbiased' way – for every data subset *Si* in a cross-validation procedure, first features *Fi* are selected from the set *S–Si* (using some of the above-discussed methods, e.g. SNR) and then a model is created based on the feature set *Fi*; the model *Mi* is validated on *Si* using features *Fi*. The leave-one-version of this procedure is outlined in Box 1.2.

Box 1.2. Leave-one-out cross validation procedure

For *i* :=1 to *N* do
Take out sample *Si* from the data set *S*
Use the rest (*S–Si*) samples for feature selection *Fi* (optional)
Train a model *Mi* on *S–Si* using features *Fi*
Test the model *Mi* on the left-out-sample *Si*, evaluate error *Ei*
end
Evaluate the overall mean error
Evaluate the features used, their frequency of selection in the iterations.
Train a final model *M* on all data and on the most frequently selected features

Example

The unbiased leave-one-out procedure is illustrated on another benchmark dataset that is used further in the book, the leukaemia classification problem of AML/ALL classes (Golub *et al.*, 1999). The dataset consists of 38 samples for training a model and 34 test samples, each having 7129 variables representing the expression of genes in two classes each of the samples from the class of AML and class of ALL leukaemia types.

Figure 1.16a shows the result of the unbiased feature selection and model validation procedure, where only the top four genes are selected on each of the 38 runs of the procedure and a k-NN model is used, $k = 3$. The overall accuracy is above 92% and the top selected four genes are shown in the diagram with their gene numbers.

The selected-above top four genes are used to build a (final) MLR model and to test it on the test data of 34 samples using the same four variables. The results, in the form of a confusion table, are shown in Fig. 1.16b. The coefficients of each of the regression formulas (shown in a box) represent the participation of each of the variables in terms of positive or negative and in terms of importance. This is important knowledge contained in the MLR model that needs to be further analysed.

A transductive modelling approach can be applied when for every vector from the test data, the closest K samples are selected from the training data using the already-selected four genes and an individual MLR model is created for this sample after which it can be used to test the model.

1.8 Exercise

Specification:

1) Select a classification or a prediction problem and a data set for it (e.g. from http://www.kedri.info, or from the repository of machine learning databases.: http://www.ics.uci.edu/~mlearn/MLRepository.html UC Irvine).
2) Select features using some of the methods from this chapter (e.g. SNR, t-test).
3) Create a global statistical model using MLR through inductive learning.
4) Validate the model and evaluate its accuracy in a leave-one-out cross-validation mode.
5) Create individual models through transductive learning and evaluate their average accuracy.
6) Answer the following questions.

Q1. Which of the models is adaptive to new data?
Q2. What knowledge can be learned from the models?

1.9 Summary and Open Problems

This chapter introduces the basic concepts in CI modelling and some benchmark datasets that are used in the rest of the chapters.

(a)

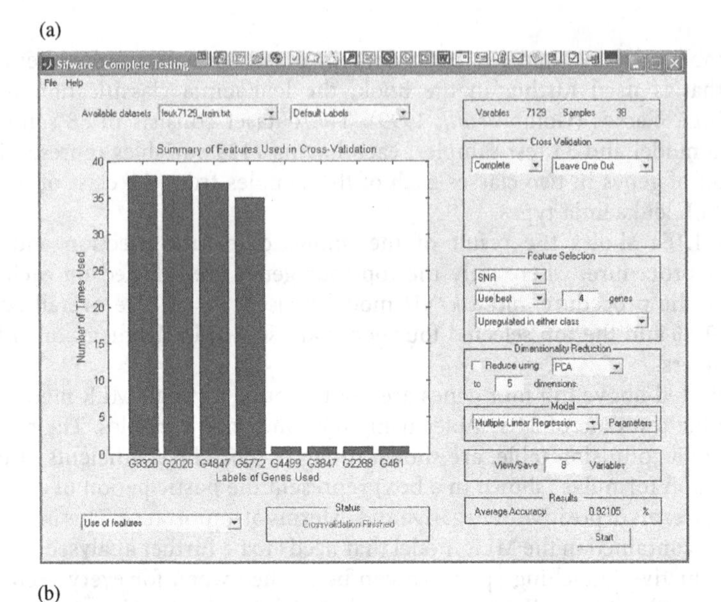

(b)

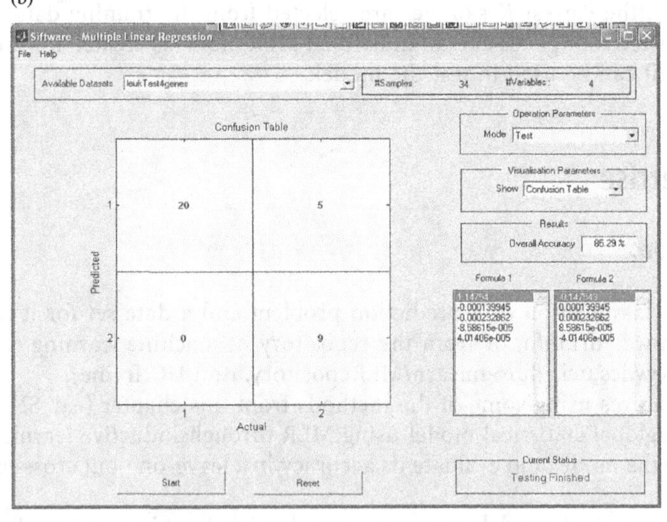

Fig. 1.16 (a) The result of an unbiased feature selection and model validation procedure, where only the top four genes are selected at each of the 38 runs of the procedure using the SNR method to rank the variables and a multiple linear regression (MLR) model for classification. The overall accuracy is above 92% and the top selected four genes are shown in the diagram with their gene numbers. (b) The four genes selected in (a – top) are used to build the final inductive MLR model and to test it on the test data of 32 samples using the same four variables. The results are shown as a confusion table (a proprietary software is used, SIFTWARE, www.peblnz.com). See www.kedri.info for colour figures.

This chapter also raises some open questions, such as:

How do we identify the problem space and the dimensionality in which a process is evolving having only a limited data collected?
Thus far, Euclidean space has predominantly been used, but is it appropriate to use it for all cases?
Most of the machine learning models use time as a linear variable, but is that the only way to present it?
How do we define the best model for the purpose of modelling an evolving process?
Prediction modelling in an open problem space: how is it verified and evaluated?
In an EIS it may be important how fast the 'intelligence' emerges within a learning system and within a population of such systems. How do we make this process faster for both machines and humans?
Can a system become faster and more efficient than humans in acquiring intelligence, e.g. in learning multiple languages?

The rest of the chapters in this part present evolving connectionist methods for incremental, adaptive, knowledge-based learning. The methods are illustrated using several benchmark datasets, some of them presented in this chapter. These methods are applied to real-world problems from life sciences and engineering in Part II of the book. All these applications deal with complex, evolving, continuous, dynamically changing processes.

1.10 Further Reading

- *Statistical Learning* (Vapnik, 1998; Cherkassky and Mulier, 1998)
- *Incremental LDA Feature Selection and Modelling* (Pang *et al.*, 2005, 2006)
- *Incremental PCA Feature Selection and Modelling* (Ozawa *et al.*, 2005, 2006)
- *SVM* (Vapniak, 1998)
- *Chaotic Processes* (Barndorff-Nielsen *et al.*, 1993; Gleick, 1987; Hoppensteadt, 1989; McCauley, 1994)
- *Emergence and Evolutionary Processes* (Holland, 1998)
- *Introduction to the Principles of Artificial Neural Networks* (Aleksander, 1989; Aleksander and Morton, 1990; Amari, 1967, 1990; Arbib, 1972, 1987, 1995, 2002; Bishop, 1995; Feldman, 1989; Hassoun, 1995; Haykin, 1994; Hecht-Nielsen, 1987; Hertz *et al.*, 1991; Hopfield, 1982; Kasabov, 1996; Rumelhart *et al.*, 1986; Werbos, 1990; Zurada, 1992)
- *Principles and Classification of online Learning Connectionist Models* (Murata *et al.*, 1997; Saad, 1999)
- *ANN and MLP for Data Analysis* (Gallinari *et al.*, 1988)
- *Catastrophic Forgetting in Multiplayer Perceptrons and other ANN* (Robins, 1996)
- *Time-series Prediction* (Weigend *et al.*, 1990; Weigend and Gershefeld, 1993)
- *Local Learning* (Bottu and Vapnik, 1992; Shastri, 1999)
- *Emerging Intelligence (EI) in Autonomous Robots* (Nolfi and Floreano, 2000)

- *Integrating ANN with AI and Expert Systems* (Barnden and Shrinivas, 1991; Giacometti *et al.*, 1992; Hendler and Dickens, 1991; Hinton, 1990; Kasabov, 1990; Medsker, 1994; Morasso *et al.*, 1992; Touretzky and Hinton, 1985, 1988; Towell and Shavlik, 1993, 1994; Tresp *et al.*, 1993)
- *Integrating ANN with Fuzzy Logic* (Furuhashi *et al.*, 1993; Hayashi, 1991; Kasabov, 1996; Kosko, 1992; Takagi, 1990; Yamakawa and Tomoda, 1989)
- *Incremental PCA and LDA* (Pang *et al.*, 2005a, 2005b; Ozawa *et al.*, 2004, 2005)
- *Transductive Learning and Reasoning* (Vapniak, 1998; Cherkassky and Mulier, 1998; Song and Kasabov, 2005, 2006)
- *Comparison Among Local, Global, Inductive, and Transductive Modelling* (Kasabov, 2007b)

2. *Evolving Connectionist Methods for Unsupervised Learning*

Unsupervised learning methods utilise data that contain input vectors only. Evolving unsupervised learning methods are about learning from a data stream of unlabelled data e.g. financial market, biological data, patient medical data, weather data, mobile telephone calls, or radioastronomy signals from the universe. They develop their structure to model the incoming data in an incremental, continuous learning mode. They learn statistical patterns such as clusters, probability distribution, and so on.

This chapter presents various methods for unsupervised adaptive incremental learning that include clustering, prototype learning, and vector quantisation, along with their generic applications for data analysis, filling missing values in data, classification, transductive learning, and reasoning. The learned clusters, categories, and the like represent new knowledge. The emphasis here is put on the model adaptability - they are evolving, and on their features to facilitate rule extraction and pattern/knowledge discovery, which are the main objectives of the knowledge engineering approach that we take in this book. The chapter material is presented in the following sections.

- Unsupervised learning from data; distance measure
- Clustering
- Evolving clustering. ECM.
- Vector quantisation. SOM. ESOM
- Prototype learning. ART
- Generic applications of unsupervised learning methods
- Exercise
- Summary
- Further readings

2.1 Unsupervised Learning from Data. Distance Measure

2.1.1 General Notions

As pointed out in Chapter 1, many real-world information systems use data streams. Such data streams are, for example: financial data such as stock market indexes; video streams transferred across the Internet; biological information, made available in an increasing volume, such as DNA and protein data; patient data; climate information; radioastronomy signals; etc. To manipulate a large amount of data in an adaptive mode and to extract useful information from it, adaptive, knowledge-based methods are needed.

Evolving, unsupervised learning methods are concerned with learning statistical and other information characteristics and knowledge from a continuous stream of data. The distribution of the data in the stream may not be known in advance. Such unsupervised methods are adaptive clustering, adaptive vector quantisation, and adaptive prototype learning presented in the next sections. The similarity and the difference among clustering, quantisation, and prototyping is schematically illustrated in Fig. 2.1. The time line and 'time-arrow' on the figure show the order in which the data vectors are presented to the learning system. Different methods for unsupervised evolving connectionist systems are presented and illustrated in the rest of the chapter.

2.1.2 Measuring Distance in Unsupervised Learning Techniques

In the context of clustering, quantisation, and prototype learning, we can assume that we have a data manifold X of dimension d; i.e., $X \subseteq R^d$. We aim at finding a set of vectors $\{c_1, \ldots, c_n\}$, that encodes the data manifold with small quantisation error. Vector quantisation usually utilizes a competitive rule; i.e. a new input vector $\boldsymbol{x}$ is represented by the best matching unit $\boldsymbol{c}_i$, that satisfies the conditions:

$$||\boldsymbol{x} - \boldsymbol{c}_i|| \leq ||\boldsymbol{x} - \boldsymbol{c}_j||, \forall j \neq i, i, j \in [1, n] \quad (2.1)$$

where $||\boldsymbol{x}\text{-}\boldsymbol{c}_i||$ measures a distance.

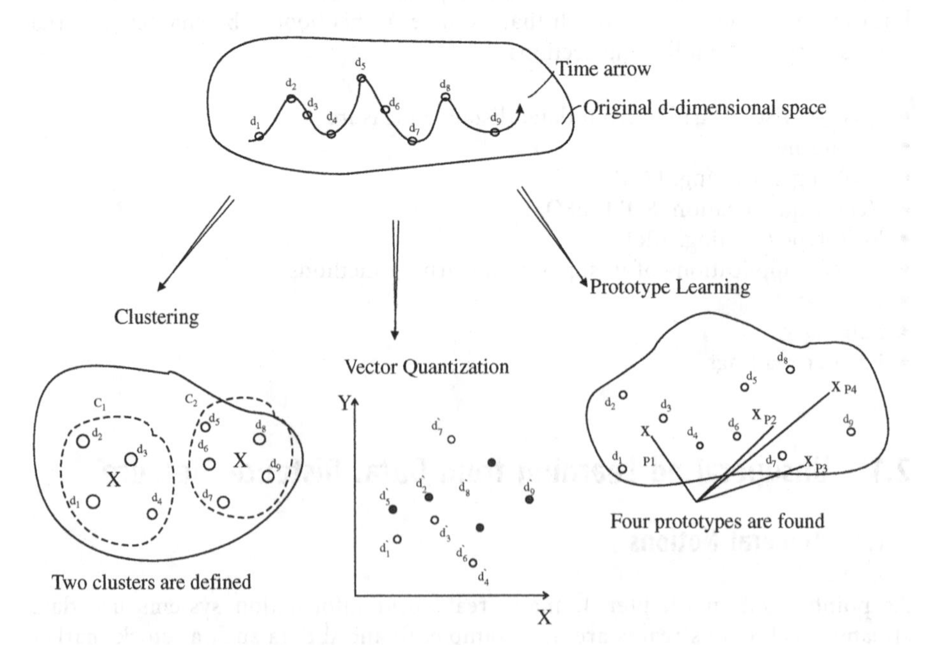

Fig. 2.1 Clustering, vector quantization, and prototype learning as unsupervised learning techniques.

The goal is to minimize the reconstruction error

$$E = \sum_{x \in X} [\Pi(x)(x - c_{i(x)})^2] \tag{2.2}$$

Here $\Pi(x)$ is the probability distribution of data vectors over the manifold X.

Measuring distance is a fundamental issue in all the above-listed methods. The following are some of the most used methods for measuring distance, illustrated on two n-dimensional data vectors $\boldsymbol{x} = (x_1, x_2, \ldots, x_n)$ and $\boldsymbol{y} = (y_1, y_2, \ldots, y_n)$:

- Euclidean distance:

$$D(\boldsymbol{x}, \boldsymbol{y}) = \sqrt{\left[\left(\sum_{i=1,..,n} (x_i - y_i)^2\right) / n\right]} \tag{2.3}$$

- Hamming distance:

$$D(\boldsymbol{x}, \boldsymbol{y}) = \left(\sum_{i=1,..,n} |x_i - y_i|\right) / n\} \tag{2.4}$$

where absolute values of the difference between the two vectors are used.

- Local fuzzy normalized distance (see Chapter 3; also Kasabov (1998)):

A local normalised fuzzy distance between two fuzzy membership vectors $\boldsymbol{x}_f$ and $\boldsymbol{y}_f$ that represent the membership degrees to which two real vector data $\boldsymbol{x}$ and $\boldsymbol{y}$ belong to predefined fuzzy membership functions is calculated as:

$$D(\boldsymbol{x}_f, \boldsymbol{y}_f) = ||\boldsymbol{x}_f - \boldsymbol{y}_f|| / ||\boldsymbol{x}_f + \boldsymbol{y}_f|| \tag{2.5}$$

where $||\boldsymbol{x}_f - \boldsymbol{y}_f||$ denotes the sum of all the absolute values of a vector that is obtained after vector subtraction (or summation in case of $||\boldsymbol{x}_f + \boldsymbol{y}_f||$) of two vectors $\boldsymbol{x}_f$ and $\boldsymbol{y}_f$ of fuzzy membership values; / denotes division.

- Cosine distance:

$$D = 1 - SUM\left(\sqrt{x_i y_i} \middle/ \sqrt{x_i^2}\sqrt{y_i^2}\right) \tag{2.6}$$

- Correlation distance:

$$D = 1 - \sum_{i=1}^{n} (x_i - \overline{x_i})(y_i - \overline{y_i}) \middle/ \sum_{i=1}^{n} (x_i - \overline{x_i})^2 (y_i - \overline{y_i})^2 \tag{2.7}$$

where $\overline{x_i}$ is the mean value of the variable x_i.

Some examples of measuring distance are shown in Fig. 2.2, which illustrates both Euclidean and fuzzy normalized distance. Using Euclidean distance may require normalization beforehand as illustrated in the figure. In this figure $x1$ is in the range of [0,100] and $x2$ is in the range of [0,1]. If $x1$ is not normalised, then the Euclidean distance $D\ (A, B)$ is greater than the distance $D(C, D)$. Otherwise, it will

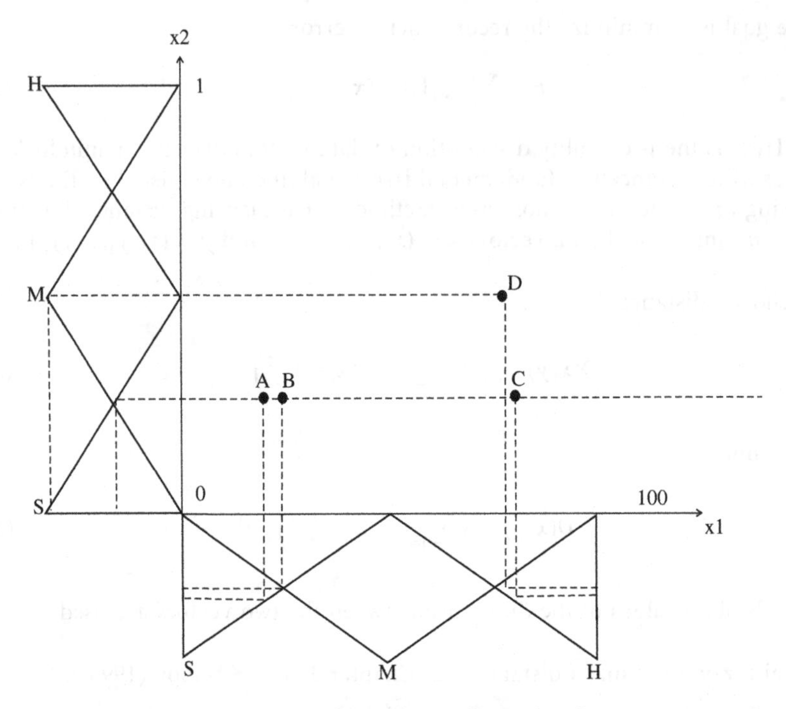

Fig. 2.2 Euclidean versus fuzzy normalized and fuzzy distance illustrated on four points in a two-dimensional space ($x1$, $x2$). If the variable values are not normalised, the Euclidean distance between A and B will be greater than the distance between D and C as the range of variable $x1$ is 100 times larger than the range of the variable $x2$. If either normalised or fuzzified (three membership functions, denoted S for small, M for medium, and H for high) values are used, the relative distance between D and C will be greater than the distance between A and B.

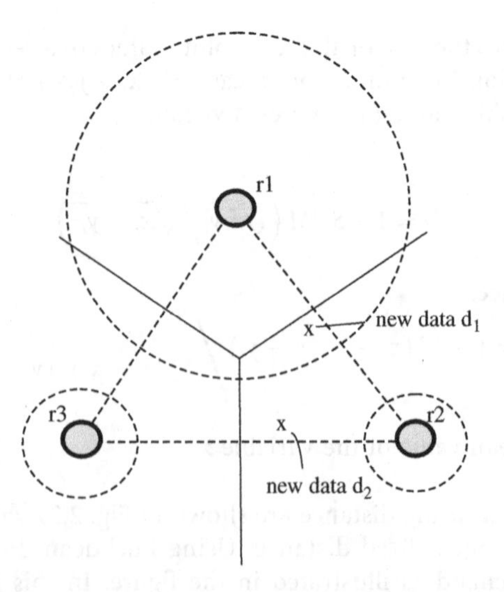

Fig. 2.3 Voronoi tessellation (the straight solid lines) versus hypersphere separation (the circles) of a hypothetical problem space separating three clusters – r_1, r_2, and r_3 – defined by their centres and hyperspheres.

be the opposite. For the fuzzy normalised distance $D(A, B) < D(C, D)$ is always held. In the example, three membership functions are used: Small (S), Medium (M), and High (H) for each of the two variables.

Figure 2.3 illustrates two ways of space partitioning among three nodes $r1$, $r2$, and $r3$: Voronoi tessellation (see Okabe *et al.* (1992)), the straight lines, and hyperspheres. The latter is described in detail in Chapter 3 for the EFuNN model. When using Voronoi tessellation a new data vector $\boldsymbol{d}1$ will be allocated to node $r2$, whereas if using hyperspheres, it will be allocated to $r1$. A new data point $d2$ will be allocated to $r2$ in the first case, but there will not be a clear allocation in the second case.

2.2 Clustering

2.2.1 Batch-Mode versus Evolving Clustering

Clustering is the process of defining how data are grouped together based on similarity. Clustering results in the following outcomes.

- Cluster centres: These are the geometrical centers of the data grouped together; their number can be either predefined (batch-mode clustering) or not defined a priori but evolving.
- Membership values, defining for each data vector to what cluster it belongs. This can be either a crisp value of 1 (the vector belongs to a cluster) or 0 (it does not belong to a cluster, as it is in the k-means method), or a fuzzy value between 0 and 1 showing the level of belonging; in this case the clusters may overlap (fuzzy clustering).

Evolving, adaptive clustering is the process of finding how data from a continuous input stream $z(t)$, $t = 0, 1, 2, \ldots$ are grouped (clustered) together at any time moment t. It requires finding the cluster centres, the cluster area occupied by each cluster in the data space, and the membership degrees to which data examples belong to these clusters.

New data, entered into the system, are either associated with existing clusters and the cluster characteristics are changed, or new clusters are created. Based on the current p vectors from an input stream $\boldsymbol{x}_1, \boldsymbol{x}_2, \boldsymbol{x}_3, \ldots, \boldsymbol{x}_p, \ldots$, n clusters are defined in the same input space, so that $n << p$. The cluster centres can be represented as points in the input space X of the data points. Adaptive evolving clustering assumes that each input data vector from the continuous input data stream is presented once to the system as it is assumed that it will not be accessible again. Adaptive clustering is a type of incremental learning, so each new data example $\boldsymbol{x}_i$ contributes to the changes in the clusters and this process can be traced over time. Through tracing an adaptive clustering procedure, it can be observed and understood how the modelled process has developed over time.

In contrast to the adaptive incremental clustering, off-line batch mode clustering methods are usually iterative, requiring many iterations to find the cluster centres that optimise an objective function. Such a function minimizes the distance between the data elements and their clusters, and maximizes the distance between the cluster centres. Such methods for example are k-means clustering (MacQueen, 1967), hierarchical clustering, and fuzzy C-means clustering (Bezdek, 1981, 1987, 1993).

2.2.2 *K*-Means Clustering

A popular clustering method is the K-means algorithm (MacQueen, 1967), which finds K disjoint groups of data (clusters) and their cluster centres as the mean of data vectors within a cluster. This procedure minimises the sum of the distances for each data vector and its closest cluster centre. Usually, this is done in a batch mode through many iterations, starting with K randomly selected cluster centres (Lloyd, 1982). The adaptive version of the K-means algorithm (MacQueen, 1967; Moody and Darken, 1989), applied without prior knowledge of the data distribution, is a stochastic gradient descent on Eq. (2.2). Starting with K randomly selected cluster centres, c_i, $I = 1, 2, \ldots, K$, for each new data vector x the closest cluster centre is updated as follows.

$$\Delta c_i = x - c_i, \text{ if } c_i \text{ is the closest cluster centre for } x;\ \Delta c_i = 0 \text{ otherwise (for } j \neq i) \tag{2.8}$$

This learning rule is also referred as the 'local k-means algorithm'. It is of the winner-takes-all type and can operate in a dynamic environment with continuously arriving data. But it can also suffer from confinement to a local minimum (Martinetz *et al.*, 1993). To avoid this problem some 'soft' computing schemes are proposed to modify reference vectors (cluster centres), in which not only the 'winner' prototype is modified, but all reference vectors are adjusted depending on their proximity to the input vector.

In both batch mode and adaptive mode of K-means clustering, the number of clusters is predefined in advance. The K-means clustering method uses an iterative algorithm that minimizes the sum of distances from each sample to its cluster centre over all clusters until the sum cannot be decreased further. The control of the minimisation procedure is done through choosing the number of clusters, the starting positions of the clusters (otherwise they will be randomly positioned), and number of iterations.

As the data vectors are grouped together in a predefined number of clusters based on similarity measure, if Euclidean distance is used, the clustering procedure may result in different cluster centers if data are normalised (scaled into a given interval, e.g. [0,1], either in a linear or in an nonlinear fashion), versus nonnormalised data; see Fig. 2.4a,b.

Another method for clustering is the DCA (dynamic clustering algorithm; Bruske and Sommer (1995)). The method does require a predefined number of clusters. This algorithm is used for dynamic fMRI cortical activation analysis data.

2.2.3 Hierarchical Clustering

The hierarchical clustering procedure finds similarity (distance) between each pair of samples using correlation analysis, and then represents this similarity as a

(a)

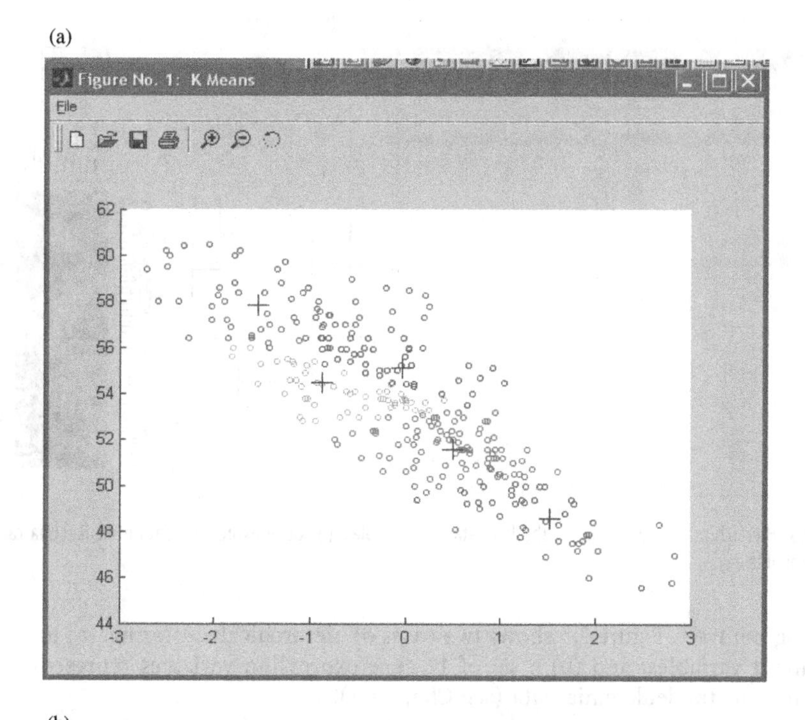

(b)

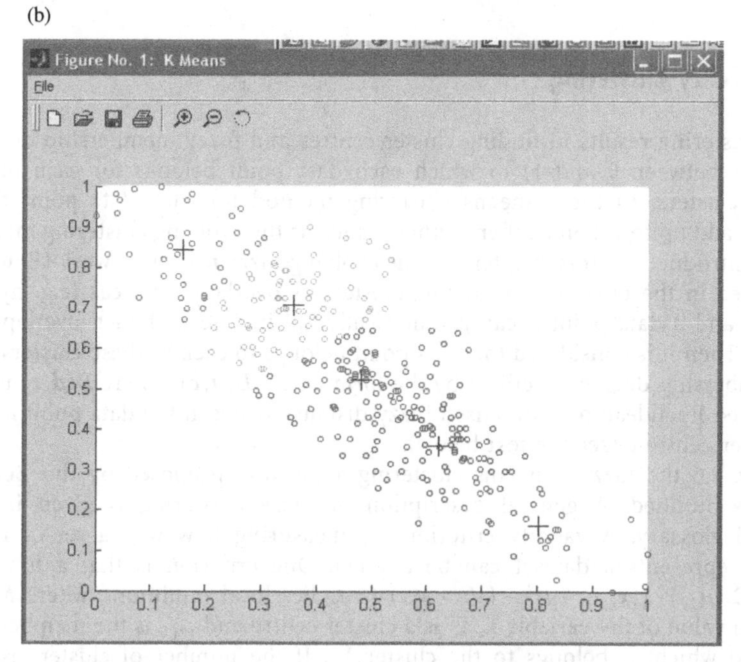

Fig. 2.4 An illustration of k-means clustering on the case study of gas-furnace data (see Fig. 1.3). The procedure results in different cluster centres and membership values for the data vectors that are not normalised, shown in (a), versus linearly normalised in the interval [0,1] data as shown in (b).

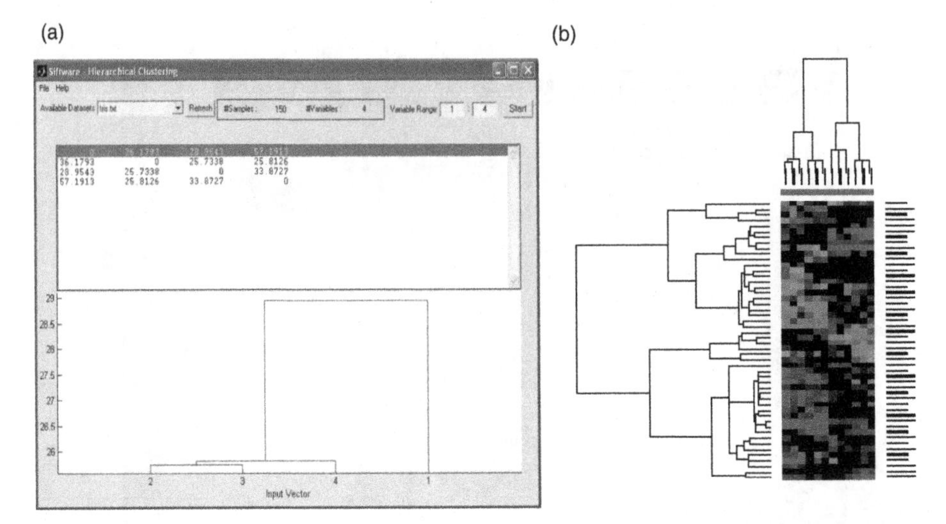

Fig. 2.5 Hierarchical clustering: (a) of the Iris data – 4 variables; (b) Gene expression data of Leukaemia cancer – 12 variables.

dendogram tree . Figure 2.5 shows two cases of hierarchical clustering: (a) the four Iris input variables, and (b) a set of 12 gene expression variables represented as columns for the leukaemia data (see Chapter 1).

2.2.4 Fuzzy Clustering

Fuzzy clustering results in finding cluster centres and fuzzy membership degrees (numbers between 0 and 1) to which each data point belongs for each of the existing clusters. In the *C*-means clustering method for each data point these numbers add up to 1. Some other methods, such as the evolving clustering method (ECM) introduced in this chapter and the evolving fuzzy neural network (EFuNN) introduced in the next chapter, define clusters as overlapping areas (e.g. hyper-spheres) and a data point $\boldsymbol{d}$ can geometrically fall into several such overlapping clusters. Then it is considered that this point belongs to each of these clusters and the membership degree is defined by the formula $1 - D(\boldsymbol{d}, \boldsymbol{c})$, where $D(\boldsymbol{d}, \boldsymbol{c})$ is the normalised Euclidean or normalised fuzzy distance between the data point $\boldsymbol{d}$ and the cluster centre $\boldsymbol{c}$ (see the text below).

In Fig. 2.6 the fuzzy *C*-means clustering algorithm, proposed by Jim Bezdek (1981) is outlined. A general description of fuzzy clustering is given in the extended glossary. A validity criterion for measuring how well a set of fuzzy clusters represents a dataset can be applied. One criterion is that a function $J(c) = \Sigma\Sigma(\mu_{i,k})^2\ [(x_k - Vi)^2 - (Vi - Mx)^2]$ reaches local minimum, where Mx is the mean value of the variable x, V_i is a cluster centre and $\mu_{i,k}$ is the membership degree to which x_k belongs to the cluster V_i. If the number of clusters is not defined, then the clustering procedure should be applied until a local minimum of $J(c)$ is found, which means that c is the optimal number of clusters. One of the advantages of the *C*-means fuzzy clustering algorithm is that it always converges

1. Initialise c fuzzy cluster centers $V_1, V_2, \ldots, V_c$ arbitrarily and calculate the membership degrees $\mu_{i,k}$ $i = 1,2,\ldots,c$, $k = 1,2,\ldots,n$ such that the general conditions are met.
2. Calculate the next values for cluster centres:

$$V_i = \left(\sum_{k=1}^{n} (\mu_{i,k})^2 . x_k\right) / \left(\sum_{k=1}^{n} (\mu_{i,k})^2\right), \quad for \quad i = 1, 2, \ldots, c$$

3. Update the fuzzy degrees of membership:

$$\mu_{i,k} = \frac{1}{\left[\sum_{j=1}^{c} \frac{d_{ik}}{d_{jk}}\right]}, for \quad d_{i,k} > 0, \forall i, k$$

where: $d_{i,k} = (x_k - V_i)^2$, $d_{j,k} = (x_k - V_j)^2$ (Euclidean distance)

4. If the currently calculated values V_i for the cluster centers are not different from the values calculated at the previous step (subject to a small error ε), then stop the procedure, otherwise go to step 2.

Fig. 2.6 A general algorithm of Bezdek's fuzzy *C*-means clustering (Bezek, 1981, from Kasabov (1996), MIT Press, reproduced with permission).

to a strict local minimum. A possible deficiency is that the shape of the clusters is ellipsoidal, which may not be the most suitable form for a particular dataset.

In most of the clustering methods (k-means, fuzzy C-means, ECM, etc.) the cluster centres are geometrical points in the input space; e.g. c is ($x = 3.7$, $y = -2.3$). But in some other methods such as the EFuNN, not only may each data point belong to the clusters to different degrees (fuzzy), but the cluster centres are defined as fuzzy coordinates and a geometrical area associated with this cluster. For example, a cluster centre c can be defined as (x is Small to a degree of 0.7, and y is Medium to a degree of 0.3; radius of the cluster area $R = 0.3$). Such clustering techniques are called fuzzy-2 clustering in this book.

Fuzzy clustering is an important data analysis technique. It helps to represent better the ambiguity in data. It can be used to direct the way other techniques for information processing are used afterwards. For example, the structure of a neural network to be used for learning from a dataset can be defined a great deal after knowing the optimal number of fuzzy clusters.

Fuzzy clustering is applied on gene expression data in Chapter 8 and in Futschik and Kasabov (2002).

2.3 Evolving Clustering Method (ECM)

2.3.1 ECM

Here, an evolving clustering method, ECM, is introduced that allows for adaptive clustering of continuously incoming data. This method performs a simple evolving, adaptive, maximum distance-based clustering (Kasabov and Song, 2002). Its extension, ECMc, evolving clustering method with constrained optimisation, to implement scatter partitioning of the input space for the purpose of deriving fuzzy inference rules, is also presented. The ECM is specially designed for adaptive evolving clustering, whereas ECMc involves some additional tuning of the cluster centres, more suitable for combined adaptive and off-line tasks (combined learning; see Chapter 1). The ECMc method takes the results from the ECM as initial values, and further optimises the clusters in an off-line mode with a predefined objective function $J(C, X)$ based on a distance measure between data X and cluster centres C, given some constraints.

The adaptive evolving clustering method, ECM, is a fast one-pass algorithm for dynamic clustering of an input stream of data (Kasabov and Song, 2002), where there is no predefined number of clusters. It is a distance-based clustering method where the cluster centres are represented by evolved nodes in an adaptive mode. For any such cluster, the maximum distance *MaxDist*, between an example point $\boldsymbol{x}_i$ and the closest cluster centre, cannot be larger than a threshold value *Dthr*, that is, a preset clustering parameter. This parameter would affect the number of the evolved clusters. The threshold value *Dthr* can be made adjustable during the adaptive clustering process, depending on some optimisation and self-tuning criteria, such as current error, number of clusters, and so on.

During the clustering process, data examples come from a data stream and this process starts with an empty set of clusters. When a new cluster C_j is created, its cluster centre Cc_j is defined and its cluster radius Ru_j is initially set to zero. With more examples presented one after another, some already created clusters will be updated through changing their centres' positions and increasing their cluster radii. Which cluster will be updated and how much it will be changed depends on the position of the current data example in the input space. A cluster C_j will not be updated any more when its cluster radius Ru_j has reached the value equal to the threshold value *Dthr*. Figure 2.7 shows an illustration of the ECM clustering process in a 2D space.

The ECM Algorithm

Step 0: Create the first cluster C_1 by simply taking the position of the first example from the input data stream as the first cluster centre Cc_1, and setting a value 0 for its cluster radius Ru_1 (see Fig. 2.7a).

Step 1: If all examples from the data stream have been processed, the clustering process finishes. Else, the current input example $\boldsymbol{x}_i$ is taken and the normalised Euclidean distance D_{ij} between this example and all n already created cluster centres Cc_j, $D_{ij} = ||\boldsymbol{x}_i - Cc_j||$, $j = 1, 2, \ldots, n$, is calculated.

Step 2: If there is a cluster C_m with a centre Cc_m, a cluster radius Ru_m, and distance value D_{im} such that:

$$\text{(i) } D_{im} = ||\boldsymbol{x}_i - Cc_m|| = \min\{D_{ij}\} = \min\{||\boldsymbol{x}_i - Cc_j||\}, \text{ for } j = 1, 2, \ldots, n; \text{ and} \tag{2.9}$$

$$\text{(ii) } D_{im} < Ru_m$$

the current example $\boldsymbol{x}_i$ is considered as belonging to this cluster. In this case neither a new cluster is created, nor any existing cluster updated (e.g. data vectors $\boldsymbol{x}_4$ and $\boldsymbol{x}_6$ in Fig. 2.7). The algorithm then returns to Step 1.

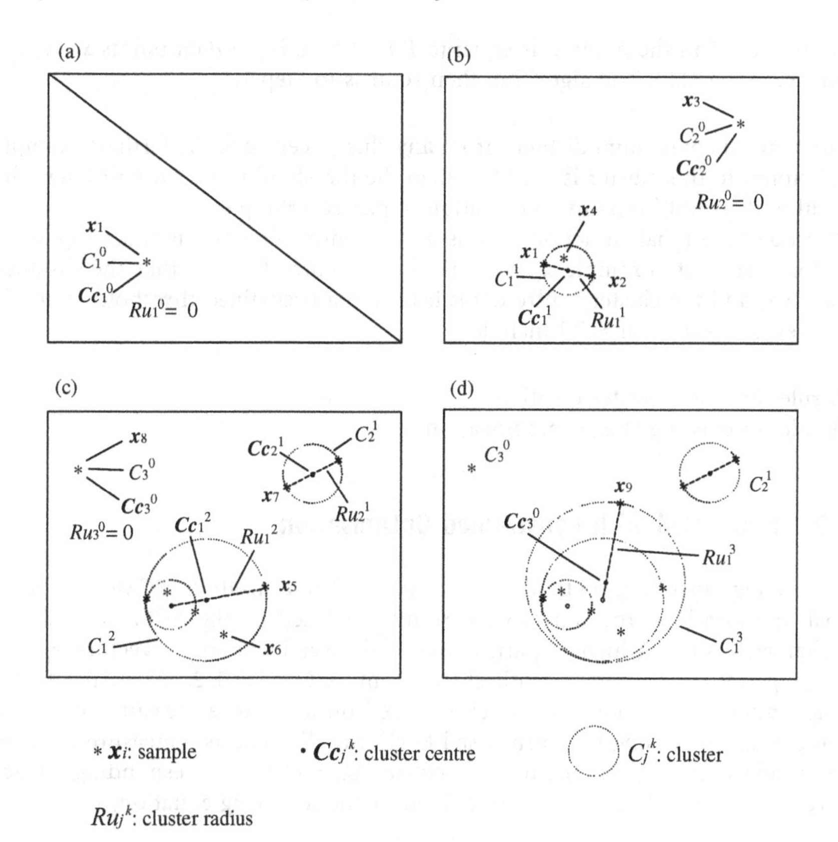

Fig. 2.7 An evolving clustering process using ECM with consecutive examples x_1 to x_9 in a 2D space (from Kasabov and Song (2002)): x_1 causes the ECM to create a new cluster $C_1{}^0$; x_2 to update cluster $C_1{}^0 \rightarrow C_1{}^1$; x_3 to create a new cluster $C_2{}^0$; x_4 to do nothing; x_5 to update cluster $C_1{}^1 \rightarrow C_1{}^2$; x_6 to do nothing; x_7 to update cluster $C_2{}^0 \rightarrow C_2{}^1$; x_8 to create a new cluster $C_3{}^0$; x_9 to update cluster $C_1{}^2 \rightarrow C_1{}^3$.

Else:

Step 3: Find a cluster C_a (with a centre Cc_a, a cluster radius Ru_a, and a distance value D_{ia}, which cluster has a minimum value S_{ia}:

$$S_{ia} = D_{ia} + Ru_a = \min\{S_{ij}\}, j = 1, 2, \ldots, n. \tag{2.10}$$

Step 4: If S_{ia} is greater than $2 \times Dthr$, the example x_i does not belong to any existing cluster. A new cluster is created in the same way as described in Step 0 (e.g. input data vectors x_3 and x_8 in Figure 2.7c). The algorithm then returns to Step 1.

Else:

Step 5: If S_{ia} is not greater than $2 \times Dthr$, the cluster C_a is updated by moving its centre Cc_a and increasing its radius value Ru_a. The updated radius $Ru_a{}^{new}$ is set to be equal to $S_{ia}/2$ and the new centre $Cc_a{}^{new}$ is located on the line connecting input vector x_i and the old cluster centre Cc_a, so that distance from the new centre Cc_a^{new} to the point x_i is equal to Ru_a^{new} (e.g. input data points x_2, x_5, x_7, and x_9 in Fig. 2.7). The algorithm then returns to Step 1.

In this way, the maximum distance from any cluster centre to the farthest example that belongs to this cluster is kept less than the threshold value *Dthr* although the algorithm does not keep any information of passed examples.

The objective (goal) function here is a very simple one and it is set to ensure that for every data example x_i there is cluster centre *Cj* such that the distance between x_i and the cluster centre *Ccj* is less than a predefined threshold *Dthr*.

The evolving rules of ECM include:

- A rule for a new cluster creation
- A rule for existing cluster modification

2.3.2 ECMc: ECM with Constrained Optimisation

The evolving clustering method with constrained optimisation, ECMc, applies a global optimisation procedure to the result produced by the ECM. In addition to what ECM does, which is partitioning a dataset including p vectors x_i, $i = 1, 2, \ldots, p$, into n clusters C_j with cluster centres Cc_j, $j = 1, 2, \ldots, n$, the ECMc further minimises an objective function based on a distance measure subject to given constraints. Using the normalised Euclidean distance as a measure between an example vector x_k, belonging to a cluster C_j, and the corresponding cluster centre Cc_j, the objective function is defined by the following equation,

$$J = \Sigma_{j=1,\ldots,n} J_j \tag{2.11}$$

where $J_j = \sum_{k, xk\in Cj} ||x_k - Cc_j||$ is the objective function within a cluster C_j, for each $j = 1, 2, \ldots, n$, and the constraints are defined as

$$||x_k - Cc_j|| \leq Dthr \tag{2.12}$$

where $x_k \in C_j$ for $j = 1, 2, \ldots, n$.

The clusters can be represented as a $p \times n$ binary membership matrix U, where the element u_{ij} is 1 if the ith data point x_i belongs to C_j, and 0 otherwise. Once the cluster centres Cc_j are defined, the values u_{ij} are derived as

$$\text{if} ||x_i - Cc_j|| \leq ||x_i - Cc_k||, \text{ for } k = 1, \ldots, n, j \neq k; \text{ then } u_{ij} = 1, \text{ else } u_{ij} = 0 \tag{2.13}$$

The ECMc algorithm works in an off-line iterative mode on a batch of data repeating the steps shown in Fig. 2.8.

Combined alternative adaptive clustering with ECM and off-line optimisation with ECMc can be used in a mode as follows. After the ECM is applied to a certain sequence of data vectors, the ECMc optimisation is applied to the latest data from a data window. After that, the system continues to work in an adaptive mode with the ECM, and so on.

ECMc evolving clustering with constraint optimisation

Step 1: Initialise the cluster centres Ccj, j = 1, 2, ..., n, that are produced through the adaptive evolving clustering method ECM.

Step 2: Determine the membership matrix *U*

Step 3: Employ the constrained minimisation method to modify the cluster centres.

Step 4: Calculate the objective function *J*

Step 5: Stop, if: (1) the result is below a certain tolerance value, or (2) the improvement of the result when compared with the previous iteration is below a certain threshold, or (3) the iteration number of minimizing operation is over a certain value. Else, the algorithm returns to Step2.

Fig. 2.8 The ECMc evolving clustering algorithm with constraint optimisation (from Kasabov and Song (2002)).

2.3.3 Comparative Analysis of ECM, ECMc, and Traditional Clustering Techniques

Here, the gas-furnace time-series data is used as a benchmark dataset. A benchmark process used widely so far is the burning process in a gas furnace (Box and Jenkins, 1970). The gas methane is fed continuously into a furnace and the produced CO_2 gas is measured every minute. This process can theoretically run forever supposing that there is a constant supply of methane and the burner keeps mechanically intact. The process of CO_2 emission is an evolving process. In this case it depends on the quantity of the methane supplied and on the parameters of the environment. For simplicity, only 292 values of CO_2 are taken in the well-known gas-furnace benchmark problem. Given the values of methane at a moment $(t-4)$ and the value of CO_2 at the moment (t) the task is to predict what the value for the CO_2 at the moment $(t+1)$ will be. The CO_2 data from Box and Jenkins (1970) along with some of their statistical characteristics, are plotted in Fig. 1.3. It shows the 292 points from the time series, the 3D phase space, the histogram, and the power spectrum of the frequency characteristics of the process. Figure 2.9 displays a snapshot from the evolving clustering process of the 2D input data (methane $(t-4)$, $CO_2(t)$) with the ECM algorithm.

For the purpose of comparative analysis, the following clustering methods are applied to the same dataset.

1. ECM, evolving clustering method (adaptive, one pass)
2. SC, subtractive clustering (off-line, one pass; see Bezdek (1993))
3. ECMc, evolving clustering with constrained optimisation (off-line)
4. FCMC, fuzzy *C*-means clustering (off-line; Bezdek (1981, 1987))
5. KMC, *K*-means clustering (off-line; MacQueen (1967))

Each of them partitions the data into a fixed number of clusters; in this case this number was chosen to be 15. The maximum distance *MaxD*, between an example and the corresponding cluster centre, as well as the value of the objective function *J* are measured for comparison as shown in Table 2.1.

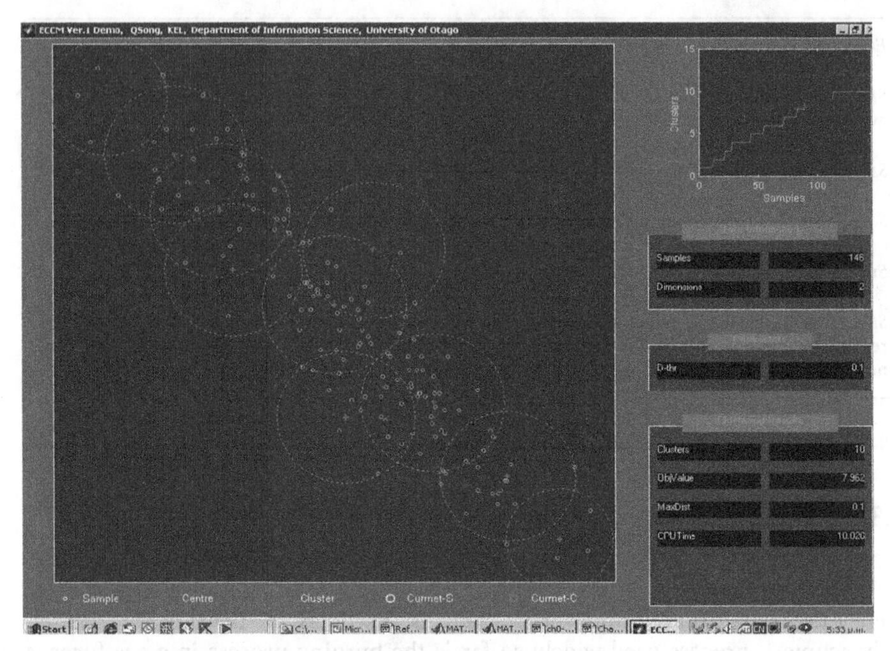

Fig. 2.9 A snapshot of the clustering process: cluster centres and their cluster radii when the ECM algorithm is applied for online clustering on 146 gas-furnace data examples (see Fig. 1.3).

Table 2.1 Comparative results of clustering the gas-furnace data set into 15 clusters by using different clustering methods.

Methods	*MaxD*	*Objective value: J*
ECM (online, one-pass)	0.1	12.9
SC (off-line, one-pass)	0.15	11.5
ECMc (off-line)	0.1	11.5
FCM (off-line)	0.14	12.4
KM (off-line)	0.12	11.8

Figure 2.10 displays the data points from the gas-furnace time series and the cluster centres obtained through the use of different clustering techniques.

Both ECM (adaptive, one pass) and ECMc (optimized through objective function, multiple passes) obtain minimum values of *MaxD*, which indicates that these methods partition the dataset more uniformly than the other methods. We can also predict that if all these clustering methods obtained the same value for *MaxD*, then the ECM and the ECMc would result in a smaller number of partitions.

Considering that the ECM clustering is a 'one-pass' adaptive process, the objective function value J for ECM simulation is acceptable as it is comparable with the J value for the other methods. With more data presented to the clustering system from the data stream, the values for the objective functions for both adaptive ECM and ECMc become closer; after a certain number of data points from a time series the two methods will eventually produce the same results provided

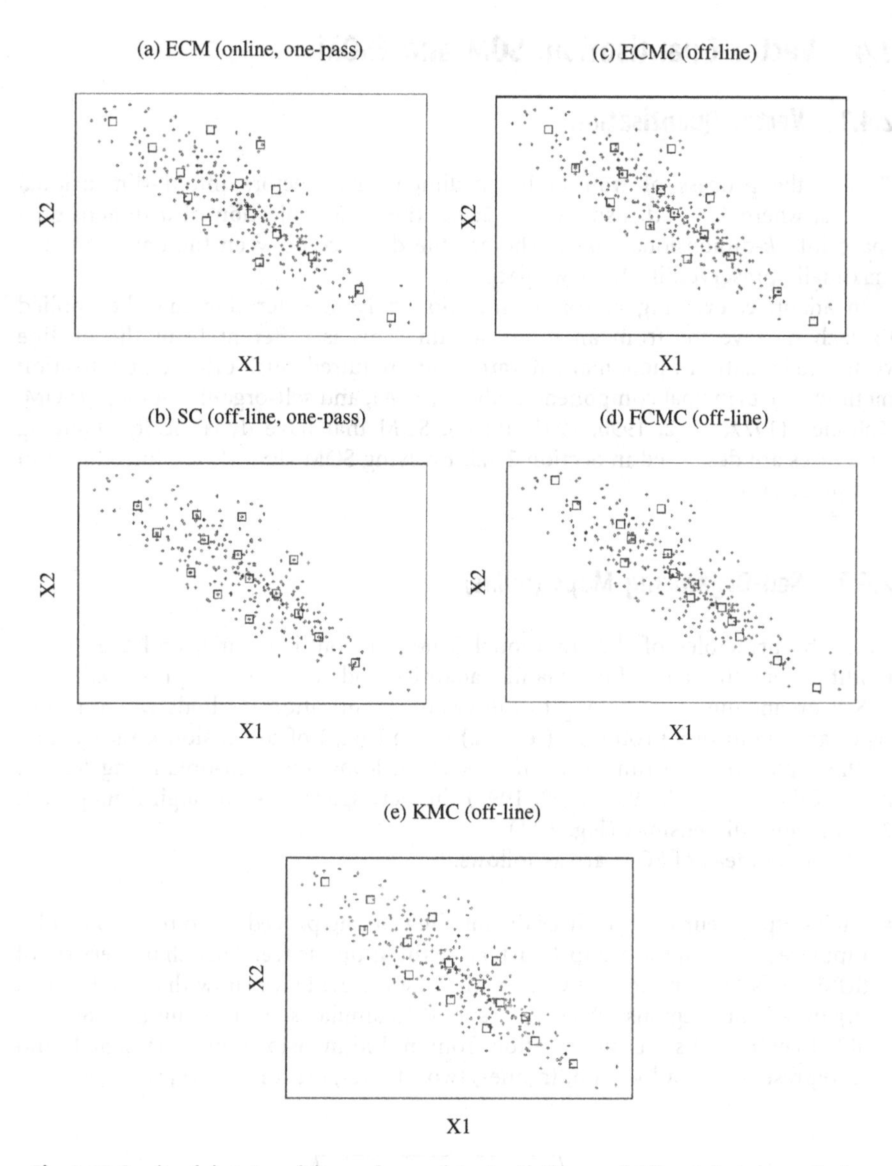

Fig. 2.10 Results of clustering of the gas-furnace dataset with the use of different clustering methods.

that the data are drawn from a closed space and the probability distribution of the data stream does not change after a certain data point in the stream.

The advantages of the ECM online clustering technique can be summarised as

1. ECM allows for unsupervised, life-long, adaptive modelling of evolving processes.
2. ECM is much faster than the off-line clustering techniques.

2.4 Vector Quantisation. SOM and ESOM

2.4.1 Vector Quantisation

This is the process of transferring d-dimensional vectors into k-dimensional vectors, where $k << d$, usually $k = 2$; i.e. this is a projection of d-dimensional space into k-dimensional space whereas the distance between the data points is maximally preserved in the new space.

In adaptive, evolving vector quantisation only one iteration may be applied to each data vector from an input stream. This is different from the off-line vector quantisation where many iterations are required. Such off-line quantisation methods are principal component analysis (PCA), and self-organizing maps (SOM; Kohonen (1977, 1982, 1990, 1993, 1997)). SOM that have dynamically changing structures are described in Section 2.4.3. Evolving SOM (ESOM) are introduced in Section 2.4.4.

2.4.2 Self-Organizing Maps (SOMs)

Here, the principles of the traditional SOMs are outlined first, and then some modifications that allow for dynamic, adaptive node creation are presented.

Self-organizing maps belong to the vector quantisation methods where prototypes are found in a prototype (feature) space (map) of dimension k rather than in the input space of dimension d, $k < d$. In Kohonen's self-organizing feature map (Kohonen, 1977, 1982, 1990, 1997) the new space is a topological map of 1, 2, 3, or more dimensions (Fig. 2.11).

The main ideas of SOM are as follows.

- Each output neuron specializes during the training procedure to react to similar input vectors from a group (cluster) of the input space. This characteristic of SOM tends to be biologically plausible as some evidence show that the brain is organised into regions which correspond to similar sensory stimuli. A SOM is able to extract abstract information from multidimensional primary signals and to represent it as a location, in one-, two-, three-, etc. dimensional space.

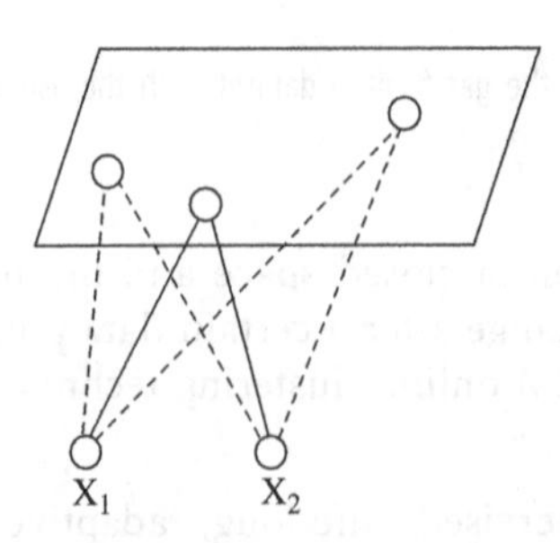

Fig. 2.11 A schematic diagram of a simple, hypothetical two-input, 2D-output SOM system (from Kasabov (1996), MIT Press, reproduced with permission).

- The neurons in the output layer are competitive ones. Lateral interaction between neighbouring neurons is introduced in such a way that a neuron has a strong excitatory connection to itself, and less excitatory connections to its neighbouring neurons in a certain radius; beyond this area, a neuron either inhibits the activation of the other neurons by inhibitory connections, or does not influence it. One possible neighbouring rule that implements the described strategy is the so-called '*Mexican hat*' rule. In general, this is a winner-takes-all scheme, where only one neuron is the winner after an input vector is fed, and a competition between the output neurons has taken place. The fired neuron represents the class, the group (cluster), the label, or the feature to which the input vector belongs.
- SOMs transform or preserve similarity between input vectors from the input space into topological closeness of neurons in the output space represented as a topological map. Similar input vectors are represented by near points (neurons) in the output space.

The unsupervised algorithm for training a SOM, proposed by Teuvo Kohonen, is outlined in Fig. 2.12. After each input pattern is presented, the winner is established and the connection weights in its neighbourhood area N_t increase, and the connection weights outside the area are kept unchanged. α is a learning parameter. Training is done through a number of training iterations so that at each iteration the whole set of input data is propagated through the SOM and the connection weights are adjusted.

SOMs learn statistical features. The synaptic weight vectors tend to approximate the density function of the input vectors in an orderly fashion. Synaptic vectors w_j converge exponentially to centres of groups of patterns and the nodes of the output map represent to a certain degree the distribution of the input data. The weight vectors are also called reference vectors, or reference codebook vectors. The whole weight vector space is called a reference codebook

In SOM the topology order of the prototype nodes is predetermined and the learning process is to 'drag' the ordered nodes onto the appropriate positions in the low-dimensional feature map (see Fig. 2.11b, upper figure). As the original input manifold can be complicated and an inherent dimension larger than that of the feature map (usually set as two for visualization purposes), the dimension

K0. Assign small random numbers to the initial weight vectors wj(t=0), for every neuron j from the output map.

K1. Apply an input vector x ($x_1, x_2, ..., x_n$) at the consecutive time moment t.

K2. Calculate the distance d_j (in n-dimensional space) between x and the weight vectors Wj(t) of each neuron j.

K3. The neuron K which is closest to X is declared winner. It becomes a center of a neighbourhood area Nt.

K4. Change all the weight vectors within the neighbourhood area:
$w_j(t+1) = w_j(t) + \alpha.(x - w_j(t))$, if $j \in Nt$,
$w_j(t+1) = w_j(t)$, if j is not from the area Nt

All of the steps from K1 to K4 are repeated for all training instances. Nt and α decrease in time. The training procedure is repeated again with the same training instances until convergence is achieved.

Fig. 2.12 The SOM training algorithm (from Kasabov (1996), MIT Press, reproduced with permission).

reduction in SOM may become inappropriate for complex data analysis tasks. The SOM have been extended for supervised learning to LVQ (Learning vector Quantisation) (Kohonen, 1997).

2.4.3 Dynamic SOMs

The constraints of a low-dimensional mapping topology of SOM are removed in Martinez and Schulten (1991), where a neural gas model is proposed with a learning rule similar to SOM, but the prototype vectors are organized in the original manifold of the input space. Each time the prototype weights are updated the neighbourhood rank, i.e. the matching rank of prototypes, needs to be computed. Unfortunately, this brings the time complexity of the algorithm to the scale of ($n \log n$) in a serial implementation, whereas searching for the best matching unit in the K-means algorithm or in the SOM algorithm takes only n steps.

Fritzke (1995) proposed a growing structure neural gas (GNG) which uses a fixed topology for reference vector space, but there is no predefined layout order for map nodes. The map creates new nodes whenever input data are not closely matched by existing reference vectors, and sets up connections between neighbouring nodes. One of the goals of the method is to insert more nodes in the model where the density of the data in that subspace is higher, thus keeping the entropy at its maximum value. If a node has more data associated with it, the node gets split and a new one is created as illustrated in Fig. 2.13a,b. It is statistical knowledge that is accumulated in the model and used to optimise its structure.

Bruske and Sommer (1995) presented another similar model, dynamic cell structure (DCS), slightly differing from GNG in the node insertion part. GNG, and DCS need to calculate local resources for prototypes, which introduces extra computational effort and reduces their efficiency.

SOM and its derivatives are unsupervised learning methods. The SOM algorithm was further extended to the learning vector quantisation (LVQ) algorithm for

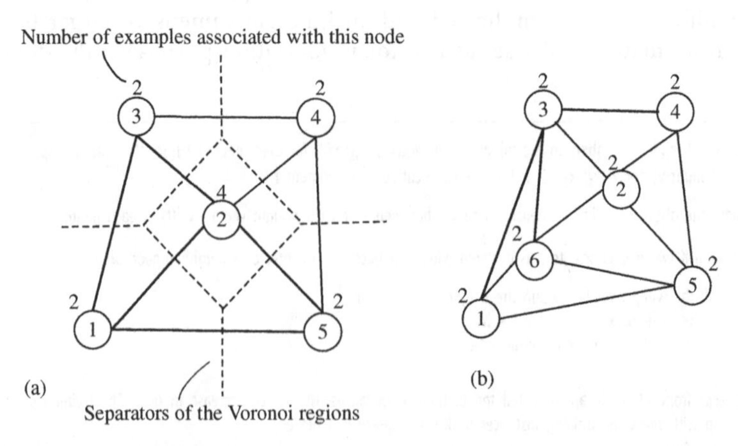

Fig. 2.13 An example of splitting neurons in the growing neural gas structure (see Fritzke (1995)): (a) initial structure; (b) the structure after node #2 was split and a new node #6 was created.

learning supervised pattern classification data (Kohonen, 1990). Vesanto (1997) incorporated a local linear regression model on the top of a SOM map for a time-series prediction problem. This method constructs local prototype vectors and uses linear regression models on these vectors. Strictly speaking, this is not an incremental learning approach and the complexity of the model usually is larger than the scale of the number of prototype vectors.

2.4.4 Evolving Self-Organizing Maps (ESOM)

Several methods, such as: dynamic topology representing networks (Si *et al.*, 2000) and evolving self-organizing maps (ESOM; Deng and Kasabov (2000)) further develop the principles of SOM. These methods allow the prototype nodes to evolve in the original data space X, and at the same time acquire and keep a topology representation. The neighbourhood of the evolved nodes (neurons) is not predefined as it is in a SOM. It is decided in an online mode according to the current distances between the nodes. These methods are free of the rigid topological constraints in a SOM. They do not require searching for neighbourhood ranking as in the neural gas algorithm, thus improving the speed of learning.

Here, the ESOM method is explained in more detail.

Given an input vector $\boldsymbol{x}$, the activation on the ith node in ESOM is defined as:

$$a_i = e^{-||x-w_i||^2/\varepsilon^2} \tag{2.14}$$

where ε is a radial. Here a_i can be regarded as a matching score for the ith prototype vector w_i onto the current input vector $\boldsymbol{x}$. The closer they are, the bigger the matching score is.

The following online stochastic approximation of the error minimization function is used.

$$E_{\text{app}} = \Sigma_{i=1,n} a_i ||x - w_i||^2 \tag{2.15}$$

where n is the current number of nodes in ESOM upon arrival of the input vector $\boldsymbol{x}$.

To minimize the criterion function above, weight vectors are updated by applying a gradient descent algorithm. From Eq. (2.15) it follows

$$\partial E_{\text{app}}/\partial w_i = a_i(w_i - \mathbf{x}) + ||x - w_i||^2 \partial a_i/\partial w_i \tag{2.16}$$

For the sake of simplicity, we assume that the change of the activation will be rather small each time the weight vector is updated, so that a_i can be treated as a constant. This leads to the following simplified weight-updating rule.

$$\Delta w_i = \gamma a_i (x - w_i), \; for \; i = 1, 2, \ldots, n \tag{2.17}$$

Here γ is a learning rate held as a small constant.

The likelihood of assigning the current input vector x onto the ith prototype w_i is defined as

$$Pi(x, w_i) = a_i / \Sigma_{k=1,2,\dots,n}(a_k) \tag{2.18}$$

Evolving the Feature Map

During online learning, the number of prototypes in the feature map is usually unknown. For a given dataset the number of prototypes may be optimum at a certain time but later it may become inapropriate as when new samples arrive the statistical characteristics of the data may change. Hence it is highly desirable for the feature map to be dynamically adaptive to the incoming data.

The approach here is to start with a null map, and gradually allocate new prototype nodes when new data samples cannot be matched well onto existing prototypes. During learning, when old prototype nodes become inactive for a long time, they can be removed from the dynamic prototype map.

If for a new data vector x none of the prototype nodes is within a distance threshold, then a new node w_{new} is inserted representing exactly the poorly matched input vector $w_{new} = x$, resulting in a maximum activation of this node for x.

The ESOM evolving algorithm is given in Box 2.1.

Box 2.1. The ESOM evolving self-organised map algorithm:

Step 1: Input a new data vector x.
Step 2: Find a set S of prototypes that are closer to x than a predefined threshold.
Step 3: If S is null, go to step 4 (insertion), otherwise calculate the activations a_i of all nodes from S and go to step 5 (updating).
Step 4 (insertion): Create a new node w_i for x and make a connection between this node and its two closest nodes (nearest neighbours) that will form a set S.
Step 5 (updating):Modify all prototypes in S according to (2.17) and recalculate the connections $s(i,j)$ between the winning node i (or the newly created one) and all the nodes j in the set S: $s(i,j) = a_i a_j / \max\{a_i, a_j\}$.
Step 6: After a certain number of input data are presented to the system, prune the weakest connections. If isolated nodes appear, prune them as well.
Step 7: Go to step 1.

Visualising the Feature Map

Sammon projection is used here to visualise the evolving nodes at each time when it is necessary. In addition to the node projection in a 2D space, the topology of node connections is also shown as links between neighbouring nodes. This is a significant difference between the ECM presented in Section 2.2 and the ESOM (see examples on Fig. 2.17b, Fig. 2.19).

2.5 Prototype Learning. ART

2.5.1 Adaptive Prototype Learning

This is a similar technique to the adaptive clustering methods, but here instead of n cluster centres and membership degrees, n prototypes of data points are found that represent to a certain degree of accuracy the whole data stream up to the current point in time. The d-dimensional space, with p examples currently presented, is transformed into n prototypes in the same space.

SOMs and ESOMs form prototypes as nodes that are placed in the original data space. Each prototype gets activated if an example from the prototype area is presented to the system. This is explained later in this chapter.

2.5.2 Adaptive Resonance Theory

Here, a brief outline of one of the historically first, and computationally simplest, adaptive prototyping systems - ART1 and ART2 - is given (Carpenter and Grossberg, 1987).

Adaptive resonance theory (ART) makes use of two terms from brain behaviour, i.e. stability and plasticity. The stability/plasticity dilemma is the ability of a system to preserve the balance between retaining previously learned patterns and learning new patterns. Two layers of neurons are used to realize the idea: a 'top' layer, an output concept layer, and a 'bottom' layer, an input feature layer. Two sets of weights between the neurons in the two layers are used. The top-down weights represent learned prototype patterns, expectations. The bottom-up weights represent a scheme for new inputs to be accommodated in the network.

Patterns, associated with an output node j, are collectively represented by the weight vector of this node $\boldsymbol{t}_j$ (top-down weight vector, prototype). The reaction of the node j to a particular new input vector is defined by another weight vector $\boldsymbol{b}_j$ (bottom-up weight). The key element in the ART realisation of the stability/plasticity dilemma is the control of the partial match between new feature vectors and already learned ones achieved by using a parameter, called *vigilance*, or *vigilance factor*. Vigilance controls the degree of mismatch between the new patterns and the learned (stored) patterns which the system can tolerate.

Figure 2.14a shows a diagram of a simple ART architecture (Carpenter and Grossberg, 1987). It consists of two sets of neurons: input (feature) neurons (first layer) and output neurons (second layer). The bottom-up connections b_{ij} from each input i to every output j and the top-down connections t_{ji} from the outputs back to the inputs are shown in the figure. Each of the output neurons has a strong excitatory connection to itself and a strong inhibitory connection to each of the other output neurons.

The ART1 learning algorithm for binary inputs and outputs is given in Fig. 2.14b. It consists of two major phases. The first one is presenting the input pattern and calculating the activation values of the output neurons. The winning neuron is defined. The second phase is for calculating the mismatch between the input pattern and the pattern currently associated with the winning neuron. If the

(a)

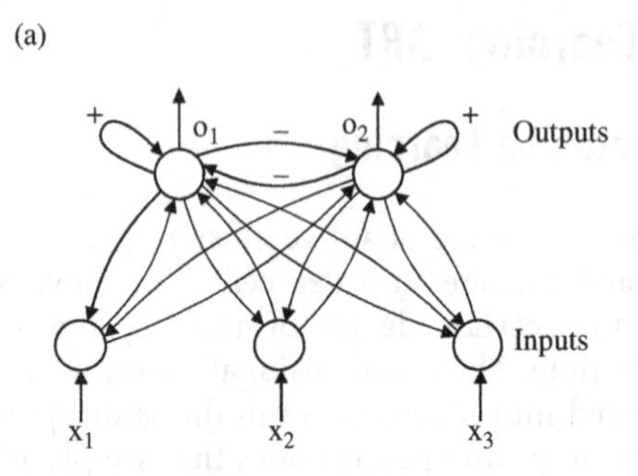

(b)

A1. Weight coefficients are initialized:
$t_{ij}(0)$:1, b_{ij}:=1/(1+n), for each i=1,2,...,n; j=1,2,...m

A2. A coefficient of similarly r, a so-called vigiliance factor, is defined, 0<=r<=1. The greater the value of r, the more similar the patterns ought to be in order to activate the same output neuron representing a category, a class, or a concept.

A3. WHILE (there are input vectors) DO
(a) a new input vector x(t) is fed at moment t, x = $(x_1, x_2, \ldots, x_n)(t)$
(b) the outputs are calculated:
$O_j = \sum b_{ij}(t).x_i(t)$, for j=1,2..., m
(c) an output o_j* with the highest value is defined;
(d) the similarly of the associated to j* input pattern is defined:

IF (number of "1"s in the intersection of the vector **x**(t) and t_j*(t)) divided to the number of "1"s in **x**(t) is greater than the vigilance r) THEN GO TO (f)

ELSE
(e) the output j* is abandoned and the procedure returns to (b) in order to calculate another output to be associated with **x**(t);
(f) the pattern **x**(t) is associated with the vector t_j*(t), therefore the pattern t_j*(t) is changed using its intersection with x(t):
$t_{ij}^*(t+1) := t_{ij}^*(t).x_i(t)$, for i=1,2,...,n
(g) the weights b_{ij} are changed:
$b_{ij}^*(t+1) := b_{ij}^*(t) + t_{ij}^*(t).x_i / (0.5 + \sum t_{ij}^*(t).\, x_i(t))$

Fig. 2.14 (a) A schematic diagram of ART1; (b) the ART1 learning algorithm presented for n inputs and m outputs (from Kasabov (1996), MIT Press, reproduced with permission).

mismatch is below a threshold (vigilance parameter) this pattern is updated to accommodate the new one. But if the mismatch is above the threshold, the procedure continues to either find another output neuron, or to create a new one

An example of applying the algorithm for learning a stream of three patterns is presented in Fig. 2.15. The network associates the first pattern with the first output neuron, the second pattern with the same output neuron, and the third input pattern with a newly created second output neuron. If the network associates a new input pattern with an old one, it changes the old one respectively. For binary inputs, the simple operation of binary intersection (multiplication) is used.

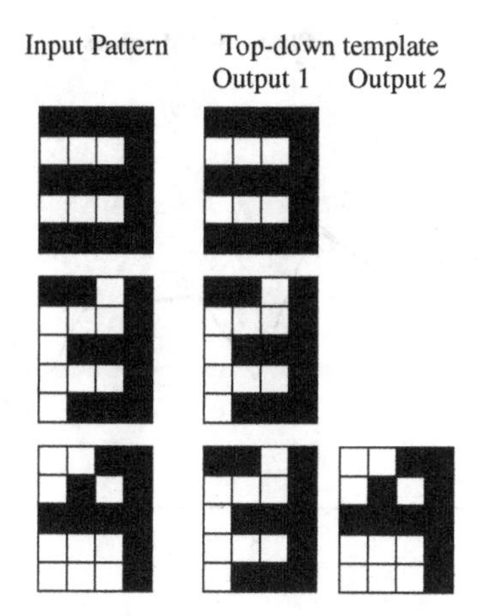

Fig. 2.15 Patterns presented to an ART1 system and learned as two prototypes at three consecutive time moments. If a new pattern did not match an existing prototype above the vigilance parameter value, a new output node was created to accommodate this pattern as a new prototype (from Kasabov (1996), MIT Press, reproduced with permission).

ART1 was further developed into ART2 (continuous values for the inputs; Carpenter and Grossberg (1987)), ART3 (Carpenter and Grossberg, 1990), Fuzzy ARTMAP (Carpenter *et al.*, 1991). The latter is an extension of ART1 when input nodes represent not 'yes or no' features, but membership degrees, to which the input data belong to these features, for example, a set of features {sweet, fruity, smooth, sharp, sour} used to categorise different samples of wines based on their taste. A particular sample of wine can be represented as an input vector consisting of membership degrees, e.g. (0.7, 0.3, 0.9, 0.2, 0.5). The fuzzy ARTMAP allows for continuous values in the interval of [0,1] for both the inputs and for the top-down weights. It uses fuzzy operators MIN and MAX to calculate intersection and union between the fuzzy input patterns $\boldsymbol{x}$ and the continuous-value weight vectors $\boldsymbol{t}$.

2.6 Generic Applications of Unsupervised Learning Methods

2.6.1 Data Analysis. Time-Series Data Analysis

Clustering of data may reveal important patterns that can lead to knowledge discovery in various application areas. Data can be either static, or dynamic time-series data as illustrated in Fig. 2.16, where three gene expression variables measured over time are clustered together based on their similarity of values over the time of measurement. The mean time series (the temporal cluster centre) is also shown.

The clustered genes together suggest that these genes may have a similar function in a cell, or may co-regulate each other which is important information for the understanding of the interaction between these genes.

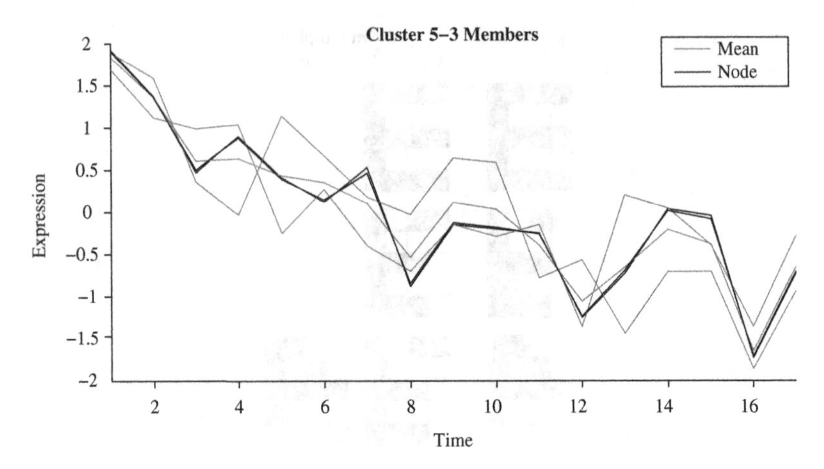

Fig. 2.16 A cluster of three data time series that have similar temporal profiles: three genes which expression level is measured at several time moments. The mean value of the time series (the temporal cluster centre) is also shown.

2.6.2 Filling Missing Values

In the case of datasets that have missing values for some variables and some samples, data can be clustered according to the available variable values and the missing values can be assigned based on similarity of the samples with missing values to other samples that have no missing values as shown in Box 2.2.

> **Box 2.2. Using clustering for filling missing values:**
>
> 1. Assume that the value x_{im} is missing in a vector (sample) $Sm = (x_{1m}, x_{2m}, \ldots, x_{im}, \ldots, x_{nm})$.
> 2. Find the closest K samples to sample Sm based on the distance, measured with the use of the only available variables (x_i is not included) - set Smk.
> 3. Substitute $x_{im} = \Sigma_{j=1,K}(1 - dj)\ x_{ij}/\Sigma_{j=1,K}(1 - dj)$ where dj is the distance between sample Sm and sample Sj from the set Smk.
> 4. For every new input vector $\boldsymbol{x}$, find the closest K samples to build a model (the new vector $\boldsymbol{x}$ is a centre of a cluster and find the K closest members of this cluster)

2.6.3 Evolving Clustering Method for Classification (ECMC)

Generally speaking, if there are class labels associated with the examples used for training a system, i.e. the examples are of the type $z = (x, y)$, where y is a class label, clustering methods can be used for classification purposes. The procedure is the following one.

1. Apply the clustering algorithm to data pairs $(\boldsymbol{x},y)$, separately for each class finding separate clusters for each class.
2. A new input datum $\boldsymbol{x}'$ with unknown class label, is first clustered into one of the existing clusters based on its distance from the cluster centres, and then the class label y assigned to this cluster is assigned to the input data point as a classification result y.

Example

The two-spirals problem is used here for illustration. The 2D training dataset is generated with a density of 1 and consists of 194 data, with 97 data points for each spiral (Fig. 2.17a) The testing dataset, generated with a density of 4, is composed of 770 data with 385 data for each spiral.
(density 1, for training data):

$$\begin{cases} \gamma = (\theta + \pi/2)/\pi \\ \theta = k\pi/16, k = 0, 1, 2, \ldots, 96 \end{cases} \tag{2.19}$$

(density 4, for testing data):

$$\begin{cases} \gamma = (\theta + \pi/2)/\pi \\ \theta = k\pi/64, k = 0, 1, 2, \ldots, 384\ldots \end{cases} \tag{2.20}$$

$$\begin{cases} \text{spiral 1}: x = \gamma \ \cos(\theta) \\ \qquad\qquad y = \gamma \ \sin(\theta) \end{cases}$$

$$\begin{cases} \text{spiral2}: x = -\gamma \ \cos(\theta) \\ \qquad\qquad y = -\gamma \ \sin(\theta) \end{cases}$$

Further points of the spirals can be generated in the two-dimensional Euclidean space, thus the process of generating spiral points can be considered evolving and expanding in an open 2D space. Figure 2.17b compares the evolved structures through using SOM (the upper figure) and through using ESOM (the lower figure).

An evolved ESOM is more suitable for the spiral data clustering problem than a trained SOM, as SOM imposes a certain 2D grid that is not suitable for the problem, whereas ESOM does not assume in advance any grid of connected nodes.

Two ECMc classification models were also created here. The first one had a threshold *Dthr* of 0.955. It created 64 nodes and achieved a classification accuracy of 100% on the training dataset and 98.4% on the test set. The second model was evolved with *Dthr* = 0.98. It evolved 146 nodes and achieved 100% classification accuracy for both the training and the test sets. Figures 2.18a,b show the classification boundaries between the two classes for the two models.

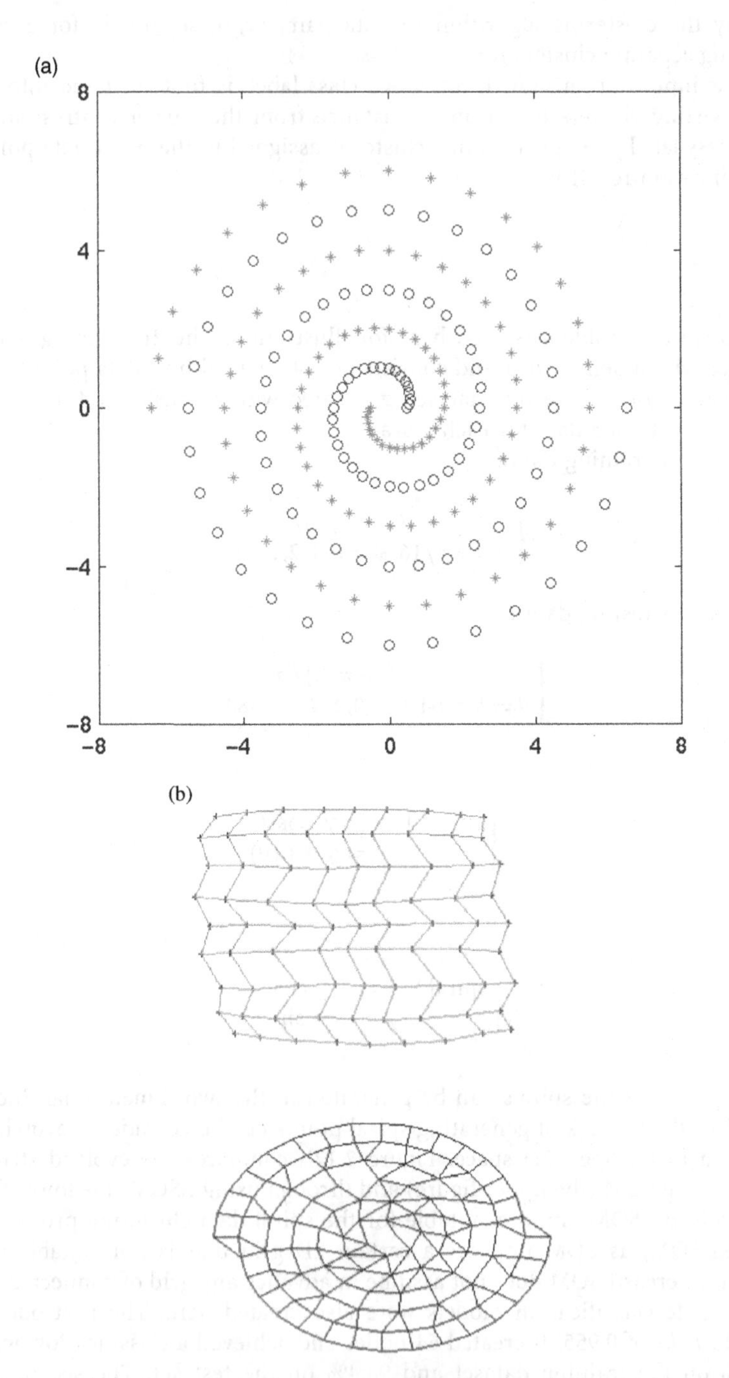

Fig. 2.17 The training data for the benchmark two-spirals problem: (a) the two-spiral benchmark data; (b) evolved structures with the use of SOM (the upper figure) and ESOM (the lower figure).

(a) 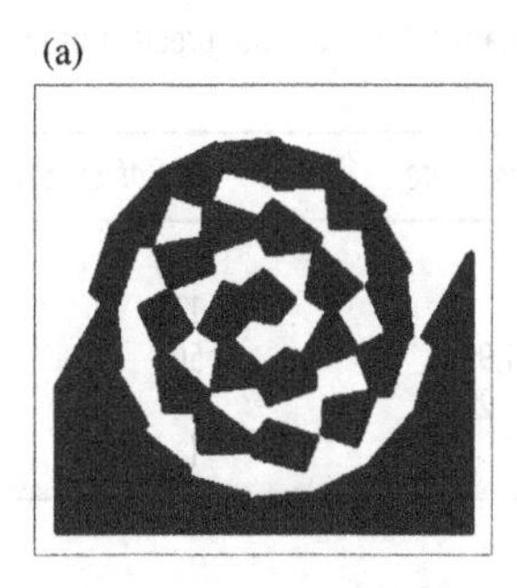 (b) 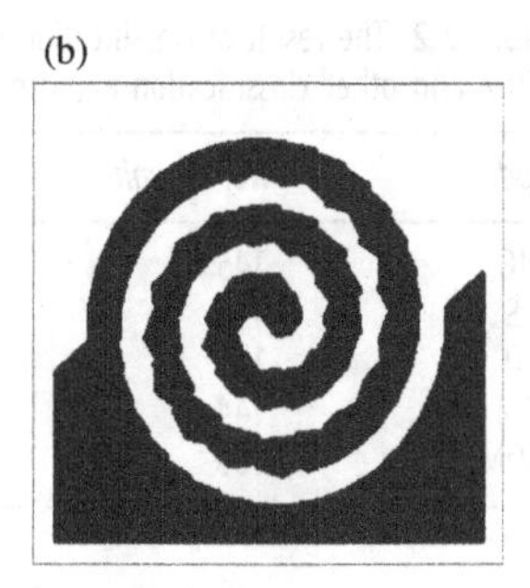

Fig. 2.18 The two-spiral problem, decision regions of ECMc: (a) decision regions for ECMc (model 1, with 64 nodes); (b) decision regions for ECMc (model 2, with 146 nodes).

The two-spiral classification problem is further used as an illustration problem in some other methods presented in this chapter. It shows that for simple classification problems unsupervised learning with clustering could provide a good solution. The problem can be solved with the use of other supervised learning schemes where the learning process takes into account the labelling of the data when the model parameters are adjusted.

In ECM, as well as in other clustering algorithms, the learning system develops in the original d-dimensional input data space X and the cluster centres Cc are points in this space. In the case of a high-dimensional X space, visualisation of the clusters becomes difficult. As a solution, the PCA or the Sammon projection algorithm can be used to project approximately the d-dimensional space into two- or three-dimensional visualisation space. This is illustrated also in Chapter 10 where ECM is used to evolve acoustic clusters from continuous speech data from multiple languages.

2.6.4 ESOM for Classification

Here we assume that data arrive in pairs $\boldsymbol{z} = (\boldsymbol{x}, y)$, where y is the class label assigned to each input data vector $\boldsymbol{x}$. When a new node $\boldsymbol{w}_j$ is created to represent an input vector $\boldsymbol{x}_i$, the node is assigned a class label y_i.

A new input vector $\boldsymbol{x}'$ with unknown class label is first mapped into the prototype nodes. A k-nearest neighbour classification is applied then on the winning node and its neighbours are linked to it through the neighbourhood links.

This is illustrated here on the benchmark two-spiral problem explained in the above section (see also Fig 2.17b). Table 2.2 shows classification results obtained with the use of different unsupervised learning models including ESOM.

ESOM is also applied on the Mackey–Glass data as explained and illustrated in Chapter 1. Five variables are used that constitute the problem space: $x(t)$, $x(t-6)$, $x(t-12)$, $x(t-18)$, and $x(t+6)$. In Fig. 2.19 the evolved nodes from 200 data points are plotted in a two-dimensional space of the first two principal components of the original 5D problem space. Except for two nodes, all of them were created during the presentation of the first 100 examples as shown in one of the windows of the snapshot of a graphical user interface in Fig. 2.19. ESOM is applied for the

Table 2.2 The result of classification of test data for the two-spiral problem using ESOM and other classification algorithms.

Model	*No. of units*	*Error rate*	*No. of Epochs*
GNG	145	0	180
DCS	135	0	135
LVQ	114	11.9%	50
SOM	144	22.2%	50
ESOM	105	0	1

analysis of gene expression data in Chapter 8, for adaptive analysis of image data in Chapter 12, and for other applications throughout this book.

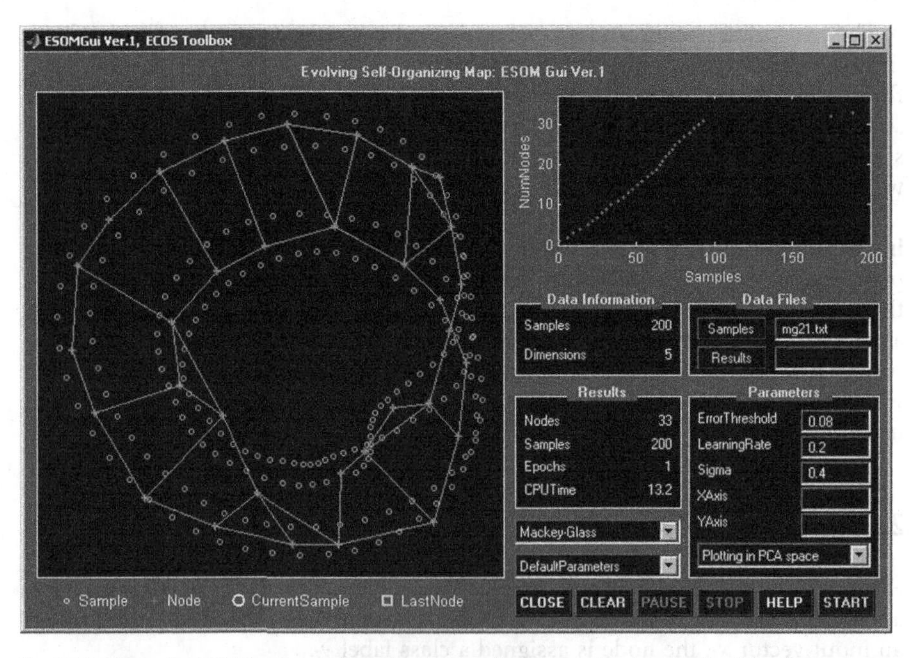

Fig. 2.19 The evolved ESOM structure from the Mackey–Glass time-series data for the following parameter values: error threshold 0.08; learning rate 0.2; sigma 0.4. Five input variables are used: $x(t)$, $x(t-6)$, $x(t-12)$, $x(t-18)$, and $x(t+6)$ for 200 examples. The evolved nodes are plotted in a two-dimensional space of the first two principal components of the problem space.

2.6.5 Evolving Clustering for Outlier Detection

Outlier detection is an important task in many applications where an unusual situation should be automatically detected from the input data and reported. If the current input vector is far from any of the clusters already created in the system, it is an outlier. The outlier may be joined by some other data vectors later in time, and thus will no longer be an outlier.

Applications of outlier detection include:

- Online processing of radio signals from a radio telescope recording signals from the universe (see, for example, SETI, Search for Extra Terrestrial Intelligence Institute at www.seti.org).
- Online processing of data from a production pipeline, indicating a good or a defect product (outlier).
- Many more.

2.7 Exercise

Assignment specification:

1. Select a dataset and run the following clustering algorithms:
   a. *k*-means, for several numbers of predefined clusters
   b. ECM, for various clustering parameter values
   c. ECMC for classification in an inductive and transductive modes
   d. Hierarchical clustering
   e. SOM
2. Analyse the results and answer the following questions.
   a. Which clustering methods are adaptive on new data?
   b. What knowledge can be learned through clustering?

2.8 Summary and Open Problems

Models for adaptive unsupervised learning and data analysis, such as the presented evolving, adaptive clustering, adaptive quantisation, and adaptive prototype creation, have the following advantages when compared with the off-line learning methods.

1. They are much faster as they require one pass of adaptive data propagation.
2. They do not require a preset number of prototypes or clusters. They create prototypes or clusters in an adaptive mode depending on the incoming data.
3. They allow for adaptive learning and accumulating of statistical knowledge.
4. They allow for the process of learning to be traced in time.

The difficulty with the adaptive unsupervised methods is that a set goal function J may not reach a minimum value, which is the case in off-line batch learning modes. This problem is overcome if a sufficient number of data examples are presented to the adaptive learning system. This is very much the case with the lifelong learning systems.

The ECM method is extended to a knowledge-based connectionist learning method DENFIS in Chapter 5.

The methods presented here are used in several applications described in Part II of the book.

This chapter also raises some open questions and problems:

1. In our everyday life we learn through using different approaches in concert, e.g. unsupervised, supervised, reinforcement, and so on. How can this flexibility be implemented in a system?
2. New ways of measuring distance between vectors, depending on their size, data distribution, and so on, are needed. Would it be always appropriate to use Euclidean distance, for example, for both measuring the distance between data points in the two-dimensional gas-furnace data space, and in the 40,000-dimensional gene expression space?
3. Would it be possible to combine different methods for measuring distance between data vectors in one model?
4. How do we measure relative distance between consecutive data points from a time series in contrast to measuring global absolute distance?
5. What is really learned through an unsupervised learning process if we do not specify goals or expectations, and if we do not analyse the results? Should we still call the process of feeding unlabelled data into a system a learning process?
6. Can unsupervised learning methods for classification perform better than supervised learning methods and when?
7. How can we integrate unsupervised learning with other sources of available information to improve the knowledge representation and discovery?

2.9 Further Reading

- *Details of the ESOM Algorithm* (Deng and Kasabov, 2002, 2003)
- *Details of the ECM Algorithm* (Kasabov and Song, 2002)
- *Clustering Algorithms – General* (Hartigan, 1975; Bezdek, 1987).
- *K-means Clustering* (MacQueen, 1967).
- *Fuzzy C-means Clustering* (Bezdek, 1981, 1987, 1993)
- *Incremental Clustering* (Fisher, 1989)
- *Adaptive Resonance Theory* (Carpenter and Grossberg, 1987, 1990, 1991)
- *Self-organizing Maps* (Kohonen, 1977, 1982, 1990, 1993, 1997)
- *Chaotic SOM* (Dingle *et al.*, 1993)
- *Neural Gas* (Martinez and Schulten, 1991).
- *Growing Neural Gas* (Fritzke, 1995)
- *Dynamic Topology Representing Networks* (Si *et al.*, 2000)
- *Kernel-based Equiprobabilistic Topographic Map Formation* (Van Hulle, 1998)
- *A Topological Neural Map for Adaptive Learning* (Gaussier and Zrehen, 1994)
- *Dynamic Cell Structures* (Bruske and Sommer, 1995).
- *Spatial Tessellation* (Okabe *et al.*, 1992)
- *Online Clustering Using Kernels* (Boubacar *et al.*, 2006)

3. Evolving Connectionist Methods for Supervised Learning

This chapter presents, as background knowledge, several well-known connectionist methods for supervised learning, such as MLP, RBF, RAN, and then introduces methods for evolving connectionist learning. These include simple evolving MLP (eMLP), evolving fuzzy neural networks (EFuNN) and other methods. The emphasis is on model adaptability and evolvability and on their facilities for rule extraction and pattern/knowledge discovery, which are the main objectives of the knowledge engineering approach that we take in this book. The chapter material is presented in the following sections.

- Connectionist supervised learning methods
- Simple evolving connectionist methods
- Evolving fuzzy neural networks - EFuNN.
- Knowledge manipulation in EFuNN - rule insertion, rule extraction, rule aggregation.
- Summary and open problems
- Exercise
- Further readings

3.1 Connectionist Supervised Learning Methods

3.1.1 General Notions

Connectionist systems for supervised learning learn from pairs of data (x,y), where the desired output vector y is known for an input vector x. If the model is incrementally adaptive, new data will be used to adapt the system's structure and function incrementally (see the classification scheme in Chapter 1).

As discussed in Chapter 1, the objective (goal) function used to optimise the structure of the learning model during the learning process can be either global, or a local goal function.

If a system is trained incrementally, the generalization error of the system on the next new input vector (or vectors) from the input stream is called here *local incrementally adaptive generalization error*. The local incrementally adaptive generalization error at the moment t,for example, when the input vector is $x(t)$, and the output vector calculated by the system is $y(t)'$, is expressed as $Err(t) = ||y(t) - y(t)'||$.

The local incrementally adaptive root mean square error, and the local incrementally adaptive nondimensional error index LNDEI(t) can be calculated at each time moment t as

$$\text{LRMSE}(t) = \sqrt{}(\Sigma i = 1, 2, \ldots, t(Err(i)^2)/\text{t})) \tag{3.1a}$$

$$\text{LNDEI}(t) = \text{LRMSE}(t)/\text{std}(y(1) : y(t)) \tag{3.1b}$$

where std($y(1)$:$y(t)$) is the standard deviation of the output data points from time unit 1 to time unit t.

In a general case, the global generalisation root mean square error RMSE and the nondimensional error index are evaluated on a set of p new (future) test examples from the problem space as follows.

$$\text{RMSE} = \sqrt{}(\Sigma i = 1, 2, \ldots, p[(yi - yi')^2]/p); \tag{3.2a}$$

$$\text{NDEI} = \text{RMSE}/\text{std}(y_1 : y_p), \tag{3.2b}$$

where std (y_1: y_p), is the standard deviation of the data from 1 to p in the test set.

After a system is evolved on a sufficiently large and representative part of the whole problem space Z, its global generalisation error is expected to become satisfactorily small, similar to the off-line, batch mode learning error.

3.1.2 Multilayer Perceptrons (MLP) and Gradient Descent Algorithms

Multilayer perceptrons (MLP) trained with a backpropagation algorithm (BP) use a global optimisation function in both incrementally adaptive (pattern mode) training, and in a batch mode training (Amari, 1967; Rumelhart *et al.*, 1986; Werbos, 1990).

The batch mode off-line training of a MLP is a typical learning method. Figure 3.1 depicts the batch mode backpropagation algorithm.

In the incremental, pattern learning mode of the backpropagation algorithm, after each training example is presented to the system and propagated through it, an error is calculated and then all connections are modified in a backward manner. This is one of the reasons for the phenomenon called catastrophic forgetting: if examples are presented only once, the model may adapt to them too much and 'forget' previously learned examples, if the model is a global model. In an incrementally adaptive learning mode, the same or very similar examples from the past need to be presented many times again, in order for the system to properly learn new examples without forgetting the "old" ones. The process of learning new examples while presenting previously used ones is called 'rehearsal' training (Robins, 1996).

MLP can be trained in an incrementally adaptive mode, but they have limitations in this respect as they have a fixed structure and the weight optimisation is a global one if a gradient descent algorithm is used for this purpose.

A very attractive feature of the MLP is that they are universal function approximators (see Cybenko (1989) and Funahashi (1989)) even though in some cases they may converge in a local minimum.

Some connectionist systems that include MLP use a local objective (goal) function to optimise the structure during the learning process. In this case when a

Forward pass:

BF1. Apply an input vector **x** and its corresponding output vector **y** (the desired output).

BF2. Propagate forward the input signals through all the neurons in all the layers and calculate the output signals.

BF3. Calculate the Err_j for every output neuron j as for example:
$Err_j = y_j - o_j$, where y_j is the jth element of the desired output vector **y**.

Backward pass:

BB1. Adjust the weights between the intermediate neurons i and output neurons j according to the calculated error:
$\Delta w_{ij}(t+1) = lrate.o_j(1-o_j).Err_j.o_i + momentum.\Delta w_{ij}(t)$

BB2. Calculate the error Err_i for neurons i in the intermediate layer:
$Err_i = \sum Err_j.w_{ij}$

BB3. Propagate the error back to the neurons k of lower level:
$\Delta w_{ki}(t+1) = lrate.o_i(1-o_i).Err_i.x_k + momentum.\Delta w_{ki}(t)$

Fig. 3.1 The backpropagation algorithm (BP) for training a multilayer perceptron (MLP) (Amari, 1967; Rumelhart *et al.*, 1986; Werbos, 1990) (from Kasabov (1996), MIT Press, reproduced with permission).

data pair (x, y) is presented, the system optimises its functioning always in a local vicinity of x from the input space X, and in the local vicinity of y from the output space Y (Saad, 1999).

3.1.3 Radial Basis Function (RBF) Connectionist Methods

Several connectionist methods for incrementally adaptive and knowledge-based learning use principles of the radial basis function (RBF) networks (Moody and Darken, 1988, 1989). The basic architecture is outlined here along with its modifications for constructive, incrementally adaptive learning.

The RBF network consists of three layers of neurons—input layer, radial basis layer, and output layer—as shown in Fig. 3.2. The radial basis layer represents clustering of the training data and is established through a clustering method. The second layer of connections is tuned through the delta rule for a global error through multiple iterations over the training data.

The input nodes are fully connected to the neurons in the second layer. A hidden node has a radial basis function as an activation function. The RBF is a symmetric function (e.g. Gaussian, belllike):

$$f(x) = \exp[-(x-M)^2/2\sigma^2] \quad (3.3)$$

where M and σ are two parameters meaning the mean and the standard deviation of the input vector x. For a particular node i, its RBF f_i is centred at the cluster centre C_i in the n-dimensional input space. The cluster centre C_i is represented by the vector $(w_{1i}, \ldots, w_{ni})$ of connection weights between the n input nodes and the hidden node i. The standard deviation for this cluster defines the range for the

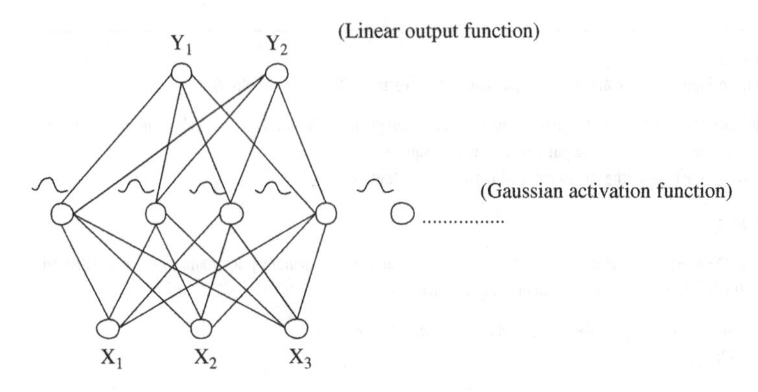

Fig. 3.2 General structure of an RBF network.

RBF f_i. The RBF is nonmonotonic, in contrast to the sigmoid function used in the MLP networks.

The second layer is connected to the output layer. The output nodes perform a simple summation function with a linear thresholding activation function.

Training of a RBFN consists of two phases:

- Adjusting the RBFs of the hidden neurons by applying a statistical clustering method; this represents an unsupervised learning phase.
- Applying gradient descent (e.g. the backpropagation algorithm) or a linear regression algorithm for adjusting the second layer of connections; this is a supervised learning phase.

During training, the following parameters of the RBFN are adjusted.

- The n-dimensional position of the centres C_i of the RBFi. This can be achieved by using the k-means clustering algorithm, for example, which finds a predefined number of hidden nodes (cluster centres and shape of the Gaussian function) that minimize the average distance between the training examples and the k-nearest centres.
- The weights of the second layer connections.

The recall procedure for the RBF network calculates the activation of the hidden nodes that represent how close an input vector $\boldsymbol{x}$ is to the centres C_i. The activation value of the closest node is propagated to the output layer.

Several methods for incrementally adaptive and constructive training of RBF networks exist. Such is the extended growing cell structure (GCS) method (see Chapter 2) called the supervised growing cell structure network (Fritzke, 1995). The method applies the growing cell algorithm on the radial basis nodes in a RBF network. The second layer of connections is tuned through the delta rule in an incrementally adaptive mode.

In Blanzieri and Katenkamp (1996) an algorithm for incrementally adaptive learning in RBF networks is presented which utilises a factorisable RBF network (F-RBFN) introduced by Poggio and Girosi (1990). Fig. 3.3 shows the structure of

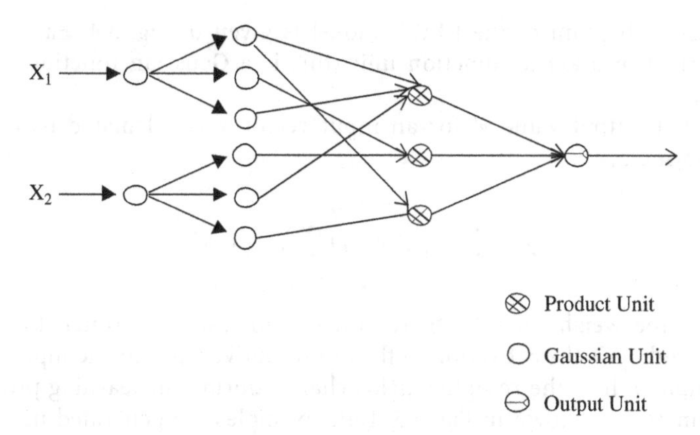

Fig. 3.3 Factorizable RBF network.

the F-RBFN. The RBF are not 'located' in the whole n-dimensional space, but on the local one-variable space for each input variable.

3.1.4 The Resource Allocation Network (RAN) Model

The resource allocation network (RAN) model was suggested by Platt (1991) and improved in other related methods presented in this section. RAN uses the same architecture as the RBF networks, but both the clustering and the second layer adjustment are performed in a two-pass incrementally adaptive mode. A RAN model allocates a new neuron for a new input example (x, y) if the input vector x is not sufficiently close to any of the already allocated radial basis neurons (centres), and also if the output error evaluation $(y - y')$, where y' is the output produced by the system for the input vector x, is above an error threshold. Otherwise, centres will be adapted to minimize the error for the example (x, y) through a gradient descent algorithm.

Some versions of RAN have been developed by Rosipal *et al.* (1997; RAN-GRD and RAN-P-GQRD). Some of these methods are used in this and the next chapters for a comparative analysis of the performance of different incrementally adaptive learning methods.

3.1.5 The Receptive Field Weighted Regression Method (RFWR)

The receptive field weighted regression method (RFWR) is a connectionist-regression technique that uses a regression formula in the form of a weighted sum of local receptive fields learned in local neurons (Schaal and Atkeson, 1998). Through learning, the receptive fields change their size and shape, but not their centre, once it is established. During learning the regression formula changes the weighting of the input variables for the purpose of incremental function approximation.

A schematic diagram of the RFWR model is given in Fig. 3.4. Each receptive field is learned in a kernel function unit (this is a Gaussian function) and in a linear unit.

A predicted output value y' for an input vector x is calculated based on the following formula,

$$y' = \sum_{k=1,K}(w_k a'_k)/\sum_{k=1,K}(w_k) \tag{3.4}$$

where w_k is the weight of the kth receptive field learned through the learning procedure, and $a_k{}'$ is the activation of the kth receptive field for the input vector $\mathbf{x}$.

An example of how the receptive fields change during the learning process of a complex function is shown in Fig. 3.5. Data examples are generated in a random manner from the following function of two variables x and y.

$$z = \max\{\exp(-10x^2), \exp(-50y^2), 1.25\exp(-5(x^2+y^2))\} + N(0, 0.01) \tag{3.5}$$

in a package of 500 examples used for one training iteration in an incrementally adaptive mode. As test data, 1681 data points are drawn from the function space and their output values are evaluated by the trained RFWR model. Figure 3.5 shows: (a) the target function; (b) the approximated function after 50 iterations of training the RFWR model; (c) receptive fields of the generated nodes after one epoch of training (that includes 500 randomly drawn examples) shown in the original input space; and (d) the receptive fields after 50 epochs of training (each epoch includes 500 randomly drawn examples from the function space). The centres of the receptive fields do not change once they are established in an incremental way. The small dots represent data examples drawn from the input space of the above function.

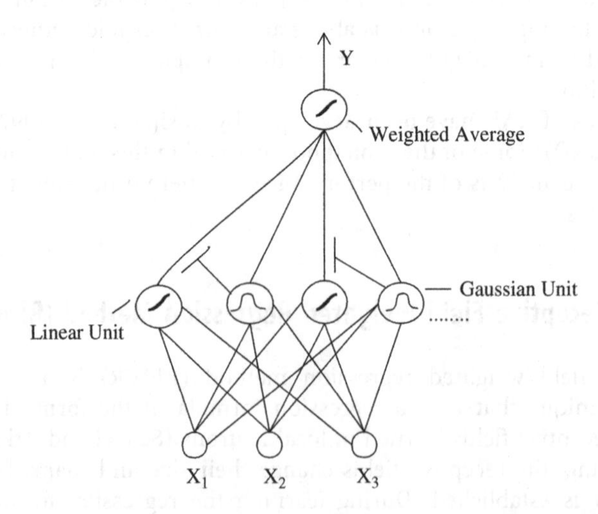

Fig. 3.4 The architecture of the receptive field weight regression network model (RFWR) (see Shaal and Atkeson (1998)).

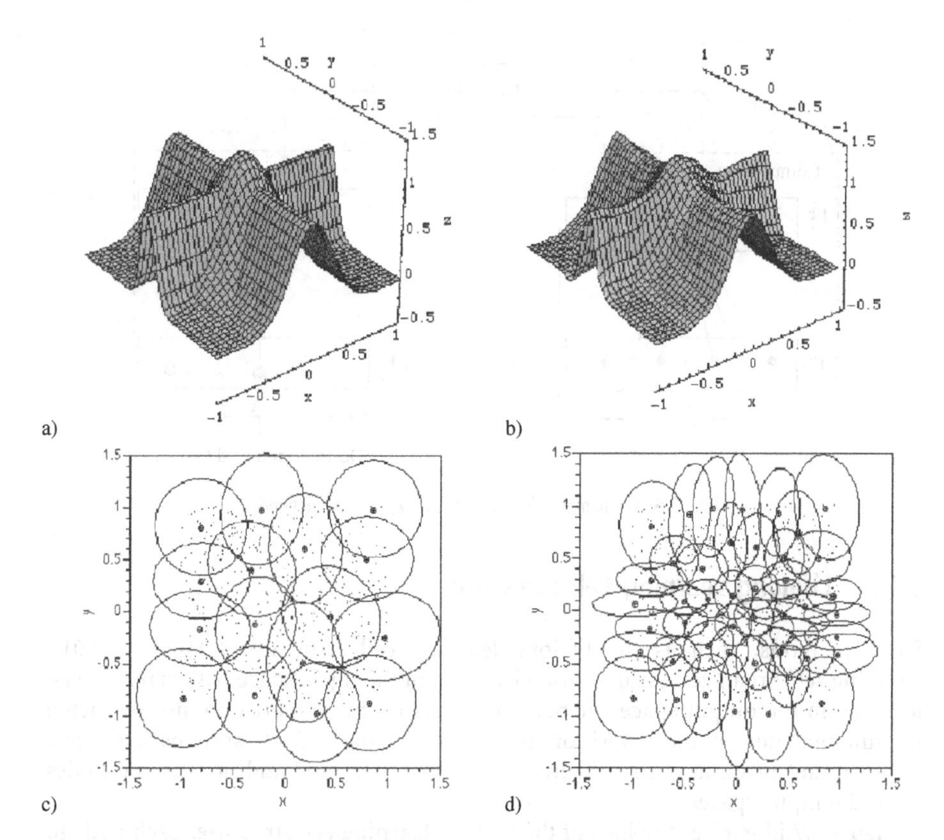

Fig. 3.5 An example of how the receptive fields change during an incremental learning process of a complex function: (a) the original function; (b) the learned in the model function; (c) receptive fields at the beginning of the learning process with a small number of examples; (d) receptive fields after more examples are added (from Schaal and Atkeson (1998); reproduced with permission).

RFWR methods are similar to the mixture of experts methods (Jordan and Jacobs, 1994) as each receptive field here represents one 'local expert'. All receptive fields cover the function under approximation. The system can cope with the changing dynamics of the function and changing probability distribution over time through creating new receptive fields and pruning old ones in an incremental way.

3.1.6 FuzzyARTMAP

FuzzyARTMAP (Carpenter *et al.*, 1991) is an incremental learning connectionist model that associates fuzzy clusters from an input space with an output space. It consists of two parts – FuzzyARTa, and FuzzyARTb – each of them being type ART2 networks that deal with fuzzy input features and fuzzy outputs (see Fig. 3.6). At each time of the learning process, rules that associate input patterns with output classes can be extracted from a FuzzyARTMAP network. A map field maps the activated node in ARTa with the desired output node from ARTb. The mapping process for each input–output training pair is iterative.

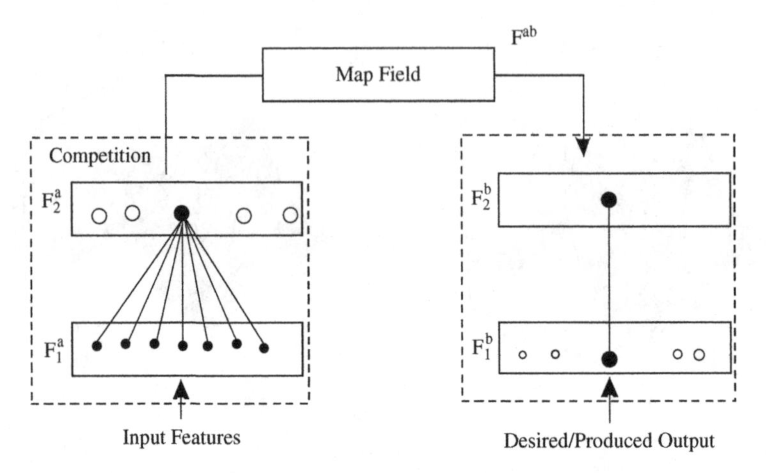

Fig. 3.6 A schematic diagram of a FuzzyARTMAP network.

3.1.7 Lifelong Learning Cell Structures

Some methods, such as the lifelong learning cell structures (Hamker, 2001), combine self-organized unsupervised learning in the input space with error-driven learning in the output space whereas the nodes in the first area are not restricted in numbers nor in their position in the input space. Such systems can grow 'forever' and pruning is also involved to remove the 'old,' and not 'useful' nodes from the input space.

Figure 3.7 illustrates the idea of the lifelong learning cell structure. Each node in the self-organised area has several parameters attached to it: centre and width of the Gaussian activation functions associated with the node, error counter, inherited error, insertion threshold, and age. These parameters are used in the evolving algorithm that defines when a node should be considered as sufficiently activated, how to link this node with the neighbouring nodes, how to update these links (see ESOM for a similar approach, Chapter 2), when to create a new node, when to prune a node, and so on.

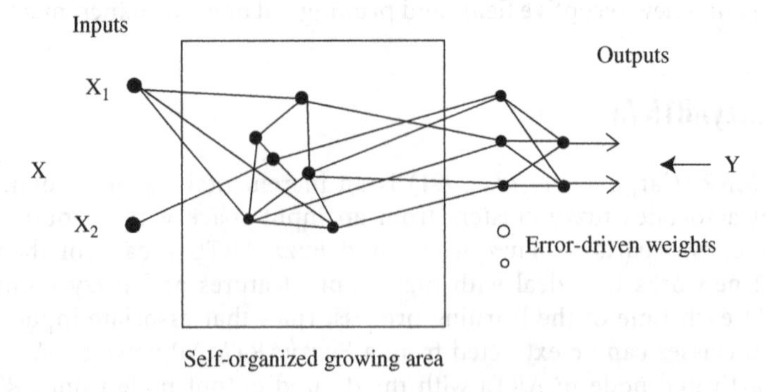

Fig. 3.7 A schematic diagram of a life-long learning cell structure.

3.2 Simple Evolving Connectionist Methods

3.2.1 Simple Evolving MLP and RBF

A representative of this class of methods and systems is ZISC (the Zero Instruction Set Computer) (ZISC Manual, 2001). ZISC is a supervised learning system in a chip that realises a growing RBF network. Each hidden node has a receptive field that is initially maximally large. A node is linked to an output class of yes/no type depending on the example that is presented. If during the learning process a new example is close to this node but belongs to another class, the radius of the field of this node is reduced to exclude this example and a new node is created. The distance between a new example and all nodes is computed in parallel.

Another simple evolving MLP method is called here eMLP and presented in Fig. 3.8 as a simplified graphical representation. An eMLP consists of three layers of neurons, the input layer, with linear or other transfer functions, an evolving layer, and an output layer with a simple saturated linear activation function. It is a simplified version of the evolving fuzzy neural network (EFuNN), presented later in this chapter (Kasabov, 2001).

The evolving layer is the layer that will grow and adapt itself to the incoming data, and is the layer with which the learning algorithm is most concerned. The meaning of the incoming connections, activation, and forward propagation algorithms of the evolving layer all differ from those of classical connectionist systems.

If a linear activation function is used, the activation A of an evolving layer node n is determined by Eq. (3.6),

$$A_n = 1 - D_n \tag{3.6}$$

where A_n is the activation of the node n and D_n is the normalised distance between the input vector and the incoming weight vector for that node.

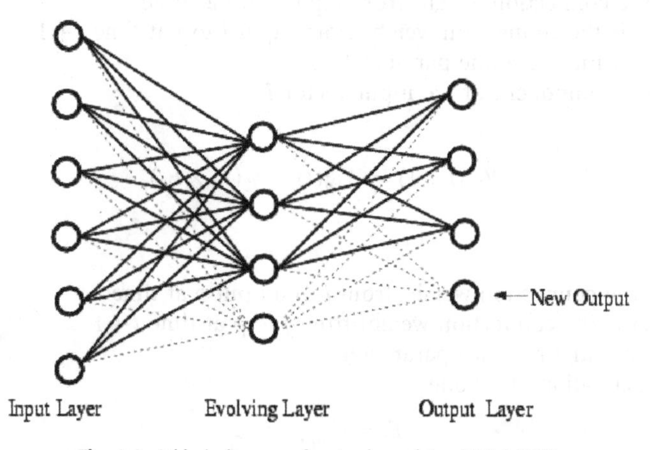

Fig. 3.8 A block diagram of a simple evolving MLP (eMLP).

Other activation functions, such as a radial basis function could be used. Thus, examples which exactly match the exemplar stored within the neurons' incoming weights will result in an activation of 1 whereas examples that are entirely outside the exemplars region of input space will result in an activation of near 0.

The preferred form learning algorithm is based on accommodating, within the evolving layer, new training examples by either modifying the connection weights of the evolving layer nodes, or by adding a new node. The algorithm employed is described below.

Box 3.1. eMLP learning algorithm

1. Propagate the input vector I through the network.
   IF the maximum activation A_{max} of a node is less than a coefficient called sensitivity threshold S_{thr}:
2. Add a node, ELSE
3. Evaluate the error between the calculated output vector O_c and the desired output vector O_d.
4. IF the error is greater than an error threshold E_{thr} OR the desired output class node is not the most highly activated,
5. Add a node, ELSE
6. Update the connections to the winning node in the evolving layer.
7. Repeat the above procedure for each training vector.

When a node is added, its incoming connection weight vector is set to the input vector I, and its outgoing weight vector is set to the desired output vector O_d.

The incoming weights to the winning node j are modified according to Eq. (3.7), whereas the outgoing weights from node j are modified according to Eq. (3.8)

$$W_{i,j}(t+1) = W_{i,j}(t) + \eta_1(I_i - W_{i,j}(t)) \tag{3.7}$$

where:

$W_{i,j}(t)$is the connection weight from input i to j at time t
$W_{i,j}(t+1)$ is the connection weight from input i to j at time $t+1$
η_1 is the learning rate one parameter
I_i is the ith component of the input vector I

$$W_{j,p}(t+1) = W_{j,p}(t) + \eta_2(A_j \times E_p) \tag{3.8}$$

where:

$W_{j,p}(t)$ is the connection weight from j to output p at time t
$W_{i,p}(t+1)$ is the connection weight from j to p at time $t+1$
η_2 is the learning rate two parameter
A_j is the activation of a node j

$$E_p = O_{d(p)} - O_{c(p)} \tag{3.9}$$

where E_p is the error at p; $O_{d(p)}$ is the desired output at p; and $O_{c(p)}$ is the calculated output at p.

The distance measure D_n in Eq. (3.6) above is preferably calculated as the normalised Hamming distance, as shown in Eq. (3.10):

$$D_n = \frac{\sum_i^K |I_i - W_i|}{\sum_i^K |I_i + W_i|} \tag{3.10}$$

where K is the number of input nodes in the eMLP, I is the input vector, and W is the input to the evolving layer weight matrix.

The eMLP architecture is similar to the Zero Instruction Set Computer architecture. However, ZISC is based on RBF ANN and requires several training iterations over input data.

Aggregation of nodes in the evolving layer can be employed to control the size of the evolving layer during the learning process. The principle of aggregation is to merge those nodes which are spatially close to each other. Aggregation can be applied for every (or after every n) training example. It will generally improve the generalisation capability of eMLP. The aggregation algorithm is as follows.

FOR each rule node $r_{j,} j = 1 : n$, where n is the number of nodes in the evolving layer and $W1$ is the connection weight matrix between the input and evolving layer and $W2$ is the connection weight matrix between the evolving and output layer.

- Find a subset R of nodes in the evolving layer for which the normalised Euclidean distances $D(W1_{rj}, W1_{ra})$ and $D(W2_{rj}, W2_{ra}) r_j, r_a \in R$ are below a threshold W_{thr}.
- Merge all the nodes from the subset R into a new node r_{new} and update $W1_{r_{new}}$ and $W2_{r_{new}}$ using the following formulas,

$$W1_{r_{new}} = \frac{\sum r_a \in R(W1_{r_a})}{m} \tag{3.11}$$

$$W2_{r_{new}} = \frac{\sum r_a \in R(W2_{r_a})}{m} \tag{3.12}$$

  where m denotes the number of nodes in the subset R.
- Delete the nodes $r_a \in R$.

Node aggregation is an important regularisation that is not present in ZISC. It is highly desirable in some application areas, such as speech and image recognition systems. In speech recognition, the vocabulary of recognition systems needs to be customised to meet individual needs. This can be achieved by adding words to the existing recognition system or removing words from the existing vocabulary.

eMLP is also suitable for online output space expansion because it uses local learning which tunes only the connection weights of the local node, so all the knowledge that has been captured in the nodes in the evolving layer will be local and only covering a 'patch' of the input–output space. Thus, adding new class outputs or new input variables does not require retraining of the whole system on both the new and old data as is required for traditional neural networks.

The task is to introduce an algorithm for online expansion and reduction of the output space in eMLP. As described above the eMLP is a three-layer network with two layers of connections. Each node in the output layer represents a particular class in the problem domain when using eMLP as a classifier. This local representation of nodes in the evolving layer enables eMLP to accommodate new classes or remove an already existing class from its output space.

In order to add a new node to the output layer, the structure of the existing eMLP first needs to be modified to encompass the new output node. This modification affects only the output layer and the connections between the output layer and the evolving layer. The graphical representation of this process is shown in Fig. 3.8. The connection weights between the new output in the output layer and the evolving layer are initialised to zero (the dotted line in Fig. 3.8.). In this manner the new output node is set by default to classify all previously seen classes as negative. Once the internal structure of the eMLP is modified to accommodate the new output class, the eMLP is further trained on the new data. As a result of the training process new nodes are created in the evolving layer to represent thenew class.

The process of adding new output nodes to eMLP is carried out in a supervised manner. Thus, for a given input vector, a new output node will be added only if it is indicated that the given input vector is a new class. The output expansion algorithm is as follows.

FOR every new output class:

1. Insert a new node j into the output layer;
2. FOR every node in the evolving layer $r_i, i = 1 : n$, where n is the number of nodes in the evolving layer, modify the outgoing connection weights $W2$ from the evolving to output layer by expanding $W2_{i,j}$ with set of zeros to reflect the zero output.
3. Insert a new node in the evolving layer to represent the new input vector and connect it to the new output node j.

This is equivalent to allocating a part of the problem space for data that belong to new classes, without specifying where this part is in the problem space.

It is also possible to remove a class from an eMLP. It only affects the output and evolving layer of eMLP architecture:

FOR every output class o to be removed,

1. Find set of nodes S in the evolving layer which are connected to that output o.
2. Modify the incoming connections $W1$ from input layer to evolving layer by deleting $S_i, i = 1 : n$, where n is the number of nodes in the set S connected to output o.
3. Modify the outgoing connection weights $W2$ from the evolving to output layer by deleting output node o.

The above algorithm is equivalent to deallocating a part of the problem space which had been allocated for the removed output class. In this manner, there will be no space allocated for the deleted output class in the problem space. In other words the network is unlearning a particular output class. The eMLP is further studied and applied in Watts and Kasabov (2002) and Watts (2006).

3.2.2 Evolving Classification Function (ECF)

Another simple evolving connectionist method for classification is the evolving classifier function ECF presented here (see Fig. 3.9). The learning and the recall algorithms of ECF are shown in Box 3.2. Internal nodes in the ECF structure capture clusters of input data that belong to a same class. For each input variable there are fuzzy membership functions define as in Fig. 2.2

Box 3.2a. Learning algorithm of ECF:

1. Enter the current input vector from the dataset (stream) and calculate the distances between this vector and all nodes already evolved (rule) using Euclidean distance (by default). If there is no node created, create the first one that has the co-ordinates of the first input vector attached as input connection weights.
2. If all calculated distances between the new input vector and the existing rule nodes are greater than a max-radius parameter *R*max, a new rule node is created. The position of the new rule node is the same as the current vector in the input data space and the radius of its receptive field is set to the min-radius parameter *R*min; the algorithm goes to step 1; otherwise it goes to the next step.
3. If there is a rule node with a distance to the current input vector less than or equal to its radius and its class is the same as the class of the new vector, nothing will be changed; go to step 1; otherwise:
4. If there is a rule node with a distance to the input vector less than or equal to its radius and its class is different from that of the input vector, its influence field should be reduced. The radius of the new field is set to the larger value from the two numbers: distance minus the min-radius; min-radius. New node is created as in step 2 to represent the new data vector.
5. If there is a rule node with a distance to the input vector less than or equal to the max-radius, and its class is the same as that of the input vector's, enlarge the influence field by taking the distance as a new radius only if such enlarged field does not cover any other rule nodes which belong to a different class; otherwise, create a new rule node in the same way as in step 2, and go to step 1.

Box 3.2.b Recall procedure (classification of a new input vector) in a trained ECF:

1. Enter the new vector in the ECF trained system; if the new input vector lies within the field of one or more rule nodes associated with one class, the vector is classified in this class;

2. If the input vector lies within the fields of two or more rule nodes associated with different classes, the vector will belong to the class corresponding to the closest rule node.
3. If the input vector does not lie within any field, then take m highest activated by the new vector rule nodes, and calculate the average distances from the vector to the nodes with the same class; the vector will belong to the class corresponding to the smallest average distance.

Two main characteristics of ECF are demonstrated in the following example that uses the Iris case study data set.

- Incrementally adaptive learning
- Rule/knowledge extraction

Example

Figure 3.10 shows: (a) an ECF model trained on 90% of the Iris data (135 samples) creating 18 clusters (rules), and afterwards adapted incrementally to the other 10% of data (class 3 only, 15 samples), updating the rules and creating a new one, #19; (b) the 19 rules that represent the adapted 18 clusters of data from the first 90% of the Iris data and the new rule, #19.

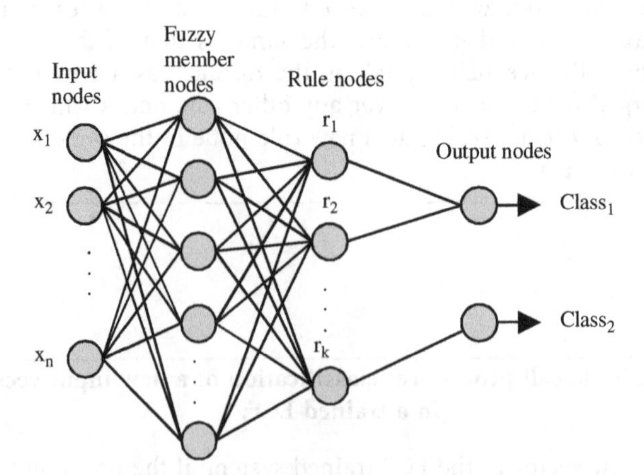

Fig. 3.9 A simplified structure of an evolving classifier function ECF. For every input variable different number and type of fuzzy membership functions can be defined or evolved see Fig. 2.2.

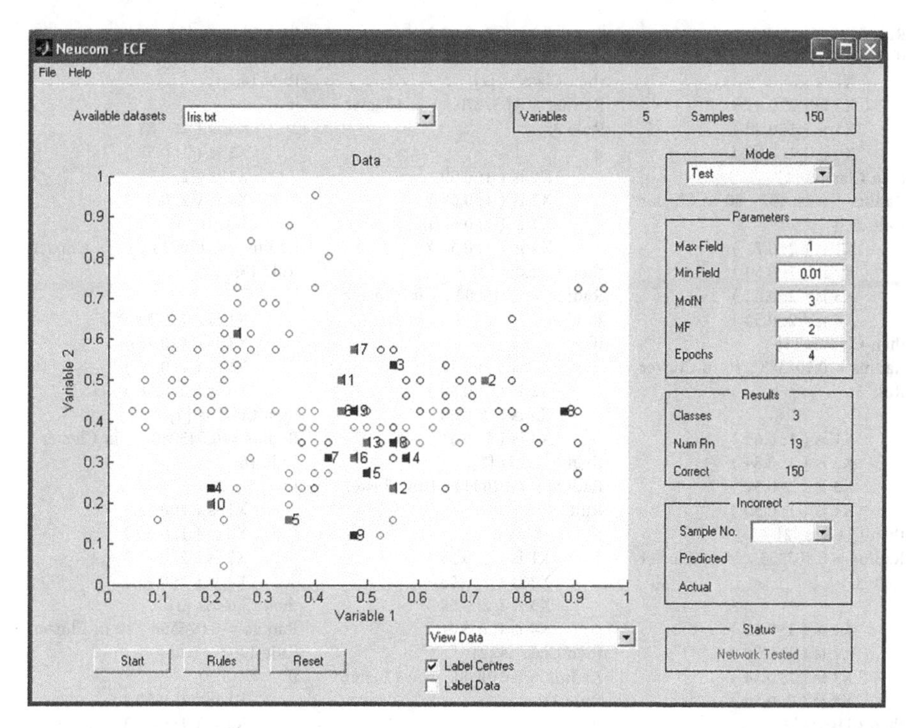

Fig. 3.10 (a) An ECF model trained on 90% of the Iris data (135 samples) creating 18 clusters (rules), and afterwards adapted incrementally to the other 10% of data (class 3 only, 15 samples), updating the rules and creating a new one, #19; (*Continued overleaf*)

3.3 Evolving Fuzzy Neural Networks (EFuNN)

Fuzzy neural networks are connectionist structures that can be interpreted in terms of fuzzy rules (Yamakawa *et al.*, 1992; Furuhashi *et al.*, 1993; Lin and Lee, 1996; Kasabov, 1996). Fuzzy neural networks are NN, with all the NN characteristics of training, recall, adaptation, and so on, whereas neuro-fuzzy inference systems (Chapter 5) are fuzzy rule-based systems and their associated fuzzy inference mechanisms that are implemented as neural networks for the purpose of learning and rule optimisation. The evolving fuzzy neural network (EFuNN) presented here is of the former type, whereas the HyFIS and DENFIS systems presented in Chapter 5 are of the latter type. Some authors do not separate the two types that make the transition from one to the other type more flexible and also broaden the interpretation and the application of each of these systems.

3.3.1 The EFuNN Architecture

EFuNNs have a five-layer structure (Fig. 3.11). Here nodes and connections are created/connected as data examples are presented. An optional short-term memory layer can be used through a feedback connection from the rule (also called case)

Rule 1:
if
X1 is (1: 0.75)
X2 is (2: 0.61)
X3 is (1: 0.89)
X4 is (1: 0.92)
then Class is [1]
Radius = 0.240437 , 50 in Cluster
Rule 2: if
X1 is (2: 0.73)
X2 is (1: 0.50)
X3 is (2: 0.62)
X4 is (2: 0.54)
then Class is [2]
Radius = 0.102388 , 10 in Cluster
Rule 3:
if
X1 is (1: 0.65)
X2 is (1: 0.84)
X3 is (2: 0.51)
X4 is (1: 0.50)
then Class is [2]
Radius = 0.107233 , 15 in Cluster
Rule 4:
if
X1 is (1: 0.55)
X2 is (1: 0.58)
X3 is (2: 0.54)
X4 is (2: 0.58)
then Class is [2]
Radius = 0.073327 , 17 in Cluster
Rule 5:
if
X1 is (1: 0.80)
X2 is (1: 0.80)
X3 is (1: 0.60)
X4 is (1: 0.61)
then Class is [2]
Radius = 0.078333 , 3 in Cluster
Rule 6: if
X1 is (1: 0.55)
X2 is (1: 0.50)
X3 is (2: 0.63)
X4 is (2: 0.69)
then Class is [2]
Radius = 0.038928 , 1 in Cluster
Rule 7:
if
X1 is (2: 0.55)
X2 is (1: 0.77)
X3 is (2: 0.65)
X4 is (2: 0.58)
then Class is [2]
Radius = 0.057870 , 1 in Cluster
Rule 8:
if
X1 is (1: 0.50)
X2 is (1: 0.65)
X3 is (2: 0.62)
X4 is (1: 0.54)
then Class is [2]
Radius = 0.010000 , 1 in Cluster
Rule 9:
if
X1 is (1: 0.53)
X2 is (1: 0.69)
X3 is (2: 0.68)
X4 is (2: 0.61)
then Class is [2]
Radius = 0.010000 , 1 in Cluster
Rule 10:
if
X1 is (1: 0.53)
X2 is (2: 0.58)
X3 is (2: 0.58)
X4 is (2: 0.61)
then Class is [2]
Radius = 0.010000 , 1 in Cluster
Rule 11:
if
X1 is (2: 0.55)
X2 is (2: 0.54)
X3 is (2: 0.82)
X4 is (2: 0.95)
then Class is [3]
Radius = 0.169390 , 20 in Cluster
Rule 12:
if
X1 is (1: 0.80)
X2 is (1: 0.77)
X3 is (2: 0.58)
X4 is (2: 0.65)
then Class is [3]
Radius = 0.089745 , 1 in Cluster
Rule 13:
if
X1 is (1: 0.58)
X2 is (1: 0.69)
X3 is (2: 0.68)
X4 is (2: 0.73)
then Class is [3]
Radius = 0.061177 , 4 in Cluster
Rule 14:
if
X1 is (2: 0.88)
X2 is (1: 0.58)
X3 is (2: 0.91)
X4 is (2: 0.80)
then Class is [3]
Radius = 0.170023 , 11 in Cluster
Rule 15:
if
X1 is (1: 0.53)
X2 is (1: 0.88)
X3 is (2: 0.66)
X4 is (2: 0.58)
then Class is [3]
Radius = 0.045060 , 1 in Cluster
Rule 16:
if
X1 is (2: 0.58)
X2 is (1: 0.69)
X3 is (2: 0.71)
X4 is (2: 0.73)
then Class is [3]
Radius = 0.076566 , 10 in Cluster
Rule 17:
if
X1 is (1: 0.50)
X2 is (1: 0.73)
X3 is (2: 0.75)
X4 is (2: 0.54)
then Class is [3]
Radius = 0.010000 , 1 in Cluster
Rule 18:
if
X1 is (2: 0.55)
X2 is (1: 0.65)
X3 is (2: 0.68)
X4 is (2: 0.58)
then Class is [3]
Radius = 0.010000 , 1 in Cluster
Rule 19:
if
X1 is (1: 0.53)
X2 is (1: 0.58)
X3 is (2: 0.63)
X4 is (2: 0.69)
then Class is [3]
Radius = 0.010000 , 1 in Cluster

Fig. 3.10 (*continued*) (b) The 19 rules that represent the adapted 19 clusters of data, obtained after further training of the ECF model from Fig. 2.10a on the other 10% of the Iris data. Rule #19 is a new one, as a new cluster #19 was created as a result of the adaptation of the model from (a) to the new 10% of the data. The cluster centers in each rule are defined by the membership degree (between 0 and 1) to which each variable belongs to a fuzzy membership function (here "1" indicates small value, and "2" indicates large value fuzzy membership function.

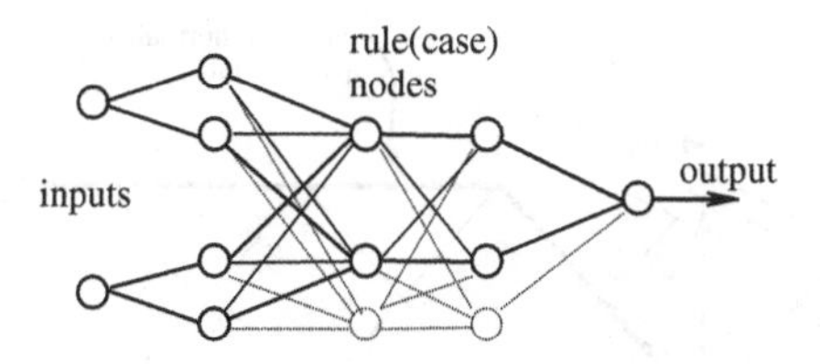

Fig. 3.11 Evolving fuzzy neural network EFuNN: an example of a simplified standard feedforward EFuNN system (from Kasabov (2001a,b), PCT patent WO 01/78003).

node layer (see Fig. 3.12). The layer of feedback connections could be used if temporal relationships of input data are to be memorized structurally.

The input layer represents input variables. The second layer of nodes (fuzzy input neurons or fuzzy inputs) represents fuzzy quantisation of each input variable space (similar to the ECF model and to the factorisable RBF networks; see Section 3.1). For example, two fuzzy input neurons can be used to represent 'small' and 'large' fuzzy values. Different membership functions (MF) can be attached to these neurons (triangular, Fig. 2.2, Fig. 3.13, Gaussian, etc.).

The number and the type of MF can be dynamically modified. The task of the fuzzy input nodes is to transfer the input values into membership degrees to which they belong to the corresponding MF. The layers that represent fuzzy MF are optional, as a nonfuzzy version of EFuNN can also be evolved with only three layers of neurons and two layers of connections as in the eMLP and also used in Chapter 6.

The third layer contains rule (case) nodes that evolve through supervised and/or unsupervised learning. The rule nodes represent prototypes (exemplars, clusters) of input–output data associations that can be graphically represented as associations of hyperspheres from the fuzzy input and the fuzzy output spaces. Each rule node r is defined by two vectors of connection weights, $W1(r)$ and $W2(r)$, the latter being adjusted through supervised learning based on the output error, and

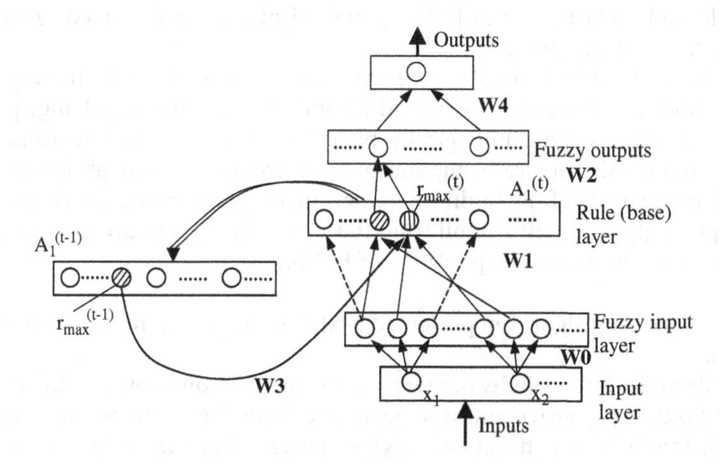

Fig. 3.12 An example of an EFuNN with a short-term memory realised as a feedback connection (from Kasabov (2001a,b), PCT patent WO 01/78003).

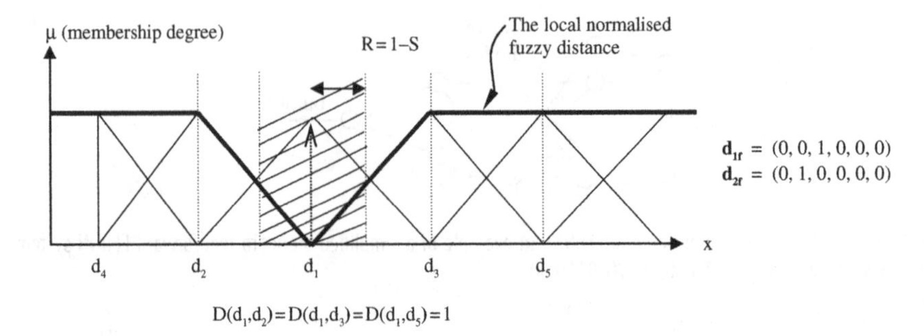

Fig. 3.13 Triangular membership functions (MF) and the local, normalised, fuzzy distance measure (from Kasabov (2001a,b)).

the former being adjusted through unsupervised learning based on a similarity measure within a local area of the problem space. A linear activation function, or a Gaussian function, is used for the neurons of this layer.

The fourth layer of neurons represents fuzzy quantisation of the output variables, similar to the input fuzzy neuron representation. Here, a weighted sum input function and a saturated linear activation function is used for the neurons to calculate the membership degrees to which the output vector associated with the presented input vector belongs to each of the output MFs. The fifth layer represents the values of the output variables. Here a linear activation function is used to calculate the defuzzified values for the output variables.

A partial case of EFuNN would be a three-layer network without the fuzzy input and the fuzzy output layers (e.g. eMLP, or an evolving simple RBF network). In this case slightly modified versions of the algorithms described below are applied, mainly in terms of measuring Euclidean distance and using Gaussian activation functions.

The evolving learning in EFuNNs is based on either of the following assumptions:

1. No rule nodes exist prior to learning and all of them are created (generated) during the evolving process; or
2. There is an initial set of rule nodes that are not connected to the input and output nodes and become connected through the learning (evolving) process. The latter case is more biologically plausible as most of the neurons in the human brain exist before birth, and become connected through learning, but still there are areas of the brain where new neurons are created during learning if 'surprisingly' different stimuli from those previously seen are presented. (See Chapter 1 for biological inspirations of ECOS.)

The EFuNN evolving algorithm presented next does not differentiate between these two cases.

Each rule node, for example, r_j, represents an association between a hypersphere from the fuzzy input space and a hypersphere from the fuzzy output space (see Fig. 3.14), the $W1(r_j)$ connection weights representing the co-ordinates of the centre of the sphere in the fuzzy input space, and the $W2$ (r_j) the co-ordinates in the fuzzy output space. The radius of the input hypersphere of a rule node r_j

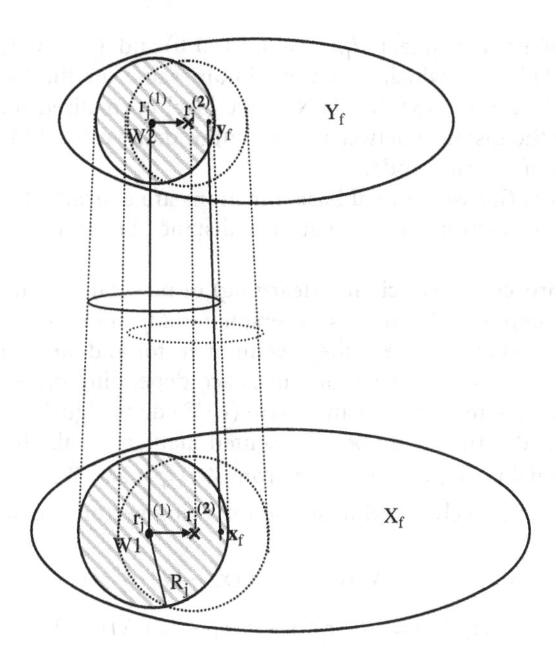

Fig. 3.14 Adaptive learning in EFuNN: a rule node represents an association of two hyperspheres from the fuzzy input space and the fuzzy output space; the rule node r_j 'moves' from a position $r_j^{(1)}$ to $r_j^{(2)}$ to accommodate a new input–output example ($\boldsymbol{x}_f$, $\boldsymbol{y}_f$) (from Kasabov (2001a,b)).

is defined as $R_j = 1 - S_j$, where S_j is the sensitivity threshold parameter defining the minimum activation of the rule node r_j to a new input vector $\boldsymbol{x}$ from a new example $(\boldsymbol{x}, \boldsymbol{y})$ in order for the example to be considered for association with this rule node.

The pair of fuzzy input–output data vectors $(\boldsymbol{x}_f, \boldsymbol{y}_f)$ will be allocated to the rule node r_j if $\boldsymbol{x}_f$ falls into the r_j input receptive field (hypersphere), and $\boldsymbol{y}_f$ falls in the r_j output reactive field hypersphere. This is ensured through two conditions: that a local normalised fuzzy difference between $\boldsymbol{x}_f$ and $W1(r_j)$ is smaller than the radius R_j, and the normalised output error $Err = ||y - y'||/N_{\text{out}}$ is smaller than an error threshold E. N_{out} is the number of the outputs and y' is the produced by EFuNN output. The error parameter E sets the error tolerance of the system.

Definition

A local normalised fuzzy distance between two fuzzy membership vectors $\boldsymbol{d}_{1f}$ and $\boldsymbol{d}_{2f}$ that represent the membership degrees to which two real-value vector data $\boldsymbol{d}_1$ and $\boldsymbol{d}_2$ belong to predefined MFs, is calculated as

$$D(\boldsymbol{d}_{1f}, \boldsymbol{d}_{2f}) = ||\boldsymbol{d}_{1f} - \boldsymbol{d}_{2f}|| / ||\boldsymbol{d}_{1f} + \boldsymbol{d}_{2f}|| \tag{3.13}$$

where $||\boldsymbol{x} - \boldsymbol{y}||$ denotes the sum of all the absolute values of a vector that is obtained after vector subtraction (or summation in case of $||\boldsymbol{x} + \boldsymbol{y}||$) of two vectors $\boldsymbol{x}$ and $\boldsymbol{y}$;

/ denotes division. For example, if $d_{1f} = (0, 0, 1, 0, 0, 0)$ and $d_{2f} = (0, 1, 0, 0, 0, 0)$, then $D(d_1, d_2) = (1+1)/2 = 1$, which is the maximum value for the local normalised fuzzy difference (see Fig. 3.13). In EFuNNs the local normalised fuzzy distance is used to measure the distance between a new input data vector and a rule node in the local vicinity of the rule node.

In RBF networks Gaussian radial basis functions are allocated to the nodes and used as activation functions to calculate the distance between the node and the input vectors.

Through the process of associating (learning) of new data points (vectors) to a rule node r_j, the centres of this node's hyperspheres adjust in the fuzzy input space depending on the distance between the new input vector and the rule node through a learning rate l_j, and in the fuzzy output space depending on the output error through the Widrow–Hoff least mean square (LMS) delta algorithm (Widrow and Hoff, 1960). This adjustment can be represented mathematically by the change in the connection weights of the rule node r_j from $W1(r_j^{(t)})$ and $W2(r_j^{(t)})$ to $W1(r_j^{(t+1)})$ and $W2(r_j^{(t+1)})$, respectively, employing the following vector operations.

$$W1(r_j^{(t+1)}) = W1(r_j^{(t)}) + lj.(x_f - W1(r_j^{(t)})) \quad (3.14)$$

$$W2(r_j^{(t+1)}) = W2(r_j^{(t)}) + lj.(y_f - A2).A1(r_j^{(t)})$$

where $A2 = f_2(W2.A1)$ is the activation vector of the fuzzy output neurons in the EFuNN structure when $\mathbf{x}$ is presented; $A1(r_j^{(t)}) = f_2(D(W1(r_j^{(t)}), \mathbf{x}_f))$ is the activation of the rule node $r_j^{(t)}$; a simple linear function can be used for f_1 and f_2; for example, $A1(r_j^{(t)}) = 1 - D(W1(r_j^{(t)}), x_f))$; l_j is the current learning rate of the rule node r_j calculated, for example, as $l_j = 1/\text{Nex}(r_j)$, where $\text{Nex}(r_j)$ is the number of examples currently associated with rule node r_j.

The statistical rationale behind this is that the more examples are currently associated with a rule node, the less it will 'move' when a new example has to be accommodated by this rule node; that is, the change in the rule node position is proportional to the number of already associated examples with the new single example.

When a new example is associated with a rule node r_j not only its location in the input space changes, but also its receptive field expressed as its radius Rj, and its sensitivity threshold Sj:

$$Rj^{(t+1)} = Rj^{(t)} + D(W1(rj^{(t+1)}), W1(rj(t))) \quad (3.15)$$

respectively,

$$Sj^{(t+1)} = Sj^{(t)} - D(W1(rj^{(t+1)}), W1(rj^{(t)})) \quad (3.16)$$

The learning process in the fuzzy input space is illustrated in Fig. 3.15 on four data points d_1, d_2, d_3, and d_4. Figure 3.15 shows how the centre $r_j^{(1)}$ of the rule node r_j adjusts (after learning each new data point) to its new positions $r_j^{(2)}$, $r_j^{(3)}$, $r_j^{(4)}$ when one-pass learning is applied. Figure 3.16 shows how the rule node position would move to new positions $r_j^{(2(2)}$, $r_j^{(3(2)}$, and $r_j^{(4(2)}$, if another pass of learning

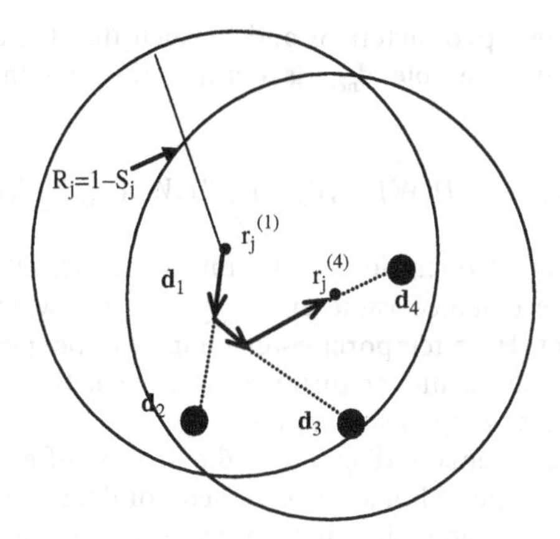

Fig. 3.15 Evolving adaptive learning in EFuNN illustrated on the example of learning four new input data vectors (points) in a rule node r_j (from Kasabov (2001a,b)).

were applied. If the two learning rates l_1 and l_2 have zero values, once established, the centres of the rule nodes will not move.

The weight adjustment formulas (3.14) define the standard EFuNN that has the first part updated in an unsupervised mode, and the second part in a supervised mode similar to the RBF networks. But here the formulas are applied once for each example (x, y) in an incrementally adaptive mode, that is similar to the RAN model (Platt, 1991) and its modifications. The standard supervised/unsupervised learning EFuNN is denoted EFuNN-s/u. In two other modifications of EFuNN, namely double-pass learning EFuNN (EFuNN-dp), and gradient descent learning EFuNN (EFuNN-gd), slightly different update functions are used as explained in the next subsection.

The learned temporal associations can be used to support the activation of rule nodes based on temporal pattern similarity. Here, temporal dependencies are learned through establishing structural links. These dependencies can be further investigated and enhanced through synaptic analysis (at the synaptic memory level) rather than through neuronal activation analysis (at the behavioural level). The ratio spatial similarity/temporal correlation can be balanced for different

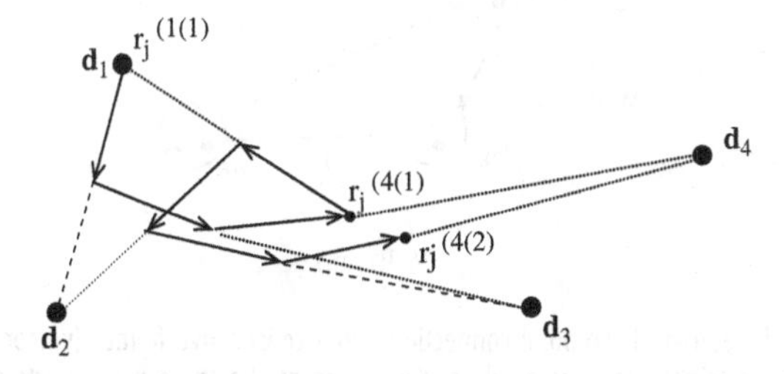

Fig. 3.16 Two-pass learning of four input data vectors (points) that fall in the receptive and the reactive fields of the rule node r_j (from Kasabov (2001a,b)).

applications through two parameters Ss and Tc such that the activation of a rule node r for a new data example $\boldsymbol{d}_{\text{new}}$ is defined through the following vector operations.

$$A1(r) = |1 - Ss.D(W1(r), \boldsymbol{d}_{\text{newf}}) + Tc.W3(r_{\max}{}^{(t-1)}, r)|_{[0,1]} \tag{3.17}$$

where $|.|_{[0,1]}$ is a bounded operation in the interval [0,1]; $D(W1(r), \boldsymbol{d}_{\text{newf}})$ is the normalised local fuzzy distance value and $r_{\max}{}^{(t-1)}$is the winning neuron at the previous time moment. Here temporal connections can be given a higher importance in order to tolerate a higher distance. If $T_c = 0$, then temporal links are excluded from the functioning of the system.

Figure 3.17 shows a schematic diagram of the process of evolving of three rule nodes and setting the temporal links between them for data taken from consecutive frames of a spoken word 'eight' similar to HMM – see chapter 1.

The EFuNN system was explained thus far with the use of one-rule node activation (the winning rule node for the current input data). The same formulas are applicable when the activation of m rule nodes is propagated and used (the so-called '*many-of-n*' mode, or '*m*-of-*n*' for short). By default, $m = 3$, but it is subject to optimisation for different data sets.

The supervised learning in EFuNN is based on the above-explained principles, so when a new data example $\boldsymbol{d} = (\boldsymbol{x},\boldsymbol{y})$ is presented, the EFuNN either creates a new rule node r_n to memorize the two input and output fuzzy vectors $W1(r_{n)} = \boldsymbol{x}_f$ and $W2(r_n) = \boldsymbol{y}_f$, or adjusts an existing rule node r_j.

After a certain time (when a certain number of examples have been presented) some neurons and connections may be pruned or aggregated.

Different pruning rules can be applied for a successful pruning of unnecessary nodes and connections. One of them is given below:

IF (Age(rj) > OLD) AND (the total activation $TA(rj)$ is less than a pruning parameter Pr times Age (rj)) THEN prune rule node rj,

where Age(r_j) is calculated as the number of examples that have been presented to the EFuNN after r_j have been first created; OLD is a predefined age limit; Pr is a pruning parameter in the range [0,1], and the total activation $TA(r_j)$ is calculated

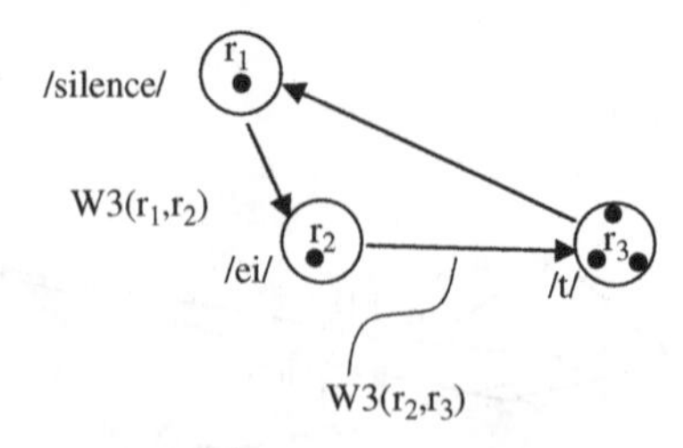

Fig. 3.17 The process of creation of temporal connections from consecutive frames (vectors) taken from speech data of a pronounced word 'eight'. The three rule nodes represent the three major parts of the speech signal, namely the phonemes /silence/, /ei/, /t/. The black dots represent data points (frame vectors) allocated to the rule nodes (from Kasabov (2001a,b)).

as the number of examples for which r_j has been the correct winning node (or among the m winning nodes in the m-of-n mode of operation).

The above pruning rule requires that the fuzzy concepts of OLD, HIGH, and so on are defined in advance. As a partial case, a fixed value can be used; e.g. a node is OLD if it has existed during the evolving process from more than p examples. The pruning rule and the way the values for the pruning parameters are defined, depend on the application task.

3.3.2 EFuNN Evolving Supervised Learning Rules and Algorithms

Three supervised learning algorithms are outlined here that differ in the weight adjustment formulas.

(a) EFuNN-s/u Learning Algorithm

Set initial values for the system parameters: number of membership functions; initial sensitivity threshold (default $S = 0.9$); error threshold E; aggregation parameter N_{agg}, a number of consecutive examples after which an aggregation is performed (explained in a later section); pruning parameters OLD and Pr; a value for m (in m-of-n mode); thresholds T_1 and T_2 for rule extraction.

Set the first rule node to memorize the first example $(\mathbf{x},\mathbf{y})$:

$$W1(r_0) = x_f \text{ and } W2(r_0) = y_f \tag{3.18}$$

Loop over presentations of input–output pairs $(\mathbf{x},\mathbf{y})$

{

Evaluate the local normalised fuzzy distance D between $\mathbf{x}_f$ and the existing rule node connections $W1$ (formulas (3.5)).

Calculate the activation $A1$ of the rule node layer. Find the closest rule node r_k (or the closest m rule nodes in case of m-of-n mode) to the fuzzy input vector $\mathbf{x}_f$

if $A1(r_k) < S_k$(sensitivity threshold for the node r_k); create a new rule node for $(\mathbf{x}_f,\mathbf{y}_f)$

else

Find the activation of the fuzzy output layer $A2 = W2.A1$ and the output error $Err = ||y - y'||/N_{out}$.

if $Err > E$

create a new rule node to accommodate the current example $(\mathbf{x}_f,\mathbf{y}_f)$

else

Update $W1(r_k)$ and $W2(r_k)$ according to (3.4) (in the case of *m-of-n* EFuNN update all the m rule nodes with the highest $A1$ activation).

Apply *aggregation* procedure of rule nodes after each group of N_{agg} examples is presented.

Update the parameters S_k, R_k, Age(r_k), TA (r_k) for the rule node r_k.

Prune rule nodes if necessary, as defined by pruning parameters.

Extract rules from the rule nodes (as explained in a later subsection)

} End of the main loop.

The two other learning algorithms presented next are exceptions and if it is not explicitly mentioned otherwise, the denotation EFuNN means EFuNN-s/u.

(b) EFuNN-dp Learning Algorithm

This is different from the EFuNN-s/u in the weight adjustment formula for $W2$ that is a modification of (3.14) as follows.

$$W2(r_j^{(t+1)}) = W2(r_j^{(t)}) + lj.(y_f - A2)A1(r_j^{(t+1)}), \quad (3.19)$$

meaning that after the first propagation of the input vector and error *Err* calculation, if the weights are going to be adjusted, $W1$ weights are adjusted first with the use of (3.14) and then the input vector $\boldsymbol{x}$ is propagated again through the already adjusted rule node r_j to its new position $r_j^{(t+1)}$ in the input space, a new error *Err* is calculated, and after that the $W2$ weights of the rule node r_j are adjusted. This is a finer weight adjustment than the adjustment in EFuNN-s/u that may make a difference in learning short sequences, but for learning longer sequences it may not cause any difference in the results obtained through the simpler and faster EFuNN-s/u.

(c) EFuNN-gd Learning Algorithm

This algorithm is different from the EFuNN-s/u in the way the $W1$ connections are adjusted, which is no longer unsupervised, but here a one-step gradient descent algorithm is used similar to the RAN model (Platt, 1991):

$$W1(r_j^{(t+1)}) = W1(r_j^{(t)}) + lj.(x_f - W1(r_j^{(t)}))(y_f - A2)A1(r_j^{(t)})W2(r_j^{(t)}) \quad (3.20)$$

Formula (3.20) should be extended when the *m*-of-*n* mode is applied. The EFuNN-gd algorithm is no longer supervised/unsupervised and the rule nodes are no longer allocated at the cluster centres of the input space.

An important characteristic of EFuNN learning is the local element tuning. Only one (or *m*, in the *m*-of-*n* mode) rule node will be either updated or created for each data example. This makes the learning procedure very fast (especially in the case when linear activation functions are used). Another advantage is that learning a new data example does not cause forgetting of old ones. A third advantage is that new input and new output variables can be added during the learning process, thus making the EFuNN system more flexible to accommodate new information, once such becomes available, without disregarding the already learned information.

The use of MFs and membership degrees (layer two of neurons), and also the use of the normalised local fuzzy difference, makes it possible to deal with missing values. In such cases, the fuzzy membership degree of all MFs will be 0.5 indicating that the value, if it existed, may belong to any of them. Preference, in terms of which fuzzy MF the missing value might belong to, can also be represented through assigning appropriate membership degrees, e.g. 0.7 degrees to 'Small' means that the value is more likely to be small rather than 'Medium,' or 'Large.'

The supervised learning algorithms above allow for an EFuNN system to always evolve and learn when a new input–output pair of data becomes available. This is an active learning mode.

(d) EFuNN Sleep-Learning Rules

In another mode, passive or sleep learning, learning is performed when there is no input pattern presented. This may be necessary to apply after an initial learning has been performed. In this case existing connections that store previously fed input patterns are used as an 'echo' to reiterate the learning process. This type of learning may be applied in the case of a short initial presentation of the data, when only a small portion of data is learned in one-pass, incremental adaptive mode, and then the training is refined through the sleep-learning method when the system consolidates what it has learned before.

Sleep learning in EFuNN and in some other connectionist models is further developed by Yamauchi and Hayami (2006).

(e) One- Pass Versus Multiple-Passes Learning

The best way to apply the above learning algorithms is to draw examples randomly from the problem space, propagate them through the EFuNN and tune the connection weights and the rule nodes, change and optimise the parameter values, and so on, until the error becomes a desirably small one. In a fast learning mode, each example is presented only once to the system. If it is possible to present examples two or more times, the error may become smaller, but that depends on the parameter values of the EFuNN and on the statistical characteristics of the data.

3.3.3 EFuNN Inference and Recall

The evolved EFuNN can perform inference when recalled on new input data. The EFuNN inference method consists of calculating the output activation value when a new input vector is applied. This is part of the EFuNN supervised learning method when only an input vector $\boldsymbol{x}$ is propagated through the EFuNN. If the new input vector falls in the receptive field of the winning rule node (the closest rule node to the input vector) *one*-of-n mode of inference is used that is based on the winning rule node activation (one rule inference). If the new input vector does not fall in the receptive field of the closest to it rule node, then the m-of-n mode is used, where m rule nodes (rules) are used in the EFuNN inference process, with an usual value of m being 3.

3.3.4 Strategies for Allocating Rule Nodes in the EFuNN Rule Node Space

There are different ways to allocate in a model space the EFuNN rule nodes evolved over time as illustrated in Fig. 3.18 and explained below:

(a) A simple consecutive allocation strategy, i.e. each newly created rule (case) node is allocated next to the previous, and to the following ones, in a linear fashion. That represents a time order.
(b) Preclustered location, i.e. for each output fuzzy node (e.g. NO, YES) there is a predefined location where the rule nodes supporting this predefined concept are located. At the centre of this area the nodes that fully support this concept (error 0) are placed; every new rule node's location is defined based on the fuzzy output error and the similarity with other nodes. In a nearest activated node insertion strategy, a new rule node is placed nearest to the highly activated node the activation of which is still less than its sensitivity threshold. The side (left or right) where the new node is inserted is defined by the highest activation of the two neighbouring nodes.
(c) As in (b) but temporal feedback connections are set as well. New connections are set that link consecutively activated rule nodes through using the short-term memory and the links established through the $W3$ weight matrix;

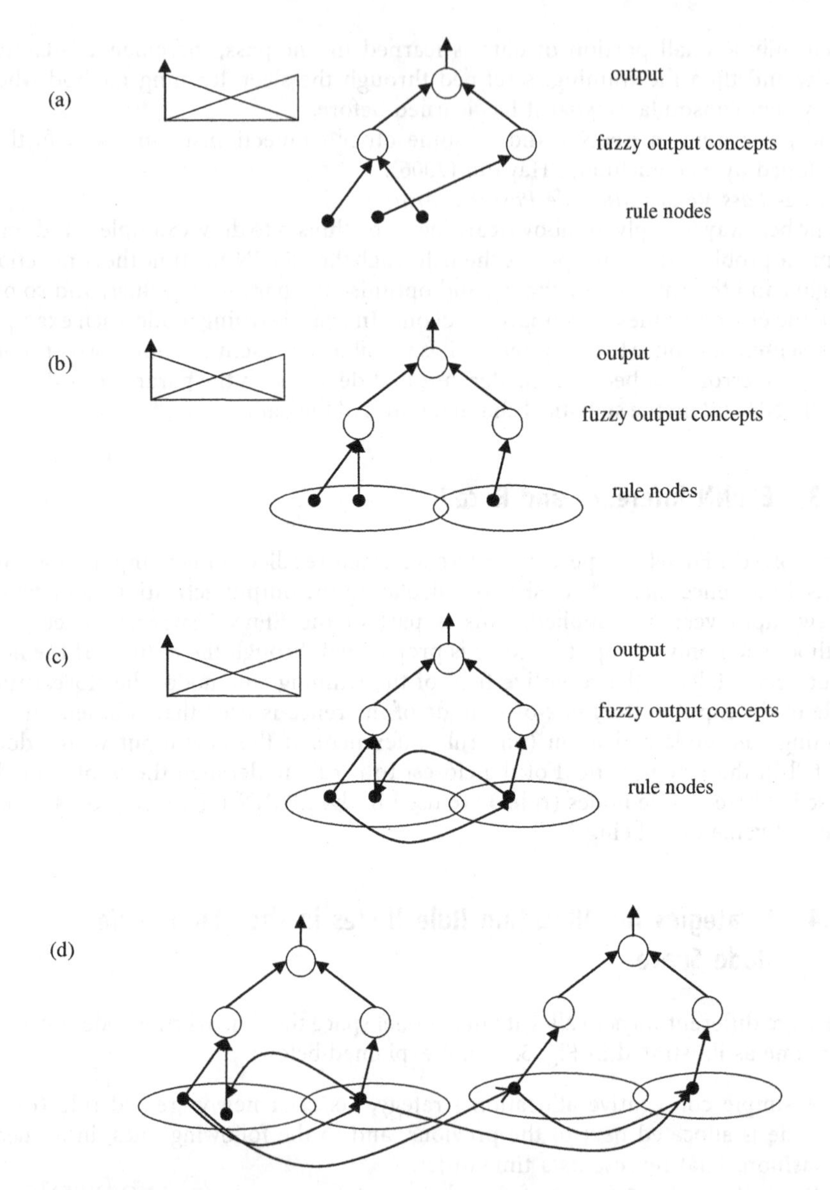

Fig. 3.18 Rule node allocation strategies: (a) simple consecutive allocation strategy; (b) a preclustered location; (c) temporal feedback connections are evolved; (d) connections are evolved between rule nodes from different EFuNN modules. For simplicity, only two membership functions (fuzzy output concepts) are used for the output variable (from Kasabov (2001a,b,c)).

that will allow for the evolving system to react properly to a series of data points starting from a certain point that is not necessarily the beginning of the series.

(d) The same as above, but in addition, new connections are established between rule nodes from different EFuNN modules that become activated simultaneously (at the same time moment). This would make it possible for an ECOS to learn a correlation between conceptually different variables, e.g. correlation between speech sound (left module) and lip movement (right module).

3.3.5 EFuNNs Evolve Using Some Evolving Rules. What Are They in a Summary?

An EFuNN model evolves its structure and functionality based on the following evolving rules, defined mathematically above.

- A rule for a new node creation
- Rules for local receptive field modifications (incremental learning rules)
- A rule for node aggregation (consolidation rule)
- A rule for node deletion (a forgetting rule)
- A sleep-learning rule
- Other rules as shown above

3.4 Knowledge Manipulation in Evolving Fuzzy Neural Networks (EFuNNs) – Rule Insertion, Rule Extraction, Rule Aggregation

It is important for an ECOS that learns in a lifelong learning mode, not only to adjust its structure and functionality, but also to 'explain' at any system operation time the essence and the 'knowledge' the system has learned. Without this ability the chances of using such systems in areas such as financial decision making, complex process control, or gene discovery and drug design, are very slim. The EFuNN architecture, and some other architectures presented thus far, are knowledge-based indeed as they manipulate knowledge in terms of rules, both inserting existing knowledge before the evolving process has started, and extracting refined knowledge from an evolving system. Here more details and analysis of the knowledge-based character of evolving connectionist systems are given with some new architectures reviewed and introduced.

3.4.1 Rule Extraction from EFuNNs

At any time (phase) of the evolving (learning) process of an EFuNN fuzzy or exact rules can be inserted and extracted. Insertion of fuzzy rules is achieved through setting a new rule node r_j for each new rule, such that the connection weights $W1(r_j)$ and $W2(r_j)$ of the rule node represent this rule. For example, the fuzzy rule(*IF* x_1 *is Small and* x_2 *is Small THEN y is Small*) can be inserted into an EFuNN structure by setting the connections of a new rule node to the fuzzy condition nodes x_1-Small and x_2-Small and to the fuzzy output node y-Small to a value of 1 each. The rest of the connections are set to a value of zero. Similarly, an exact rule can be inserted into an EFuNN structure: e.g. *IF* x_1 *is 3.4 and* x_2 *is 6.7 THEN y is 9.5.* Here the membership degrees to which the input values $x_1 = 3.4$ and $x_2 = 6.7$, and the output value $y = 9.5$ belong to the corresponding fuzzy values are calculated and attached to the corresponding connection weights. Each rule node r_j can be expressed as a fuzzy rule, for example:

Rule *rj*: IF *x*1 is Small 0.85 and *x*1 is Medium 0.15 and *x*2 is Small 0.7 and *x*2 is Medium 0.3 (Radius of the receptive field $Rj = 0.1$, maxRadius$j = 0.75$) THEN *y* is Small 0.2 and *y* is Large 0.8 (20 out of 175 examples associated with this rule),

where the numbers attached to the fuzzy labels denote the degree to which the centres of the input and the output hyperspheres belong to the respective MF. The degrees associated with the condition elements are the connection weights from the matrix $W1$. Only values that are greater than a threshold $T1$ are left in the rules as the most important ones. The degrees associated with the conclusion part are the connection weights from $W2$ that are greater than a threshold of $T2$. An example of rules extracted from a benchmark dynamic time-series data is given in Section 3.5. The two thresholds $T1$ and $T2$ are used to disregard the connections from $W1$ and $W2$ that represent small and insignificant membership degrees (e.g. less than 0.1). A set of simple rules extracted from ECF was shown in Fig. 3.10.

3.4.2 Rule Aggregation in EFuNNs

Another knowledge-based technique applied to EFuNNs is rule node aggregation. Through this technique several rule nodes are merged into one as shown in Fig. 3.19a,b,c on an example of three rule nodes r_1, r_2, and r_3 (only the input space is shown there).

For the aggregation of three rule nodes $r_{1,}r_2$, and r_3 the following two aggregation rules can be used to calculate the new aggregated rule node r_{agg} $W1$ connections (the same formulas are used to calculate the $W2$ connections):

(a) As a geometrical centre of the three nodes:

$$W1(r_{agg}) = (W1(r1) + W1(r2) + W1(r3))/3 \quad (3.21)$$

(b) As a weighted statistical centre:

$$W2(r_{agg}) = (W2(r1).\mathrm{Nex}(r1) + W2(r2)\mathrm{Nex}(r2) + W2(r3).\mathrm{Nex}(r3))/\mathrm{Nsum} \quad (3.22)$$

where

$$\mathrm{Nex}(\mathrm{r}_{agg}) = \mathrm{Nsum} = \mathrm{Nex}(r1) + \mathrm{Nex}(r2) + \mathrm{Nex}(r3)$$

The three rule nodes will aggregate only if the radius of the aggregated node receptive field is less than a predefined maximum radius *R*max:

$$Rr_{agg} = D(W1(r_{agg}), W1(r_j)) + R_j <= R\mathrm{max},$$

r_j is the rule node from the three nodes that have a maximum distance from the new node r_{agg} and *Rj* is its radius of the receptive field.
(see Fig. 3.19c).

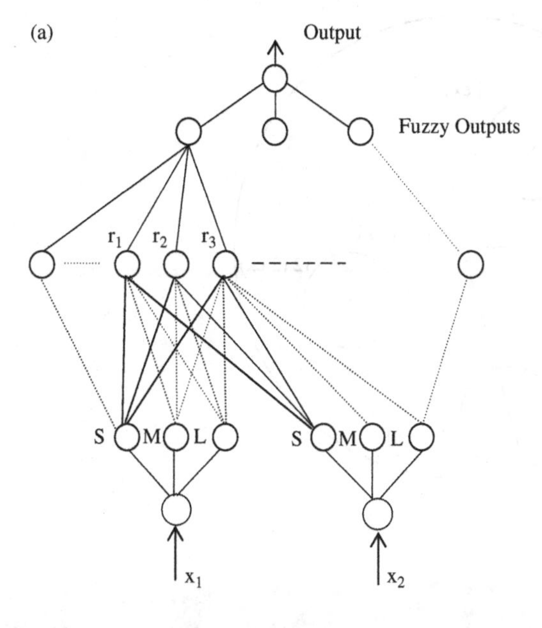

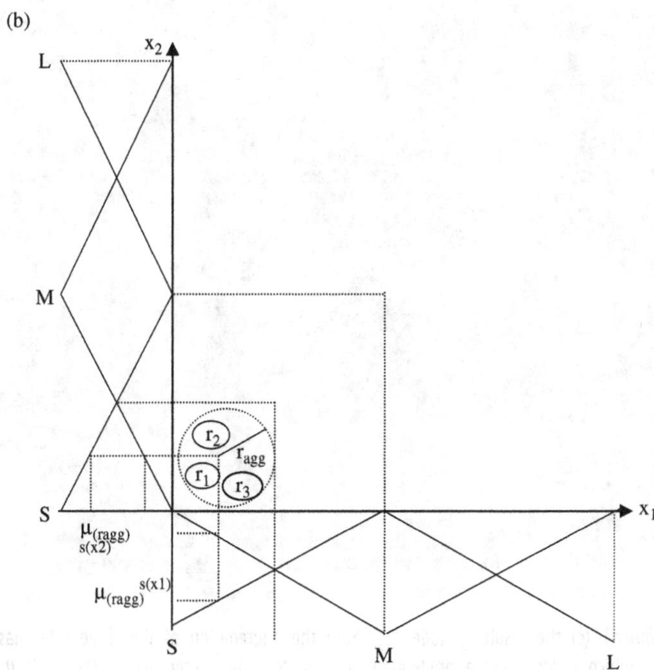

Fig. 3.19 Aggregation of rule nodes in an EFuNN: (a) an example of an evolved EFuNN structure; (b) the process of aggregation of three rule nodes r_1, r_2, and r_3 into one cluster node $r_{agg.}$. (*Continued overleaf*)

(c)

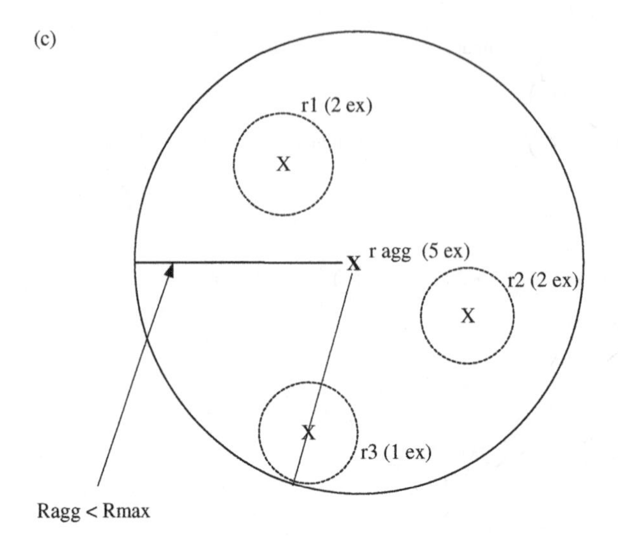

(d)

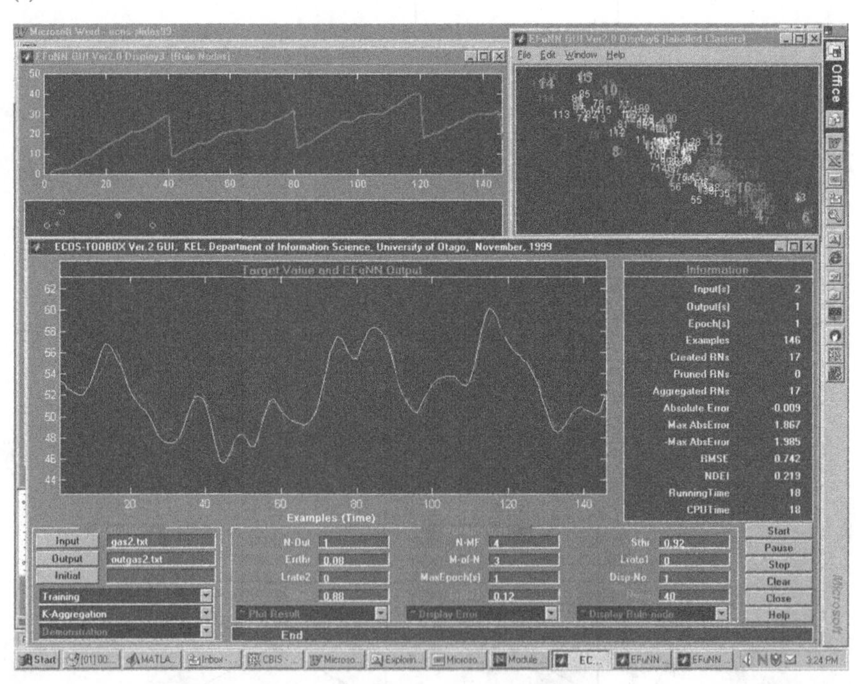

Fig. 3.19 (*continued*) (c) the resulting node r_{agg} from the aggregation of the three rules has a receptive field radius R_{agg} which is less than a predefined (as a system parameter) value R_{max}; (d) the process of aggregation in time, shown for the example of gas-furnace data; the number of rule nodes is aggregated after every 40 examples; the picture also shows the resulting rule node allocation, their corresponding clusters, the desired and the approximated gas-furnace function in time, and other EFuNN parameter values (from Kasabov (2001a,b)).

In order for a given node r_j to aggregate with other nodes, two subsets of nodes are formed: the subset of nodes r_k that if activated to a degree of 1 will produce an output value $y'(r_k)$ that is different from $y'(r_j)$ in less than the error threshold E, and the subset of nodes that cause output values different from $y'(r_k)$ in more than E. The $W2$ connections define these subsets. Then all the rule nodes from the first subset that are closer to r_j in the input space than the closest to r_j node from the second subset in terms of $W1$ distance, get aggregated if the radius of the new node r_{agg} is less than the predefined limit Rmax for a receptive field (Fig. 3.19c).

Figure 3.19d shows the process of incrementally adaptive learning and aggregation (after every 40 examples) from the gas-furnace time series. Through aggregation after the 146th example, 17 rule nodes are created. These rule nodes also represent cluster centres in the input space. The data points that belong to each of these clusters are shown in different colours.

For classification, instead of aggregating all rule nodes that are closer to a rule node r_j than the closest node from the other class, it is possible to keep the closest to the other class node from the aggregation pool out of the aggregation procedure - as a separate node - a 'guard' (see Fig. 3.20), thus preventing a possible misclassification of new data on the bordering area between the two classes. "Guard" vectors are conceptually similar to support vector in SVM (see chapter 1).

Through node creation and their consecutive aggregation, an EFuNN system can adjust over time to changes in the data stream and at the same time preserve its generalisation capabilities.

Through analysis of the weights $W3$ of an evolved EFuNN, temporal correlation between time consecutive exemplars can be expressed in terms of rules and conditional probabilities, e.g.:

$$\text{IF } r_1(t-1) \text{ THEN } r_2(t)(0.3) \tag{3.23}$$

The meaning of the above rule is that some examples that belong to the rule (prototype) r_2 follow in time examples from the rule prototype r_1 with a relative conditional probability of 0.3.

3.4.3 Evolving Membership Functions in EFuNN

Changing membership functions is another knowledge-based operation that may be needed for a refined performance after a certain time moment of the EFuNN's operation. Changing the shape of the MF in a fuzzy neural structure such as FuNN through a gradient descent algorithm is suggested in Kasabov *et al.* (1997). The same algorithm, but in a one-epoch incrementally adaptive version, can be used in EFuNNs. Changing the number of the MFs may also be needed. For example, instead of three MFs, the system may perform better if it had five MFs for some of the variables. In traditional fuzzy neural networks this change is difficult to implement.

In EFuNNs there are several possibilities to implement such dynamical changes of MF, two of them graphically illustrated in Fig. 3.21a,b. These are: (a) new MFs

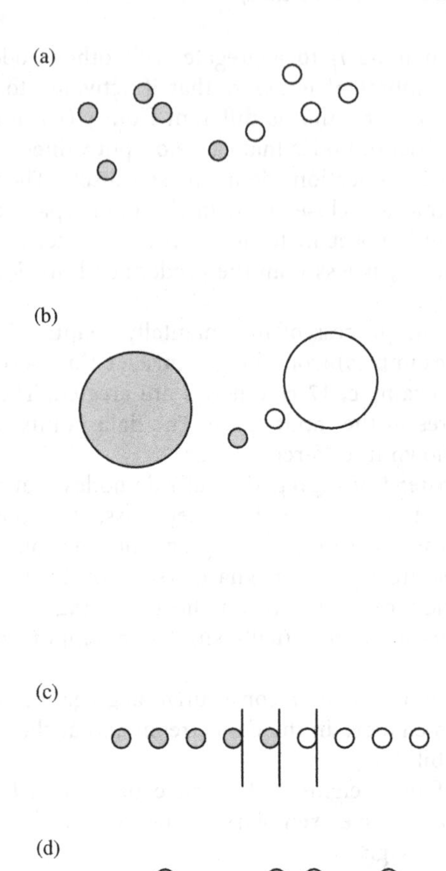

Fig. 3.20 Aggregation of rule nodes ('support vector') with the use of a 'guard' node strategy. Rule nodes are presented as circles, the radius of which define their receptive fields and the colour representing the class that the nodes support (two classes are used): (a) before aggregation; (b) after aggregation, the receptive fields of the new rule nodes have changed, but the receptive fields of the unchanged nodes, the 'guard' nodes, are unchanged; (c) and (d) the process of aggregation as in (a) and (b) but here presented in one-dimensional space of the ordered rule nodes where spatial allocation of nodes is applied (from Kasabov (2001a,b)).

are created (inserted) without a need for the old ones to be changed. The degree to which each cluster centre (each rule node) belongs to the new MF can be calculated through defuzzifying the centres. (b) All MFs change in order for new ones to be introduced. For example, all stored fuzzy exemplars in $W1$ and $W2$ that had three MFs, are defuzzified (e.g., through the centre of gravity defuzzification technique) and subsequently used to evolve a new EFuNN structure that has five MFs (Fig. 3.21a,b).

Adjustment of MF based on x^2 criterion in the incrementally adaptive learning context of EFuNN can be applied as follows.

1. Initialise the EFuNN with a standard number of MF before the learning begins, based on some expected rule representation and based on the context of the data.

(a)

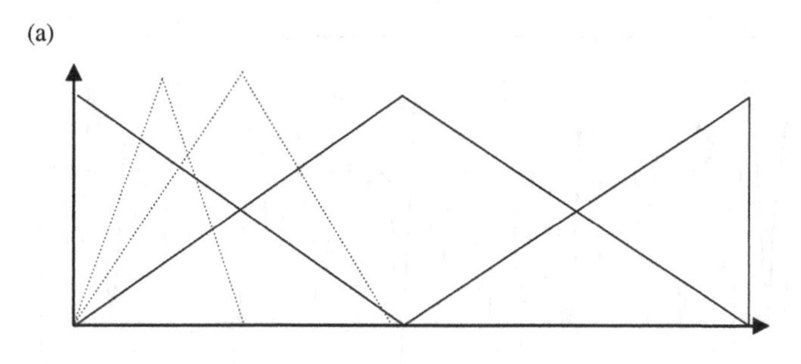

(b)

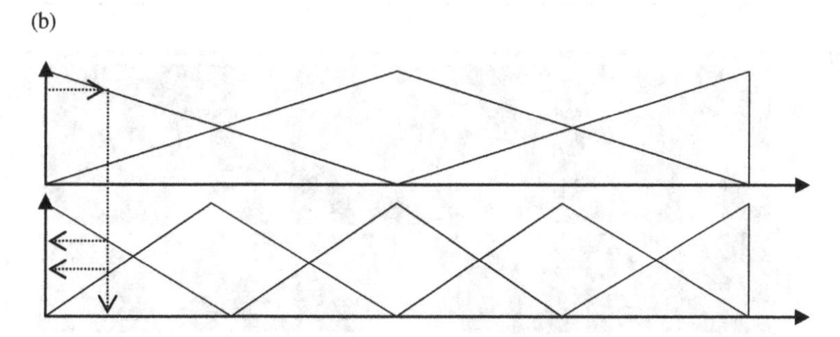

Fig. 3.21 Online membership function modification: (a) new MF are inserted without modifying the existing ones; (b) five new MF are created that substitute the three old MF (from Kasabov (2001a,b)).

2. During the evolving process, calculate the number of examples that fall in each of the areas of the already defined MF and their class (output) values.
3. Regularly, after a sufficient number of examples are presented (e.g. 100), start merging the neighbouring fuzzy intervals and evaluate new ones based on the X^2criteria. The final fuzzy intervals will define the optimal fuzzy MF of a certain type, e.g. triangular.

3.4.4 Case Study Examples: Learning, Aggregation, and Rule Extraction from the Mackey–Glass Time Series

Example 1

The following values for the EFuNN parameters were set: initial value for sensitivity threshold S of 0.9; error threshold $E = 0.1$; a maximum radius $R\max = 0.2$; a rule extraction threshold of 0.5; aggregation is performed after each consecutive group of 50 examples is presented; m-of-n mode, where $m = 1$, is used; the number of membership functions MF is 5; and 1000 consecutive data examples are used. Some experimental results of the incrementally adaptive evolving of an EFuNN are presented in Fig. 3.22a–d, as follows: (a) the desired versus the predicted six-steps-ahead values through one-pass incrementally adaptive learning; (b) the

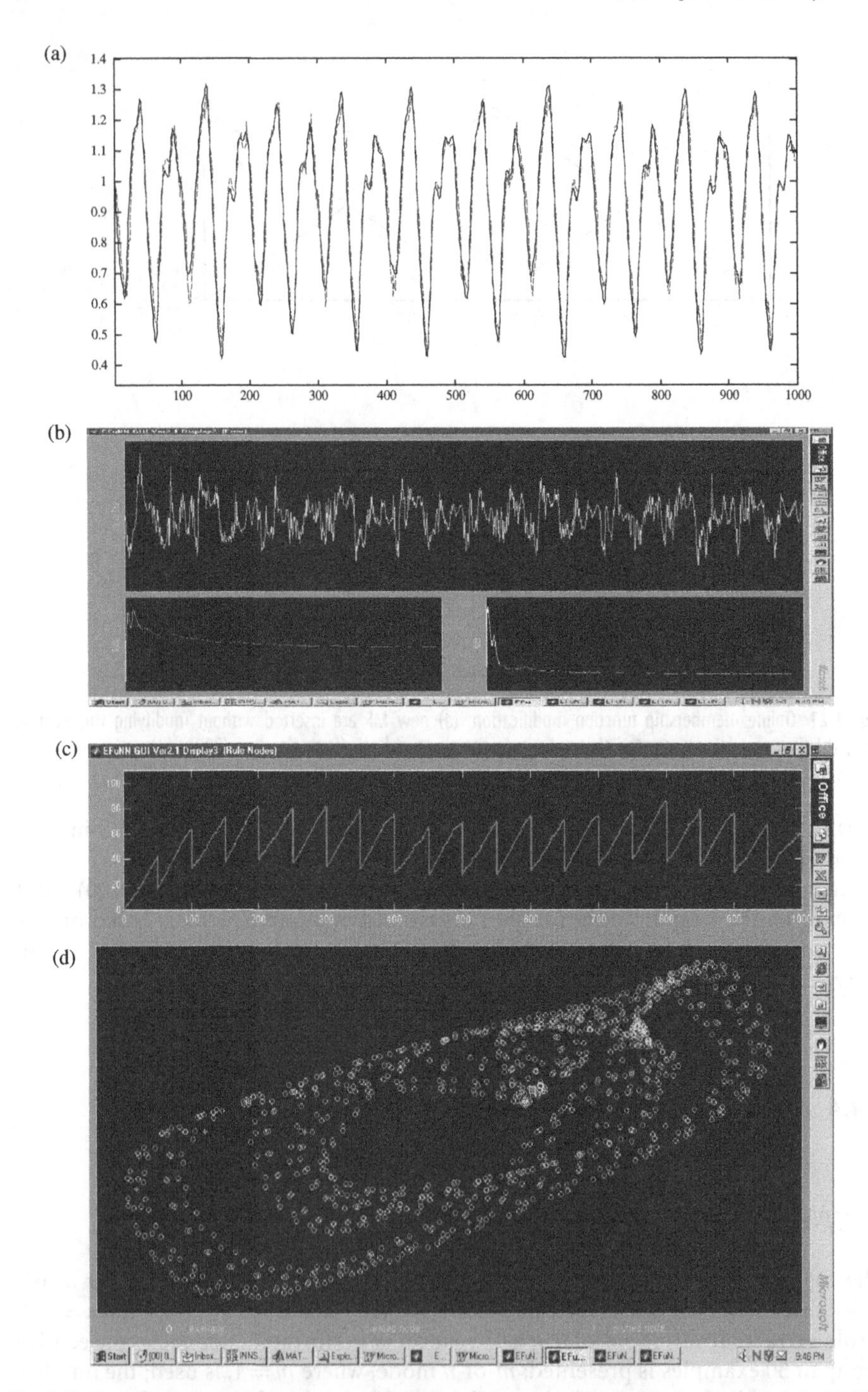

Fig. 3.22 Experiments for online evolving of an EFuNN from the Mackey–Glass chaotic time-series data. An EFuNN is evolved on 1000 data examples from the Mackey–Glass time series (four inputs: $x(t)$, $x(t-6)$, $x(t-12)$, and $x(t-18)$, and one output $x(t+6)$; see http://legend.gwydion.cs.cmu/neural-bench/benchmarks/mackey-glass.html): (a) the desired versus the predicted six-steps-ahead value through

Table 3.1 Some of the fuzzy rules extracted from the evolved from the Mackey–Glass data EFuNN (see Fig. 3.22).

Rule 1: If [x1 is (3 0.658) AND [x2 is (4 0.884)] AND [x3 is (4 0.822)] AND
[x4 is (4 0.722)] [Radius of the receptive field R1 = 0.086]
then [y is (4 0.747)[accommodated training examples Nex(r1) = 6]
Rule 2: If [x1 is (3 0.511)] AND [x2 is (4 0.774)] AND [x3 is (4 0.852)] AND
[x4 is (4 0.825)] [Radius of the receptive field R2 = 0.179]
then [y is (3 0.913)][accommodated training examples Nex(r2) = 2]
....
Rule 16: If [x1 is (2 0.532)]AND [x2 is (2 0.810)] AND [x3 is (3 0.783)] AND
[x4 is(4 0.928)] [Radius of the receptive field R16 = 0.073)
then [y is (5 0.516)] [accommodated training examples Nex(r16) = 12]

Notation: The fuzzy values are denoted with numbers as follows: 1,very small; 2, small; 3, medium; 4, large; 5, very large; the antecedent and the consequent weights are rounded to the third digit after the decimal point; smaller values than 0.5 are ignored as 0.5 is used as a threshold $T1 = T2$ for rule extraction.

absolute, the local incrementally adaptive RMSE (LRMSE), and the local incrementally adaptive NDEI (LNDEI) error over time as described below; (c) the number of the rule nodes created and aggregated over time; (d) a plot of the input data vectors (circles) and the evolved rule nodes (the $W1$ connection weights, crosses) projected in the two-dimensional input space of the first two input variables $x(t)$ and $x(t-6)$.

For different values of the EFuNN parameters, a different number of rule nodes are evolved, each of them represented as one rule through the rule extraction procedure (some of the rules are shown in Table 3.1).

After a certain time moment, the LRMSE and LNDEI converge to constant values subject to a small error. Generally speaking, in the case of compact and bounded problem space the error can be made sufficiently small subject to appropriate selection of the parameter values for the EFuNN.

The example here demonstrates that EFuNN can learn a complex chaotic function through incrementally adaptive evolving from one-pass data propagation. But the real strength of the EFuNNs is in learning processes that change their dynamics through time, e.g. changing values for the parameter τ of the Mackey–Glass equation. Time-series processes with changing dynamics could be of different origin, e.g. biological, financial, environmental, industrial processes, or control.

EFuNNs can also be used for off-line training and testing similar to other standard NN techniques. This is illustrated in another example shown in Fig. 3.23a,b.

Fig. 3.22 one-pass online learning and consecutive prediction; (b) the absolute, the local online RMSE, and the local online NDEI over time; (c) the process of creation and aggregation of rule nodes over time; (d) the input data vectors (circles) and the rule node co-ordinates ($W1$ connection weights; crosses) projected in the two-dimensional input space of the first two input variables $x(t)$ and $x(t-6)$. Some of the extracted rules are shown in Table 3.1.

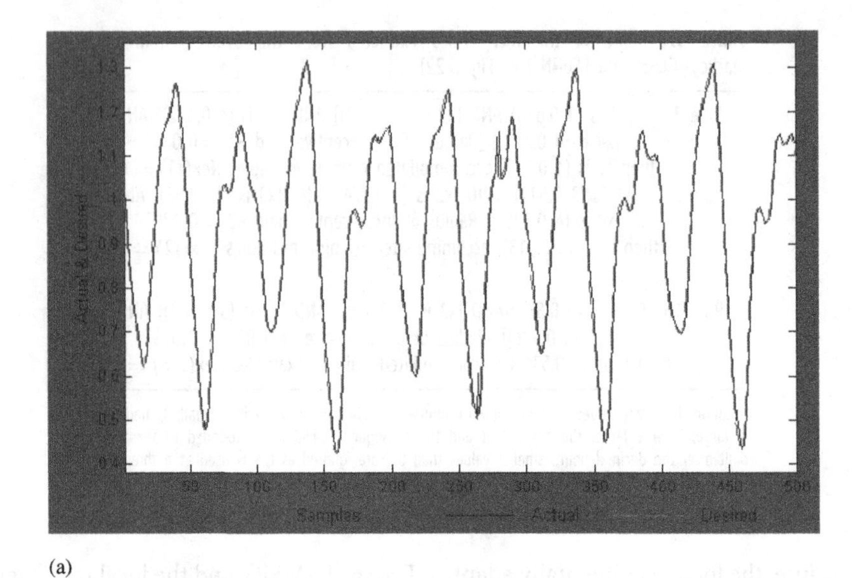

(a)

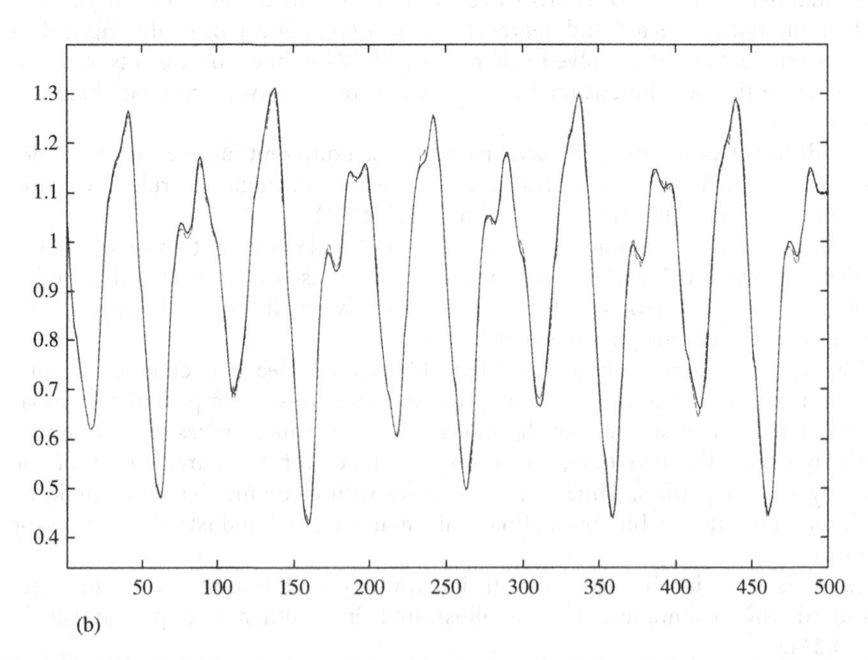

(b)

Fig. 3.23 (a) An EFuNN is evolved on 500 data examples from the Mackey–Glass time series (four inputs: $x(t)$, $x(t-6)$, $x(t-12)$, and $x(t-18)$, and one output $x(t+6)$; the figure shows the desired versus the predicted online values of the time series; (b) after the EFuNN is evolved, it is tested for a global generalisation on a chunk of 500 future data.

Example 2

The following parameter values are set before an EFuNN is evolved: MF = 5; initial $S = 0.9$; $E = 0.01$; $m = 1$; $R\max = 0.2$. The EFuNN is evolved on the first 500 data examples from the same Mackey–Glass time series. Figure 3.23a shows the desired versus the predicted online values of the time series. The following are the parameter values: rule nodes, 62; created nodes, 452; pruned nodes, 335; aggregated nodes, 55; RMSE, 0.034; and NDEI, 0.149.

After the EFuNN is evolved, it is tested for a global generalisation on the second 500 examples. Figure 3.23b shows the desired versus the predicted by the EFuNN values in an off-line mode.

As pointed out before, after having evolved an EFuNN on a small but representative part of the whole problem space, its global generalisation error can become satisfactorily small.

The EFuNN was also tested for incrementally adaptive test error on the test data while further training on it was performed. The incrementally adaptive local test error was slightly smaller.

Generally speaking, if continuous and incremental learning is possible (in many cases of time-series prediction it is) EFuNNs will be continuously evolved all the time through adaptive lifelong learning, always improving their performance. Typical applications of EFuNN would be modelling and predicting of continuous financial time series, modelling of large DNA data sequences, adaptive spoken word classification, and many others (see applications in Part II of this book).

3.4.5 Evolving Fuzzy-2 Clustering in EFuNN

The rule nodes in EFuNN represent cluster centres and have areas associated with them as the cluster area. If a data point $\boldsymbol{d}$ falls in a cluster area, the membership degree belonging to the cluster is defined by the formula $1 - D(\boldsymbol{d}, \boldsymbol{c})$, where $D(\boldsymbol{d}, \boldsymbol{c})$ is the normalised fuzzy distance between the data point $\boldsymbol{d}$ and the cluster centre $\boldsymbol{c}$ (see Chapter 2).

The clustering that is performed in EFuNN is called here fuzzy-2 clustering as not only each data point may belong to several clusters to different degrees (fuzzy), but a cluster centre is defined as fuzzy co-ordinates and a geometrical area associated with this cluster. For example, a cluster centre c is defined as (x is Small to a degree of 0.7, and y is Medium to a degree of 0.3; radius of the cluster area $Rc = 0.3$).

This is illustrated in Fig. 3.24 where two random number input variables x and y are mapped into the same variables as outputs (here EFuNN is used as a replicator); 1000 data points were generated and 98 cluster centres (rule nodes) were evolved for the following initial parameter values: $Sthr = 0.9$; $Errthr = 0.1$; $lr1 = lr2 = 0.1$; 3 MF.

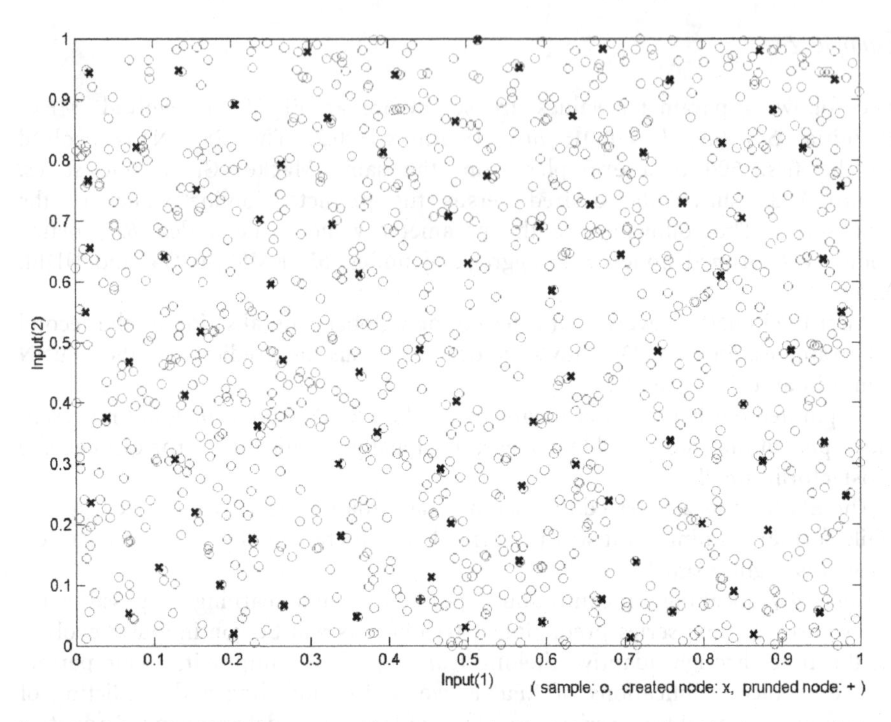

Fig. 3.24 Rule nodes in EFuNN represent cluster centres in the input space of randomly generated two-dimensional input vectors. o, raw data points in a 2D space; x, rule nodes; +, pruned nodes. EFuNN is used here as a replicator (two inputs, two outputs, that have the same values as the corresponding inputs).

3.4.6 Comparative Analysis of EFuNNs and Other ANN and AI Techniques for Incrementally Adaptive, Knowledge-Based Learning

EFuNNs are learning models that can learn in an incrementally adaptive mode any dataset, regardless of the problem (function approximation, time-series prediction, classification, etc.) in a supervised, unsupervised, or hybrid learning mode, subject to appropriate parameter values selected and a certain minimum number of examples presented. Some well-established NN and AI techniques have difficulties when applied to incrementally adaptive, knowledge-based learning. For example, the multilayer perceptrons (MLP) and the backpropagation learning algorithm have the following problems: catastrophic forgetting (Robins, 1996; Hopfield, 1982), local minima problem, difficulties to extract rules (Duch *et al.*, 1998), not being able to adapt to new data without retraining on old ones, and too long training when applied to large datasets.

The radial basis function RBF neural networks require clustering to be performed first and then the backpropagation algorithm applied. They are not efficient for incrementally adaptive learning unless they are significantly modified.

Many neurofuzzy systems, such as ANFIS (Jang, 1993), FuNN (Kasabov *et al.*, 1997), and neofuzzy neuron (Yamakawa *et al.*, 1992) cannot update the learned rules through continuous training on additional data without suffering catastrophic forgetting.

A comparative analysis of different incrementally adaptive learning methods on the Mackey–Glass time series is shown in Table 3.2. Each model is evolved on 3000 data examples from the Mackey–Glass time series (4 inputs: $x(t)$, $x(t-6)$, $x(t-12)$, and $x(t-18)$, and one output $x(t+85)$, from the CMU data: *http://legend.gwydion.cs.cmu/neural-bench/benchmarks/mackey-glass.html).*

The analysis of Table 3.2 shows that the EFuNN evolving procedure leads to a similar local incrementally adaptive error as RAN and its modifications, but EFuNNs allow for rules to be extracted and inserted at any time of the operation of the system thus providing knowledge about the problem and reflecting changes in its dynamics. In this respect the EFuNN is a flexible, incrementally adaptive, knowledge engineering model.

One of the advantages of EFuNN is that rule nodes in EFuNN represent dynamic fuzzy-2 clusters.

Despite the advantages of EFuNN, there are some difficulties when using them:

(a) EFuNNs are sensitive to the order in which data are presented and to the initial values of the parameters.
(b) There are several parameters that need to be optimised in an incrementally adaptive mode. Such parameters are: error threshold *Err*; number, shape, and type of the membership functions; type of learning; aggregation threshold and number of iterations before aggregation, etc. In Chapter 6 parameters related to the simple ECF model, such as max*R*, min*R*, *m*-of-*n*, and number of membership functions, are optimised using a genetic algorithm (GA).

Table 3.2 Comparative analysis of different online learning models on the Mackey–Glass time series. Each model is evolved on 3000 data examples from the Mackey–Glass time series (four inputs: $x(t)$, $x(t-6)$, $x(t-12)$, and $x(t-18)$, and one output $x(t+85)$, from the CMU data: *http://legend.gwydion.cs.cmu/neural-bench/benchmarks/mackey-glass.html*).

Model	*Parameter values*	*Number of Centres (Rule nodes in EFuNN)*	*Online LNDEI after learning 3000 examples*
RAN	$\varepsilon = 0.02$	113	0.373
RAN-GRD	$\varepsilon = 0.01$	50	0.165
RAN-P-GQRD	$\varepsilon = 0.02$	31	0.160
EFuNN-su	$E = 0.05$; *R*max = 0.2	91	0.115
EFuNN-dp	$E = 0.05$; *R*max = 0.2	93	0.113
EFuNN-gd	$E = 0.05$; *R*max = 0.2	102	0.103

3.4.7 EFuNNs as Universal Classifiers and Universal Function Approximators

When issues such as applicability of the EFuNN model, learning accuracy, generalisation, and convergence are discussed for different tasks, two cases must be distinguished.

Case A

The incoming data are from a compact and bounded problem space. In this case the more data vectors are presented to an evolving EFuNN, the better its generalisation is on the whole problem space. After a time moment T, if appropriate values for the EFuNN parameters are used, each of the fuzzy input and the fuzzy output spaces (they are compact and bounded) will be covered by hyperspheres of the evolved rule nodes that will have different receptive fields in the general case.

We can assume that by a certain time moment T a sufficient number of examples from the stream will have been presented and rule node hyperspheres cover the problem space to a desired accuracy. The local incrementally adaptive error will saturate at this time because any two associated compact and bounded fuzzy spaces X_f and Y_f that represent a problem space can be fully covered by a sufficient number of associated (possibly overlapping) fuzzy hyperspheres. The number of these spheres (the number of rule nodes) depends on the error threshold E, set before the training of the EFuNN system, and on some other parameters. The error threshold can be automatically adjusted during learning.

If the task is function approximation, a theorem can be proved that EFuNNs are universal function approximators subject to the above conditions. This is analogous to the proof that MLPs with only two layers are universal function approximators (see for example Cybenko (1989), Funahashi (1989), and Kurkova (1992) and the proof that fuzzy systems are universal approximators too (see for example Kosko (1992) and Koczy and Zorat (1997)).

These proofs are based on the well-known Kolmogorov theorem (Kolmogorov, 1957), which states that:

For all $n >= 2$ there exist $n(2n+1)$ continuous, monotonously increasing, univariate functions on the domain [0,1], by which an arbitrary continuous real function f of n variables can be constructed by the following equation.

$$f(x_1, x_2, \ldots, x_n) = \sum\nolimits_{q=0,2n} \{\phi_q(\sum\nolimits_{p=1,n} \psi_{p,q}(x_p))\} \tag{3.23}$$

As a continuous function, the sigmoid function is mostly used in MLP and other ANN architectures. Linear or Gaussian functions are also used which is the case in the EFuNN architecture.

Case B

The incoming data are from an open problem space, where data dynamics and data distribution may change over time in a continuous way. In this case the local incrementally adaptive error will depend on the closeness of the new input data to the existing rule nodes.

3.4.8 Incrementally Adaptive Parameter and Feature Evaluation in EFuNNs

The performance of the EFuNN depends on its parameter values as illustrated in Table 3.3 on the Mackey–Glass data, when 3000 examples are used to evolve an EFuNN on the task of predicting the values at time moments $(t+85)$, and 500 examples are used to test the system. Different values for the sensitivity threshold *Sthr* and for the error threshold E result in a different number of rule nodes and different values for the RMSE achieved.

Once set, the values of the EFuNN parameters can be either kept fixed during the entire operation of the system, or can be adapted (optimised). Such parameters are, for example: the number of membership functions; the value m for the m-of-n parameter; the error threshold E; the maximum receptive field *Rmax*; the rule extraction thresholds $T1$ and $T2$; the number of examples for aggregation N_{agg}; and the pruning parameters *OLD* and *Pr*. Adaptation can be achieved through: (1) using relevant statistical parameters of the incoming data; (2)incrementally adaptive self-analysis of the behaviour of the system; (3) feedback connection from higher-level modules in the ECOS architecture; and (4) all the above methods.

Table 3.4 shows the results of a similar experiment to the one shown in Table 3.3, but here the two EFuNN parameters are adapted automatically after every 700 examples based on the current RMSE. If the current RMSE is higher than an expected one, the value for E decreases by a delta value δE and the value of *Sthr* increases by a delta value $\delta Sthr$. In the experiment shown in Table 3.4 delta values of 0.05 and 0.06 are used for both parameters.

Genetic algorithms (GA) and evolutionary programming techniques can also be applied to optimise the EFuNNs structural and functional parameters through evolving populations of EFuNNs over generations and evaluating each EFuNN in

Table 3.3 Number of rule nodes, training, and test errors when different EFuNNs are evolved for different parameter values of sensitivity threshold *Sthr* and error threshold E. The Mackey–Glass series is used for training for incremental training and prediction of the value of the series at time moment $t+$ 85 using 3000 initial data points (called training) and further 500 points (called testing). The error on the test 500 vectors is smaller than the training error on the first 3000 vectors, as the model is tested on each test data vector, but then it is further trained on it, in the same way as it was done for the training data.

Sensitivity threshold sthr	*Error threshold E*	*Number of evolved rule nodes Rn*	*RMSE on the training data*	*RMSE on the test data*
0.7	0.3	335	0.077	0.06
0.75	0.25	373	0.073	0.055
0.8	0.2	425	0.068	0.05
0.85	0.15	541	0.064	0.043
0.9	0.1	742	0.055	0.033
0.95	0.055	1300	0.046	0.024

Table 3.4 Self- tuning of the parameters *Sthr* and error threshold E in an EFuNN structure. EfuNNs, as from the experiment shown in Table 3.3. Delta values are used for an automatic increase/decrease of the sensitivity threshold *Sthr* and the error threshold E based on the RMSE, after every 700 examples. The experimented delta values here are 0.05 and 0.06.

Delta value	*Initial* sthr	*Final* sthr	*Initial* errthr	*Final* errthr	*Rule nodes* Rn	*RMSE on training data*	*RMSE on test data*
0.05	0.7	0.9	0.3	0.1	467	0.074	0.04
0.06	0.7	0.96	0.3	0.04	1388	0.052	0.024

Note: A desired maximum RMSE is set to 0.045; *Sthr* and *Errthr* modification after every 700 examples.

the population at certain time intervals (see Chapter 6 for an example of ECF optimisation through GA).

The evaluation of the relevance of the input variables to the task can be done in an incrementally adaptive mode. One way to achieve this is to continuously evaluate the correlation of each input variable to each output class, or to each membership function of the output variables, e.g. *Corr* ($x1$, [y is Small, Medium, High]) = [0.7, 0.4, −0.3] thus producing continuous information on the most relevant input features (see Chapter 7 for details).

In addition to optimising the set of features for an EFuNN in an incrementally adaptive mode, new features can be added, new inputs and new outputs, while the system is operating, similar to the case of the eMLP. Because EFuNN uses local normalized fuzzy distance, new input variables and new output variables can be added to the EFuNN structure at any time of its operation if new data contain these variables.

The following algorithm allows for adding new outputs to already trained EFuNN for a further training.

1. Insert new output node and its initial two fuzzy output nodes representing 'yes' and 'no' for this output.
2. Connect the 'no' output fuzzy nodes with zero connection weights to the already existing rule nodes.
3. Continue the evolving learning process as previously done.

The above simple algorithm is used in Chapter 11 to add new classes of words into an already-trained EFuNN for word recognition.

3.5 Exercise

Choose a classification problem and a dataset for it.

Select a set of features for a classification model.

Build an inductive and a transductive SVM classification model and validate their accuracy through a cross-validation technique.

Build an inductive and a transductive MLP classification model and validate their accuracy through a cross-validation technique.

Build an inductive and a transductive RBF classification model and validate their accuracy through a cross-validation technique.
Build an inductive and a transductive ECF classification model and validate their accuracy through a cross-validation technique.
Demonstrate model and rule adaptation of an ECF model on new data.
Answer the following questions:

(a) Which of the above models are adaptive to new data and under what conditions and constraints?
(b) Which models allow for knowledge extraction and what type of knowledge can be acquired from them?

3.6 Summary and Open Questions

This chapter presents dynamic supervised learning systems. A simple evolving model, EFuNN, and other dynamic supervised learning models are presented that incorporate important AI features, such as adaptive, evolving learning; nonmonotonic reasoning; knowledge manipulation in the presence of imprecision and uncertainties; and knowledge acquisition and explanation.

EFuNNs have features of knowledge-based systems, logic systems, case-based reasoning systems, and adaptive connectionist-based systems, all together. Through self-organization and self-adaptation during the learning process, they allow for solving difficult engineering tasks as well as for simulation of emerging, evolving biological and cognitive processes to be attempted. The lifelong learning mode is the natural learning mode of all biological systems.

The EFuNN models can be implemented in software or in hardware with the use of either conventional or new computational techniques.

The EFuNN applications span across several application areas of information science, life sciences, and engineering, where systems learn from data and improve continuously (Kasabov, 2000a). Some of them are presented in PartII of this book.

Despite the excellent properties of the EFuNNs and the other types of incrementally adaptive ECOS, there are several issues that need to be addressed in the future:

1. How to optimise all ECOS parameters, including choosing the best set of features in an incrementally adaptive mode. This question relates to modifying the evolving rules of EFuNN to better model the incoming data and to reflect on changes in the evolving rules of the modelled process. Only one possible answer is presented in Chapter 6.
2. How to evaluate the convergence property of an ECOS if it is working in an open space.
3. How to evaluate in an incrementally adaptive mode, which supervised model, out of several available, is the best for a given task, or for a given time period of this task.
4. How can knowledge be transferred from one connectionist model to another if the two methods use different knowledge representations?

5. How much can one rely on the labels (desired data, output values) provided with the data for supervised learning? (e.g. are the diagnostic labels associated with patients data always correct?) Would fuzzy representation help to accommodate and deal with the imprecision during data collection?
6. If wrong labels are associated with data, would unsupervised evolving learning be more precise than supervised learning? How can we make an ECOS model 'unlearn' associations between input vectors and output classes if more precise labels for the samples become available in the future?

3.7 Further Reading

A full description of the evolving connectionist architectures presented as well as of some other architectures for supervised incrementally adaptive learning can be found as follows.

- *EFuNN* (Kasabov, 1998, 2001a,b)
- *Simple ECOS, eMLP* (Watts and Kasabov, 2002; Ghobakglou *et al.*, 2003; Watts, 2006)
- *Incrementally Adaptive Learning in Multilayer Perceptron Architectures* (Amari, 1990; Saad, 1999)
- *ART Architectures and the Stability-plasticity Dilemma* (Grossberg, 1981, 1988.
- *ARTMAP* (Carpenter *et al* ., 1991)
- *FuzzyARTMAP* (Carpenter *et al.*, 1992)
- *Incrementally Adaptive Q-learning* (Rummery and Niranjan, 1994)
- *Online Learning in ZISC (Zero Instruction Set Computer)* (ZISC Manual, 2001)
- *Life-long Learning Cell Structures* (Hamker, 2001; Bruske *et al.*, 1998; Hamker and Gross, 1997)
- *Hybrid Neuro-fuzzy Systems for Adaptive and Continuous Learning* (Berenji, 1992; Lim and Harrison, 1998)
- *Incrementally Adaptive Learning in RBF Networks* (Karayiannis and Mi, 1997; Platt, 1991; Fritzke, 1995; Freeman and Saad, 1997)
- *Quantizable RBF Networks* (Poggio and Girosi, 1990)
- *Prediction of Chaotic Time-series with a Resource-allocating RBF Network* (Rosipal *et al.*, 1997)
- *Sleep Learning in EFuNN and other Connectionist Models* (Yamauchi and Hayami, 2006)

4. Brain Inspired Evolving Connectionist Models

The chapter presents some closer to the brain information processing connectionist methods, namely state-based ANN realized in recurrent connectionist structures, reinforcement learning ANN, and spiking ANN. In the state-based ANN the output signal from the model depends not only on the inputs and the connections, but on its previous states. A mathematical model describing such behaviour is finite state automata, realized here in a recurrent network structure, where connections from outputs or hidden nodes connect back to the inputs or to the hidden layer. Spiking neural networks (SNN) are brainlike connectionist methods, where the output activation is represented as a train of spikes rather then as a potential. The chapter is presented in the following sections.

- State-based ANN
- Reinforcement learning
- Evolving spiking neural networks
- Summary and open problems
- Further reading

4.1 State-Based ANN

A classical model for modelling systems described by states and their transitions is the finite automata model that has already been shown to be a good theoretical candidate for modelling brain states and their transitions (see Arbib (1972, 1987)). Here we present a new version of it - evolving finite automata - and show how the model can be realised in a recurrent evolving connectionist structure.

4.1.1 Evolving Finite Automata

A deterministic finite-state automaton is characterized by a finite number of states. It is described as a five-tuple $A = \{X, S, \delta, q, O\}$, where $S = \{s1, s2, \ldots, sn\}$, is a set of states, $s0$ is a designated initial state. $X = \{x1, x2, \ldots, xk\}$ is the alphabet of the input language.

The transition table δ: $(X \times S)$->S defines the state transitions in A. F is a set of final states (outputs) defined through an output transformation q : S->O.

A deterministic finite-state fuzzy automaton is characterized by a seven-tuple $A = \{X, FX, S, \delta, q, O, FO\}$, where S, X, and O are defined as in the nonfuzzy automaton. Fuzzy membership functions are defined as sets FX and FO for the input and the output variables, respectively. Transitions are defined as follows: $\delta : (FX \times S)\text{-}{>}S$ defines the state transitions, and $q : S\text{-}{>}FO$ defines the output fuzzy transitions.

Further in this section the concepts of evolving automata and evolving fuzzy automata are first introduced and then implemented in an evolving connectionist structure.

In an evolving automaton, the number of states in the set S is not defined a priori; rather it increases and decreases in a dynamic way, depending on the incoming data. New transitions are added to the transition table. The number of inputs and outputs can change over time.

In an evolving fuzzy automaton, the number of states is not defined a priori as is the case for the nonfuzzy automata. New transitions are added to the transition table as well as new output fuzzy transitions. The number of inputs and outputs can change over time.

4.1.2 Recurrent Evolving Neural Networks and Evolving Automata

Recurrent connectionist architectures, having the feature to capture time dependencies, are suitable techniques to implement finite automata. In Omlin and Giles (1994) recurrent MLPs that have fixed structures are used to implement finite automata. In a reverse task, a finite automaton is extracted from a trained recurrent MLP.

Recurrent connectionist systems have feedback connections from a hidden or output layer of neurons back to the inputs or to the hidden layer nodes.

There are two main types of recurrent connectionist architectures of EFuNN that are derivatives of the main EFuNN architecture. They are depicted in Fig. 4.1a,b.

1. The feedback connections are from the hidden rule nodes to the same nodes but with a delay of some time intervals, similar to the recurrent MLP (Elman, 1990).
2. The feedback connections are from the output nodes to the hidden nodes, similar to the proposed system in Lawrence *et al.* (1996)

Recurrent connectionist structures capture temporal dependencies between the presented data examples from the data stream (Grossberg, 1969; Fukuda *et al.*, 1997). Sometimes these dependencies are not known in advance. For example, a chaotic time series function with changing dynamics may have different autocorrelation characteristics at different times. This implies a different dependency between the predicted signal in the future and the past data values. The number of the time-lags cannot be determined in advance. It has to be learned and built in the system's structure as the system operates.

Figure 4.2a,b illustrates the autocorrelation characteristics of a speech signal, phoneme /e/ in English, pronounced by a male speakers of New Zealand English: (a) the raw signal in time; (b) the autocorrelation. The autocorrelation analysis

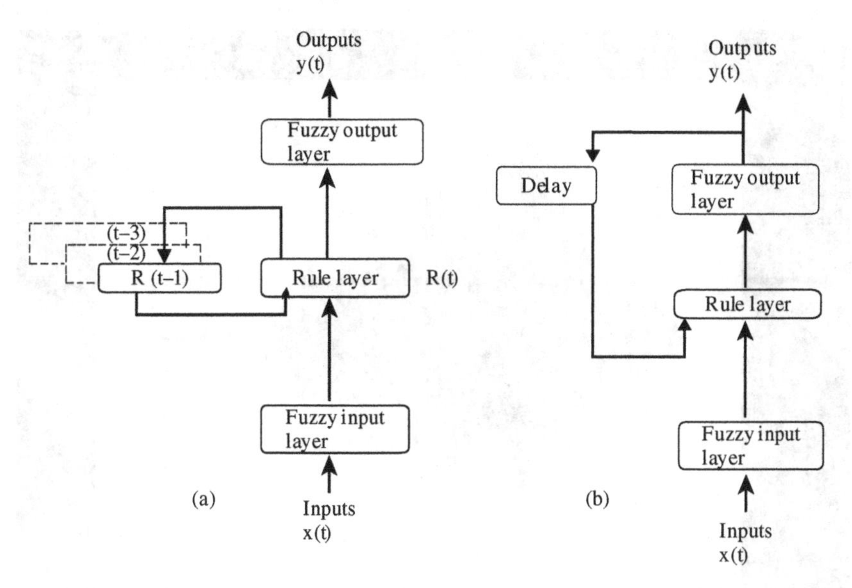

Fig. 4.1 Two types of recurrent EFuNN structures: (a) recurrent connections from the rule layer; (b) recurrent connections from the output layer.

shows that there is correlation between the signal at a current time moment ($t = 0$) and the signal at previous moments. These time dependencies are very difficult to know in advance and a preferred option would be that they are learned in an online mode.

Autocorrelation and other time-dependency characteristics can be captured online in the recurrent connections of an evolving connectionist structure as explained in this section.

Although the connection weights $W1$ and $W2$ capture fuzzy co-ordinates of the learned prototypes (exemplars) represented as centres of hyperspheres, the temporal layer of connection weights $W3$ of the EFuNN (Chapter 3) captures temporal dependencies between consecutive data examples. If the winning rule node at the moment $(t-1)$ (with which the input data vector at the moment $(t-1)$ is associated) is $r_{\max}{}^{(t-1)}$, and the winning node at the moment t is $r_{\max}{}^{(t)}$, then a link between the two nodes is established as follows.

$$W3(r_{\max}{}^{(t-1)}, r_{\max}{}^{(t)}) = W3(r_{\max}{}^{(t-1)}, r_{\max}{}^{(t)}) + l_3.A1(r_{\max}{}^{(t-1)})A1(r_{\max}{}^{(t)}) \qquad (4.1)$$

where $A1(r^{(t)})$ denotes the activation of the rule node r at a time moment (t); and l_3 defines the degree to which the EFuNN associates links between rule nodes (clusters, prototypes) that include consecutive data examples. If $l_3 = 0$, no temporal associations are learned in an EFuNN structure. Figure 4.3 shows a hypothetical process of rule node creation in a recurrent EFuNN for learning the phoneme /e/ from input data that are presented frame by frame.

Rather than using fixed time-lags as inputs to a time-series modeling system, the structure shown in Fig. 4.1b of a recurrent EFuNN can be used to learn temporal dependencies of a time series 'on the fly'.

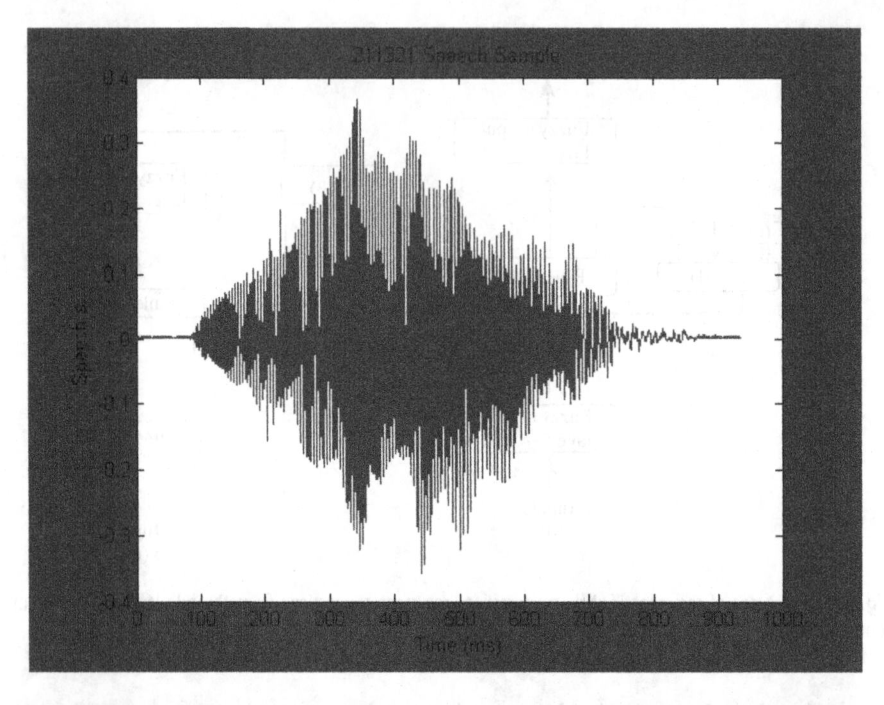

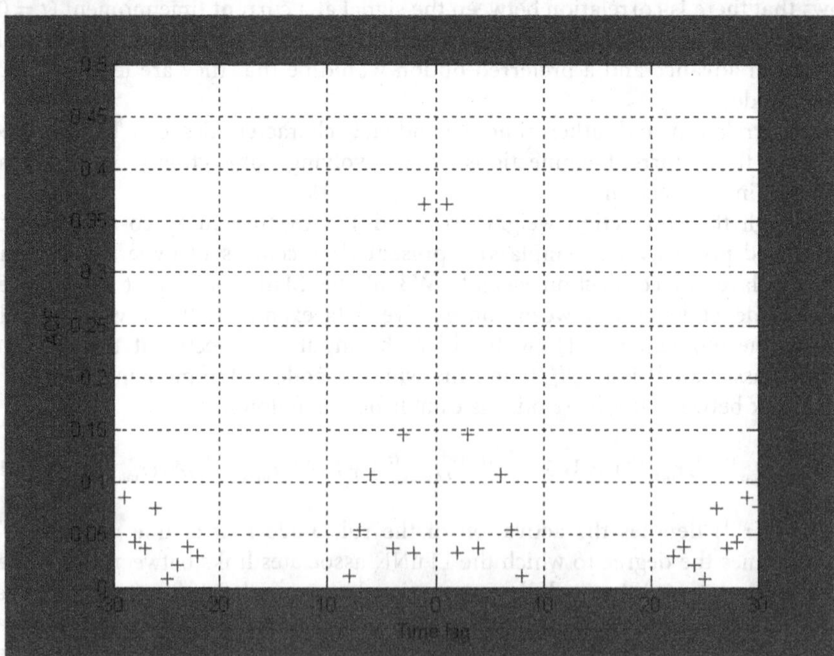

Fig. 4.2 (a) A waveform of a speech signal over time representing a pronunciation of the phoneme /e/ in English, by a male speaker; (b) the autocorrelation characteristics of the signal. The autocorrelation analysis shows that there is correlation between the signal at a time moment (indicated as 0 time), and the signal at previous moments. The middle vertical line represents the signal at a time 0.

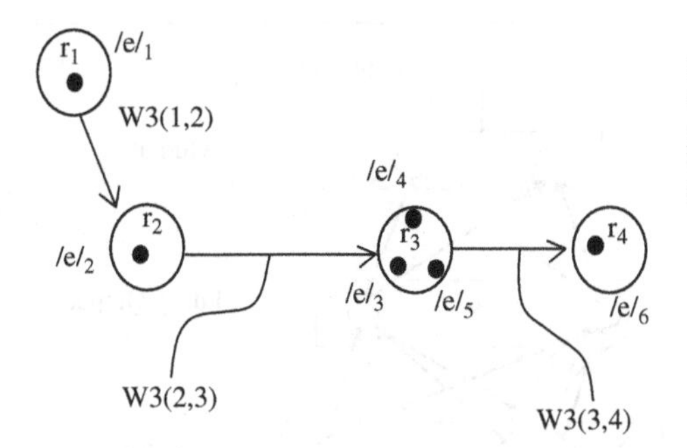

Fig. 4.3 The process of evolving nodes and recurrent connections for a pronounced phoneme /e/.

Two experiments were conducted to compare the EFuNN and the recurrent EFuNN (called REFuNN) on the two benchmark time-series data used in this book: the gas-furnace data, and the Mackey–Glass data. The results shown in Table 4.1 suggest that even for a stationary time series the REFuNN gives a slightly better result. If the dynamics of the time series change over time, the REFuNN would be much superior to the EFuNN in a longer term of the evolving process. The following parameters were used for both the EFuNN and the REFuNN systems for the gas-furnace data: 4 MF, $Sthr = 0.9$, $Errthr = 0.1$, $lr1 = lr2 = 0$, no pruning, no aggregation. For the Mackey–Glass data experiments, the following parameter values were used: 4 MF, $Sthr = 0.87$, $Errthr = 0.13$, $lr1 = lr2 = 0$, no pruning, no aggregation.

For the same values of the parameters, the recurrent EFuNN - REFuNN, achieves less error for less number of rule nodes. This is due to the contribution of the feedback connections to capture existing temporal relationship in the time series.

A recurrent EFuNN can realize an evolving fuzzy automaton as illustrated in Fig. 4.4. In this realization the rule nodes represent states and the transition function is learned in the recurrent connections. Such an automaton can start learning and operating without any transition function and it will learn this function in an incremental, lifelong way.

Table 4.1 Comparative analysis between an EFuNN architecture and a recurrent EFuNN architecture with recurrent connections from the output nodes back to the rule nodes. Two benchmark datasets are used: Mackey–Glass series and gas-furnace time series (see Chapter 1). In both cases the recurrent version of EFuNN–REFuNN, evolves less nodes and achieves better accuracy. The reason is that the REFuNN captures some temporal relationship in the time-series data.

	Number of rule nodes	RMSE
EFuNN for the gas furnace data	58	1.115
REFuNN for the gas furnace data	43	1.01
EFuNN for the Mackey–Glass data	188	0.067
REFuNN for the Mackey–Glass data	156	0.065

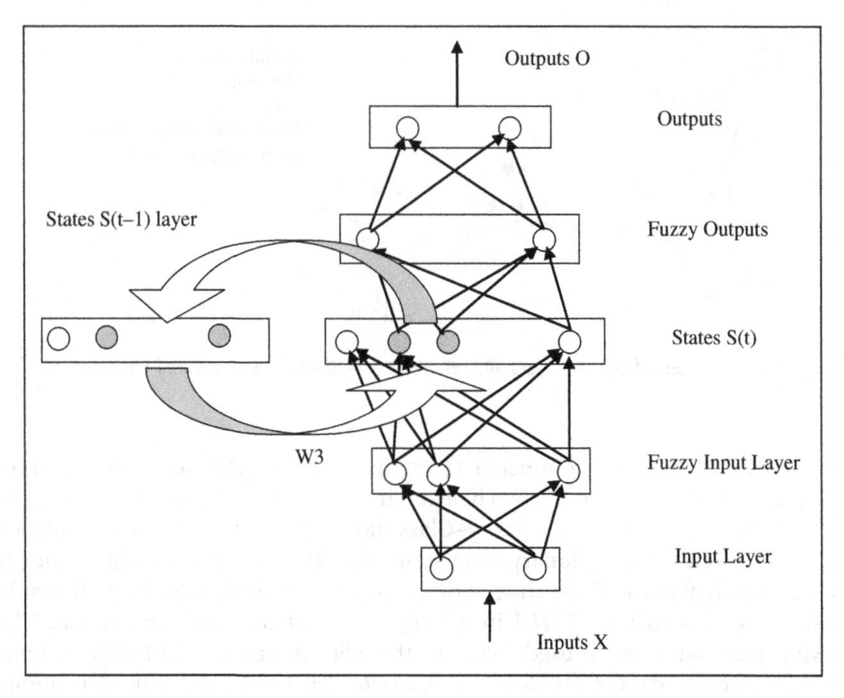

Fig. 4.4 Recurrent EFuNN realising an evolving fuzzy finite automaton. The transitions between states are captured in the short-term memory layer and in the feedback connections.

As shown in Chapter 3, at any time of the evolving process of a recurrent EFuNN, a meaningful internal representation of the network such as a set of rules or their equivalent fuzzy automaton can be extracted. The REFuNN has some extra evolving rules, such as the recurrent evolving rule defined in Eq. (4.1).

4.2 Reinforcement Learning

Reinforcement learning is based on similar principles as supervised learning, but there is no exact desired output and no calculated exact output error. Instead, feedback "hints" are given. There are several cases in a reinforcement learning procedure for an evolving connectionist architecture, such as EFuNN (see Chapter 3):

(a) There is a rule node activated (by the current input vector x) above the preset threshold, and the highest activated fuzzy output node is the same as the received fuzzy hint. In this case the example x is accommodated in the connection weights of the highest activated rule node according to the learning rules of EFuNN.
(b) Otherwise, there will be a new rule node created and new output neuron (or new module) created to accommodate this example. The new rule node is then connected to the fuzzy input nodes and to a new output node, as is the case in the supervised evolving systems (e.g. as in the EFuNN algorithm).

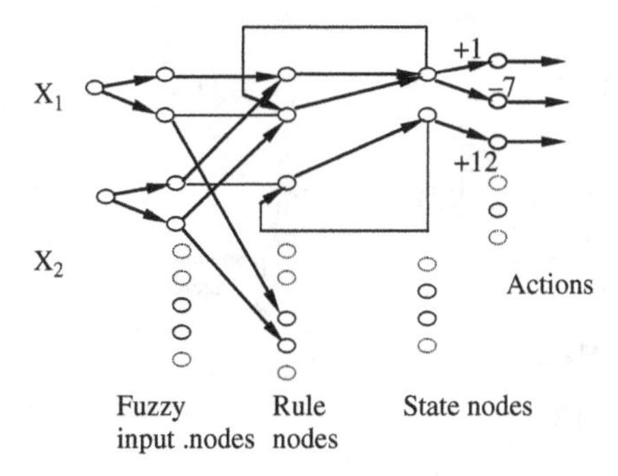

Fig. 4.5 An exemplar recurrent EFuNN for reinforcement learning.

Figure 4.5 shows an example of a recurrent EFuNN for reinforcement learning. The fuzzy output layer is called here a state node layer. The EFuNN structure has feedback connections from its fuzzy outputs back to its rule nodes.

The connection weights from the state to the action (output) nodes can be learned through reinforcement learning, where the awards are indicated as positive connection weights and the punishments as negative connection weights. This type of recurrent EFuNN can be used in mobile robots that learn and evolve as they operate. They are suitable techniques for the realization of intelligent agents when supervised, unsupervised, or reinforcement learning is applied at different stages of the system's operation.

4.3 Evolving Spiking Neural Networks

4.3.1 Spiking Neuron Models

SNN models are more biologically plausible to brain principles than any of the above ANN methods. A neuron in a spiking neural network, communicates with other neurons by means of spikes (Maass, 1996, 1998; Gerstner and Kistler, 2002; Izhikevich, 2003). A neuron *Ni* receives continuously input spikes from presynaptic neurons *Nj* (Fig. 4.6). The received spikes in *Ni* accumulate and cause the emission of output spikes forming an output spike train that is transmitted to other neurons.

This is a more biologically realistic model of a neuron that is currently used to model various brain functions, for instance pattern recognition in the visual system, speech recognition, and odour recognition.

We describe here the Spike Response Model (SRM) as a representative of spiking neuron models that are all variations of the same theme. In a SRM, the state of a neuron *Ni* is described by the state variable *ui*(t) that can be interpreted as a total somatic postsynaptic potential (PSP). The value of the state variable $u_i(t)$ is the weighted sum of all excitatory and inhibitory synaptic post synaptic potentials PSPs:

$$u_i(t) = \sum_{j \in \Gamma_i} \sum_{t_j \in F_j} W_{ij} \varepsilon_{ij}(t - t_j - \Delta_{ij}) \tag{4.2}$$

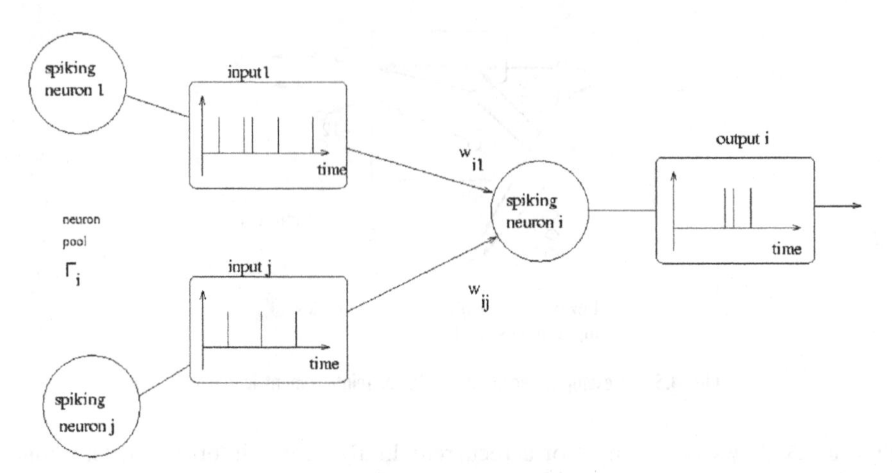

Fig. 4.6 Spiking model of a neuron sends and receives spikes to and from other neurons in the network, similar to biological neurons (from Benuskova and Kasabov (2007)).

where Γ_i is the pool of neurons presynaptic to neuron *Ni*, F_i is the set of times $t_j < t$ when presynaptic spikes occurred, and Δ_{ij} is an axonal delay between neurons *i* and *j*, which increases with the increase of the physical distance between neurons in the network. The weight of synaptic connection from neuron *Nj* to neuron *Ni* is denoted W_{ij}. It takes positive (negative) values for excitatory (inhibitory) connections, respectively. When $u_i(t)$ reaches the firing threshold $\vartheta_i(t)$ from below, neuron *Ni* fires, i.e. emits a spike.

Immediately after firing the output spike at t_i, the neuron's firing threshold $\vartheta_i(t)$ increases *k* times and then returns to its initial value ϑ_0 in an exponential fashion. In such a way, absolute and relative refractory periods are modeled:

$$\vartheta_i(t - t_i) = k \times \vartheta_0 \exp\left(-\frac{t - t_i}{\tau_\vartheta}\right) \tag{4.3}$$

where τ_ϑ is the time constant of the threshold decay. Synaptic PSP evoked on neuron *i* when a presynaptic neuron *j* from the pool Γ_i fires at time t_j is expressed by the positive kernel $\varepsilon_{ij}(t - t_j - \Delta_{ij}) = \varepsilon_{ij}(s)$ such that

$$\varepsilon_{ij}(s) = A\left(\exp\left(-\frac{s}{\tau_{decay}}\right) - \exp\left(-\frac{s}{\tau_{rise}}\right)\right) \tag{4.4}$$

where τ are time constants of the decay and rise of the double exponential, respectively, and *A* is the amplitude of PSP. To make the model more biologically realistic, each synapse be it an excitatory or inhibitory one, can have a fast and slow component of its PSP, such that

$$\varepsilon_{ij}^{type}(s) = A^{type}\left(\exp\left(-\frac{s}{\tau_{decay}^{type}}\right) - \exp\left(-\frac{s}{\tau_{rise}^{type}}\right)\right) \tag{4.5}$$

where *type* denotes one of the following: *fast_excitation*, *fast_inhibition*, *slow_excitation*, and *slow_inhibition*, respectively. These types of PSPs are based on neurobiological data (Destexhe, 1998; Deisz, 1999; Kleppe and Robinson, 1999; White *et al.*, 2000).

In each excitatory and inhibitory synapse, there can be a fast and slow component of PSP, based on different types of postsynaptic receptors.

A SNN is characterized in general by:

- An encoding scheme for the representation of the input signals as spike trains, to be entered into a spiking neuronal model
- A spiking model of a single neuron
- A learning rule of a neuron, including a spiking threshold rule
- A SNN structure
- Learning rules for the SNN including rules for changing connection weights and creation of neurons

In Bohte *et al.* (2000) a MLP architecture is used for a SNN model and the backpropagation algorithm is modified for spike signals. In Strain *et al.* (2006) this architecture is further developed with the introduction of a new rule for a dynamically adjusting firing threshold.

The evolving rules in a SNN, being a biologically plausible ANN model, can include some parameters that are directly related to genes and proteins expressed in the brain as it is presented in the computational neuro-genetic model in Chapter 9 and in Table 9.2 (see Benuskova and Kasabov (2007)). A simple, evolving rule there relates to evolving the output spiking activity of a neuron based on changes in the genetic parameters.

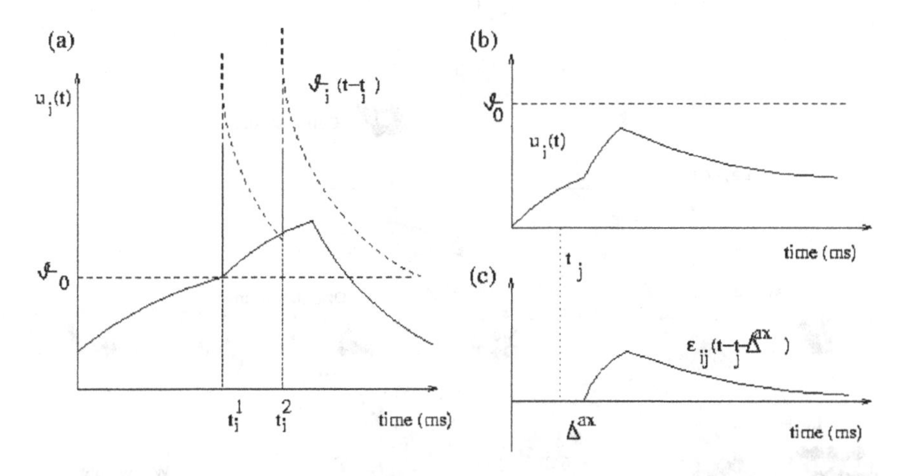

Fig. 4.7 (a) Suprathreshold summation of PSPs in the spiking neuron model. After each generation of postsynaptic spike there is a rise in the firing threshold that decays back to the resting value between the spikes. (b) Subthreshold summation of PSPs that does not lead to the generation of postsynaptic spike. (c) PSP is generated after some delay taken by the presynaptic spike to travel from neuron j to neuron i (from Benuskova and Kasabov (2007)).

4.3.2 Evolving Spiking Neural Networks (eSNN)

Evolving SNN (eSNN) are built of spiking neurons (as described above), where there are new evolving rules for:

- Evolving new neurons and connections, based on both parameter (genetic) information and learning algorithm, e.g. the Hebbian learning rule (Hebb, 1949)
- Evolving substructures of the SNN

An example of such eSNN is given in Wysoski *et al.* (2006) where new output classes presented in the incoming data (e.g. new faces in a face recognition problem) cause the SNN to create new substructures; see Fig. 4.8.

The neural network is composed of three layers of integrate-and-fire neurons. The neurons have a latency of firing that depends upon the order of spikes received. Each neuron acts as a coincidence detection unit, where the postsynaptic potential for neuron *Ni* at a time *t* is calculated as

$$PSP(i,t) = \sum \mathrm{mod}^{\,order(j)} w_{j,i} \tag{4.6}$$

where mod $\in (0,1)$ is the modulation factor, j is the index for the incoming connection, and $w_{j,i}$ is the corresponding synaptic weight.

Each layer is composed of neurons that are grouped in two-dimensional grids forming neuronal maps. Connections between layers are purely feedforward and each neuron can spike at most once on a spike arrival in the input synapses. The first layer cells represent the ON and OFF cells of the retina, basically enhancing

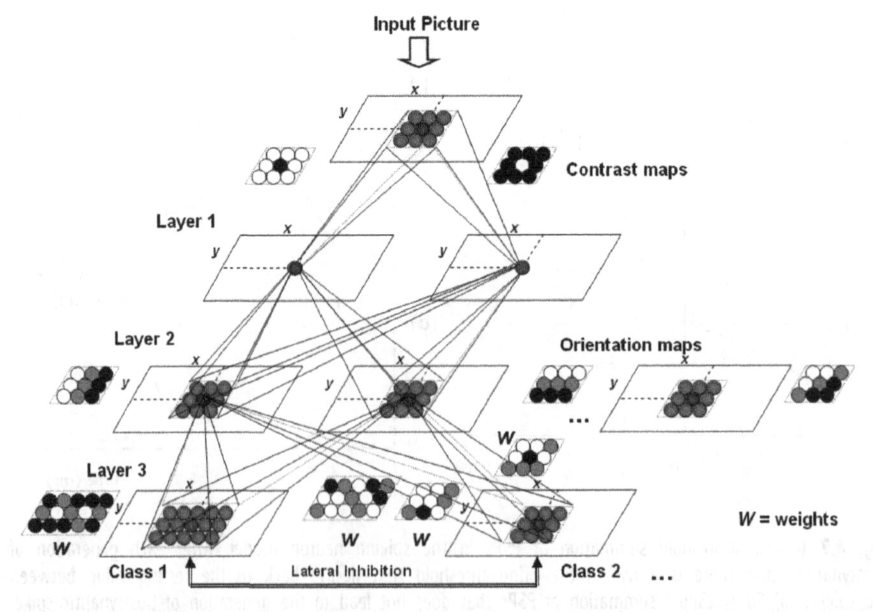

Fig. 4.8 Evolving spiking neural network (eSNN) architecture for visual pattern recognition (from Wysoski *et al.* (2006)).

the high-contrast parts of a given image (highpass filter). The output values of the first layer are encoded into pulses in the time domain. High output values of the first layer are encoded as pulses with short time delays whereas long delays are given low output values. This technique is called rank order coding (Thorpe *et al.*, 1998) and basically prioritizes the pixels with high contrast that consequently are processed first and have a higher impact on neurons' PSP.

The second layer is composed of eight orientation maps, each one selective to a different direction (0°, 45°, 90°, 135°, 180°, 225°, 270°, and 315°). It is important to notice that in the first two layers there is no learning, in such a way that the structure can be considered simply passive filters and time-domain encoders (layers 1 and 2). The theory of contrast cells and direction selective cells was first reported by Hubel and Wiesel (1962). In their experiments they were able to distinguish some types of cells that have different neurobiological responses according to the pattern of light stimulus.

The third layer is where the learning takes place and where the main contribution of this work is presented. Maps in the third layer are to be trained to represent classes of inputs. In Thorpe *et al.* (1998) the learning is performed off-line using the rule:

$$\Delta w_{j,i} = \frac{\mathrm{mod}^{\,order(a_j)}}{N} \tag{4.7}$$

where $w_{j,i}$ is the weight between neuron j of the second layer and neuron i of the third layer, mod $\in (0,1)$ is the modulation factor, $order(a_j)$ is the order of arrival of a spike from neuron j to neuron i, and N is the number of samples used for training a given class.

In this rule, there are two points to be highlighted: (a) the number of samples to be trained needs to be known a priori; and (b) after training, a map of a class will be selective to the average pattern.

In Wysoski *et al.* (2006) a new approach is proposed for learning with structural adaptation, aiming to give more flexibility to the system in a scenario where the number of classes and/or class instances is not known at the time the training starts. Thus, the output neuronal maps need to be created, updated, or even deleted online, as the learning occurs. To implement such a system the learning rule needs to be independent of the total number of samples because the number of samples is not known when the learning starts. In the implementation of Equation (4.7) in Delorme *et al.* (1999, 2001) the outcome is the average pattern. However, the new equation in Wysoski *et al.* (2006) calculates the average dynamically as the input patterns arrive as explained below.

There is a classical drawback to learning methods when, after training, the system responds to the average pattern of the training samples. The average does not provide a good representation of a class in cases where patterns have high variance (see Fig. 4.9). A traditional way to attenuate the problem is the divide-and-conquer procedure. We implement this procedure through the structural modification of the network during the training stage. More specifically, we integrate into the training algorithm a simple clustering procedure: patterns within a class that comply with a similarity criterion are merged into the same neuronal map. If the similarity criterion is not fulfilled, a new map is generated.

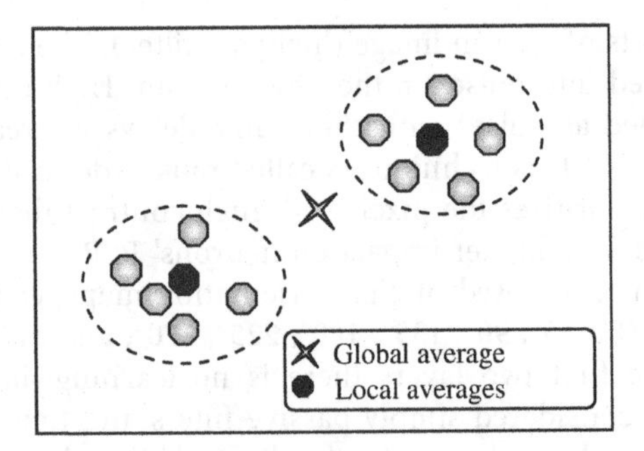

Fig. 4.9 Divide-and-conquer procedure to deal with high intraclass variability of patterns in the hypothetical space of class K. The use of multiple maps that respond optimally to the average of a subset of patterns provides a better representation of the classes than using a global average value.

The entire training procedure follows four steps described next and summarized in the flowchart of Fig. 4.10.

The new learning procedure can be described in these sequential steps:

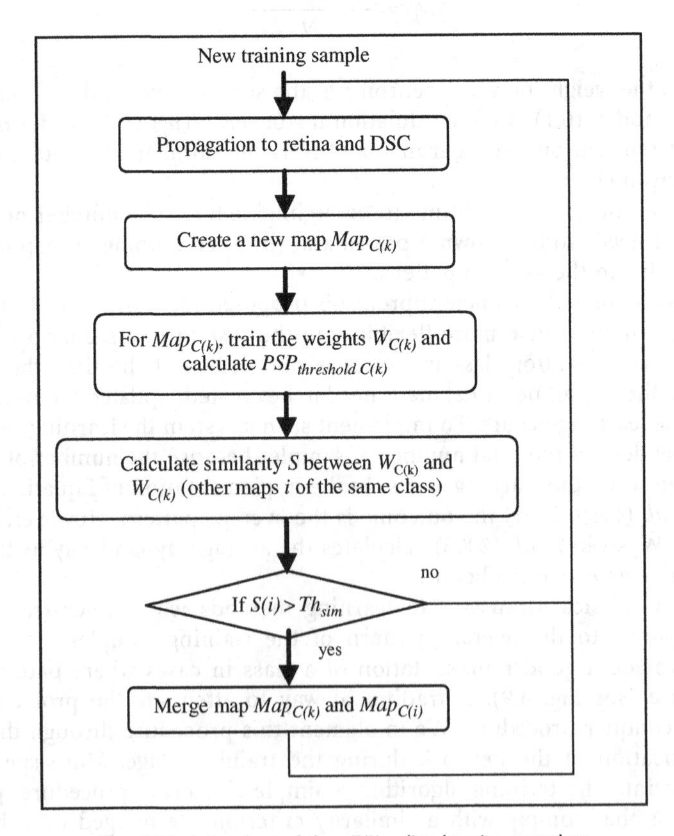

Fig. 4.10 A flowchart of the eSNN online learning procedure.

Propagate a sample k of class K for training into the layer 1 (retina) and layer 2 (direction selective cells, DSC);
Create a new map $Map_{C(k)}$ in layer 3 for sample k and train the weights using the equation:

$$\Delta w_{j,i} = \mathrm{mod}^{\,order(a_j)} \tag{4.8}$$

where $w_{j,i}$ is the weight between neuron j of layer 2 and neuron i of layer 3, mod $\in$ (0,1) is the modulation factor, and $order(a_j)$ is the order of arrival of spike from neuron j to neuron i.
The postsynaptic threshold ($PSP_{threshold}$) of the neurons in the map is calculated as a proportion $c \in [0,1]$ of the maximum postsynaptic potential (PSP) created in a neuron of map $Map_{C(k)}$ with the propagation of the training sample into the updated weights, such that:

$$PSP_{threshold} = c \max(PSP) \tag{4.9}$$

The constant of proportionality c expresses how similar a pattern needs to be to trigger an output spike. Thus, c is a parameter to be optimized in order to satisfy the requirements in terms of false acceptance rate (FAR) and false rejection rate (FRR).
Calculate the similarity between the newly created map $Map_{C(k)}$ and other maps belonging to the same class $Map_{C(K)}$. The similarity is computed as the inverse of the Euclidean distance between weight matrices.
If one of the existing maps for class K has similarity greater than a chosen threshold $Th_{simC(K)} > 0$, merge the maps $Map_{C(k)}$ and $Map_{C(K\mathrm{similar})}$ using the arithmetic average as expressed in

$$W = \frac{W_{Map_{C(k)}} + N_{samples} W_{Map_{C(Ksimilar)}}}{1 + N_{samples}} \tag{4.10}$$

where matrix W represents the weights of the merged map and $N_{samples}$ denotes the number of samples that have already being used to train the respective map. In similar fashion the $PSP_{threshold}$ is updated:

$$PSP_{threshold} = \frac{PSP_{Map_{C(k)}} + N_{samples} PSP_{Map_{C(Ksimilar)}}}{1 + N_{samples}} \tag{4.11}$$

4.4 Summary and Open Questions

The methods presented in this chapter only indicate the potential of the evolving connectionist systems for learning in a reinforcement mode, for learning temporal dependencies, for the realisation of evolving finite automata, and for the implementation of more biologically plausible ANN models such as the SNN and the evolving SNN (eSNN).

The topics covered in the chapter also raise some issues such as:

1. How to optimize the time lags of output values or inner states that the system should use in order to react properly on future input vectors in an online mode and an open problem space.
2. How can ECOS handle fuzzy reinforcement values if these values come from different sources, each of them using different, unknown to the system, fuzzy membership functions?
3. How can evolving fuzzy automata be evaluated in terms of the generalisation property?
4. How biologically plausible should a SNN model be in order to model a given brain function?
5. How biologically plausible should the SNN be in order to be used for solving complex problems of CI, such as speech recognition, image recognition, or multimodal information processing?
6. How to develop an eSNN that evolves its rules for:
   - Firing threshold adjustment
   - Neuronal connection creation and connection deletion
   - Neuron aggregation
7. How to develop eSNN automata, where the whole SNN at a certain time is represented as a state. The transition between states may be interpreted then as learning knowledge representation.

4.5 Further Reading

Some more on the subject of this chapter can be read in the following references.

- *Recurrent Structures of NN* (Elman, 1990; Arbib, 1995, 2002)
- *Finite Automata and their Realization in Connectionist Architectures* (Arbib, 1987; Omlin and Giles, 1994)
- *Introduction to the Theory of Automata* (Zamir,1983; Hopkin and Moss, 1976).
- *Symbolic Knowledge Representation in Recurrent Neural Networks* (Omlin and Giles, 1994)
- *Reinforcement Learning* (Sutton and Barto,1998)
- *Spiking MLP and Backpropagation Algorithm* (Bohte *et al.*, 2000)
- *Spiking Neural Networks* (Maass, 1996, 1998; Gerstner and Kistler, 2002; Izhikevich, 2003)
- *SNN with a Firing Threshold Adjustment* (Strain *et al.*, 2006)
- *Using SNN for Pattern Recognition* (Delorme *et al.*, 1999; Delorme and Thorpe, 2001; Wysoski *et al.*, 2006; Thorpe *et al.*, 1998)
- *Evolving SNN (eSNN)* (Wysoski *et al.*, 2006)
- *Computational Neuro-genetic Modeling Using SNN* (Benuskova and Kasabov, 2007)

5. *Evolving Neuro-Fuzzy Inference Models*

Some knowledge-based fuzzy neural network models for adaptive incremental (possibly online) learning, such as EFuNN and FuzzyARTMAP, were presented in the previous chapter. Fuzzy neural networks are connectionist models that are trained as neural networks, but their structure can be interpreted as a set of fuzzy rules. In contrast to them, neuro-fuzzy inference systems consist of a set of rules and an inference method that are embodied or combined with a connectionist structure for a better adaptation. Evolving neuro-fuzzy inference systems are such systems, where both the knowledge and the inference mechanism evolve and change in time, with more examples presented to the system. In the models here knowledge is represented as both fuzzy rules and statistical features that are learned in an online or off-line, possibly, in a lifelong learning mode. In the last three sections of the chapter different types of fuzzy rules, membership functions, and receptive fields in ECOS (that include both evolving fuzzy neural networks and evolving neuro-fuzzy inference systems) are analysed and introduced. The chapter covers the following topics.

- Knowledge-based neural networks
- Hybrid neuro-fuzzy inference system: HyFIS
- Dynamic evolving neuro-fuzzy inference system (DENFIS).
- TWNFI: Transductive weighted neuro-fuzzy inference systems for 'personalised' modelling
- Other neuro-fuzzy inference systems
- Exercise
- Summary and open problems
- Further reading

5.1 Knowledge-Based Neural Networks

5.1.1 General Notions

Knowledge (e.g. rules) is the essence of what a knowledge-based neural network (KBNN) has accumulated during its operation (see Cloete and Zurada (2000). Manipulating rules in a KBNN can pursue the following objectives.

1. Knowledge discovery, i.e. understanding and explanation of the data used to train the KBNN. The extracted rules can be analysed either by an expert, or by the system itself. Different methods for reasoning can be subsequently applied to the extracted set of rules.
2. Improvement of the KBNN system, e.g. maintaining an optimal size of the KBNN that is adequate to the expected accuracy of the system. Reducing the structure of a KBNN can be achieved through regular pruning of nodes and connections thus allowing for knowledge to emerge in the structure, or through aggregating nodes into bigger rule clusters. Both approaches are explored in this chapter.

Types of Rules Used in KBNN

Different KBNNs are designed to represent different types of rules, some of them listed below.

1. *Simple propositional rules* (e.g. IF $x1$ is A AND/OR $x2$ is B THEN y is C, where A, B, and C are constants, variables, or symbols of true/false type) (see, for example, Feigenbaum (1989), Gallant (1993), and Hendler and Dickens (1991)). As a partial case, interval rules can be used, for example:
   IF $x1$ is in the interval [$x1$min, $x1$max] AND $x2$ is in the interval [$x2$min, $x2$max] THEN y is in the interval [$y$min, $y$max], with $Nr1$ examples associated with this rule.
2. *Propositional rules with certainty factors* (e.g., IF $x1$ is A (CF1) AND $x2$ is B (CF2) THEN y is C (CFc)), (see, e.g. Fu (1989)).
3. *Zadeh–Mamdani fuzzy rules* (e.g., IF $x1$ is A AND $x2$ is B THEN y is C, where A, B, and C are fuzzy values represented by their membership functions) (see, e.g. Zadeh (1965) and Mamdani (1977)).
4. *Takagi–Sugeno fuzzy rules* (e.g. the following rule is a first-order rule: IF $x1$ is A AND $x2$ is B THEN y is $a.x1 + b.x2 + c$, where A and B are fuzzy values and a, b, and c are constants) (Takagi and Sugeno, 1985; Jang, 1993). More complex functions are possible to use in higher-order rules.
5. *Fuzzy rules* with degrees of importance and certainty degrees (e.g. IF $x1$ is A (DI1) AND $x2$ is B (DI2) THEN y is C (CFc), where DI1 and DI2 represent the importance of each of the condition elements for the rule output, and the CFc represents the strength of this rule (see Kasabov (1996)).
6. Fuzzy rules that represent associations of clusters of data from the problem space (e.g. Rule j: IF [an input vector $\mathbf{x}$ is in the input cluster defined by its centre ($x1$ is Aj, to a membership degree of MD1j, AND $x2$ is Bj, to a membership degree of MD2j) and by its radius Rj-in] THEN [y is in the output cluster defined by its centre (y is C, to a membership degree of MDc) and by its radius Rj-out, with Nex(j) examples represented by this rule]. These are the EFuNN rules discussed in Chapter 3.
7. *Temporal rules* (e.g. IF $x1$ is present at a time moment $t1$ (with a certainty degree and/or importance factor of DI1) AND $x2$ is present at a time moment $t2$ (with a certainty degree/importance factor DI2) THEN y is C (CFc)).
8. *Temporal recurrent rules* (e.g., IF $x1$ is A (DI1) AND $x2$ is B (DI2) AND y at the time moment $(t-k)$ is C THEN y at a time moment $(t+n)$ is D (CFc)).

9. *Type-2 fuzzy rules*, that is, fuzzy rules of the form of: IF x is A~ AND y is B~ THEN z is C~ , where A~, B~, and C~ are type-2 fuzzy membership functions (see the extended glossary, and also the section in this chapter on type-2 ECOS).

Generic Methods for Rule Extraction from KBNN

There are several methods for rule extraction from a KBNN. Three of them are explained below:

1. Rule extraction through activating a trained KBNN on input data and observing the patterns of activation ("the short-term memory"). The method is not practical for online incremental learning as past data may not be available for a consecutive activation of the trained KBNN. This method is widely used in brain research (e.g. analysing MRI, fMRI, and EEG patterns and signals to detect rules of behaviour.
2. Rule extraction through analysis of the connections in a trained KBNN ("the long-term memory"). This approach allows for extracting knowledge without necessarily activating the connectionist system again on input data. It is appropriate for online learning and system improvement. This approach is not yet used in brain study as there are no established methods thus far for processing information stored in neuronal synapses.
3. Combined methods of (1) and (2). These methods make use of the above two approaches.

A seminal work on fuzzy rule extraction from KBNN is the publication by Mitra and Hayashi (2000).

Methods for Inference over Rules Extracted from KBNN

In terms of applying the extracted from a KBNN rules to infer new information, there are three types of methods used in the KBNN:

1. The rule learning and rule inference modules constitute an integral structure where reasoning is part of the rule learning, and vice versa. This is the case in all fuzzy neural networks and of most of the neuro-fuzzy inference systems.
2. The rules extracted from a KBNN are interpreted in another inference machine. The learning module is separated from the reasoning module. This is a main principle used in many AI and expert systems, where the rule base acquisition is separated from the inference machine.
3. The two options from above are possible within one intelligent system.

Figure 5.1 shows a general scheme of a fuzzy inference system. The decision-making block is the fuzzy inference engine that performs inference over fuzzy rules and data from the database. The inference can be realized in a connectionist structure, thus making the system a neuro-fuzzy inference system.

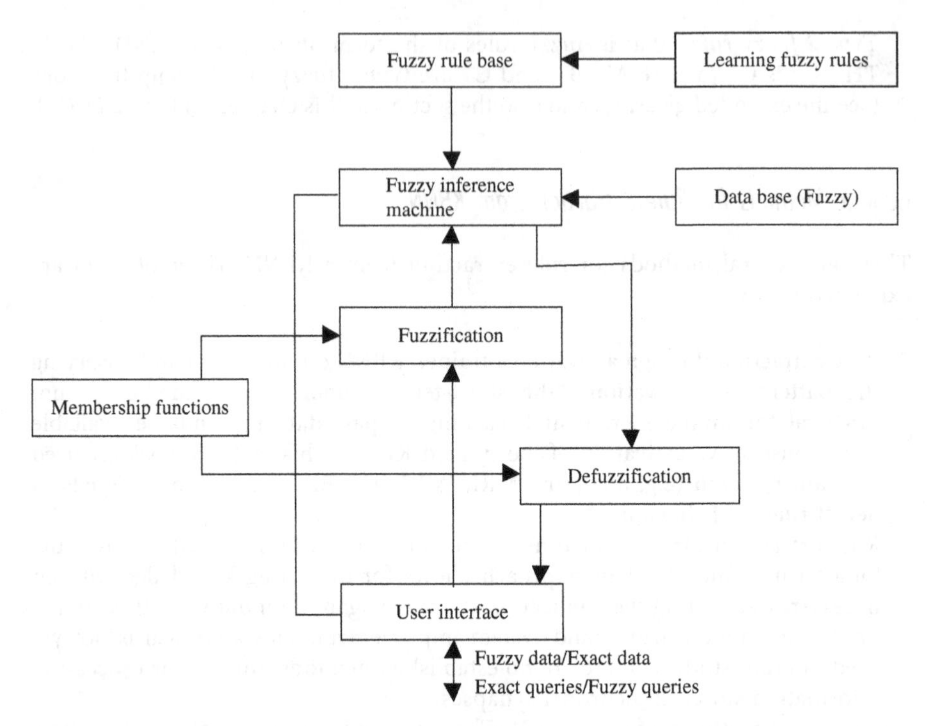

Fig. 5.1 A general diagram of a fuzzy inference system (from Kasabov (1996), ©MIT Press, reproduced with permission).

5.1.2 Adaptive Neuro-Fuzzy Inference Systems (ANFIS)

ANFIS (Jang, 1993) implements Takagi–Sugeno fuzzy rules in a five-layer MLP network. The first layer represents fuzzy membership functions. The second and the third layers contain nodes that form the antecedent parts in each rule. The fourth layer calculates the first-order Takagi–Sugeno rules for each fuzzy rule. The fifth layer – the output layer – calculates the weighted global output of the system as illustrated in Fig. 5.2a and b.

The backpropagation algorithm is used to modify the initially chosen membership functions and the least mean square algorithm is used to determine the coefficients of the linear output functions. Here, the min and the max functions of a fuzzy inference method (Zadeh, 1965) are replaced by differentiable functions.

As many rules can be extracted from a trained ANFIS as there are a predefined number of rule nodes. Two exemplar sets of fuzzy rules learned by an ANFIS model are shown below (see also Fig. 5.2):

Rule 1: If x is A_1 and y is B_1, then $f_1 = p_1x + q_1y + r_1$
Rule 2: If x is A_2 and y is B_2, then $f_2 = p_2x + q_2y + r_2$

where x and y are the input variables; A_1, A_2, B_1 and B_2 are the membership functions; f is the output; and p, q, and r are the consequent parameters.

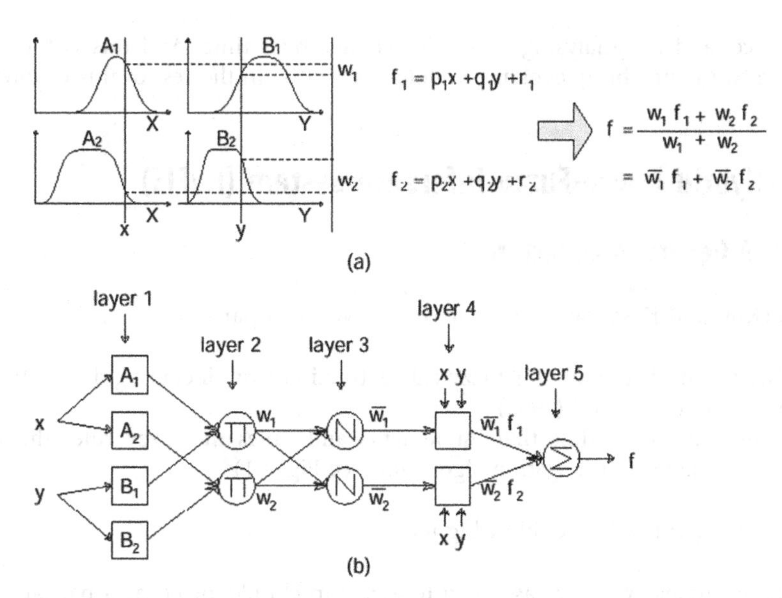

Fig. 5.2 (a) An exemplar set of two fuzzy rules and the inference over them that is performed in an ANFIS structure; (b) the exemplar ANFIS structure for these two rules (see Jang (1993) and the MATLAB Tutorial book, *Fuzzy Logic Toolbox*).

By employing a hybrid learning procedure, the proposed architecture can refine fuzzy if–then rules obtained from human experts to describe the input–output behaviour of a complex system. If human expertise is not available, reasonable initial membership functions can still be set up intuitively and the learning process can begin to generate a set of fuzzy if–then rules to approximate a desired dataset.

ANFIS employs a multiple iteration learning procedure and has a fast convergence due to the hybrid learning algorithm used. It does not require preselection of the number of the hidden nodes; they are defined as the number of combinations between the fuzzy input membership functions.

Despite the fact that ANFIS is probably the most popular neuro-fuzzy inference system thus far, in some cases it is not adequate to use. For example, ANFIS cannot handle problems with high dimensionality, for example, more than 10 variables (we are not talking about 40,000 gene expression variables) as the complexity of the system becomes incredible and the million of rules would not be comprehensible by humans. ANFIS has a fixed structure that cannot adapt to the data in hand, therefore it has limited abilities for incremental, online learning.

There can be only one output from an ANFIS. This is due to the nature of the format of the fuzzy rules it represents. Thus ANFIS can only be applied to tasks such as prediction or approximation of nonlinear functions where there is only one output. The number of membership functions associated with each input and output node cannot be adjusted, only their shape. Prior choice of membership functions is a critical issue when building the ANFIS system. There are no variations apart from the hybrid learning rule available to train ANFIS.

In contrast to ANFIS, incremental adaptive learning and local optimisation in a fuzzy-neural inference system would allow for tracing the process of knowledge emergence, and for analysing how rules change over time. And this is the case with the two neuro-fuzzy evolving systems presented in the rest of this chapter.

5.2 Hybrid Neuro-Fuzzy Inference System (HyFIS)

5.2.1 A General Architecture

HyFIS (Kim and Kasabov, 1999) consists of two main parts (Fig. 5.3):

1. A fuzzy analysis module for fuzzy rule extraction from incoming data with the use of Wang's method (1994)
2. A connectionist module that implements and tunes the fuzzy rules through applying the backpropagation algorithm (see Fig. 3.1)

The system operates in the following mode.

1. Data examples $(\boldsymbol{x}, \boldsymbol{y})$ are assumed to arrive in chunks of m (as a partial case, $m = 1$).
2. For the current chunk Ki, consisting of m_i examples, n_i fuzzy rules are extracted as described below. They have a form illustrated with the following example. IF $x1$ is Small AND $x2$ is Large THEN y is Medium (certainty 0.7)
3. The n_i fuzzy rules are inserted in the neuro-fuzzy module, thus updating the current structure of this module.
4. The updated neuro-fuzzy structure is trained with the backpropagation algorithm on the chunk of data Ki, or on a larger dataset if such is available.
5. New data $\boldsymbol{x}'$ that do not have known output vectors, are propagated through the neuro-fuzzy module for recall.

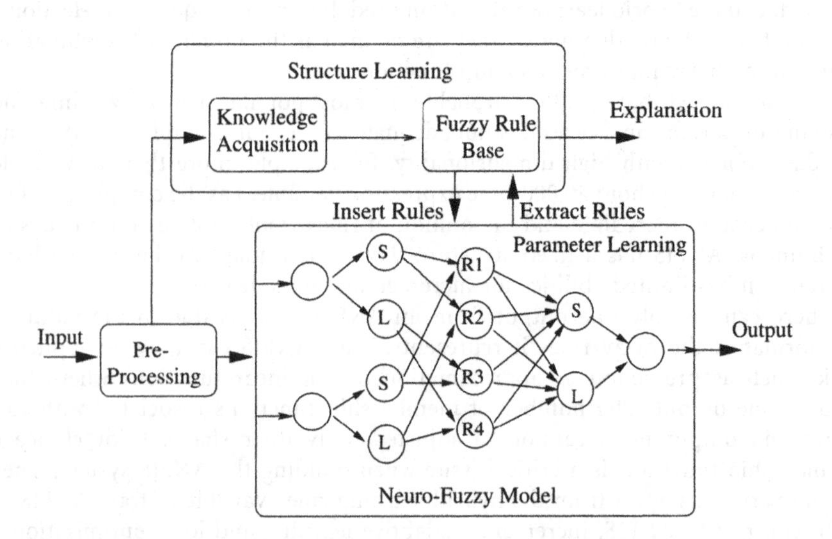

Fig. 5.3 A schematic block diagram of HyFIS (from Kim and Kasabov (1999)).

The fuzzy rule extraction method is illustrated here on the following two examples of input–output data (the chunk consists of only two examples):

$$\text{exampl} : [x1 = 0.6; x2 = 0.2; y = 0.2]$$
$$\text{examp2} : [x1 = 0.4; x2 = 0; y = 0.4]$$

These examples are fuzzified with membership functions (not shown here) defined in the neuro-fuzzy module, or initialised if this is the first chunk of data, for example:

$$\text{exampl} : [x1 = \text{Medium}(0.8); x2 = \text{Small}(0.6); y = \text{Small}(0.6)]$$
$$\text{examp2} : [x1 = \text{Medium}(0.8)\ ; x2 = \text{Small}(1); y = \text{Medium}(0.8)]$$

Here we have assumed that the range of the three variables $x1$, $x2$, and y is [0,1] and there are three membership functions uniformly distributed on this range (Small, Medium, Large).

Each of the examples can be now represented as a fuzzy rule of the Zadeh–Mamdani type:

Rule 1: IF $x1$ is Medium and $x2$ is Small THEN y is Small.
Rule 2: IF $x1$ is Medium and $x2$ is Small THEN y is Medium.

The rules now can be inserted in the neuro-fuzzy modules, but in this particular case they are contradictory rules; i.e. they have the same antecedent part but different consequent part. In this case there will be a certainty degree calculated as the product of all the membership degrees of all the variables in the rule, and only the rule with the highest degree will be inserted in the neuro-fuzzy system.

Rule 1: Certainty degree $CF1 = 0.8 * 0.6 * 0.6 = 0.288$ (5.1)
Rule 2: Certainty degree $CF2 = 0.8 * 1 * 0.8 = 0.64$

Only Rule 2 will be inserted in the connectionist structure.

5.2.2 Neuro-Fuzzy Inference Module

A block diagram of a hypothetical neuro-fuzzy module is given in Fig. 5.4. It consists of five layers: layer one is the input layer, layer two is the input membership functions layer, layer three is the rule layer, layer four is the output membership function layer, and layer five is the output layer.

Layer three performs the AND operation calculated as the *min* function on the incoming activation values of the membership function nodes. The membership functions are of a Gaussian type. Layer four performs the OR operation, calculated as the *max* function on the weighted activation values of the rule nodes connected to nodes in layer four:

$$Oj^{(4)} = \max\{Oi^{(3)} w_{i,j}\} \tag{5.2}$$

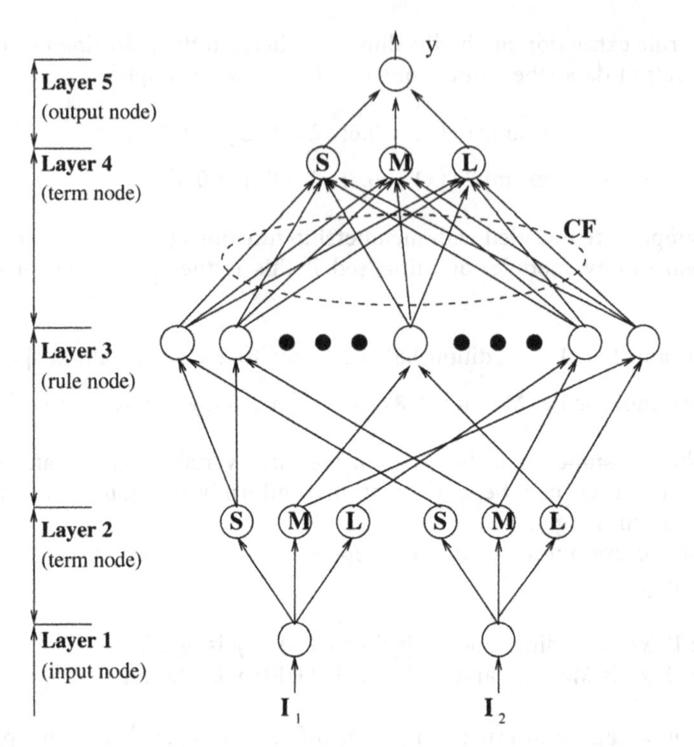

Fig. 5.4 The structure of the neuro-fuzzy module of HyFIS (from Kim and Kasabov (1999)).

Layer five performs a centre of area defuzzification:

$$Ol^{(5)} = \sum Oj^{(4)} Cj^{(4)} \sigma j^{(4)} / \sum Oj^{(4)} \sigma j^{(4)} \tag{5.3}$$

where $Ol^{(5)}$ is the activation of the lth output node; $Oj^{(4)}$ is the activation of the jth node from layer 4 that represents a Gaussian output membership function with a centre $Cj^{(4)}$ and a width of $\sigma j^{(4)}$.

Through the backpropagation learning algorithm the connection weights $w_{i,j}$ as well as the centres and the width of the membership functions are adjusted to minimize the mean square error over the training dataset (or the current chunk of data).

5.2.3 Modelling and Prediction of the Gas-Furnace Data with HyFIS

In this experiment 200 data examples were drawn randomly from the gas-furnace data time series (Box and Jankins (1970); see Chapter 1); 23 fuzzy rules were extracted from them and were inserted in an initial neuro-fuzzy structure (black circles in Fig. 5.5). Two new rules were extracted from the rest of the 92 data examples (inserted also in structure: the empty circles).

This was trained on the 200 examples and tested and incrementally evolved on the rest 92 Examples. The test results were compared with similar results obtained

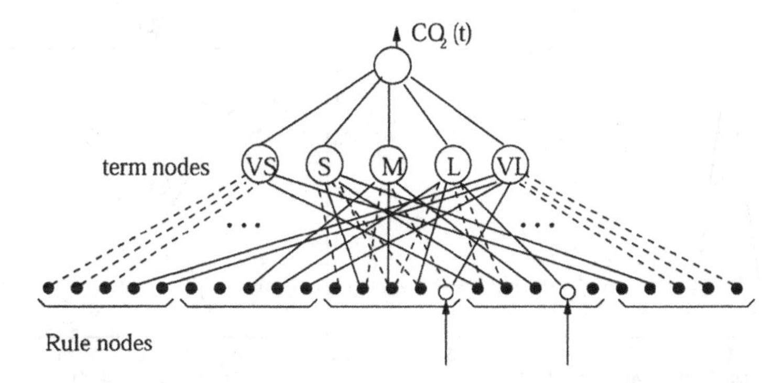

Fig. 5.5 The initial structure of HyFIS for the gas-furnace time-series prediction (the filled circles represent the initial 2 fuzzy rules inserted before training), and the resulted HyFIS structure after training is performed (the empty circles represent newly created rules; from Kim and Kasabov (1999)).

Table 5.1 A comparative analysis of the test results of different fuzzy neural networks and neuro-fuzzy inference models for the prediction of the gas-furnace time series. All the models, except HyFIS, use a prior fixed structure that does not change during training. HyFIS begins training with no rule nodes and builds nodes based on fuzzy rules extracted from data in an online mode (from Kim and Kasabov (1999)).

Model name and reference	*Number of inputs*	*Number of rules*	*Model error (MSE)*
ARMA (Box and Jenkins, 1970)	5	—	0.71
Takagi–Sugeno model (1985)	6	2	0.068
(Hauptman and Heesche, 1995)	2	10	0.134
ANFIS (Jang, 1993)	2	25	0.73×10^{-3}
FuNN (Kasabov *et al.*, 1997)	2	7	0.51×10^{-3}
HyFIS (Kim and Kasabov, 1999)	2	15	0.42×10^{-3}

with the use of other statistical connections, fuzzy, and neuro-fuzzy techniques as shown in Table 5.1.

After training with the 200 examples the updated rules can be extracted from the neuro-fuzzy structure. The membership functions are modified from their initial forms, as shown in Fig. 5.6.

5.3 Dynamic Evolving Neuro-Fuzzy Inference Systems (DENFIS)

5.3.1 General Principles

The dynamic evolving neuro-fuzzy system, DENFIS, in its two modifications – for online and for off-line learning – use the Takagi–Sugeno type of fuzzy inference

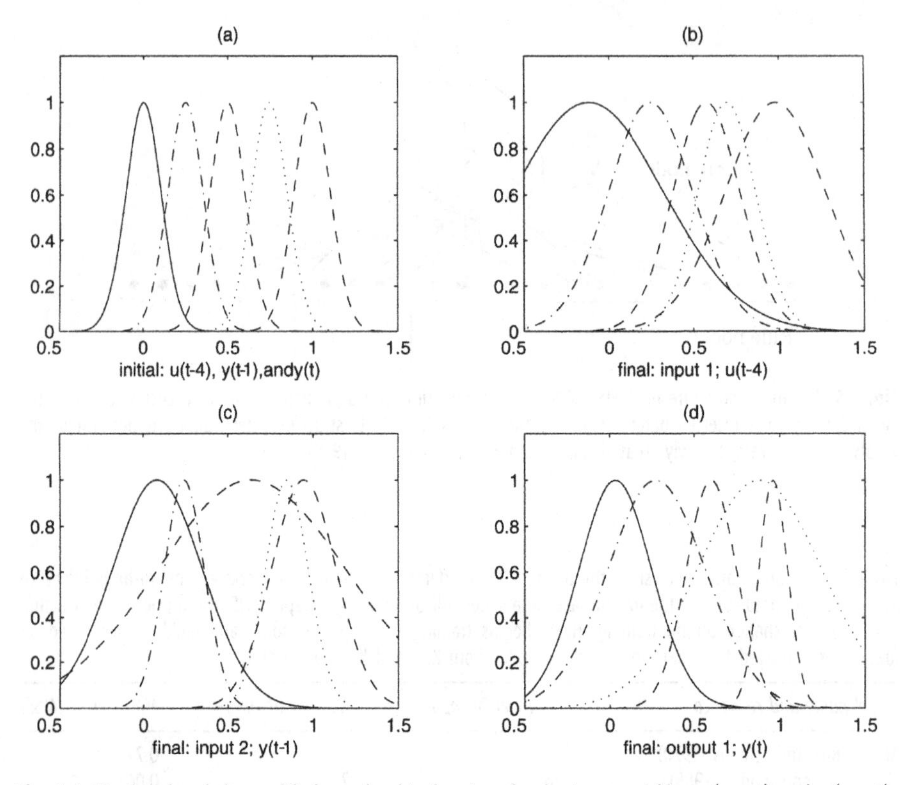

Fig. 5.6 The initial and the modified membership functions for the input variables $u(t-4)$ and $y(t-1)$ and the output variable $y(t)$ after certain epochs of training HyFIS on the gas-furnace data are performed (from Kim and Kasabov (1999)).

method (Kasabov and Song, 2002). The inference in DENFIS is performed on m fuzzy rules indicated as follows.

$$\begin{cases} \text{if } x_1 \text{ is } R_{11} \text{ and } x_2 \text{ is } R_{12} \text{ and} \ldots \text{and } x_q \text{ is } R_{1q}, \text{ then } y \text{ is } f_1(x_1, x_2, \ldots, x_q) \\ \text{if } x_1 \text{ is } R_{21} \text{ and } x_2 \text{ is } R_{22} \text{ and} \ldots \text{and } x_q \text{ is } R_{2q}, \text{ then } y \text{ is } f_2(x_1, x_2, \ldots, x_q) \\ \text{if } x_1 \text{ is } R_{m1} \text{ and } x_2 \text{ is } R_{m2} \text{ and} \ldots \text{and } x_q \text{ is } R_{mq}, \text{ then } y \text{ is } f_m(x_1, x_2, \ldots, x_q) \end{cases} \tag{5.4}$$

where x_j is R_{ij}, $i = 1, 2, \ldots, m$; $j = 1, 2, \ldots, q$, are $m \times q$ fuzzy propositions that form m antecedents for m fuzzy rules respectively; x_j, $j = 1, 2, \ldots, q$, are antecedent variables defined over universes of discourse X_j, $j = 1, 2, \ldots, q$, and R_{ij}, $i = 1, 2, \ldots, m$; $j = 1, 2, \ldots, q$ are fuzzy sets defined by their fuzzy membership functions μ_{Rij}: $X_j \to [0, 1]$, $i = 1, 2, \ldots, m$; $j = 1, 2, \ldots, q$. In the consequent parts of the fuzzy rules, y is the consequent variable, and crisp functions f_i, $i = 1, 2, \ldots, m$, are employed.

In both online and off-line DENFIS models, fuzzy membership functions triangular type can be of that depend on three parameters, a, b, and c, as given below:

$$\mu(x) = mf(x, a, b, c) = \begin{cases} 0, & x \leq a \\ \dfrac{x-a}{b-a}, & a \leq x \leq b \\ \dfrac{c-x}{c-b}, & b \leq x \leq c \\ 0. & c \leq x \end{cases} \tag{5.5}$$

where b is the value of the cluster centre on the variable x dimension; $a = b - d \times Dthr$; $c = b + d \times Dthr$; and $d = 1.2 \sim 2$; the threshold value $Dthr$ is a clustering parameter (see the evolving clustering method ECM presented in Chapter 2).

If $f_i(x_1, x_2, \ldots, x_q) = C_i$, $i = 1, 2, \ldots, m$, and C_i are constants, we call this inference a zero-order Takagi–Sugeno type fuzzy inference system. The system is called a first-order Takagi–Sugeno fuzzy inference system if $f_i(x_1, x_2, \ldots, x_q)$, $i = 1, 2, \ldots, m$, are linear functions. If these functions are nonlinear functions, it is called a high-order Takagi–Sugeno fuzzy inference system.

5.3.2 Online Learning in a DENFIS Model

In the DENFIS online model, the first-order Takagi–Sugeno type fuzzy rules are employed and the linear functions in the consequence parts are created and updated through learning from data by using the linear least-square estimator (LSE).

For an input vector $\mathbf{x}^0 = [x_1{}^0 x_2{}^0 \ldots x_q{}^0]$, the result of inference y^0(the output of the system) is the weighted average of each rule's output indicated as follows.

$$y^0 = \frac{\sum_{i=1,m} \omega_i f_i(x_1{}^0, x_2{}^0, \ldots, x_q{}^0)}{\sum_{i=1,m} \omega_i} \tag{5.6}$$

$$\text{where } \omega_i = \prod_{j=1}^{q} \mu_{\mathrm{R}ij}(xj^0); \;\; i = 1, 2, \ldots m; \;\; j = 1, 2, \ldots, q. \tag{5.7}$$

In the DENFIS online model, the first-order Takagi–Sugeno type fuzzy rules are employed and the linear functions in the consequences can be created and updated by linear least-square estimator on the learning data. Each of the linear functions can be expressed as follows.

$$y = \beta_0 + \beta_1 x_1 + \beta_2 x_2 + \cdots + \beta_q x_q. \tag{5.8}$$

For obtaining these functions a learning procedure is applied on a dataset, which is composed of p data pairs $\{([x_{i1}, x_{i2}, \ldots, x_{iq}], y_i), \; i = 1, 2, \ldots, p\}$. The least-square estimator of $\beta = [\beta_0 \; \beta_1 \; \beta_2 \ldots \beta_q]^T$, is calculated as the coefficients $\boldsymbol{b} = [b_0 \; b_1 \; b_2 \ldots b_q]^T$, by applying the following formula,

$$\boldsymbol{b} = (\mathrm{A}^T \mathrm{A})^{-1} \mathrm{A}^T \boldsymbol{y} \tag{5.9}$$

where

$$A = \begin{pmatrix} 1 & x_{11} & x_{12} & \dots & x_{1q} \\ 1 & x_{21} & x_{22} & \dots & x_{2q} \\ \cdot & \cdot & \cdot & \cdot & \cdot \\ \cdot & \cdot & \cdot & \cdot & \cdot \\ \cdot & \cdot & \cdot & \cdot & \cdot \\ 1 & x_{p1} & x_{p2} & \dots & x_{pq} \end{pmatrix}$$

and $y = [y_1\ y_2 \dots y_p]^T$. A weighted least-square estimation method is used here as follows.

$$b_w = (\mathbf{A}^T\mathbf{W}\ \mathbf{A})^{-1}\mathbf{A}^T\mathbf{W}y, \tag{5.10}$$

where

$$\mathbf{W} = \begin{pmatrix} w_1 & 0 & \dots & 0 \\ 0 & w_2 & \dots & 0 \\ \vdots & \vdots & \vdots & \vdots \\ 0 & \dots & \dots & w_p \end{pmatrix}$$

and w_j is the distance between jth example and the corresponding cluster centre, $j = 1, 2, \dots, p$.

We can rewrite Eqs. (5.9) and (5.10) as follows.

$$\begin{cases} P = (A^T A)^{-1} \\ b = P\ A^T y \end{cases} \tag{5.11}$$

$$\begin{cases} P_w = (A^T W\ A)^{-1} \\ b_w = P_w A^T W\ y \end{cases} \tag{5.12}$$

Let the kth row vector of matrix A defined in Eq. (5.9) be $a_k{}^T = [1\ x_{k1}\ x_{k2}\ \dots\ x_{kq}]$ and the kth element of y be y_k; then b can be calculated iteratively as follows.

$$\begin{cases} b_{k+1} = b_k + \mathbf{P}_{k+1}\mathbf{a}_{k+1}(y_{k+1} - \mathbf{a}_{k+1}{}^T\ b_k) \\ \mathbf{P}_{k+1} = \mathbf{P}_k - \dfrac{\mathbf{P}_k\mathbf{a}_{k+1}\ \mathbf{a}_{k+1}{}^T\mathbf{P}_k}{1 + \mathbf{a}_{k+1}{}^T\mathbf{P}_k\mathbf{a}_{k+1}} \end{cases} \tag{5.13}$$

for $k = n, n+1, \dots, p-1$.

Here, the initial values of $\mathbf{P}_n$ and $\mathbf{b}_n$ can be calculated directly from Eq. (5.12) with the use of the first n data pairs from the learning dataset.

Equation (5.13) is the formula of recursive LSE. In the DENFIS online model, we use a weighted recursive LSE with forgetting factor defined as the following equations

$$b_{k+1} = b_k + w_{k+1}\mathbf{P}_{k+1}\mathbf{a}_{k+1}(y_{k+1} - \mathbf{a}_{k+1}{}^T b_k),$$

$$\mathbf{P}_{k+1} = \frac{1}{\lambda}\left[\mathbf{P}_{k-} \frac{w_{k+1}\mathbf{P}_k\mathbf{a}_{k+1}\ \mathbf{a}_{k+1}{}^T}{\lambda + \mathbf{a}_{k+1}{}^T\mathbf{P}_k\mathbf{a}_{k+1}}\mathbf{P}_k\right] \quad k = n, n+1, \ldots, p-1 \tag{5.14}$$

where w is the weight defined in Eq. (5.10) and λ is a forgetting factor with a typical value between 0.8 and 1.

In the online DENFIS model, the rules are created and updated at the same time with the input space partitioning using the online evolving clustering method (ECM) and Eqs. (5.8) and (5.14). If no rule insertion is applied, the following steps are used for the creation of the first m fuzzy rules and for the calculation of the initial values **P** and b of the functions.

1. Take the first n_0 learning data pairs from the learning dataset.
2. Implement clustering using ECM with these n_0 data to obtaining m cluster centres.
3. For every cluster centre C_i, find p_i data points whose positions in the input space are closest to the centre, $i = 1, 2, \ldots, m$.
4. For obtaining a fuzzy rule corresponding to a cluster centre, create the antecedents of the fuzzy rule using the position of the cluster centre and Eq. (5.8). Using Eq. (5.12) on p_i data pairs calculate the values of **P** and b of the consequent function. The distances between p_i data points and the cluster centre are taken as the weights in Eq. (5.12). In the above steps, m, n_0, and p are the parameters of the DENFIS online learning model, and the value of p_i should be greater than the number of input elements, q.

As new data pairs are presented to the system, new fuzzy rules may be created and some existing rules updated. A new fuzzy rule is created if a new cluster centre is found by the ECM. The antecedent of the new fuzzy rule is formed by using Eq. (5.8) with the position of the cluster centre as a rule node. An existing fuzzy rule is found based on the rule node that is closest to the new rule node; the consequence function of this rule is taken as the consequence function for the new fuzzy rule. For every data pair, several existing fuzzy rules are updated by using Eq. (5.14) if their rule nodes have distances to the data point in the input space that are not greater than $2 \times Dthr$ (the threshold value, a clustering parameter). The distances between these rule nodes and the data point in the input space are taken as the weights in Eq. (5.14). In addition to this, one of these rules may also be updated through changing its antecedent so that, if its rule node position is changed by the ECM, the fuzzy rule will have a new antecedent calculated through Eq. (5.8).

5.3.3 Takagi–Sugeno Fuzzy Inference in DENFIS

The Takagi–Sugeno fuzzy inference system utilised in DENFIS is a dynamic inference. In addition to dynamically creating and updating fuzzy rules the DENFIS online model has some other major differences from the other inference systems.

First, for each input vector, the DENFIS model chooses m fuzzy rules from the whole fuzzy rule set for forming a current inference system. This operation depends on the position of the current input vector in the input space. In the case of two input vectors that are very close to each other, especially in the DENFIS off-line model, the inference system may have the same fuzzy rule inference group. In the DENFIS online model, however, even if two input vectors are exactly the same, their corresponding inference systems may be different. This is because these two input vectors are presented to the system at different time moments and the fuzzy rules used for the first input vector might have been updated before the second input vector has arrived.

Second, depending on the position of the current input vector in the input space, the antecedents of the fuzzy rules chosen to form an inference system for this input vector may vary. An example is illustrated in Fig. 5.7a,b where two different groups of fuzzy inference rules are formed depending on two input vectors $\boldsymbol{x}_1$ and $\boldsymbol{x}_2$, respectively, in a 2D input space. We can see from this example that, for instance, the region C has a linguistic meaning 'large', in the X_1 direction for the Fig. 5.7a group, but for the group of rules from Fig. 5.7b it denotes a linguistic meaning of 'small' in the same direction of $X1$. The region C is defined by different membership functions, respectively, in each of the two groups of rules.

5.3.4 Time-Series Modelling and Prediction with the DENFIS OnLine Model

In this section the DENFIS online model is applied to modelling and predicting the future values of a chaotic time series: the Mackey–Glass (MG) dataset (see Chapter 1), which has been used as a benchmark example in the areas of neural networks, fuzzy systems, and hybrid systems (see Jang (1993)). This time series is created with the use of the MG time-delay differential equation defined below:

$$\frac{dx(t)}{dt} = \frac{0.2x(t-\tau)}{1+x^{10}(t-\tau)} - 0.1x(t) \tag{5.15}$$

To obtain values at integer time points, the fourth-order Runge–Kutta method was used to find the numerical solution to the above MG equation. Here we assume that the time step is 0.1, $x(0) = 1.2, \tau = 17$, and $x(t) = 0$ for $t < 0$. The task is to predict the values $x(t+85)$ from input vectors $[x(t-18)\,x(t-12)\,x(t-6)x(t)]$ for any value of the time t. For the purpose of a comparative analysis, we also use some existing online learning models applied to the same task. These models are neural gas, resource-allocating network (RAN), evolving self-organising maps (ESOM; see Chapter 2) and evolving fuzzy-neural network (EFuNN; Chapter 3). Here, we estimate the nondimensional error index (NDEI) which is defined as the root mean square error divided by the standard deviation of the target series.

The following experiment was conducted: 3000 data points, for $t = 201$ to 3200, are extracted from the time series and used as learning (training) data; 500 data

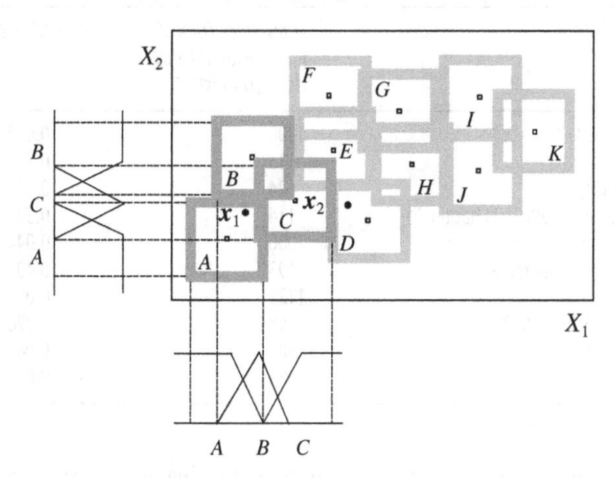

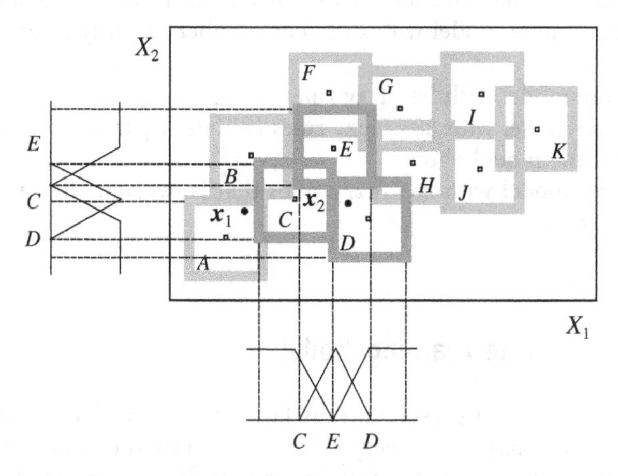

Fig. 5.7 Two fuzzy rule groups are formed by DENFIS to perform inference for an input vector $\boldsymbol{x}_1$ (a), and for an input vector $\boldsymbol{x}_2$ (b) that is entered at a later time moment, all represented in the 2D space of the first two input variables $X1$ and $X2$ (from Kasabov and Song (2002)).

points, for $t = 5001$ to 5500, are used as testing data. For each of the online models mentioned above the learning data are used for the online learning processes, and then the testing data are used with the recalling procedure.

Table 5.2 lists the prediction results (NDEI on test data after online learning) and the number of rules (nodes, units) evolved (used) in each model.

In another experiment the properties of rule insertion and rule extraction were utilised where we first obtained a group of fuzzy rules from the first half of the training data (1500 samples), using the DENFIS off-line model I (introduced in the next section); then we inserted these rules into the DENFIS online model and

Table 5.2 Prediction results of online learning models on the Mackey–Glass test data.

Methods	*Fuzzy rules (DENFIS)* *Rule nodes (EFuNN)* *Units (Others)*	*NDEI for testing data*
Neural gas (Fritzke,1994)	1000	0.062
RAN (Platt,1991)	113	0.373
RAN (other parameters)	24	0.17
ESOM (Deng and Kasabov, 2002) (Chapter 2)	114	0.32
ESOM (other parameters)	1000	0.044
EFuNN (Kasabov, 2001) (Chapter 3)	193	0.401
EFuNN (other parameters)	1125	0.094
DENFIS (Kasabov and Song, 2002)	58	0.276
DENFIS (other parameters)	883	0.042
DENFIS with rule insertion	883	0.033

let it learn continuously from the next half of the learning data (1500 samples). Then, we tested the model on the test data.

Figures 5.8a,b,c display the test errors (from the recall processes on the test data) of DENFIS online model with different number of fuzzy rules:

DENFIS online model with 58 fuzzy rules
DENFIS online model with 883 fuzzy rules (different parameter values are used from those in the model above)
DENFIS online model with 883 fuzzy rules that is evolved after an initial set of rules was inserted

5.3.5 DENFIS Off-Line Learning Model

The DENFIS online model presented thus far can be used also for off-line, batch mode training, but it may not be very efficient when used on comparatively small datasets. For the purpose of batch training the DENFIS online model is extended here to work efficiently in an off-line, batch training mode.

Two DENFIS models for off-line learning are developed and presented here: a linear model, model I, and a MLP-based model, model II.

A first-order Takagi–Sugeno type fuzzy inference engine, similar to the DENFIS online model, is employed in model I, and an extended high-order Takagi–Sugeno fuzzy inference engine is used in model II. The latter employs several small-size, two-layer (the hidden layer consists of two or three neurons) multilayer perceptrons to realise the function f in the consequent part of each fuzzy rule instead of using a function that has a predefined type.

The DENFIS off-line learning process is implemented in the following way.

- Cluster (partition) the input space to find n cluster centres (n rule nodes, $n$rules) by using the off-line evolving clustering method with constrained optimisation (ECMc; see Chapter 2).

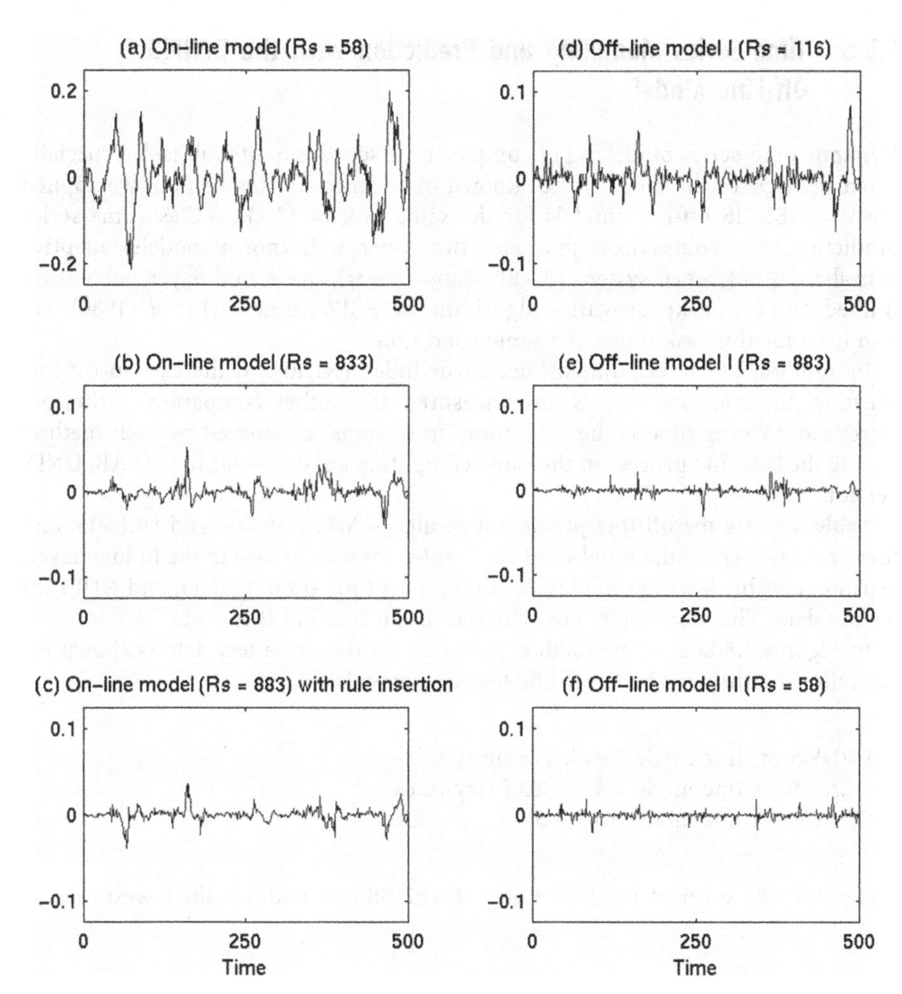

Fig. 5.8 Prediction error of DENFIS online (a)(b)(c) and off-line (d)(e)(f) models on test data taken from the Mackey–Glass time series (from (Kasabov and Song (2002)).

- Create the antecedent part for each fuzzy rule using Eq. (5.8) and also the current position of the cluster centre (rule node).
- Find n datasets, each of them including one cluster centre and p learning data pairs that are closest to the centre in the input space. In the general case, one data pair can belong to several clusters.
- For model I, estimate the functions f to create the consequent part for each fuzzy rule using Eq. (5.10) or Eq. (5.12) with n datasets; the distance between each data point and its corresponding centre is represented as a connection weight.
- For model II, each consequent function f of a fuzzy rule (rule node, cluster centre) is learned by a corresponding MLP network after training it on the corresponding dataset with the use of the backpropagation algorithm.

5.3.6 Time-Series Modelling and Prediction with the DENFIS Off-Line Model

Dynamic time-series modelling of complex time series is a difficult task, especially when the type of the model is not known in advance. In this section, we applied the two DENFIS off-line models for the same task of Mackey–Glass time-series prediction. For comparison purposes two other well-known models, adaptive neural-fuzzy inference system (ANFIS; Jang (1993)) and a multilayer perceptron trained with the backpropagation algorithm (MLP-BP; Rumelhart *et al.* (1986)) are also used for this task under the same conditions.

In addition to the nondimensional error index (NDEI), in the case of off-line learning, the learning time is also measured as another comparative criterion. Here, the learning time is the CPU-time (in seconds) consumed by each method during the learning process in the same computing environment (MATLAB, UNIX version 5.5).

Table 5.3 lists the off-line prediction results of MLP, ANFIS, and DENFIS, and these results include the number of fuzzy rules (or rule nodes) in the hidden layer, learning epochs, learning time (CPU-time), NDEI for training data, and NDEI for testing data. The best results are achieved in the DENFIS II model.

In Figures 5.8d,e,f the prediction test error on the same test data is shown for the following three DENFIS off-line learning models,

DENFIS off-line mode I with 116 fuzzy rules
DENFIS off-line mode I with 883 fuzzy rules
DENFIS off-line mode II with 58 fuzzy rules

The prediction error of DENFIS model II with 58 rule nodes is the lowest one.

5.3.7 Rule Extraction from DENFIS Models

DENFIS allow for rules to be extracted at any time of the system operation. The rules are first-order Takagi–Sugeno rules.

Table 5.3 Prediction results of off-line learning models on Mackey–Glass training and test data.

Methods	*Neurons or rules*	*Epochs*	*Training time (s)*	*Training NDEI*	*Testing NDEI*
MLP-BP	60	50	1779	0.083	0.090
MLP-BP	60	500	17928	0.021	0.022
ANFIS	81	50	23578	0.032	0.033
ANFIS	81	200	94210	0.028	0.029
DENFIS I	116	2	352	0.068	0.068
DENFIS I	883	2	1286	0.023	0.019
DENFIS II	58	100	351	0.017	0.016

This is illustrated on the gas-furnace time-series dataset. A DENFIS system is trained on part of the data. The system approximates this data with a RMSE of 0.276 and NDEI of 0.081; see Fig. 5.9a. There are 11 rule nodes created that partition the input space as shown in Fig. 5.9b. Eleven rules are extracted, each of them representing the 11 rule nodes as shown in Fig. 5.9c.

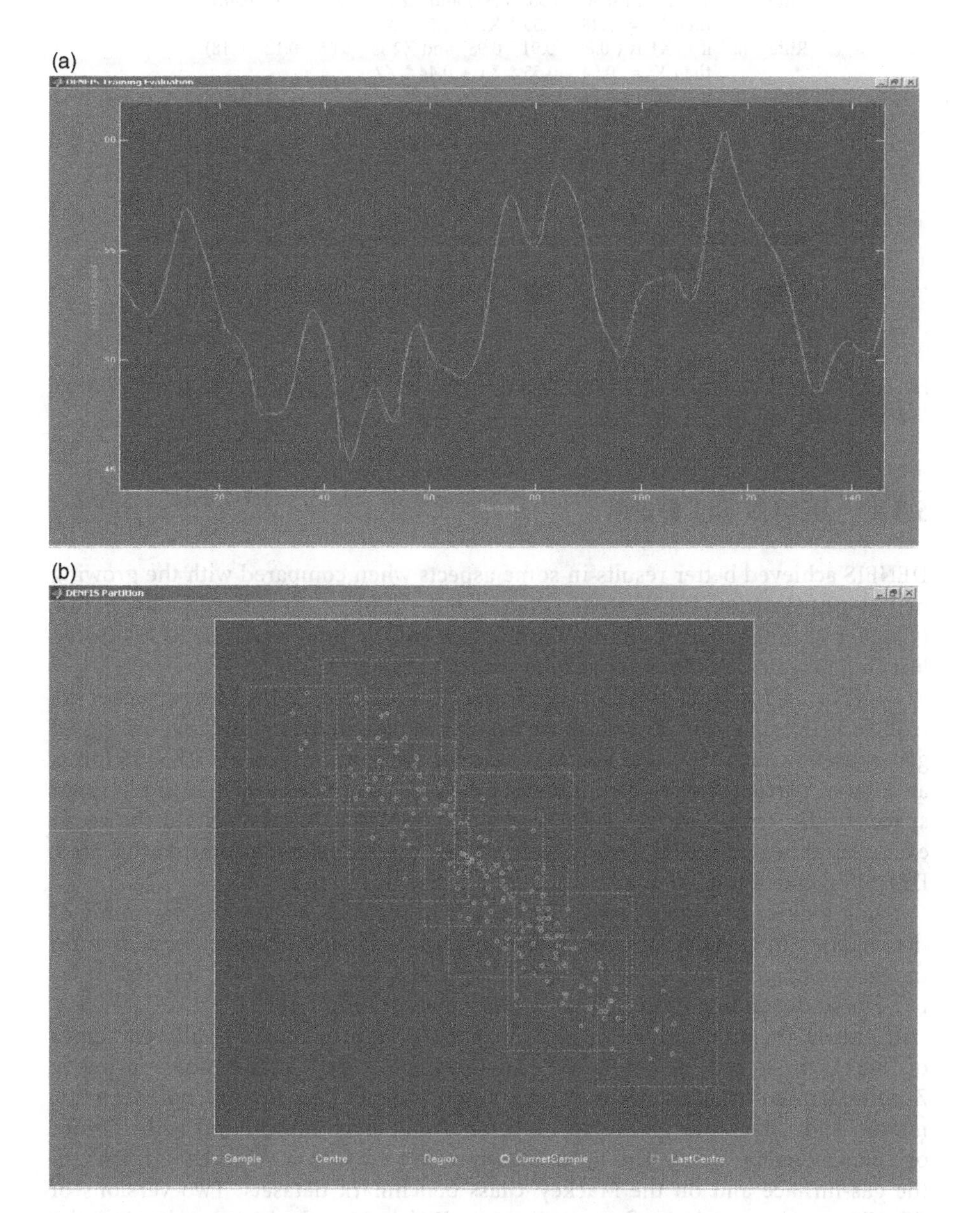

Fig. 5.9 (a) The trained DENFIS approximates the gas-furnace time-series data; (b) partitioning of the input space by the 11 evolved rule nodes in DENFIS; (*Continued overleaf*)

```
Rule  1: if   X1 is ( 0.44  0.50  0.57) and X2 is ( 0.45  0.52  0.58)
         then Y  =  0.53 - 0.58 * X1 + 0.53 * X2
Rule  2: if   X1 is ( 0.23  0.29  0.36) and X2 is ( 0.63  0.69  0.76)
         then Y  =  0.52 - 0.52 * X1 + 0.51 * X2
Rule  3: if   X1 is ( 0.65  0.71  0.78) and X2 is ( 0.25  0.32  0.38)
         then Y  =  0.45 - 0.49 * X1 + 0.60 * X2
Rule  4: if   X1 is ( 0.63  0.70  0.76) and X2 is ( 0.14  0.20  0.27)
         then Y  =  0.41 - 0.43 * X1 + 0.60 * X2
Rule  5: if   X1 is ( 0.50  0.56  0.63) and X2 is ( 0.33  0.39  0.46)
         then Y  =  0.48 - 0.53 * X1 + 0.59 * X2
Rule  6: if   X1 is ( 0.85  0.91  0.98) and X2 is ( 0.05  0.12  0.18)
         then Y  =  0.54 - 0.55 * X1 + 0.44 * X2
Rule  7: if   X1 is ( 0.26  0.32  0.39) and X2 is ( 0.51  0.58  0.64)
         then Y  =  0.54 - 0.59 * X1 + 0.50 * X2
Rule  8: if   X1 is ( 0.23  0.29  0.36) and X2 is ( 0.74  0.81  0.87)
         then Y  =  0.51 - 0.52 * X1 + 0.52 * X2
Rule  9  if   X1 is ( 0.51  0.57  0.64) and X2 is ( 0.55  0.62  0.68)
         then Y  =  0.59 - 0.61 * X1 + 0.46 * X2
Rule 10: if   X1 is ( 0.01  0.08  0.14) and X2 is ( 0.77  0.83  0.90)
         then Y  =  0.53 - 0.51 * X1 + 0.49 * X2
Rule 11: if   X1 is ( 0.19  0.26  0.32) and X2 is ( 0.83  0.90  0.96)
         then Y  =  0.53 - 0.51 * X1 + 0.50 * X2
```

Fig. 5.9 (*continued*) (c) Takagi–Sugeno fuzzy rules extracted from a trained DENFIS model on the gas-furnace time-series dataset.

5.3.8 DENFIS and EFuNN

DENFIS achieved better results in some aspects when compared with the growing neural gas (Fritzke, 1995), RAN (Platt, 1991), EFuNN (Chapter 3), and ESOM (Chapter 2) in the case of online learning of a chaotic time series. DENFIS off-line learning produces comparable results with ANFIS and MLP.

DENFIS uses a local generalisation, like EFuNN and CMAC neural networks (Albus, 1975), therefore it needs more training data than the models that use global generalisation such as ANFIS and MLP. During the learning process DENFIS forms an area of partitioned regions, but these regions may not cover the whole input space. In the recall process, DENFIS would give satisfactory results if the recall examples appeared inside these regions. In the case of examples outside this area, DENFIS is likely to produce results with a higher error rate.

Using a different type of rules (see the list of types of rules at the beginning of the chapter) in an ECOS architecture may lead to different results depending on the task in hand. ECOS allow for both fuzzy and propositional (e.g. interval) rules to be used depending on if there is a fuzzy membership layer.

If the ECOS architecture deals with a fuzzy representation, different types of fuzzy rules can be exploited. For example, for classification purposes, Zadeh–Mamdani fuzzy rules may give a better result, but for function approximation and time series prediction a better result may be achieved with the use of Takagi–Sugeno rules. The latter is demonstrated with a small experiment on the gas-furnace and on the Mackey–Glass benchmark datasets. Two versions of EFuNN are compared: the first version uses Zadeh–Mamdani fuzzy rules, and the second version, Takagi–Sugeno fuzzy rules (Table 5.4).

Table 5.4 Comparing two versions of EFuNN: the first version uses Zadeh–Mamdani fuzzy rules, and the second version Takagi–Sugeno fuzzy rules on the gas-furnace times-series data, and on the Mackey–Glass time-series data.

	EFuNN with Zadeh–Mamdani fuzzy rules	*EFuNN with Takagi–Sugeno fuzzy rules*
Gas-furnace time-series data: number of rule nodes	51	51
Gas-furnace time-series data: online testing NDEI	0.283	0.156
Mackey–Glass time-series data: number of rule nodes	151	151
Mackey–Glass time-series data: online testing NDEI	0.147	0.039

For the same number of rule nodes evolved in the two types of EFuNNs, the EFuNN that uses the Takagi–Sugeno fuzzy rules gives better results on the time-series prediction problem for the two benchmark time-series data.

5.4 Transductive Neuro-Fuzzy Inference Models

5.4.1 Principles and Structure of the TWNFI

TWNFI is a dynamic neural-fuzzy inference system with a local generalization, in which, either the Zadeh–Mamdani type fuzzy inference engine is used, or the Takagi–Sugeno fuzzy inference is applied. Here, the former case is introduced. The local generalisation means that in a subspace of the whole problem space (local area) a model is created that performs generalisation in this area. In the TWNFI model, Gaussian fuzzy membership functions are applied in each fuzzy rule for both the antecedent and the consequent parts. A steepest descent (BP) learning algorithm is used for optimizing the parameters of the fuzzy membership functions. The distance between vectors $\boldsymbol{x}$ and $\boldsymbol{y}$ is measured in TWNFI in weighted normalized Euclidean distance defined as follows (the values are between 0 and 1),

$$\|\boldsymbol{x}-\boldsymbol{y}\| = \left[\frac{1}{P}\sum_{j=1}^{P} w_j \left|x_j - y_j\right|^2\right]^{\frac{1}{2}} \tag{5.16}$$

where: $\boldsymbol{x}, \boldsymbol{y} \in \boldsymbol{R}^P$, and w_j are weights.

To partition the input space for creating fuzzy rules and obtaining initial values of fuzzy rules, the ECM (evolving clustering method) is applied (Chapter 2) and the cluster centres and cluster radii are respectively taken as initial values of the centres and widths of the Gaussian membership functions. Other clustering techniques can be applied as well. A block diagram of the TWNFI is shown in Fig. 5.10.

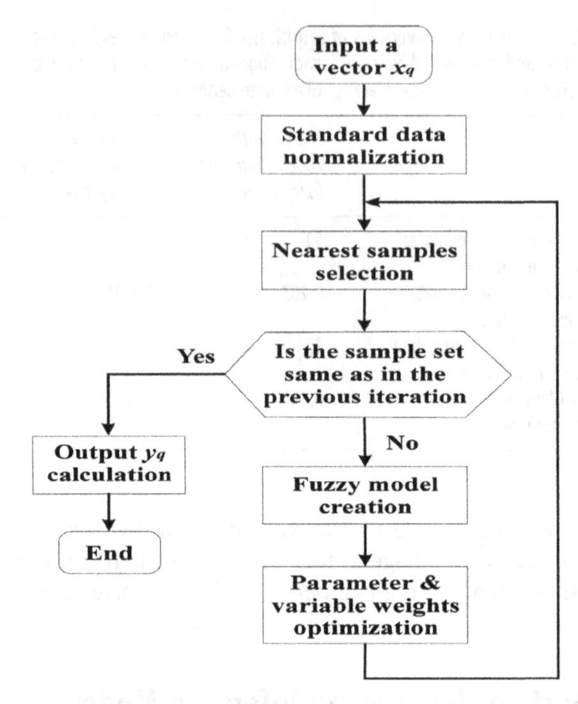

Fig. 5.10 A block diagram of the TWNFI method (from Song and Kasabov (2006)).

5.4.2 The TWNFI Learning Algorithm

For each new data vector $\boldsymbol{x}_q$ an individual model is created with the application of the following steps.

Normalize the training dataset and the new data vector $\boldsymbol{x}_q$ (the values are between 0 and 1) with value 1 as the initial input variable weights.

Search in the training dataset in the input space to find N_q training examples that are closest to $\boldsymbol{x}_q$ using weighted normalized Euclidean distance. The value of N_q can be predefined based on experience, or optimised through the application of an optimisation procedure. Here we assume the former approach.

Calculate the distances d_i, $i = 1, 2, \ldots, N_q$, between each of these data samples and $\boldsymbol{x}_q$. Calculate the vector weights $v_i = 1 - (d_i - \min(\boldsymbol{d}))$, $i = 1, 2, \ldots, N_q$, $\min(\boldsymbol{d})$ is the minimum value in the distance vector $\boldsymbol{d} = [d_1, d_2, \ldots, d_{Nq}]$.

Use the ECM clustering algorithm to cluster and partition the input subspace that consists of N_q selected training samples.

Create fuzzy rules and set their initial parameter values according to the ECM clustering procedure results; for each cluster, the cluster centre is taken as the centre of a fuzzy membership function (Gaussian function) and the cluster radius is taken as the width.

Apply the steepest descent method (backpropagation) to optimize the weights and parameters of the fuzzy rules in the local model M_q following the equations given below.

Search in the training dataset to find N_q samples (the same as Step 2); if the same samples are found as in the last search, the algorithm goes to Step 8, otherwise, to Step 3.

Calculate the output value y_q for the input vector $\boldsymbol{x}_q$ applying fuzzy inference over the set of fuzzy rules that constitute the local model M_q.

End of the procedure.

The weight and parameter optimization procedure is described below:

Consider the system having P inputs, one output and M fuzzy rules defined initially through the ECM clustering procedure, the lth rule has the form of:

$$\mathrm{R}_l: \quad \text{If } x_1 \text{ is } \mathrm{F}_{l1} \text{ and } x_2 \text{ is } \mathrm{F}_{l2} \text{ and} \ldots x_\mathrm{P} \text{ is } \mathrm{F}_{l\mathrm{P}}, \text{ then } y \text{ is } \mathrm{G}_l \tag{5.17}$$

Here, F_{lj} are fuzzy sets defined by the following Gaussian type membership function,

$$GaussianMF = \alpha \exp\left[-\frac{(x-m)^2}{2\sigma^2}\right] \tag{5.18}$$

and G_l are logarithmic regression functions:

$$\mathrm{G}l = b_{l0} x_1^{b_{l1}} x_2^{b_{l2}} \cdots x_p^{b_{lp}} \tag{5.19}$$

Using the modified centre average defuzzification procedure the output value of the system can be calculated for an input vector $\boldsymbol{x}_i = [x_1, x_2, \ldots, x_\mathrm{P}]$ as follows.

$$f(\boldsymbol{x}_i) = \frac{\sum\limits_{\mathrm{l}=1}^{\mathrm{M}} G_l \prod\limits_{\mathrm{j}=1}^{\mathrm{P}} \alpha_{lj} \exp\left[-\frac{w_j^2 \left(x_{ij} - m_{lj}\right)^2}{2\sigma_{\mathrm{lj}}^2}\right]}{\sum\limits_{\mathrm{l}=1}^{\mathrm{M}} \prod\limits_{\mathrm{j}=1}^{P} \alpha_{lj} \exp\left[-\frac{w_j^2 \left(x_{ij} - m_{lj}\right)^2}{2\sigma_{\mathrm{lj}}^2}\right]} \tag{5.20}$$

Here, w_j are weights of the input variables.

Suppose the TWNFI is given a training input–output data pair $[\boldsymbol{x}_i, t_i]$, the system minimizes the following objective function (a weighted error function):

$$b_{l0}(k+1) = b_{l0}(k) - \frac{\eta_b}{b_{l0}(k)} G_l(k) v_i \left[f^{(k)}(\boldsymbol{x}_i) - t_i\right] \Phi(\boldsymbol{x}_i) \tag{5.21}$$

$$\mathrm{E} = \frac{1}{2} v_i \left[f(x_i) - t_i\right]^2$$

(v_i are defined in Step 3).

The steepest descent algorithm (BP) is used then to obtain the formulas for the optimisation of the parameters b_{lj}, α_{lj}, m_{lj}, σ_{lj}, and w_j such that the error E is minimised:

$$b_{lj}(k+1) = b_{lj}(k) - \eta_b G_l(k) \ln\left(x_{ij}\right) v_i \left[f^{(k)}(\boldsymbol{x}_i) - t_i\right] \Phi(\boldsymbol{x}_i) \tag{5.22}$$

$$\alpha_{lj}(k+1) = \alpha_{lj}(k) - \frac{\eta_\alpha v_i \Phi(xi)}{\alpha_{lj}(k)} \left[f^{(k)}(\boldsymbol{x}_i) - t_i\right]\left[G_l(k) - f^{(k)}(\boldsymbol{x}_i)\right] \tag{5.23}$$

$$m_{lj}(k+1) = m_{lj}(k) - \frac{\eta_m w_j^2(k) v_i \Phi(\boldsymbol{x}_i)}{\sigma_{lj}^2(k)} \left[f^{(k)}(\boldsymbol{x}_i) - t_i\right]\left[G_l(k) - f^{(k)}(\boldsymbol{x}_i)\right]\left[x_{ij} - m_{lj}(k)\right] \tag{5.24}$$

$$\sigma_{lj}(k+1) = \sigma_{lj}(k) - \frac{\eta_\sigma w_j^2(k) v_i \Phi(\boldsymbol{x}_i)}{\sigma_{lj}^3(k)} \left[f^{(k)}(\boldsymbol{x}_i) - t_i\right]\left[G_l(k) - f^{(k)}(\boldsymbol{x}_i)\right]\left[x_{ij} - m_{lj}(k)\right]^2 \tag{5.25}$$

$$w_j(k+1) = w_j(k) - \frac{\eta_w w_j(k) v_i \Phi(\boldsymbol{x}_i)}{\sigma_{lj}^2(k)} \left[f^{(k)}(\boldsymbol{x}_i) - t_i\right]\left[f^{(k)}(\boldsymbol{x}_i) - G_l(k)\right]\left[x_{ij} - m_{lj}(k)\right]^2 \tag{5.26}$$

$$\Phi(\boldsymbol{x}_i) = \frac{\prod_{j=1}^{P} \alpha_{lj} \exp\left\{-\frac{w_j^2(k)\left[x_{ij} - m_{lj}(k)\right]^2}{2\sigma_{lj}^2(k)}\right\}}{\sum_{l=1}^{M}\prod_{j=1}^{P} \alpha_{lj} \exp\left\{-\frac{w_j^2(k)\left[x_{ij} - m_{lj}(k)\right]^2}{2\sigma_{lj}^2(k)}\right\}} \tag{5.27}$$

where η_b, η_α, η_m, η_σ, and η_w and are learning rates for updating the parameters b_{lj}, α_{lj}, m_{lj}, σ_{lj}, and w_j, respectively.

In the TWNFI training-simulating algorithm, the following indexes are used.

Training data samples: $i = 1, 2, \ldots, N$.
Input variables: $j = 1, 2, \ldots, P$.
Fuzzy rules: $l = 1, 2, \ldots, M$.
Training epochs: $k = 1, 2, \ldots$.

5.4.3 Applications of TWNFI for Time-Series Prediction

In this section several evolving NFI systems are applied for modelling and predicting the future values of a chaotic time series: Mackey–Glass (MG) dataset (see Chapter 1), which has been used as a benchmark problem in the areas of neural networks, fuzzy systems, and hybrid systems. This time series is created with the use of the MG time-delay differential equation defined below:

$$\frac{dx(t)}{dt} = \frac{0.2x(t-\tau)}{1+x^{10}(t-\tau)} - 0.1x(t) \tag{5.28}$$

To obtain values at integer time points, the fourth-order Runge–Kutta method was used to find the numerical solution to the above MG equation. Here we assume that: the time step is 0.1; $x(0) = 1.2$; $\tau = 17$; and $x(t) = 0$ for $t < 0$. The task is to predict the values $x(t + 6)$ from input vectors $[x(t-18), x(t-12), x(t-6), x(t)]$ for any value of the time t. The following experiment was conducted: 1000 data points, from $t = 118$ to 1117, were extracted; the first 500 data were taken as the training data and another 500 as testing data. For each of the testing data sample a TWNFI model was created and tested on this data. Figure 5.11 displays the target data and Table 5.5 lists the testing results represented by NDEI, nondimensional error index, that is defined as the root mean square error (RMSE) divided by the standard deviation of the target series. For the purpose of a comparative analysis, we have quoted the prediction results on the same data produced by some other methods, which are also listed in Table 5.5. The TNFI method there is the same as the TWNFI method described in Section 5.2, but there is no weight optimization (all they are assumed equal to one and do not change during the model development).

The TWNFI model performs better than the other models. This is a result of the fine tuning of each local model in TWNFI for each tested example, derived according to the TWNFI learning procedure. The finely tuned local models achieve a better local generalisation. For each individual model, created for each test sample, input variable weights w_{1q}, w_{2q}, w_{3q}, and w_{4q} are derived. Table 5.1 shows the average weight for each variable across all test samples. The weights suggest a higher importance of the first, second, and fourth input variables, but not the third one.

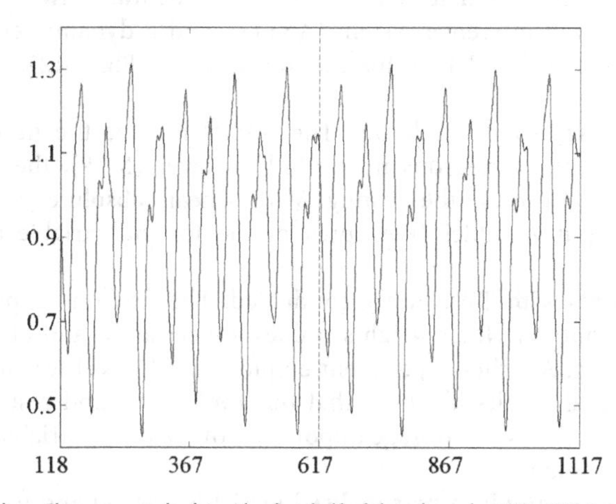

Fig. 5.11 The Mackey–Glass case study data: the first half of data (samples 118–617) is used as training data, and the second half (samples 618–1117) as testing data. An individual TWNFI prediction model is created for each test data vector, based on the nearest data vectors (from Song and Kasabov (2006)).

Table 5.5 Comparative analysis of test accuracy of several methods on the MG series.

Model	Testing NDEI	Weights of input variables			
		w1	w2	w3	w4
CC–NN model	0.06	1	1	1	1
Sixth-order polynomial	0.04	1	1	1	1
MLP (BP)	0.02	1	1	1	1
HyFIS	0.01	1	1	1	1
ANFIS	0.007	1	1	1	1
DENFIS	0.006	1	1	1	1
TNFI	0.008	1	1	1	1
TWNFI	0.005	0.95	1.0	0.46	0.79

5.4.4 Applications of TWNFI for Medical Decision Support

A real dataset from a medical institution is used here for experimental analysis (Marshal *et al.*, 2005) The dataset has 447 samples, collected at hospitals in New Zealand and Australia. Each of the records includes six variables (inputs): age, gender, serum creatinine, serum albumin, race, and blood urea nitrogen concentrations, and one output: the glomerular filtration rate value (GFR).

All experimental results reported here are based on tencross-validation experiments with the same model and parameters and the results are averaged. In each experiment 70% of the whole dataset is randomly selected as training data and another 30% as testing data.

For comparison, several well-known methods are applied on the same problem, such as the MDRD logistic regression function widely used in the renal clinical practice (MDRD, see Marshall *et al.*, 2005), MLP neural network (Chapter 3), adaptive neural fuzzy inference system (ANFIS), and a dynamic evolving neural fuzzy inference system (DENFIS; this chapter, and also Fig. 5.12), along with the TWNFI.

Results are presented in Table 5.6. The results include the number of fuzzy rules (fuzzy models), or neurons in the hidden layer (MLP), the testing RMSE (root mean square error), the testing MAE (mean absolute error), and the weights of the input variables (the upper bound for the variable normalization range).

Two experiments with TWNFI are conducted. The first one applies the transductive NFI without WDN: all weights' values are set as '1' and are not changed during the learning. Another experiment employs the TWNFI learning algorithm.

The experimental results illustrate that the TWNFI method results in a better accuracy and also depicts the average importance of the input variables represented as the calculated weights.

For every patient sample, a personalised model is created and used to evaluate the output value for the patient and to also estimate the importance of the variables for this patient as shown in Table 5.6. The TWNFI not only results in a better

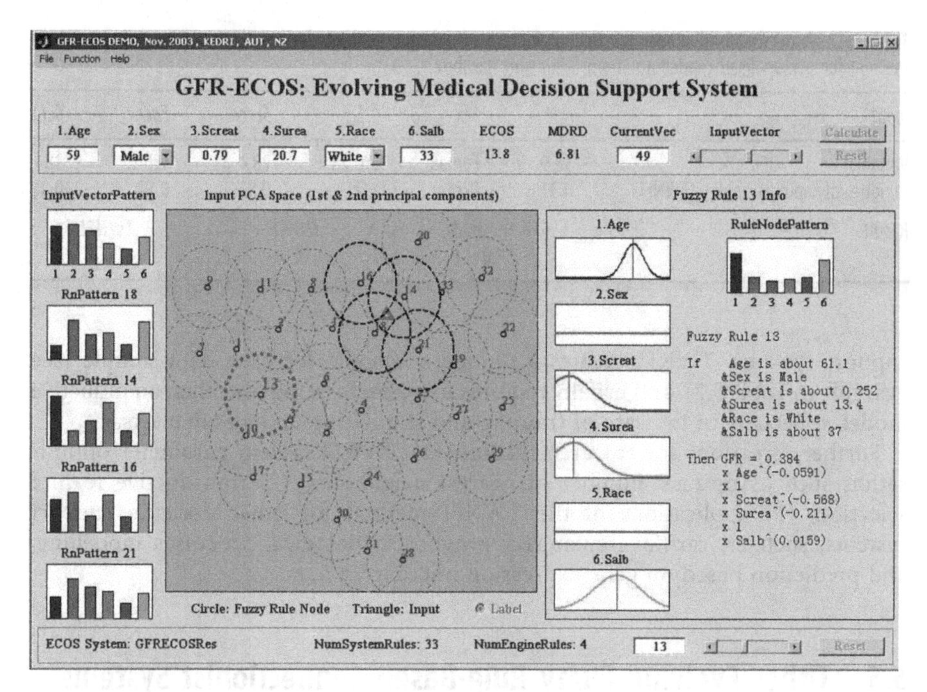

Fig. 5.12 The interface of an adaptive, kidney function evaluation decision support system: GFR-ECOS (from Marshall *et al.* (2005)).

accuracy for this patient, but shows the importance of the variables for her or him that may result in a more efficient personalised treatment.

The transductive neuro-fuzzy inference with weighted data normalization method (TWNFI) performs a better local generalisation over new data as it develops an individual model for each data vector that takes into account the new input vector location in the space, and it is an adaptive model, in the sense that input–output pairs of data can be added to the dataset continuously and immediately made available for transductive inference of local models. This type of modelling can be called 'personalised', and it is promising for medical decision

Table 5.6 Experimental results on GFR data using different methods (from Song *et al.* (2005)).

Model	*Neurons or rules*	*Test RMSE*	*Test MAE*	*Weights of Input Variables*					
				Age w1	*Sex w2*	*Scr w3*	*Surea w4*	*Race w5*	*Salb w6*
MDRD	__	7.74	5.88	1	1	1	1	1	1
MLP	12	8.38	5.71	1	1	1	1	1	1
ANFIS	36	7.40	5.43	1	1	1	1	1	1
DENFIS	27	7.22	5.21	1	1	1	1	1	1
TNFI	6.8 (average)	7.28	5.26	1	1	1	1	1	1
TWNFI	**6.8 (average)**	**7.08**	**5.12**	**0.87**	**0.70**	**1**	**0.93**	**0.40**	**0.52**

Table 5.7 A personalised model for GFR prediction of a single patient derived with the use of the TWNFI method for personalised modelling (from Song and Kasabov (2006)).

Input	*Age*	*Sex*	*Scr*	*Surea*	*Face*	*Salb*
Variables	58.9	Female	0.28	28.4	White	38
Weights of input variables (TWNFI)	0.91	0.73	1	0.82	0.52	0.46
Results	GFR (desired) 18.0		MDRD 14.9		**TWNFI 16.6**	

support systems. TWNFI creates a unique submodel for each data sample (see example in Table 5.7), and usually requires more performing time than an inductive model, especially in the case of training and simulating on large datasets.

Further directions for research include: (1) TWNFI system parameter optimisation such as optimal number of nearest neighbours; (2) transductive feature selection; (3) applications of the TWNFI method for other decision support systems, such as: cardiovascular risk prognosis, biological processes modelling, and prediction based on gene expression microarray data.

5.5 Other Evolving Fuzzy Rule-Based Connectionist Systems

5.5.1 Type-2 Neuro-Fuzzy Systems

In all models and ECOS presented thus far in this part of the book, we assumed the following.

- A connection weight is associated with a single number.
- Each neuron's activation has a numerical value.
- The degree to which an input variable belongs to a fuzzy membership function is a single numerical value.

In brief, each structural element of an ECOS is associated with a single numerical value.

The above assumptions may not be appropriate when processing complex information such as noisy information (with a random noise added to each data item that may vary according to a random distribution function).

Here, the ECOS paradigm presented thus far is extended further to using higher-order representation, for example association of a function rather than a single value with a connection, with a neuron, or with a membership function. Such an extension can be superior when:

- Data are time varying (e.g. changing dynamics of a chaotic process).
- Noise is nonstationary.
- Features are nonstationary (they change over time).
- Dealing with inexact human knowledge that changes over time and varies across humans. Humans, especially experts, change their minds during the process of

learning and understanding phenomena. For various people the same concepts may have different meaning (e.g. the concept of a small salary points to different salary scales in the United States, Bulgaria, and Sudan).

As a theoretical basis for type-2 ECOS, some principles from the theory of type-2 fuzzy systems can be used (for a detailed description see Mendel (2001)). The type-2 fuzzy system theory is based on several concepts as explained below.

Zadeh in 1967. This is in contrast to the type-1 fuzzy sets where every element belongs to the set to a certain membership degree that is represented as a single number between 0 and 1.

An example of a type-2 fuzzy membership function is given in Fig. 5.13.

Type-2 MF can be used to represent MF that may change over time, or MF that are interpreted differently by different experts.

Type-2 fuzzy rules are fuzzy rules of the form of: IF x is A~ AND y is B~ THEN z is C~ , where A~ , B~, and C~ are type-2 fuzzy membership functions. These rules deal with interval values rather than with single numerical values.

Type-2 fuzzy inference systems are fuzzy rule-based systems that use type-2 fuzzy rules. The inference process consists of the following steps.

Step 1. Fuzzyfication of input data with the use of type-2 fuzzy MF.
Step 2. Fuzzy inference (decision making) with the use of type-2 fuzzy rules. The inference produces type-2 output MF.
Step 3. Defuzzification that would include a step of type reduction which transforms the type 2 output MF into type 1 MF, and a step of calculating a single output value from a type-1 MF.

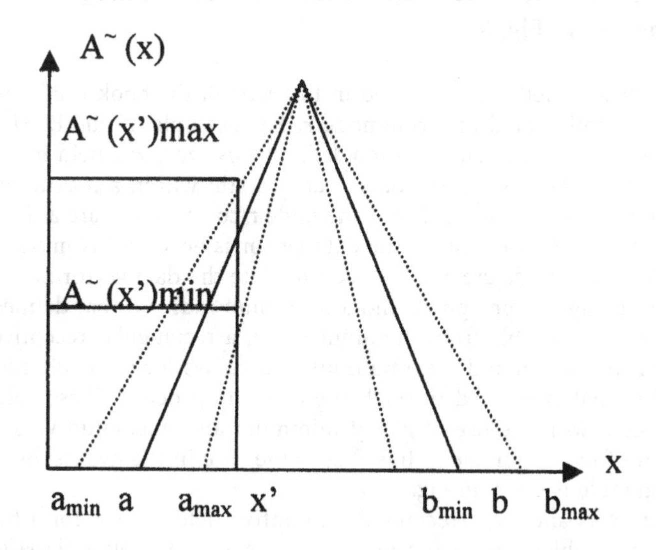

Fig. 5.13 An example of type-2 triangular fuzzy membership function (MF) of a variable x. Although the centre of the MF is fixed, the left and the right sides may vary. A fuzzification procedure produces min–max interval membership values as illustrated on a value x' for the variable x.

The move from type-1 systems to type-2 systems is equivalent to moving from a 2D space to a 3D space of representation and reasoning.

Type-2 EFuNN

In type-2 ECOS each connection has either a function or a numerical min–max interval value associated with it. For an EFuNN for example that interval could be formed by the value of the connection weight **Wi,j**, that represents the fuzzy coordinate of the *i*th cluster (rule node r_i), plus/minus the radius of this node *Ri*: $Wi,j(Ri) = [Wi,j - Ri, Wi,j + Ri]$.

A type-2 EFuNN uses type-2 fuzzy MF defined a priori and modified during the system operation. For each input vector $\mathbf{x}$, the fuzzification procedure produces interval membership values. For example, for an input variable x_k, we can denote its membership interval to a type-2 MF A~ as $A^{\sim}(x_k) = [\min(A^{\sim}(x_k)), \max(A^{\sim}(x_k))]$.

Having interval values in a connectionist system requires interval operations to be applied as part of its inference procedure. The interval operations that are defined for type-2 fuzzy systems (see Mendel (2001)) can be applied here. The distance between the input interval $A^{\sim}(x_k)$ and the rule node r_i interval *Wi,j(Ri)* is defined as an interval operation, so as the activation of the rule node r_i. The activation of the output MF nodes is calculated also based on multiplication operation of interval values and scalars, and on summation of interval values. In Mendel (2001) several defuzzification operations over type-2 fuzzy MF are presented that can be used in a type-2 EFuNN.

5.5.2 Interval-Based ECOS and Other Ways of Defining Receptive Fields

Most of the ECOS methods presented in this part of the book used hyperspheres to define a receptive field of a rule node *r*. For example, in the ECM and in the EFuNN models, if a new data vector $\mathbf{x}$ falls in this receptive field it is associated with this node. Otherwise, if the data vector is still within a maximum radius *R* distance from the receptive field of *r*, the node *r* co-ordinates are adjusted and its receptive field is increased, but if the data vector is beyond this maximum radius there will be a new node created to accommodate the data vector $\mathbf{x}$.

Instead of using a hypersphere and a maximum radius *R* that defines the same distance for each variable from the input space, a rectangular receptive field can be used with minimum and maximum interval values for each rule node and for each variable that is derived through the evolving process. These values can be made restricted with the use of global minimum and maximum values Min and Max instead of using a single radius *R* as is the case in the hypersphere receptive field. An example is given in Fig. 5.14.

Using intervals and hyperrectangular receptive fields allows for a better partitioning of the problem space and in many cases leads to better classification and prediction results.

Interval-based receptive fields can be used in both a fuzzy version and a nonfuzzy version of ECOS. Figure 5.14 applies to both. The extracted rules from a trained

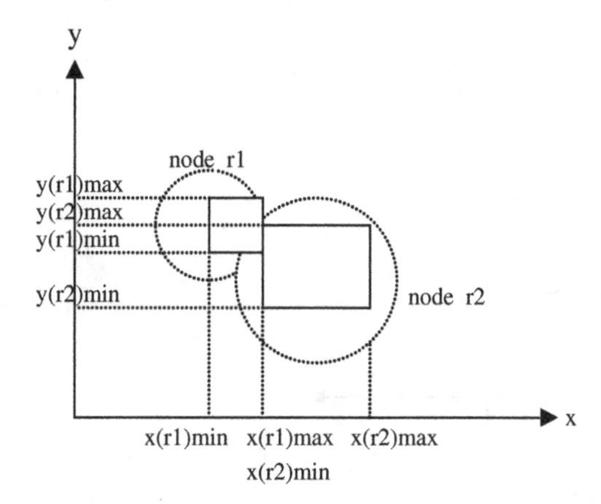

Fig. 5.14 Using intervals, hyperrectangles (solid lines) for defining receptive fields versus using hyperspheres (the dotted circles) on a case study of two rule nodes $r1$ and $r2$ and two input variables x and y.

ECOS will be interval-based. For example, the rule that represents rule node 1 from Fig. 5.14 will read:

IF x is in the interval [$x(r1)$min, $x(r1)$max] AND y is in the interval [$y(r1)$min, $y(r1)$max] THEN class will be (...), with $Nr1$ examples associated with this rule.

Divide and Split Receptive Fields

Online modification of the receptive fields is crucial for successful online learning. Some times a receptive field is created for a rule node, but within this receptive field there is a new example that belongs to a different class. In this case the new example will be assigned a new node that divides the previous receptive field into several parts. Each of these parts will be assigned new nodes that will have the same class label as the 'mother node'. This approach is very efficient when applied for online classification in complex multiclass distribution spaces.

Figure 5.15. shows an example where a new class example divides the receptive field of the existing node from (a), drawn in a 2D space into six new regions represented by six nodes.

Example

Here a classical classification benchmark problem was used – the classification of Iris flower samples into three classes – Setosa, Versicolor, and Virginica, based on four features – sepal length, petal length, sepal width, and petal width. A dataset of 150 examples, 50 examples of each of the three classes was used (see Fisher (1936) and also Duda and Hart (1973)). Three types of EFuNNs were evolved through one-pass learning on the whole dataset of 150 examples. Then each EFuNN was tested for classification results on the same dataset. The EFuNN with the

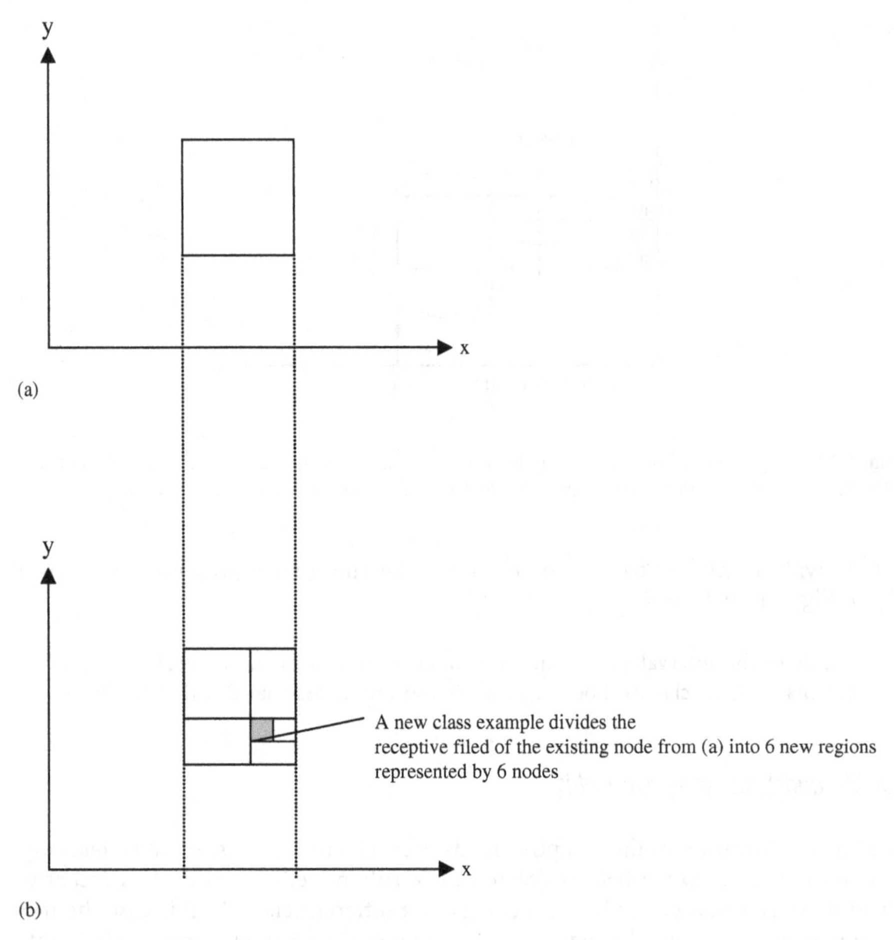
y
x
(a)
y
A new class example divides the
receptive filed of the existing node from (a) into 6 new regions
represented by 6 nodes
x
(b)

Fig. 5.15 (a) An existing receptive field of a rule node in a 2D input space; (b) a new class example divides the receptive field of the existing node from (a) into six new regions represented by six nodes.

hyperspherical receptive field produced three misclassified examples, the EFuNN with hyperrectangular fields produced two errors, and the EFuNN with the divide and split receptive fields resulted in no misclassified examples.

5.5.3 Evolving Neuro-Fuzzy System: EFS

In Angelov (2002) an evolving neuro-fuzzy system framework, called EFS, is introduced that is represented as a connectionist architecture in Fig. 5.16. The framework is used for adaptive control systems and other applications. Here a brief presentation of the structure and the learning algorithm of EFS is presented adopted from Angelov (2002).

The first layer consists of neurons corresponding to the membership functions of fuzzy sets. The second layer represents the antecedent parts of the fuzzy rules.

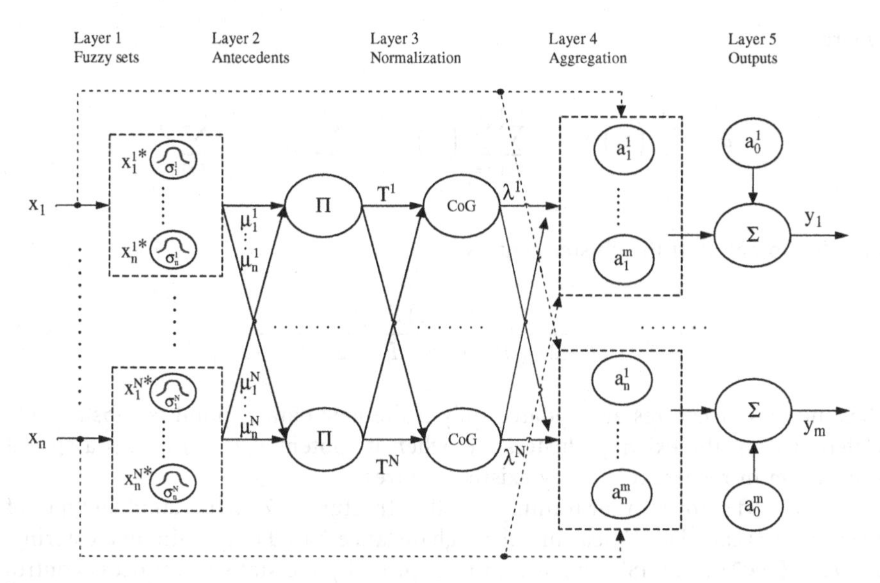

Fig. 5.16 Neuro-fuzzy interpretation of the evolving fuzzy system (Angelov, 2002). The structure is not predefined and fixed; rather it evolves 'from scratch' by learning from data simultaneously with the parameter adjustment/adaptation (reproduced with permission).

It takes as inputs the membership function values and gives as output the firing level of the ith rule. The third layer of the network takes as inputs the firing levels of the respective rule and gives as output the normalized firing level. The fourth layer aggregates the antecedent and the consequent part that represents the local subsystems (singletons or hyperplanes). Finally, the last, fifth, layer forms the total output of the evolving fuzzy system performing a weighed summation of local multiple model outputs.

There are two main algorithms for training an EFS. The first one, called eTS, is based on combining unsupervised learning with respect to the antecedent part of the model and supervised learning in terms of the consequent parameters, where the fuzzy rules are of the Takagi–Sugeno type, similar to the DENFIS training algorithm presented in Section 5.3 (Kasabov and Song, 2002). An unsupervised clustering algorithm is employed to continuously analyze the input–output data streams and identifies emerging new data structures. It clusters the input–output space into N fuzzy regions. The clusters define a fuzzy partitioning of the input space into subsets that are obtained by projecting the cluster centres onto the space of input (antecedent) variables. The learning algorithm also assigns to each of the clusters a linear subsystem. The eTS learning algorithm (Angelov, 2002) is a density-based clustering that stems from the Mountain clustering method and an extension called the Subtractive clustering.

The eTS clustering is based on the recursive calculation of the potential $P_t(z_t)$ of each data point z_t in the input–output space $z_t = R^{n+m}$:

$$P_t(z_t) = \frac{t-1}{(t-1)(a_t+1) - 2c_t + b_t}, \tag{5.29}$$

where

$$a_t = \sum_{j=1}^{p} \left(z_t^j\right)^2;\ b_t = \sum_{i=1}^{t-1}\sum_{j=1}^{p} \left(z_i^j\right)^2;\ c_t = \sum_{j=1}^{p} z_t^j f_t^j;\ f_t^j = \sum_{i=1}^{t-1} z_i^j$$

and the potential of the cluster centres

$$P_t(z^*) = \frac{(t-1)\,P_{t-1}(z^*)}{(t-2) + P_{t-1}(z^*) + P_{t-1}(z^*)||z^* - z_{t-1}||^2} \quad (5.30)$$

Existing cluster centres are updated only if the new data z_t point is substantially different from the existing clusters, i.e. when its potential $P_t(z_t)$ brings a spatial innovation in respect to already existing centres.

The second algorithm for training the EFS structure of Zadeh–Mamdani types of fuzzy rules is called fSPC (learning through distance-based output–input clustering; Angelov (2002)). The fSPC algorithm is inspired by the statistical process control (SPC), a method for process variability monitoring. The SPC procedure naturally clusters the system output data into granules (clusters) that relate to the same process control conditions and are characterized with similar system behavior. The output granules induce corresponding granules (clusters) in the input domain and define the parameters of the rule antecedents:

$$close =: \exp(-0.5(x - x_i^*)' C_{xi}^{-1} (x - x_i^*)) \quad (5.31)$$

where x_i^* is the vector of input means and C_{xi} is the input covariance matrix.

For outputs that belong to an output granule, i.e. satisfy condition (5.31), the rule parameters associated with the respective granule are recursively updated through exponential smoothing.

5.5.4 Evolving Self-Organising Fuzzy Cerebellar Model Articulation Controller

The cerebellar model articulation controller (CMAC) is an associative memory ANN that is inspired by some principles of the information processing in the part of the brain called cerebellum (Albus, 1975).

In Nguyen *et al.* (2006) an evolving self-organising fuzzy cerebellar model articulation controller (ESOF-CMAC) is proposed. The method applies an unsupervised clustering algorithm similar to ECM (see Chapter 2), and introduces a novel method for evolving the fuzzy membership functions of the input variables. The connectionist structure implements Zadeh–Mamdani fuzzy rules, similar to the EFuNN structure. A good performance is achieved with the use of cluster aggregation and fuzzy membership function adaptation in an evolving mode of learning.

5.6 Exercise

Choose a time-series prediction problem and a dataset for it.
Select a set of features for the model.
Create an inductive ANFIS model and validate its accuracy.
Create an inductive HyFIS ANFIS model and validate its accuracy.
Create an inductive DENFIS model and validate its accuracy.
Create a transductive TWNFI model and validate its accuracy.
Answer the following questions.

(a) Which of the above models are adaptive to new data and under what conditions and constraints?
(b) Which models allow for knowledge extraction and what type of knowledge can be acquired from them?

5.7 Summary and Open Problems

The chapter presents evolving neuro-fuzzy inference systems for both off-line and online learning from data, rule insertion, rule extraction, and inference over these rules. ANFIS is not flexible in terms of changing the number of membership functions and rules over time, according to the incoming data. HyFIS and DENFIS can be used as both off-line and online knowledge-based learning systems. Whereas HyFIS and EFuNN use Zadeh–Mamdani simple fuzzy rules, DENFIS exploits Takagi–Sugeno first-order fuzzy rules. Each of the above systems has its strengths and weaknesses when applied to different tasks.

Some major issues need to be addressed in a further study:

1. How to choose dynamically the best inference method in an ECOS for the current time interval depending on the data flow distribution and on the task in hand.
2. What type of fuzzy rules and receptive fields are most appropriate for a given problem and a given problem space? Can several types of rules be used in one system, complementing each other?
3. How to test the accuracy of a neuro-fuzzy inference system in an online mode if data come from an open space.
4. How to build transductive, personalised neuro-fuzzy inference systems without keeping all data samples in a database, but only prototypes and already built transductive systems in the past.

5.8 Further Reading

More details on related to this chapter issues can be found as follows.

- *Fuzzy Sets and Systems* (Zadeh, 1965; Terano *et al.*, 1992; Dubois and Prade, 1980, 1988; Kerr, 1991; Bezdek, 1987; Sugeno, 1985)

- *Zadeh–Mamdani Fuzzy Rules* (Zadeh, 1965; Mamdani, 1977)
- *Takagi–Sugeno Fuzzy Rules* (Takagi and Sugeno, 1985; Jang, 1995)
- *Type-2 Fuzzy Rules and Fuzzy Inference Systems* (Mendel, 2001)
- *Neuro-fuzzy Inference Systems and Fuzzy Neural Networks* (Yamakawa *et al.*, 1992; Furuhashi, 1993; Hauptmann and Heesche, 1995; Lin and Lee, 1996; Kasabov, 1996; Feldcamp, 1992; Gupta, 1992; Gupta and Rao, 1994)
- *ANFIS* (Jang, 1993)
- *HyFIS* (Kim and Kasabov, 1999)
- *DENFIS* (Kasabov and Song, 2002)
- *Rule Extraction from Neuro-fuzzy Systems* (Hayashi, 1991; Mitra and Hayashi, 2001; Duch *et al.*, 1997)
- *Neuro-fuzzy Systems as Universal Approximators* (Hayashi, 1991)
- *Evolving Fuzzy Systems* (Angelov, 2002)
- *TNFI: Transductive Neuro-fuzzy Inference Systems* (Song and Kasabov, 2005)
- *TWNFI: Transductive Weighted Neuro-fuzzy Inference Systems* (Song and Kasabov, 2006)

6. Population-Generation-Based Methods: Evolutionary Computation

Nature's diversity of species is tremendous. How does mankind evolve in the enormous variety of variants? In other words, how does nature solve the optimisation problem of perfecting mankind? An answer to this question may be found in Charles Darwin's theory of evolution (1858). Evolution is concerned with the development of generations of populations of individuals governed by fitness criteria. But this process is much more complex, as individuals, in addition to what nature has defined for them, develop in their own way: they learn and evolve during their lifetime. This chapter is an attempt to apply the principle of nature-nurture duality to evolving connectionist systems. While improving its performance through adaptive learning, an individual evolving connectionist system (ECOS) can improve (optimise) its parameters and features through evolutionary computation (EC). The chapter is presented in the following sections.

- A brief introduction to EC
- Genetic algorithms (GA) and evolutionary strategies (ES)
- Traditional use of EC as learning and optimisation techniques for ANN
- EC for parameter and feature optimisation of adaptive, local learning models
- EC for parameter and feature optimisation of transductive, personalised models
- Particle swarm intelligence
- Artificial life systems
- Exercise
- Summary and open questions
- Further reading

6.1 A Brief Introduction to EC

Charles Darwin favoured the 'Mendelian heredity' explanation that states that features are passed from generation to generation.

In the early 1800s Jean-Baptiste Lamarck had expounded the view that changes in individuals over the course of their lives were passed on to their progeny. This perspective was adopted by Herbert Spencer and became an established view along with the Darwin's theory of evolution.

The evolution in nature inspired computational methods called evolutionary computation (EC). ECs are stochastic search methods that mimic the behaviour of natural biological evolution. They differ from traditional optimisation techniques in that they involve a search from a population of solutions, not from a single point, and carry this search over generations. So, EC methods are concerned with population-based search and optimisation of individual systems through generations of populations (Goldberg, 1989; Koza, 1992; Holland, 1992, 1998).

Several different types of evolutionary methods have been developed independently. These include genetic programming (GP) which evolves programs, evolutionary programming (EP), which focuses on optimising continuous functions without recombination, evolutionary strategies (ES), which focus on optimising continuous functions with recombination, and genetic algorithms (GAs), which focus on optimising general combinatorial problems, the latter being the most popular technique. EC has been applied thus far to the optimisation of different structures and processes, one of them being the connectionist structures and connectionist learning processes (Fogel *et al.*, 1990; Yao, 1993). Methods of EC include in principle two stages (see Fig. 6.1):

1. A stage of creating new population of individuals
2. A stage of development of the individual systems, so that a system develops and evolves through interaction with the environment, which is also based on the genetic material embodied in the system

The process of individual (internal) development has been ignored or neglected in many EC methods as insignificant from the point of view of the long process of generating hundreds of generations, each of them containing hundreds and thousands of individuals.

But my personal concern as an individual - and also as the author of the book - is that it matters to me not only how much I have contributed to the improvement of the genetic code of the population that is going to live, possibly, 2,000,000 years after me, but also how I can improve myself during my lifetime, and how I evolve as an individual in a particular environment, making the best out of my genetic material.

ECOS deal with the process of interactive off-line or online adaptive learning of a single system that evolves from incoming data. The system can either have

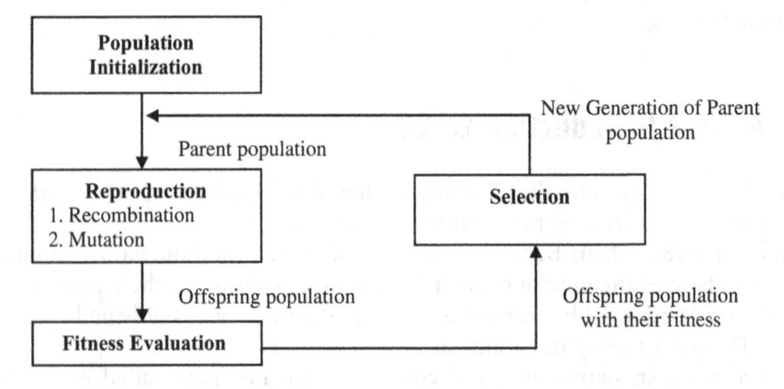

Fig. 6.1 A schematic algorithmic diagram of evolutionary computation (EC).

its parameters (genes) predefined, or it can be self-optimised during the learning process starting from some initial values. ECOS should also be able to improve their performance and adapt better to a changing environment through evolution, i.e. through population-based improvement over generations.

There are several ways in which EC and ECOS can be interlinked. For example, it is possible to use EC to optimise the parameters of an ECOS at a certain time of its operation, or to use the methods of ECOS for the development of the individual systems (individuals) as part of the global EC process.

Before we discuss methods for using EC for the optimisation of connectionist systems, a short introduction to two of the most popular EC techniques - genetic algorithms and evolutionary strategies - is given below.

6.2 Genetic Algorithms and Evolutionary Strategies

6.2.1 Genetic Algorithms

Genetic algorithms were introduced for the first time in the work of John Holland in 1975. They were further developed by him and other researchers (Holland, 1992, 1998; Goldberg, 1989; Koza, 1992). The most important terms used in GA are analogous to the terms used to explain the evolution processes. They are:

- *Gene:* A basic unit that defines a certain characteristic (property) of an individual
- *Chromosome:* A string of genes; used to represent an individual or a possible solution to a problem in the population space
- *Crossover (mating) operation*: Substrings of different individuals are taken and new strings (offspring) are produced
- *Mutation:* Random change of a gene in a chromosome
- *Fitness (goodness) function:* A criterion which evaluates how good each individual is
- *Selection:* A procedure of choosing a part of the population which will continue the process of searching for the best solution, while the other individuals 'die'

A simple genetic algorithm consists of steps shown in Fig. 6.2. The process over time has been 'stretched' in space. Whereas Fig. 6.2 shows graphically how a GA searches for the best solution in the solution space, Fig. 6.3 gives an outline of the GA.

When using the GA method for a complex multioption optimisation problem, there is no need for in-depth problem knowledge, nor is there a need for many data examples stored beforehand. What is needed here is merely a 'fitness' or 'goodness' criterion for the selection of the most promising individuals (they may be partial solutions to the problem). This criterion may require a mutation as well, which is a heuristic approach of trial-and-error type. This implies keeping (recording) the best solutions at each stages.

Many complex optimisation problems find their way to a solution through genetic algorithms. Such problems are, for example, the travelling salesman problem (TSP): finding the cheapest way to visit n towns without visiting a town

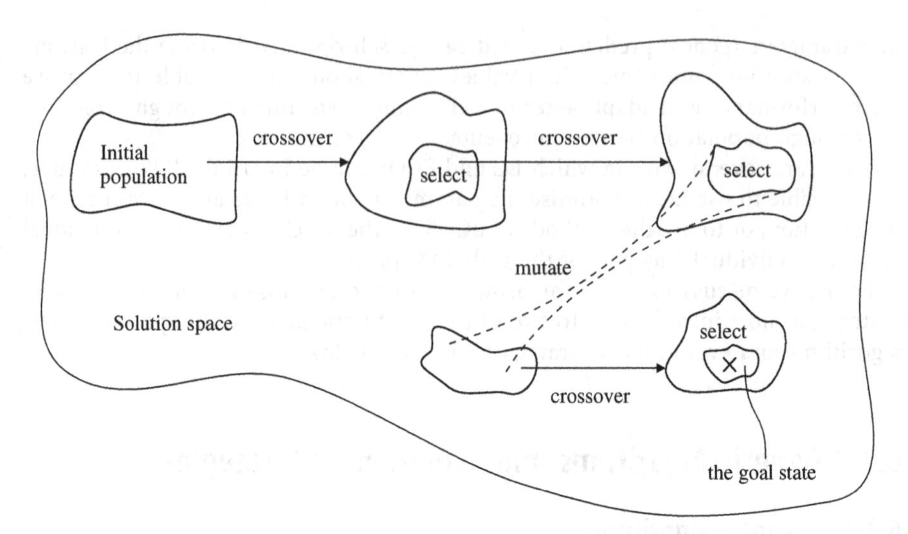

Fig. 6.2 A schematic diagram of how a genetic algorithm (GA) works in time (from Kasabov (1996), ©MIT Press, reproduced with permission).

```
1. Initialize population of possible solutions

2. WHILE a criterion for termination is not reached DO
      {
           2a. Crossover two specimens ("mother and father") and generate new
individuals;
           2b. Select the most promising ones, according to a fitness function;
           2c. Development (if at all);
           2d. Possible mutation (rare) }
      }
```

Fig. 6.3 A general representation of the GA (from Kasabov (1996), ©MIT Press, reproduced with permission).

twice; the min cut problem: cutting a graph with minimum links between the cut parts; adaptive control; applied physics problems; optimisation of the parameters of complex computational models; optimisation of neural network architectures (Fogel *et al.*, 1990); and finding fuzzy rules and membership functions (Furuhashi *et al.*, 1994).

The main issues in using genetic algorithms relate to the choice of genetic operations (crossover, selection, mutation). In the case of the travelling salesman problem the crossover operation can be merging different parts of two possible roads ('mother' and 'father' roads) until new usable roads are obtained. The criterion for the choice of the most prospective ones is minimum length (or cost).

A GA offers a great deal of parallelism: each branch of the search tree for a best individual can be utilised in parallel with the others. That allows for an easy realisation on parallel architectures. GAs are search heuristics for the 'best' instance in the space of all possible instances. A GA model requires the specification of the following features.

- *Encoding scheme*: How to encode the problem in terms of the GA notation: what variables to choose as genes, how to construct the chromosomes, etc.
- *Population size:* How many possible solutions should be kept in a population for their performance to be further evaluated
- *Crossover operations:* How to combine old individuals and produce new, more prospective ones
- *Mutation heuristic:* When and how to apply mutation

In short, the major characteristics of a GA are the following. They are heuristic methods for search and optimisation. In contrast to the exhaustive search algorithms, GAs do not evaluate all variants in order to select the best one. Therefore they may not lead to the perfect solution, but to one which is closest to it taking into account the time limits. But nature itself is imperfect too (partly due to the fact that the criteria for perfection keep changing), and what seems to be close to perfection according to one 'goodness' criterion may be far from it according to another.

6.2.2 Selection, Crossover, and Mutation Operators in EC

The theory of GA and the other EC techniques includes different methods for selection of individuals from a population, different crossover techniques, and different mutation techniques.

Selection is based on fitness that can employ several strategies. One of them is proportional fitness, i.e. 'if A is twice as fit as B, A has twice the probability of being selected.' This is implemented as roulette wheel selection and gives chances to individuals according to their fitness evaluation as shown in an example in Fig. 6.4.

Other selection techniques include tournament selection (e.g. at every time of selection the roulette wheel is turned twice, and the individual with the highest fitness is selected), rank ordering, and so on (Fogel *et al.*, 1990). An important feature of the selection procedure is that fitter individuals are more likely to be selected.

The selection procedure may also involve keeping the best individuals from previous generations (if this principle was used by nature, Leonardo Da Vinci would still be alive, as he was one of the greatest artists ever, presumably having the best artistic genes). This operation is called *elitism.*

After the best individuals are selected from a population, a crossover operation is applied between these individuals. The crossover operator defines how individuals (e.g. 'mother' and 'father') exchange genes when creating the offspring. Different crossover operations can be used, such as one-point crossover (Fig. 6.5), two-point crossover, etc.

Individual	1	2	3	4	5	6	7	8	9	10
Fitness	2.0	1.8	1.6	1.4	1.2	1.0	0.8	0.6	0.4	0.2
Selection Chance	0.18	0.16	0.15	0.13	0.11	0.09	0.07	0.06	0.03	0.02

Fig. 6.4 An example of a roulette selection strategy. Each of the ten individuals has its chance to 'survive' (to be selected for reproduction) based on its evaluated fitness.

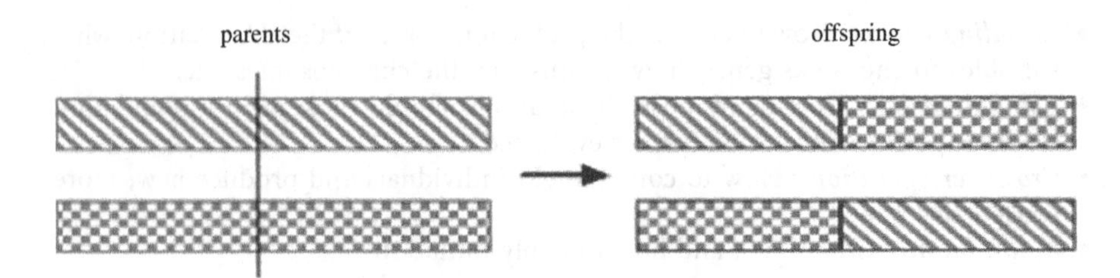

Fig. 6.5 One-point crossover operation.

Mutation can be performed in several ways, e.g.

- For a binary chromosome, just randomly 'flip' a bit (a gene allele).
- For a more complex chromosome structure, randomly select a site, delete the structure associated with this site, and randomly create a new substructure.

Some EC methods just use mutation and no crossover ('asexual reproduction'). Normally, however, mutation is used to search in a 'local search space', by allowing small changes in the genotype (and therefore it is hoped in the phenotype) as it is in the evolutionary strategies.

6.2.3 Evolutionary Strategies (ES)

Another EC technique is called evolutionary strategies (ES). These techniques use only one chromosome and a mutation operation, along with a fitness criterion, to navigate in the solution (chromosomal) space.

In the reproduction phase, the current population, called the parent population, is processed by a set of evolutionary operators to create a new population called the offspring population. The evolutionary operators include two main operators: mutation and recombination; both imitate the functions of their biological counterparts. Mutation causes independent perturbation to a parent to form an offspring and is used for diversifying the search. It is an asexual operator because it involves only one parent. In GA, mutation flips each binary bit of a parent string at a small, independent probability p_m (which is typically in the range [0.001, 0.01]) to create an offspring. In ES, mutation is the addition of a zero-mean Gaussian random number to a parent individual to create the offspring. Let $\mathbf{s}_{PA}$ and $\mathbf{s}_{OF}$ denote the parent and offspring vector; they are related through the Gaussian mutation

$$\mathbf{s}_{OF} = \mathbf{s}_{PA} + \mathbf{z} \qquad \mathbf{z} \sim N(0, s) \tag{6.1}$$

where $N(a, s)$ represents a normal (Gaussian) distribution with a mean a and a covariance s and $\sim$ denotes sampling from the corresponding distribution.

ES uses the mutation as the main search operator.

The selection operator is probabilistic in GA and deterministic in ES. Many heuristic designs, such as the rank-based selection that assigns to the individuals a survival probability proportional (or exponentially proportional) to their ranking, have also been studied. The selected individuals then become the new generation of parents for reproduction. The entire evolutionary process iterates until some stopping criteria are met. The process is essentially a Markov chain; i.e. the outcome of one generation depends only on the last. It has been shown that under certain design criteria of the evolutionary operators and selection operator, the average fitness of the population increases and the probability of discovering the global optimum tends towards unity. The search could, however, be lengthy.

6.3 Traditional Use of EC for Learning and Optimisation in ANN

Before we present in the next section the use of EC for the optimisation of ECOS, here a brief review of different approaches for using EC for the optimisation of MLP and other traditional, not evolving, ANN models is presented. Such reviews have been published by several authors (Yao, 1993; Fogel *et al.*, 1990; Watts and Kasabov, 1998, 1999).

6.3.1 ANN Topology Determination by GA

The number of layers within a connectionist structure, e.g. MLP, and the number of nodes within each layer can often have a significant effect upon the performance of the network. Too many nodes in the hidden layers of the network may cause the network to overfit the training data, whereas too few may reduce its ability to generalize.

In Schiffman et al. (1993) the length of the chromosome determined the number of nodes present, as well as the connectivity of the nodes. This approach to ANN design was tested in the above paper on a medical classification problem, that of identifying thyroid disorders and provided networks that were both smaller and more accurate than manually designed ANNs were.

6.3.2 Selection of ANN Parameters by GA

In addition to the selection of the ANN topology, it is also possible to select the training parameters for the network. This has been investigated by many authors: see for example Choi and Bluff (1995), who used a GA to select the training parameters for the backpropagation training of an MLP. In this work the chromosome encoded the learning rate, momentum, sigmoid parameter, and number of training epochs to be used for the backpropagation training of the network. This technique was tested with several different datasets, including bottle classification data, where a glass bottle is classified as either being suitable for reuse or suitable for recycling, and breast cancer data, which classifies tissue samples as malignant or benign. For each of the test datasets, the genetically derived network outperformed those networks whose control parameters were manually set, often by a significant margin.

6.3.3 Training of ANN via GA

For some problems, it is actually more efficient to abandon the more conventional training algorithms, such as backpropagation, and train the ANN via GA. For some problems GA training may be the only way in which convergence can be reached. GA-based training of ANNs has been extensively investigated by Hugo de Garis (1989, 1993). The networks used in these experiments were not of the MLP variety, but were instead fully self-connected synchronous networks. These 'GenNets' were used to attempt tasks that were time-dependent, such as controlling the walking action of a pair of stick legs. With this problem the inputs to the network were the current angle and angular velocity of the 'hip' and 'knee' joints of each leg. The outputs were the future angular acceleration of each of the joints. Both the inputs and outputs for this problem were time-dependent.

Conventional training methods proved to be incapable of solving this problem, whereas GenNets solved it very easily. Other applications of GenNets involved creating 'production rule' GenNets that duplicated the function of a production rule system. These were then inserted into a simulated artificial insect and used to process inputs from sensor GenNets. The outputs of the production rule GenNets sent signals to other GenNets to execute various actions, e.g. eat, flee, or mate.

A similarly structured recurrent network was used in Fukuda *et al.* (1997) to attempt a similar problem. The application area in this research was using the genetically trained network to control a physical biped robot. The results gained from this approach were quite impressive. Not only was the robot able to walk along flat and sloped surfaces, it was able to generalise its behaviour to deal with surfaces it had not encountered in training. Comparison of the results gained from the genetically trained network with those from networks trained by other methods showed that not only did the genetically trained network train more efficiently than the others, it was also able to perform much better than the others.

6.3.4 Neuro-Fuzzy Genetic Systems

GAs have been used to optimise membership functions and other parameters in neuro-fuzzy systems (Furuhashi et al., 1994; Watts and Kasabov, 1998, 1999). An example is the FuNN fuzzy neural network (Kasabov *et al.*, 1997). It is essentially an ANN that has semantically meaningful nodes and weights that represent input and output variables, fuzzy membership functions, and fuzzy rules. Tuning the membership functions in FuNNs is a technique intended to improve an already trained network. By slightly shifting the centres of the MFs the overall performance of the network can be improved. Because of the number of MFs in even a moderately sized network, and the degree of variation in the magnitude of the changes that each MF may require, a GA is the most efficient means of achieving the optimisation.

Much of the flexibility of the FuNN model is due to the large number of design parameters available in creating a FuNN. Each input and output may have an arbitrary number of membership functions attached. The number of combinations that these options yield is huge, making it quite impractical to search for the optimal configuration of the FuNN combinatorially. Using a GA is one method of solving this difficult problem.

Optimisation of FuNN MFs involves applying small delta values to each of the input and output membership functions. Optimisation of conventional fuzzy systems by encoding these deltas into a GA structure has been investigated (Furuhashi *et al.*, 1994) and has been shown to be more effective than manual tuning. The initial GA population is randomly initialised except for one chromosome, which has all the encoded delta values set to zero to represent the initial network. This, along with elitism, ensures that the network can only either improve in performance or stay the same, and never degrade in performance. To evaluate each individual, the encoded delta values are added to the centre of each membership function and the recall error over the training datasets is calculated. In situations where a small number of examples of one class could be overwhelmed by large numbers of other classes, the average recall error is taken over several datasets, with each dataset containing examples from one class. The fitness f of an individual is calculated by the following formula.

$$f = 1/e \tag{6.2}$$

where e is the average overall error on the test datasets.

6.3.5 Evolutionary Neuro-Genetic Systems and Cellular Automata

Other research methods combine GA and cellular automata to 'grow' cells (neurons) in a cellular automaton (de Garis, 1993). Here a cellular automaton (CA) is a simplified simulation of a biological cell, with a finite number of states, whose behaviour is governed by rules that determine its future state based upon its current state and the current state of its neighbours. CAs have been used to model such phenomena as the flocking behaviour of birds and the dynamics of gas molecules. The advantage of CAs is their ability to produce seemingly complex dynamic behaviour from a few simple rules.

Applying EC methods in an off-line mode on a closed and compact problem space of possible solutions, through generations of populations of possible solutions from this space, is time-consuming and not practical for realtime online applications. The next sections suggest methods that overcome this problem and optimise the parameters of ECOS.

6.4 EC for Parameter and Feature Optimisation of ECOS

When an ECOS is evolved from a stream of data we can expect that for a better performance of the system, along with the structural and the functional evolution of the model, its different parameters should also evolve. This problem has been discussed in numerous papers and methods have been introduced (Watts and Kasabov, 2001, 2002; Kasabov *et al.*, 2003; Minku and Ludemir, 2005).

In this and in the next two sections, we present some other methods for the optimisation of the parameters and the structures of ECOS.

6.4.1 ES for Incremental Learning Parameter Optimisation

Here, a method for adaptive incremental (possibly, online) optimisation of the parameters of a learning model using ES is presented on a simple example of optimising the learning parameters l_1 and l_2 of the two weight connection matrices $W1$ and $W2$ respectively, in an EFuNN ECOS architecture (see Chapter 3; from Chan and Kasabov (2004)).

Each individual $s^{(k)} = (s_1^{(k)}, s_2^{(k)})$ is a two-vector solution to l_1 and l_2. Because there are only two parameters to optimise, the ES method requires only a small population and a short evolutionary time. In this case, we use one parent and ten offspring and a small number of generations of $gen_{max} = 20$. We set the initial values of s to (0.2,0.2) in the first run and use the previous best solution as initial values for subsequent runs. Using a nonrandomised initial population encourages a more localised optimisation and hence speeds up convergence. We use simple Gaussian mutation with standard deviation 0.1 (empirically determined) for both parameters to generate new points. For selection, we use the high selection pressure scheme to accelerate convergence, which picks the best of the joint pool of parents and offspring to be the next-generation parents.

The fitness function (or the optimisation objective function) is the prediction error over the last n_{last} data, generated by using EFuNN model at $(t - n_{last})$ to perform incremental learning and predicting over the last n_{last} data. The smaller n_{last} is, the faster the learning rates adapts and vice versa. Because the effect of changing the learning rates is usually not expressed immediately but after a longer period, the fitness function can be noisy and inaccurate if n_{last} is too small. In this work we set $n_{last} = 50$. The overall algorithm has five steps (Fig. 6.6).

To verify the effectiveness of the proposed online ES, we first train EFuNN (see Chapter 3) with the first 500 data of the Mackey–Glass series (using $x(0) = 1.2$ and $\tau = 17$) to obtain a stable model, and then apply online ES to optimise the learning

1 Population Initialisation. Reset generation counter *gen*. Initialise a population of *p* parent EFuNNs with the best estimate (l_1', l_2') for the two learning rates of EFuNN:

$$S_{PA}{}^{(k)} = (l_1', l_2'), K = \{1, 2, \ldots, p\}$$

2 Reproduction. Rondomly select one of the *P* parents, $S_{PA}{}^{(r)}$, to undergo Gaussian mutation defined as a normal distribution function *N* to produce a new offspring

$$S_{OF}^{(i)} = S_{PA}^{(r)} + Z^{(i)} \quad \text{Where } Z^{(i)} \sim N(0, \sigma^2), i = \{1, 2, \ldots \lambda\}$$

3 Fitness Evaluation. Apply each of the offspring EFuNN models to perform incremental learning and prediction using data in the interval $[t - t_{last}, t]$, where it is the current time moment of the learned temporal process, and t_{last} is the last moment of the parameter measure and optimisation in the past. Set the respective prediction error as fitness.

4 Selection. Perform selection

5 Termination. Increment the number of generations (*gen*). Stop if $gen \geq gen_{max}$ or if no fitness improvement has been recorded over the past 3 generations; otherwise go to step (2).

Fig. 6.6 An ES algorithm for online optimisation of two parameters of an ECOS, in this case the learning rates l_1 and l_2 of an EFuNN (see Chapter 3).

rates over the next 500 data. The corresponding prediction error is recorded. Example results are shown in Figure 6.7a,b.

Figure 6.7a shows an example of the evolution of the RMSE (used as the fitness function) and the learning rates. The RMSE decreases, which is a characteristic of the selection because the best individual is always kept in the population. The optimal learning rates are achieved quickly after 14 generations.

Figure 6.7b shows the dynamics of the learning rate over the period $t = [500, 1000]$. Both learning rates l_1 and l_2 vary considerably over the entire course with only short stationary moments, showing that they are indeed dynamic parameters. The average RMSE for online prediction one-step-ahead obtained with and without online EC are 0.0056 and 0.0068, respectively, showing that online EC is effective in enhancing EFuNN's prediction performance during incremental learning.

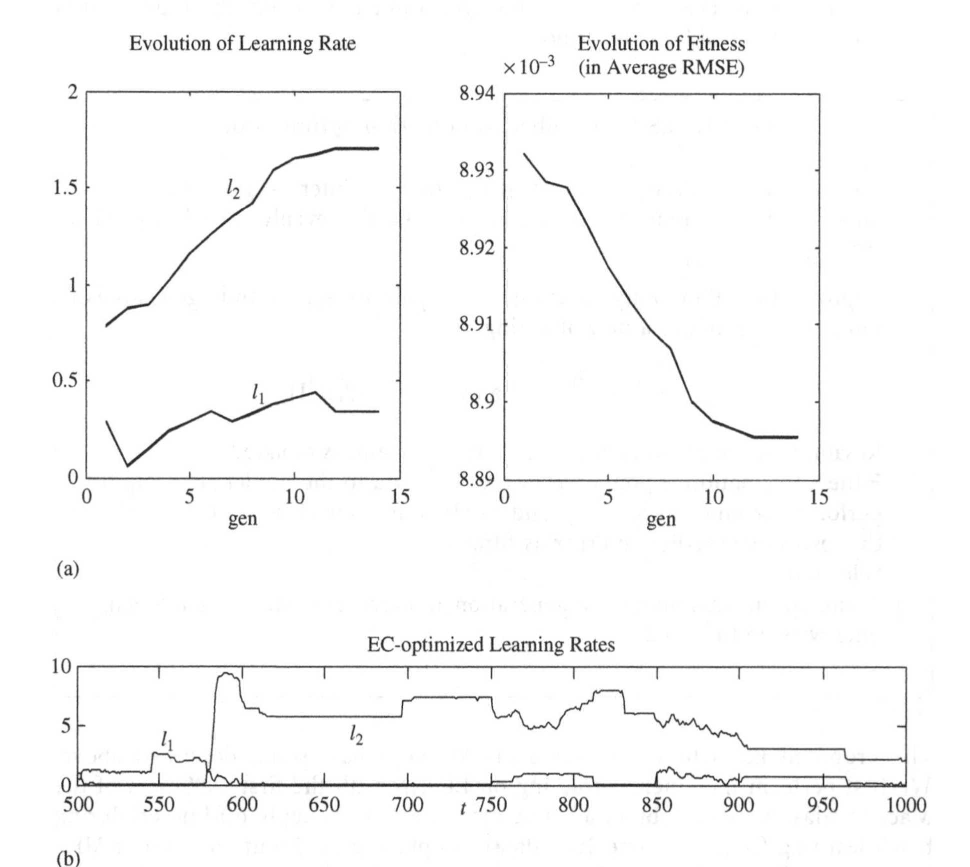

Fig. 6.7 (a) Evolution of the best fitness and the learning rates over 15 generations; (b) optimised learning rates l_1 and l_2 of an EFuNN ECOS over a period of time $t = [500, 1000]$.

6.4.2 ES for Fuzzy Membership Function Optimisation in ECOS

As an example, here, we apply ES to optimise the fuzzy input and output membership functions (MFs) at the second and fourth layers of an EFuNN (see Chapter 3) with the objective of minimising the training error (Chan and Kasabov, 2004). For both the input and output MFs, we use the common triangular function, which is completely defined by the position of the MF centre.

Given that there are p input variables and p_{MF} fuzzy quantisation levels for each input variable, m output variables and m_{MF} fuzzy quantisation levels for each output variable, there are $n_c = (p \times p_{\text{MF}} + m \times m_{\text{MF}})$ centres $c = [c_1, c_2, \ldots, c_{nc}]$ to be optimised.

The ES represents each individual as a $(p \times p_{\text{MF}} + m \times m_{\text{MF}})$ real vector solution to the positions of the MFs. We use five parents and 20 offspring and a relatively larger number of generations of $\text{gen}_{\text{max}} = 40$. Each individual of the initial population is a copy of the evenly distributed MFs within the boundaries of the variables. Standard Gaussian mutation is used for reproduction. Every offspring is checked for the membership hierarchy constraint; i.e. the value of the higher MF must be larger than that of the lower MF. If the constraint is violated, the individual is resampled until a valid one is found.

Box 6.1. ES for membership function optimisation

1) Population Initialisation. Reset generation counter - *gen*. Initialise a population of parents $[c_1', c_2', \ldots, c_{n_c}']$ with the evenly distributed MFs $\mathbf{s}_{\text{PA}}^{(k)} = [c_1', c_2', \ldots, c_{n_c}']$
2) Reproduction. Randomly select one of the parents, $s_{\text{PA}}^{(r)}$, to undergo Gaussian mutation to produce a new offspring

$$\mathbf{s}_{\text{OF}}^{(i)} = \mathbf{s}_{\text{PA}}^{(r)} + z^{(i)} \text{ and } z^{(i)} \sim N\left(0, \sigma^2 \text{I}\right)$$

   Resample if the membership hierarchy constraint is violated.
3) Fitness Evaluation. Apply each of the offspring to the model at $(t - t_{\text{last}})$ to perform incremental learning and prediction using data in $[t - t_{\text{last}}, t]$. Set the respective prediction error as fitness.
4) Selection.
5) Termination. Increment the generation number gen. Stop if $\text{gen} \geq \text{gen}_{\text{max}}$, otherwise go to step 2.

The proposed ES is tested on the same Mackey–Glass series described above. We first perform incremental training on EFuNN with the first 1000 data of the Mackey–Glass series to obtain a stable model, and then apply off-line ES during batch learning (over the same 1000 data) to optimise the input and output MFs. Example results are shown in Figure 6.8.

Figure 6.8a shows the evolution of the best fitness recorded in each generation. The unidirectional drop in prediction error shows that the optimisation of MFs has

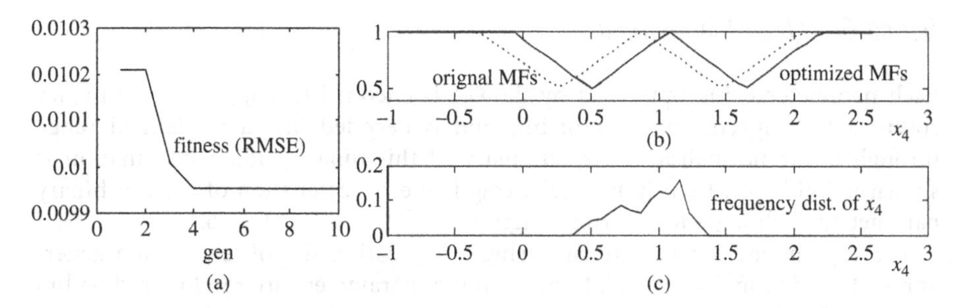

Fig. 6.8 (a) The evolution of the best fitness from the off-line ES; (b) initial membership functions and EC-optimised membership functions on an input variable; (c) the frequency distribution of the same input variable as in (b).

a positive impact on improving model performance. Figure 6.8b shows the initial MFs and the EC-optimised MFs and Figure 6.8c shows the frequency distribution of one of the input variables $\{x_4^{(t)}\}$. Clearly, the off-line ES evolves the MFs towards the high-frequency positions, which maximises the precision for fuzzy quantisation and in turn yields higher prediction accuracies. The training RMSEs are 0.1129 and 0.1160 for EFuNN with and without off-line ES optimisation, respectively, showing that the ES-optimised fuzzy MFs are effective in improving EFuNN's data tracking performance.

6.4.3 EC for Integrated Parameter Optimisation and Feature Selection in Adaptive Learning Models

As discussed in Chapter 3, a simple version of EFuNN–ECF (evolving classifier function), can be applied in both online and off-line modes. When working in an off-line mode, ECF requires accurate setting of several control parameters to achieve optimal performance. However, as with other ANN models, it is not always clear what the best values for these parameters are. EC provides a robust global optimisation method for choosing values for these parameters.

In this work, EC is applied to optimise the following four parameters of an ECF model:

- R_{max}: The maximum radius of the receptive hypersphere of the rule nodes. If, during the training process, a rule node is adjusted such that the radius of its hypersphere becomes larger than R_{max} then the rule node is left unadjusted and a new rule node is created.
- R_{min}: The minimum radius of the receptive hypersphere of the rule nodes. It becomes the radius of the hypersphere of a new rule node.
- n_{MF}: The number of membership functions used to fuzzify each input variable.
- M-of-n: If no rules are activated when a new input vector is entered, ECF calculates the distance between the new vector and the M closest rule nodes. The average distance is calculated between the new vector and the rule nodes of each class. The vector is assigned the class corresponding to the smallest average distance.

Testing GA-ECF on the Iris Data

Each parameter being optimised by the GA is encoded through standard binary coding into a specific number of bits and is decoded into a predefined range through linear normalisation. A summary of this binary coding information is shown in Table 6.1. Each individual string is the concatenation of a set of binary parameters, yielding a total string-length $l_{tot} = (5+5+3+3) = 16$.

The experiments for GA are run using a population size of ten, for ten generations. For each individual solution, the initial parameters are randomised within the predefined range. Mutation rate p_m is set to $1/l_{tot}$, which is the generally accepted optimal rate for unimodal functions and the lower bound for multimodal functions, yielding an average of one bit inversion per string. Two-point crossover is used. Rank-based selection is employed and an exponentially higher probability of survival is assigned to high-fitness individuals. The fitness function is determined by the classification accuracy. In the control experiments, ECF is performed using the manually optimised parameters, which are $R_{max} = 1$, $R_{min} = 0.01$, $n_{MF} = 1$, M-of-$N = 3$. The experiments for GA and the control are repeated 50 times and between each run the whole dataset is randomly split such that 50% of the data is used for training and 50% for testing. Performance is determined as the percentage of correctly classified test data. The statistics of the experiments are presented in Table 6.2.

The results show that there is a marked improvement in the average accuracy in the GA-optimised network. Also the standard deviation of the accuracy achieved by the 50 GA experiments is significantly less than that for the control experiments, indicating that there is a greater consistency in the experiments using the GA.

Figure 6.9 shows the evolution of the parameters over 40 generations. Each parameter converges quickly to its optimal value within the first 10 generations, showing the effectiveness of the GA implementation.

Table 6.1 Each parameter of the ECF being optimised by the GA is encoded through standard binary coding into a specific number of bits and is decoded into a predefined range through linear normalisation.

Parameter	*Range*	*Resolution*	*Number of Bits*
R_{max}	0.01–0.1	2.81e-3	5
R_{min}	0.11–0.8	2.16e-2	5
n_{MF}	1–8	1	3
M-of-N	1–8	1	3

Table 6.2 The results of the GA experiment repeated 50 times and averaged and contrasted with a control experiment.

	Average accuracy (%)	*Standard dev.*
Evolving Parameters	97.96	1.13
Control	93.68	2.39

Figure 6.10 shows the process of GA optimization of the ECF parameters and features on the case study example of cancer outcome prediction based on clinical variable (#1) and 11 gene expression variables across 56 samples (dataset is from Ship *et al.* 2002).

6.4.4 EC for Optimal Feature Weighting in Adaptive Learning Models

Feature weighting is an alternative form of feature selection. It assigns to each variable a weighting coefficient that reduces or amplifies its effect on the model based on its relevance to the output. A variable that has little relevance to the output is given a small weight to suppress its effect, and vice versa. The purpose of feature weighting is twofold: first, to protect the model from the random and perhaps detrimental influence of irrelevant variables and second, to act as a guide for pruning away irrelevant variables (feature selection) by the size of the weighting coefficients. Feature weighting/selection are generally implemented through three classes of methods: Bayesian, incremental/sequential, or stochastic methods. The first two classes are local methods that are fast and yet susceptible to local optima; the last class includes EC applications that use computationally intensive population search to search for the global optimum. Here the algorithm proposed for ECOS, called weighted data normalisation (WDN; Song and Kasabov (2006)) belongs to such a class, using the robust optimisation capability of the genetic algorithm to implement feature weighting.

The WDN method optimises the normalisation intervals (range) of the input variables and allocates weights to each of the variables from a dataset. The method consists of the following steps.

1. The training data are preprocessed first by a general normalisation method. There are several ways to achieve this: (a) normalising a given dataset so that they fall in a certain interval, e.g. [0, 1], [0, 255] or [−1, 1] etc; (b) normalising the dataset so that the inputs and targets will have means of zero

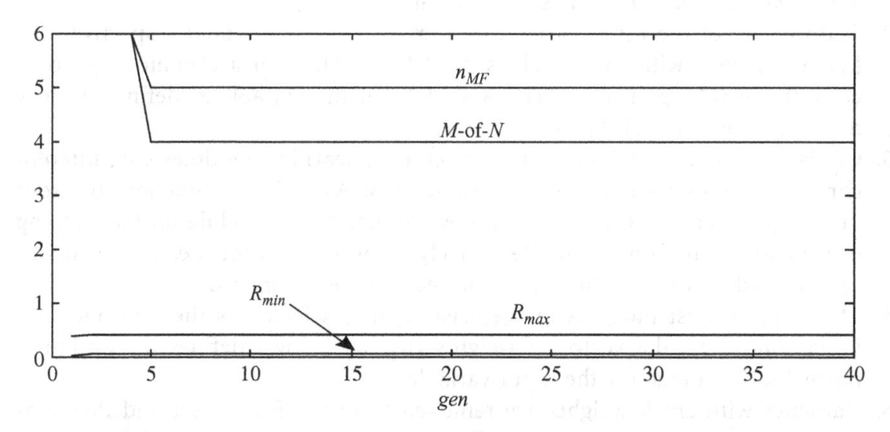

Fig. 6.9 Evolution of the parameters: number of membership functions n_{MF}; m-of-n; Rmax and Rmin of an ECF evolving model over 40 generations of a GA optimisation procedure.

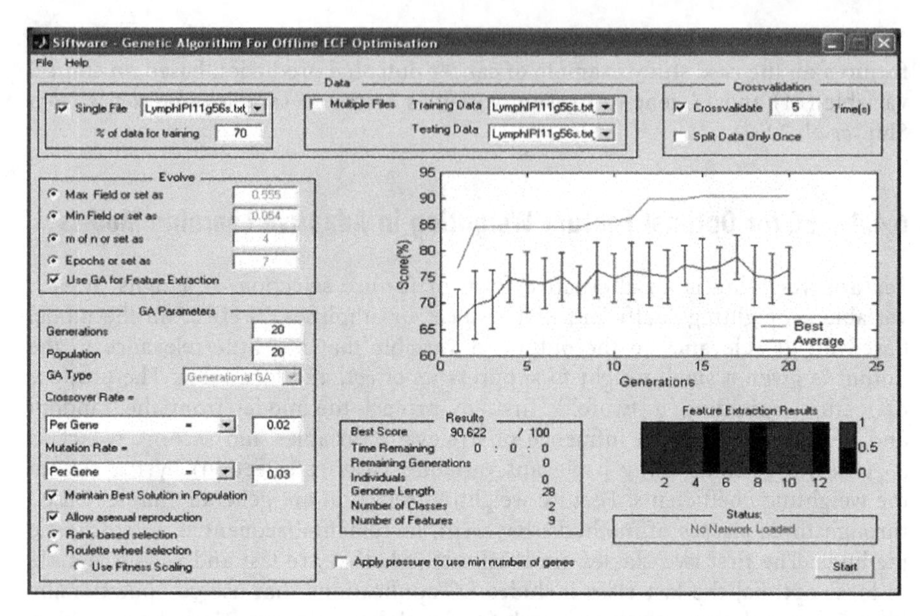

Fig. 6.10 GA for ECF parameter and feature evaluation on the case study example of lymphoma cancer outcome prediction (Ship *et al.*, 2002) based on a clinical variable (#1, IPI, International Prognostic Index) and 11 gene expression variables (#2–12) across 56 samples. Twenty generations are run over populations of 20 individual ECF models, each trained five times (five-cross validation) on 70% of the data selected for training and 30% for testing. The average accuracy of all five validations is used as a fitness function. The accuracy of the best model evolves from 76% (the first generation) to 90.622% due to optimization of the ECF parameters (Rmin, Rmax, m-of-n, number of training iterations). The best model uses 9 input variables instead of 12 (variables 5, 8, and 12 are not used).

and standard deviations of 1; (c) normalising the dataset so that the deviation of each variable from its mean is normalised by its standard deviation. In the WDN, we normalise the dataset in the interval [0, 1].

2. The weights of the input variables $[x_1, x_2, \ldots, x_n]$ represented respectively by $[w_1, w_2, \ldots, w_n]$ with initial values of [1,1, ..., 1], form a chromosome for a consecutive GA application. The weight w_i of the variable x_i defines its new normalisation interval $[0,w_i]$.
3. GA is run on a population of connectionist learning modules for different chromosome values over several generations. As a fitness function, the root mean square error (RMSE) of a trained connectionist module on the training or validation data is used, or alternatively the number of the created rule nodes can be used as a fitness function that needs to be minimised.
4. The connectionist model with the least error is selected as the best one, and its chromosome, the vector of weights $[w_1, w_2, \ldots, w_n]$ defines the optimum normalisation range for the input variables.
5. Variables with small weights are removed from the feature set and the steps from above are repeated again to find the optimum and the minimum set of variables for a particular problem and a particular connectionist model.

The above WDN method is illustrated in the next section on two case study ECOS and on two typical problems, namely EFuNN for a time-series prediction, and ECMC for classification.

Example 1: Off-Line WDN of EFuNN for Prediction

Here an EFuNN model is developed for a time-series prediction problem. Improved learning with the WDN method is demonstrated on the Mackey–Glass (MG) time-series prediction task. In the experiments, 1000 data points, from $t = 118$ to 1117, are extracted for predicting the six-steps-ahead output value. The first half of the dataset are taken as the training data and the rest as the testing data.

The following parameters are set in the experiments for the EFuNN model: $R_{max} = 0.15$; $E_{max} = 0.15$ and $n_{MF} = 3$. The following GA parameter values are used: for each input variable, the values from 0.16 to 1 are mapped onto a four-bit string; the number of individuals in a population is 12; mutation rate is 0.001; termination criterion (the maximum epochs of GA operation) is 100 generations; the RMSE on the training data is used as a fitness function. The optimised weight values, the number of the rule nodes created by EFuNN with such weights, the training and testing RMSE, and the control experiments are shown in Table 6.3.

With the use of the WDN method, better prediction results are obtained for a significantly smaller number of rule nodes (clusters) evolved in the EFuNN models. This is because of the better clustering achieved when different variables are weighted according to their relevance.

Example 2: Off-Line WDN Optimisation and Feature Extraction for ECMC

In this section, an evolving clustering method for classification ECMC (see Chapter 2) with WDN is applied to the Iris data for both classification and feature weighting/selection. All experiments in this section are repeated 50 times with the same parameters and the results are averaged. Fifty percent of the whole dataset is randomly selected as the training data and the rest as the testing data. The following parameters are set in the experiments for the ECMC model: $R_{min} = 0.02$; each of the weights for the four normalised input variables is a value from 0.1 to 1 and is mapped into a six-bit binary string.

The following GA parameters are used: number of individuals in a population 12; mutation rate $p_m = 0.005$; termination criterion (the maximum epochs of

Table 6.3 Comparison between EFuNN without weighted data normalisation (WDN) and EFuNN with WDN.

	Normalisation weights	*Number on rule nodes*	*Training RMSE*	*Testing RMSE*
EFuNN without WDN	1, 1, 1, 1	87	0.053	0.035
EFuNN with WDN	0.4, 0.8, 0.28, 0.28	77	0.05	0.031

Table 6.4 Comparison between evolving clustering method for classification ECMC (see Chapter 2) without weighted data normalisation (WDN) and ECMC with WDN.

	Normalised feature weights	*Number of rule nodes*	*Number of test errors*
4 Inputs without WN	1, 1, 1, 1	9.8	3.8
4 Inputs with WN	0.25, 0.44, 0.73, 1	7.8	3.1
3 Inputs without WN	1, 1, 1	8.8	3.7
3 Inputs with WN	0.50, 0.92, 1	8.1	2.9
2 Inputs without WN	1,1	7.7	3.2
2 Inputs with WN	1, 0.97	7.4	3.1

GA operation) 50; the fitness function is determined by the number of created rule nodes.

The final weight values, the number of rule nodes created by ECMC, and the number of classification errors on the testing data, as well as the control experiment are shown in the first two rows of Table 6.4, respectively. Results show that the weight of the first variable is much smaller than the weights of the other variables. Now using the weights as a guide to prune away the least relevant input variables, the same experiment is repeated without the first input variable. As shown in the subsequent rows of Table 6.4, this pruning operation slightly reduces test errors. However, if another variable is removed (i.e. the total number of input variables is two) test error increases. So we conclude that for this particular application the optimum number of input variables is three.

6.5 EC for Feature and Model Parameter Optimisation of Transductive Personalised (Nearest Neighbour) Models

6.5.1 General Notes

In transductive reasoning, for every new input vector $\mathbf{x}_i$ that needs to be processed for a prognostic/classification task, the Ni nearest neighbors, which form a data subset Di, are derived from an existing dataset D and a new model Mi is dynamically created from these samples to approximate the function in the locality of point x_i only. The system is then used to calculate the output value y_i for this input vector $\mathbf{x}_i$ (Vapnik (1998); also see Chapter 1).

The transductive approach has been implemented in medical decision support systems and time-series prediction problems, where individual models are created for each input data vector (i.e. specific time period or specific patient). The approach gives good accuracy for individual models and has promising applications especially in medical decision support systems. This transductive approach has also been applied using support vector machines as the base model in the area of bioinformatics and the results indicate that transductive inference performs better than inductive inference models mainly because it exploits the structural information of unlabelled data. However, there are a few open questions that need to be addressed when implementing transductive modelling (Mohan and Kasabov, 2005).

6.5.2 How Many Nearest Neighbours Should Be Selected?

A standard approach adopted to determine the number of nearest neighbours is to consider a range starting with 1, 2, 5, 10, 20, and so on and finally select the best value based on the classifier's performance. In the presence of unbalanced data distribution among classes in the problem space recommended value of nearest neighbours is to range from 1 to a maximum of number of samples in the smaller class or the square root of the number of samples in the problem space. A similar recommendation is made by Duda and Hart (1973) based on concepts of probability density estimation of the problem space. They suggest that the number of nearest neighbours to consider depends on two important factors: distribution of sample proportions in the problem space and the relationship between the samples in the problem space measured using covariance matrices.

The problem of identifying the optimal number of neighbours that improves the classification accuracy in transductive modeling remains an open question that is addressed here with the use of a GA.

6.5.3 What Distance Measure to Use and in What Problem Space?

There exist different types of distance measures that can be considered to measure the distance of two vectors in a different part of the problem/feature space such as Euclidean, Mahalanobis, Hamming, cosine, correlation, and Manhattan distance among others (see Chapter 2). It has been proved mathematically that using an appropriate distance metric can help reduce classification error when selecting neighbours without increasing number of sample vectors. Hence it is important to recognise which distance measure will best suit the data in hand.

Some authors suggest that in the case where the dataset consists of numerical data, the Euclidean distance measure should be used when the attributes are independent and commensurate with each other. However, in the case where the numerical data are interrelated, then Mahalanobis distance should be considered as this distance measure takes interdependence between the data into consideration.

If the data consist of categorical information, Hamming distance can appropriately measure the difference between categorical data. Also, in the case where the dataset consists of a combination of numerical and categorical values, for example, a medical dataset that includes the numerical data such as gene expression values and categorical data such as clinical attributes, then the distance measure considered could be the weighted sum of Mahalanobis or Euclidean for numerical data and Hamming distance for nominal data.

Keeping these suggestions in perspective, it is important to provide a wide range of options to select the distance measure based on the type of dataset in a particular part of the problem space for a particular set of features.

6.5.4 How is the Input Vector Assigned an Output Value?

There are different ways of determining the contribution of the nearest neighbours to the output class of the input vector, already discussed in Chapter 1, such as:

- The K-nearest neighbours (K-NN)
- WKNN
- WWKNN
- Other methods, such as MLR, ECOS, etc.

6.5.5 What Features are Important for a Specific Input Vector?

We discussed the issue of feature selection in Chapter 1, but here we raise the issue of which feature set Fi is most appropriate to use for each individual vector $\mathbf{x}_i$

6.5.6 A GA Optimization of Transductive Personalised Models

The proposed algorithm aims to answer all these questions by applying a transductive modeling approach that uses GA to define the optimal: (a) distance measure, (b) number of nearest neighbours, and (c) feature subset for every new input vector (Mohan and Kasabov, 2005).

The number of neighbours to be optimized lies in the minimum range of the number of features selected in the algorithm and a maximum of the number of samples available in the problem space.

As an illustration the multiple linear regression method (MLR) is used as the base model for applying the transductive approach. The model is represented by a linear equation which links the features/variables in the problem space to the output of the classification task and is represented as follows:

$$r = w0 + w1X1 + \ldots + wnXn,$$

where r represents the output, and wi represents the weights for the features/variables of the problem space which are calculated using the least square method. The descriptors Xi are used to represent the structural information of the data samples, that is, the features/variables, and n represents the number of these features/variables. The reason for selecting the MLR model is the simplicity of this model that will make the comparative analysis of the transductive approach using GA with the inductive approach easier to understand and interpret.

The main objective of this algorithm is to develop an individualized model for every data vector in a semi-supervised manner by exploiting the data vector's structural information, to identify its nearest neighbours in the problem space and finally to test the model using the neighbouring data vectors to check the effectiveness of the model created. The GA is used to locate an effective set of features that represent most of the data's significant structural information along with the optimal number of neighbours to consider and the optimal distance measure to identify the neighbours. The complete algorithm is described in two parts, the transductive setting and the GA.

1. We normalize the dataset linearly with values between 0 and 1 to ensure a standardization of all values especially when variables are represented in different units. This normalization procedure is based on the assumption that all variables/features have the same importance for output of the system for the whole problem space.
2. For every test sample *Ti*, perform the following steps: select the closest neighbouring samples, create a model, and evaluate its accuracy
   For every test sample *Ti* we select through a GA optimisation: a set of features to be considered, the number of nearest neighbours, and the distance measure to locate the neighbours.
   The accuracy of the selected set of parameters for the *Ti* model is calculated by creating a model with these parameters for each of the neighbours of the test sample *Ti* and calculating the accuracy of each of these models. The cross-validation is run in a leave-one-out manner for all neighbours of *Ti*. If, for the identified set of parameters, the neighbours of *Ti* give a high classification accuracy rate, then we assume that the same set of parameters will also work for the sample *Ti*. This criterion is used as a fitness evaluation criterion for the GA optimisation procedure.
3. Perform the set of operations in step 2 in a leave-one-out manner for all the samples in the dataset and calculate the overall classification accuracy for this transductive approach (see Chapter 1).

Experiments

We conducted experiments on various UC Irvine datasets with their characteristics represented in Table 6.5a. The tests were carried out on all the data using the leave-one-out validation technique. The datasets were selected as the ones without any missing values except for the breast cancer dataset that had four missing values. At the preprocessing stage, the four samples with missing values were deleted and the size of the breast cancer dataset reduced from 198 to 194. As the next step of preprocessing, all the datasets were normalised using linear normalisation resulting in values in the range of 0 and 1 to provide standardisation.

Table 6.5b presents cross-validation results for comparison between the inductive modelling approach and the transductive modelling approach without and with GA parameter optimisation for MLR models.

Table 6.5 (a) Different machine learning benchmark datasets; (b) classification results using: inductive multiple linear regression (MLR) model for classification; a transductive MLR without and with GA parameter optimisation of the following parameters: number of neighbouring samples K, input variables, distance measure.

(a)

Dataset	*No. of classes*	*No. of features*	*No. of data points*
Thyroid	3	5	215
Sonar	2	60	208
Breast Cancer	2	30	194
Liver	2	6	345
Glass	7	10	214

(b)

Data	*Classification accuracy of an inductive .MLR*	*Transductive MLR with fixed parameters selected manually as the best from multiple runs*	*Transductive MLR with GA optimised parameters*
Thyroid	86.51	94.88	94.88
Sonar	75.48	78.81	81.25
Breast Cancer	72.16	67.01	73.711
Liver	69.57	73.62	78
Glass	60.75	68.69	74

The results show that the transductive modelling approach significantly outperforms the inductive modelling and parameter optimisation also improves the accuracy of the individual models at average.

6.6 Particle Swarm Intelligence

In a GA optimisation procedure, a solution is found based on the best individual represented as a chromosome, where there is no communication between the individuals.

Particle swarm optimization (PSO), introduced by Kennedy and Eberhard (1995) is motivated by social behaviour of organisms such as bird flocking, fish schooling, and swarm theory. In a PSO system, each particle is a candidate solution to the problem at hand. The particles in a swarm fly in multidimensional search space, to find an optimal or suboptimal solution by competition as well as by cooperation among them. The system initially starts with a population of random solutions. Each potential solution, called a *particle*, is given a random position and velocity.

The particles have memory and each particle keeps track of its previous best position and the corresponding fitness. The previous best position is called the

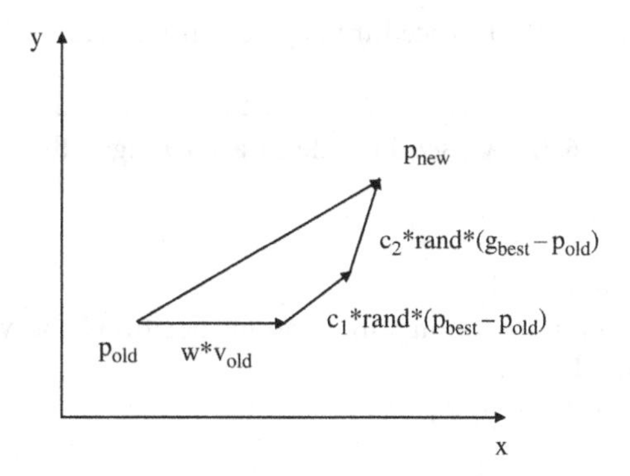

Fig. 6.11 A graphical representation of a particle swarm optimisation process (PSO) in a 2D space.

pbest. Thus, *pbest* is related only to a particular particle. The best value of all particles' *pbest* in the swarm is called the *gbest.* The basic concept of PSO lies in accelerating each particle towards its *pbest* and the *gbest* locations at each time step. This is illustrated in Fig. 6.11 for a two-dimensional space.

PSO have been developed for continuous, discrete, and binary problems. The representation of the individuals varies for the different problems. Binary particle swarm optimisation (BPSO) uses a vector of binary digit representation for the positions of the particles. The particle's velocity and position updates in BPSO are performed by the following equations.

$$v_{new} = w^* v_{old} + c_1^* rand()^* (p_{best} - p_{old}) + c_2^* rand()^* (g_{best} - p_{old}) \tag{6.3}$$

$$p_{new} = \begin{cases} 0 & if\ r \geq s(v_{new}) \\ 1 & if\ r < s(v_{new}) \end{cases} \tag{6.4}$$

where

$$s(v_{new}) = \frac{1}{1 + \exp(-v_{new})} \quad \text{and } r \sim \mathrm{U}(0, 1) \tag{6.5}$$

The velocities are still in the continuous space. In BPSO, the velocities are not considered as velocities in the standard PSO but are used to define probabilities that a bit flip will occur. The inertia parameter w is used to control the influence of the previous velocity on the new velocity. The term with c_1 corresponds to the "cognitive" acceleration component and helps in accelerating the particle towards the *pbest* position. The term with c_2 corresponds to the "social" acceleration component which helps in accelerating the particle towards the *gbest* position.

A simple version of a PSO procedure is given in Box 6.2.

Box 6.2. A pseudo-code of a PSO algorithm

begin

t ← 0 (time variable)
1) Initialize a population with random positions and velocities
2) Evaluate the fitness
3) Select the *pbest* and *gbest*

while (termination condition is not met) do

begin
t ← t + 1
4) Compute velocity and position updates
5) Determine the new fitness
6) Update the *pbest* and *gbest* if required
end

end

6.7 Artificial Life Systems (ALife)

The main characteristics of life are also the main characteristics of a modelling paradigm called artificial life (ALife), namely:

1. Self-organisation and adaptation
2. Reproducibility
3. Population/generation-based
4. Evolvability

A popular example of an ALife system is the so-called Conways' Game of Life (Adami, 1998): Each cell in a 2D grid can be in one of the two states: either *on* (alive) or *off* (dead, unborn). Each cell has eight neighbours, adjacent across the sides and corners of the square.

Whether cells stay alive, die, or generate new cells depends upon how many of their eight possible neighbours are alive and is based on the following transition rule:

Rule S23/B3, a live cell with two live neighbours, or any cell with three neigbhours, is alive at the next time step (see Fig. 6.12).

- Example 1: If a cell is *off* and has three living neighbours (out of eight), it will become alive in the next generation.
- Example 2: If a cell is *on* and has two or three living neighbours, it survives; otherwise, it dies in the next generation.

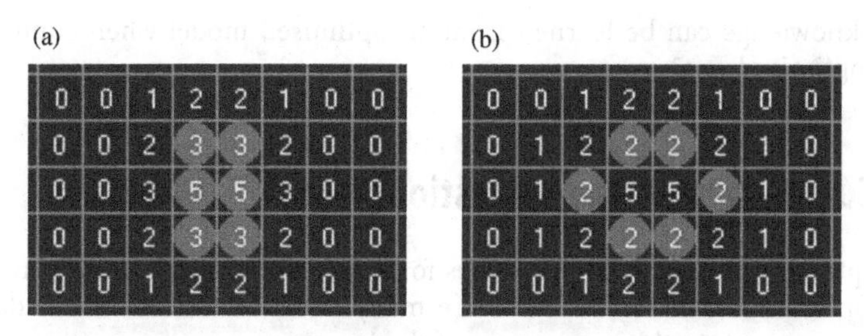

Fig. 6.12 Two consecutive states of the Game of Life according to rule S23/B3 (one of many possible rules), meaning that every cell survives if it is alive and is surrounded by two or three living cells, and a cell is born if there are three living cells in the neighbourhood, otherwise a cell dies (as a result of an either overcrowded neighbourhood of living cells, or of lack of sufficient living cells, 'loneliness').

- Example 3: A cell with less than two neighbours will die of loneliness and a cell with more then three neighbours will die of overcrowding.

In this interpretation, the cells (the individuals) never change the above rules and behave in this manner forever (until there is no individual left in the space). A more intelligent behaviour would be if the individuals were to change their rules of behaviour based on additional information they were able to collect. For example, if the whole population is likely to become extinct, then the individuals would create more offspring, and if the space became too crowded, the individual cells would not reproduce every time they are 'forced' to reproduce by the current rule. In this case we are talking about emerging intelligence of the artificial life ensemble of individuals (see Chapter 1). Each individual in the Game of Life can be implemented as an ECOS that has connections with its neighbours and has three initial exact (or fuzzy) rules implemented, but at a later stage new rules can be learned.

6.8 Exercise

Choose a classification problem and a dataset.
Create an evolving classification (ECF) or other classfication model for the problem and evaluate its accuracy.
Apply a GA for 20 generations, 20 individuals in a population, for parameter and feature optimisation of ECF and evaluate the accuracy.
Apply an ES for 20 generations, 20 individuals in a population, for parameter and feature optimisation of ECF and evaluate the accuracy.
Apply ECF with WDN (weighted data normalisation) for weighting the input variables in an optimised model.
Apply transductive modelling with MLR or other methods, with a GA optimisation of the number of the nearest samples, the input features and the distance measure.
Apply particle swarm optimisation (PSO) to the problem.
Compare the results from the above experiments.
Question: In general, what EC method (GA, ES, WDN, transductive, PSO) would be most suitable to the problem in hand and why?

What knowledge can be learned from an optimised model when compared to an unoptimised one?

6.9 Summary and Open Questions

This chapter presents several approaches for using evolutionary computation (EC) for the optimisation of ECOS. There are many issues that need to be addressed for further research in this area. Some of the issues are:

1. Online optimisation of the fitness function of an EC.
2. Using individual fitness functions for each ECOS.
3. EC helps choose the parameter values of ECOS, but how do we choose the optimal parameters for the EC method at the same time?
4. Interactions between individuals and populations that have different genetic makeups.

6.10 Further Reading

- *Generic Material on Evolutionary Computation* (Goldberg, 1989; Michaliewicz, 1992)
- *Genetic Programming* (Koza, 1992)
- *Evolutive Fuzzy Neural Networks* (Machado *et al.*, 1992)
- *Using EC Techniques for the Optimisation of Neural Networks* (Fogel, 1990; Schiffman *et al.*, 1993; Yao, 1993)
- *Using GA for the Optimisation and Training of Fuzzy Neural Networks* (Watts and Kasabov, 1998)
- *The Evolution of Connectivity: Pruning Neural Networks Using Genetic Algorithms* (Whitley and Bogart, 1990)
- *Neuronal Darwinism* (Edelman, 1992)
- *GA for the Optimisation of Fuzzy Rules* (Furuhashi *et al.*, 1994)
- *Using EC for Artificial Life* (Adami, 1998)
- *Online GA and ES Optimisation of ECOS* (Kasabov 2003; Chan *et al.* 2004)
- *Parameter Optimisation of EFuNN* (Watts and Kasabov, 2001, 2002; Minku and Ludemir, 2005)
- *Swarm Intelligence* (Kennedy and Eberhard, 1995)

7. Evolving Integrated Multimodel Systems

Chapters 2 to 6 presented different methods for creating single evolving connectionist models. This chapter presents a framework and several methods for building evolving connectionist machines that integrate in an adaptive way several evolving connectionist models to solve a given task, allowing for using different models (e.g. regression formulas, ANN), and for adding new data and new variables. The chapter covers the following topics.

- A framework for evolving multimodel systems
- Adaptive, incremental data and model integration
- Integrating kernel functions and regression formulas in knowledge-based ANN
- Ensemble learning methods for ECOS
- Integrating ECOS with evolving ontologies
- Summary and open problems
- Further reading

7.1 Evolving Multimodel Systems

7.1.1 A General Framework

Complex problems usually require a more complex intelligent system for their solution, consisting of several models. Some of these models can be evolving models. A block diagram of a framework for evolving connectionist machines, consisting of several evolving models (EM) is given in Fig. 7.1 (from Kasabov (1998a)).

The framework facilitates multilevel multimodular ensembles where many EM or ECOS are connected with inter- and intraconnections. The evolving connectionist machine does not have a 'clear' multilayer structure. It has a modular 'open' structure. The main parts of the framework are described below.

1. Input presentation part: This filters the input information, performs feature extraction and forms the input vectors. The number of inputs (features) can vary from example to example.

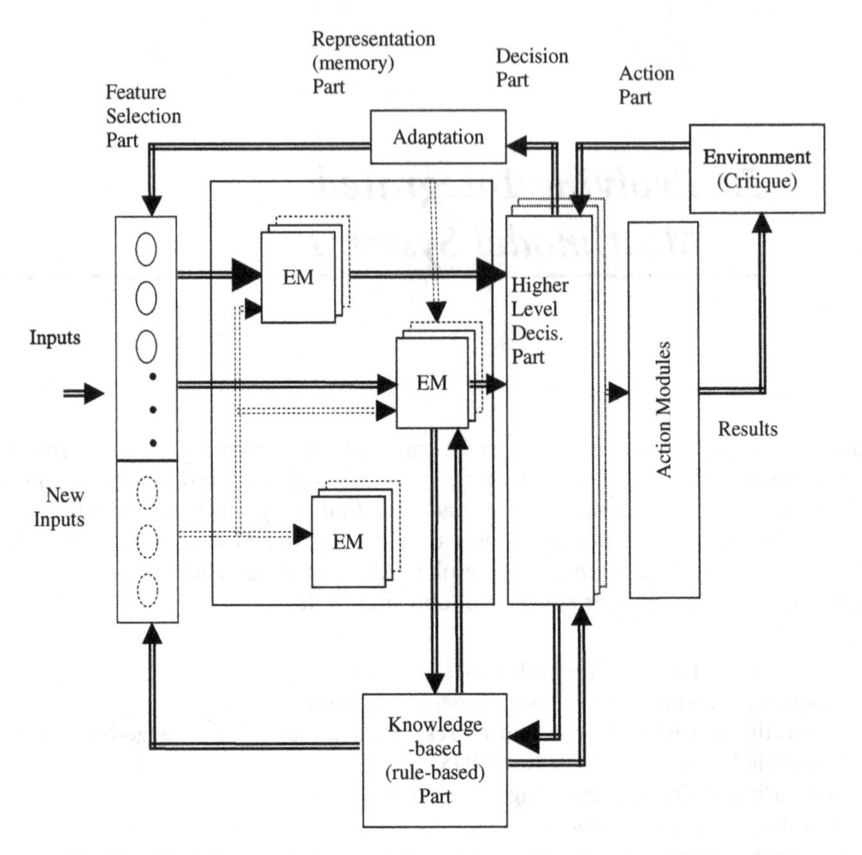

Fig. 7.1 A block diagram of a framework for evolving multimodel systems (Kasabov, 2001).

2. Representation and memory part, where information (patterns) are stored: This is a multimodular evolving structure of evolving connectionist modules and systems organised in spatially distributed groups; for example, one group can represent the phonemes in a spoken language (e.g. one ECOS representing a phoneme class in a speech recognition system).
3. Higher-level decision part: This consists of several modules, each taking decisions on a particular problem (e.g. word recognition or face identification). The modules receive feedback from the environment and make decisions about the functioning and adaptation of the whole evolving machine.
4. Action part: The action modules take output values from the decision modules and pass information to the environment.
5. Knowledge-based part: This part extracts compressed abstract information from the representation modules and from the decision modules in different forms of rules, abstract associations, etc.

Initially, an evolving machine is a simple multimodular structure, each of the modules being a mesh of nodes (neurons) with very little connection between them, predefined through prior knowledge or 'genetic' information. An initial set of rules can be inserted in this structure. Gradually, through self-organisation, the system becomes more and more 'wired'. New modules are created as the system operates. A network module stores patterns (exemplars) from the entered examples. A node in a module is created and designated to represent an individual example if it is significantly different from the previously used examples (with a level of differentiation set through dynamically changing parameters).

The functioning of the evolving multimodel machines is based on the following general principles.

1. The initial structure is defined only in terms of an initial set of features, a set of initial values for the ECOS parameters ('genes') and a maximum set of neurons, but no connections exist prior to learning (or connections exist but they have random values close to zero).
2. Input patterns are presented one by one or in chunks, not necessarily having the same input feature sets. After each input example is presented, the ECOS either associates this example with an already existing module and a node in this module, or creates a new module and/or creates a new node to accommodate this example. An ECOS module denoted in Fig. 7.1 as an evolving module (EM), or a neuron, is created when needed at any time during the functioning of the whole system.
3. Evolving modules are created as follows. An input vector x is passed through the representation module to one or more evolving modules. Nodes become activated based on the similarity between the input vector and their input connection weights. If there is no EM activated above a certain threshold a new module is created. If there is a certain activation achieved in a module but no sufficient activation of a node inside it, a new node will be created.
4. Evolving a system can be achieved in different modes, e.g. supervised, reinforcement, or unsupervised (see Chapters 2 through 5). In a supervised learning mode the final decision on which class (e.g. phoneme) the current vector x belongs to is made in the higher-level decision module that may activate an adaptation process.
5. The feedback from the higher-level decision module goes back to the feature selection and filtering part (see Chapter 1).
6. Each EM or ECOS has both aggregation and pruning procedures defined. Aggregation allows for modules and neurons that represent close information instances in the problem space to merge. Pruning allows for removing modules and neurons and their corresponding connections that are not actively involved in the functioning of the ECOS (thus making space for new input patterns). Pruning is based on local information kept in the neurons. Each neuron in ECOS keeps track of its 'age', its average activation over the whole lifespan, the global error it contributes to, and the density of the surrounding area of neurons (see, for example, EFuNN, Chapter 3).
7. The modules and neurons may be spatially organised and each neuron has relative spatial dimensions with regard to the rest of the neurons based on their reaction to the input patterns. If a new node is to be created when an input vector $\mathbf{x}$ is presented, then this node will be allocated closest to the neuron that had the highest activation to the input vector $\mathbf{x}$, even though insufficiently high to accommodate this input vector.

8. In addition to the modes of learning from (4), there are two other general modes of learning (see Chapter 1):
   (a) *Active learning mode*: Learning is performed when a stimulus (input pattern) is presented and kept active.
   (b) *Sleep learning mode*: Learning is performed when there is no input pattern presented at the input of the machine. In this case the process of further elaboration of the connections in an ECOS is done in a passive learning phase, when existing connections, that store previous input patterns, are used as eco-training examples. The connection weights that represent stored input patterns are now used as exemplar input patterns for training other modules in ECOS.
9. ECOS provide explanation information (rules) extracted from the structure of the NN modules.
10. The ECOS framework can be applied to different types of ANN (different types of neurons, activation functions, etc.) and to different learning algorithms.

Generally speaking, the ECOS machine from Fig. 7.1 can theoretically model the five levels of evolving processes as shown in Fig.I.1. We can view the functioning of an ECOS machine as consisting of the following functional levels.

1. *Gene, parameter, level:* Each neuron in the system has a set of parameters – genes – that are subject to adaptation through both learning and evolution.
2. *Neuronal level:* Each neuron in every ECOS has its information-processing functions, such as the activation function or the maximum radius of its receptive field.
3. *Ensembles of neurons:* These are the evolving neural network modules (EM) each of them comprising a single ECOS, e.g. an EFuNN.
4. *The whole ECOS machine:* This has a multimodular hierarchical structure with global functionality.
5. *Populations of ECOS machines* and their development over generations (see Chapter 6).

The following algorithm describes a scenario of the functioning of this system.

Loop 0: {Create a population of ECOS machines with randomly chosen parameter values – chromosomes (an optional higher-level loop)
Loop 1: {Apply evolutionary computation after every p data examples over the whole population of ECOS machines.
Loop 2: {Apply adaptive lifelong learning (evolving) methods to an ECOS (e.g. ECM, EFuNN, DENFIS) from the ECOS machine to learn from p examples.
Loop 3: {For each created neuron in an ECOS adapt (optimise) its parameter values (genes) during the learning process either after each example, or after a set of p examples is presented. Mutation or other operations over the set of parameters can be applied. During this process a gene interaction network (parameter dependency network) can be created, allowing for observation of how genes (parameters) interact with each other.
} end of loop 3 } end of loop 2 } end of loop 1 } end of loop 0 (optional)

The main challenge here is to be able to model both the evolving processes at each level of modelling and the interaction between these levels.

7.1.2 An Integrated Framework of Global, Local, and Personalised Models

Global models capture trends in data that are valid for the whole problem space, and local models capture local patterns, valid for clusters of data (see Chapter 1). Both models contain useful information and knowledge. Local models are also adaptive to new data as new clusters and new functions, that capture patterns of data in these clusters, can be incrementally created. Usually, both global and local modelling approaches assume a fixed set of variables and if new variables, along with new data, are introduced with time, the models are very difficult to modify in order to accommodate these new variables. This can be done in the personalised models, as they are created on the fly and can accommodate any new variables, provided that there are data for them. All three approaches are useful for complex modelling tasks and all of them provide complementary information and knowledge, learned from the data. Integrating all of them in a single multimodel system would be an useful approach and a challenging task.

A graphical representation of an integrated multimodel system is presented in Fig. 7.2. For every single input vector, the outputs of the tree models are weighted. The weights can be adjusted and optimised for every new input vector in a similar way as the parameters of a personalised model (Kasabov, 2007b).

$$y_i = w_{i,g}\, y_i(\mathbf{x}_i)^{(\text{global})} + w_{i,l}\, y_i(\mathbf{x}_i)^{(\text{local})} + w_{i,p}\, y_i(\mathbf{x}_i)^{(\text{personalised})} \tag{7.1}$$

7.1.3 Spatial and Temporal Complexity of a Multimodel ECOS

Spatial complexity of a system defines the system architecture requirements in terms of nodes, connections, etc. The spatial complexity of an ECOS could be evaluated as follows.

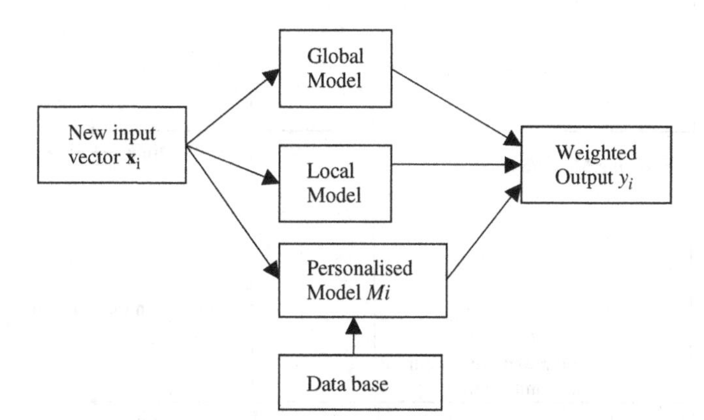

Fig. 7.2 A graphical representation of an integrated global, local, and personalised multimodel system. For every single input vector, the outputs of the tree models are weighted (Kasabov, 2007b).

- Number of features; number of NN modules; total number of neurons
- Total number of connections
- Function of growth (linear, exponential)
- Level of pruning and aggregation

As shown graphically in Fig. 7.3, for a hypothetical example, an ECOS creates modules and connects nodes all the time, but it happens more often at the beginning of the evolving process (until time *t*1). After that, the subsequent input patterns are accommodated, but this does not cause the creation of many new connections until time *t*2, when a new situation in the modelled process arises (e.g. a user speaks a different accent to the machine, or there is a new class of genes, or the stock market rises/crashes unexpectedly). Some mechanisms prevent infinite growth of the ECOS structure. Such mechanisms are pruning and aggregation. At this moment the ECOS 'shrinks' and its complexity is reduced (time *t*3 in the figure).

Time complexity is measured in the required time for the system to react to an input datum.

7.1.4 An Agent-Based Implementation of Multimodel Evolving Systems

An agent-based implementation of the framework from Fig. 7.1 is shown in Fig. 7.4. The main idea of this framework is that agents, which implement some evolving models, are created online whenever they are needed in time. Some agents collect data, some learn from these data in an online lifelong mode, and some extract and refine knowledge. The models are dynamically created (on the fly). Figure 7.4 shows

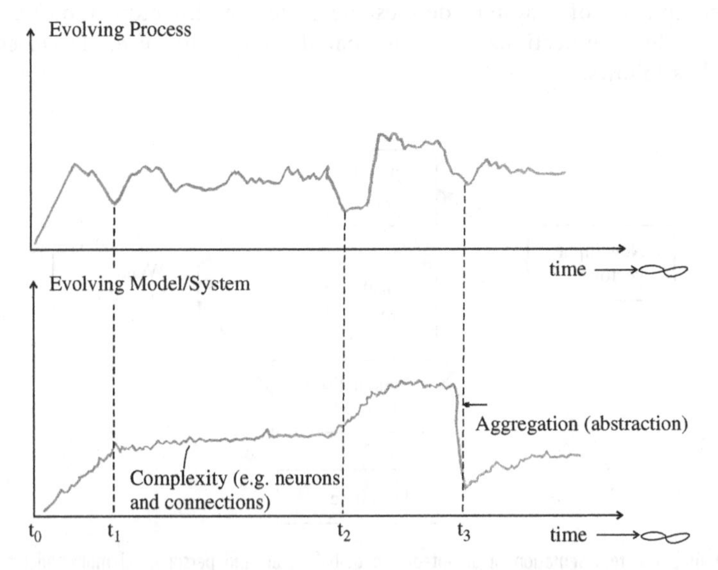

Fig. 7.3 A hypothetical example of a complexity measure in an evolving connectionist system.

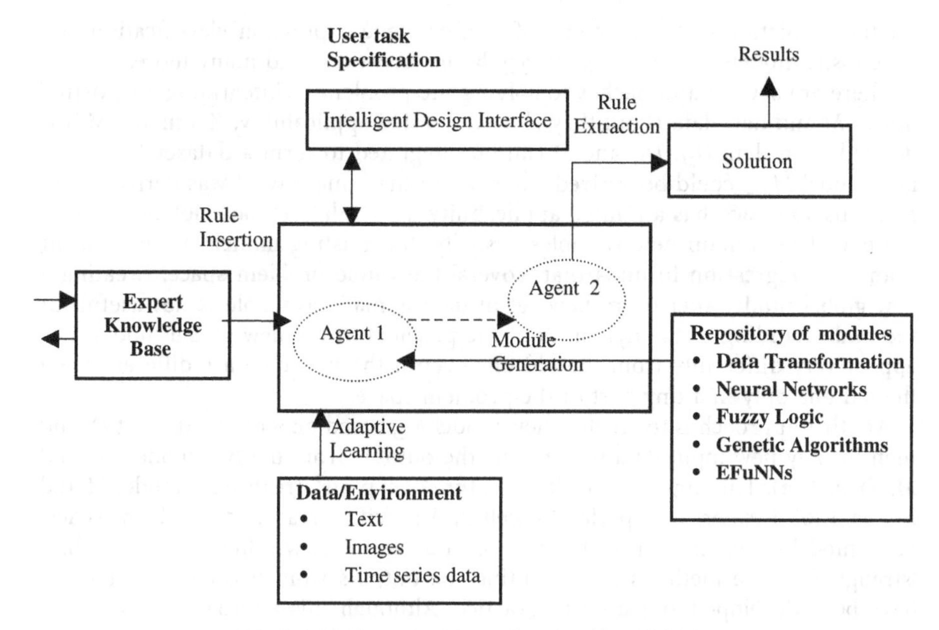

Fig. 7.4 Multiagent implementation of multimodel ECOS.

the general architecture of the framework. The user specifies the initial problem parameters and the task to be solved. Then intelligent agents (designers) create dynamic units – modules – that initially may contain no structured knowledge, but rules on how to create the structure and evolve the functionality of the module.

The modules combine expert rules with data from the environment. The modules are continuously trained with these data. Rules may be extracted from trained modules. This is facilitated by several evolving connectionist models, such as the evolving fuzzy neural network EFuNN (Chapter 3). Rules can be extracted for the purpose of later analysis or for the creation of new modules. Modules can be deleted if they are no longer needed for the functioning of the whole system.

7.2 ECOS for Adaptive Incremental Data and Model Integration

7.2.1 Adaptive Model and Data Integration: Problem Definition

Despite the advances in mathematical and information sciences, there is a lack of efficient methods to extend an existing model M to accommodate new (reliable) dataset D for the same problem, if M does not perform well on D. Examples of existing models that need to be further modified and extended to new data are numerous: differential equation models of cells and neurons, a regression formula to predict the outcome of cancer, an analytical formula to evaluate renal functions, a logistic regression formula for evaluating the risk of cardiac events, a set of rules for the prediction of the outcome of trauma, gene expression classification and prognostic models, models of gene regulatory networks, and many more.

There are several approaches to solving the problem of integrating an existing model M and new data D but they all have limited applicability. If a model M was derived from data D_M, D_M and D can be integrated to form a dataset D_{all} and a new model M_{new} could be derived from D_{all} in the same way M was derived from D_M. This approach has a limited applicability if past data D_M are not available or the new data contain new variables. Usually, the existing models are global, for example a regression formula that 'covers' the whole problem space. Creating a new global model after every new set of data is made available is not useful for the understanding of the dynamics of the problem as the new global model may appear very differently from the old one even if the new data are different from the old one only in a tiny part of the problem space.

Another approach is to create a new model M_D based only on the dataset D, and then for any new input data to combine the outputs from the two models M and M_D (Fig. 7.5a). This approach, called 'mixture of experts', treats each model M and M_D as a local expert that performs well in a specific area of the problem space. Each model's contribution to the final output vector is weighted based on their 'strengths'. Some methods for weighting the outputs from two or more models have been developed and used in practice. Although this approach is useful for some applications, the creation, the weight optimization, and the validation of several models used to produce a single output in areas where new input data are generated continuously, could be an extremely difficult task. In Kasabov (2007b) an alternative approach is proposed that is a generic solution to the problem as explained below.

7.2.2 Integrating New Data and Existing Models Through Evolving Connectionist Systems

In the method introduced here, we assume that an existing model M performs well in part of the problem space, but there is also a new dataset D that does not fit into the model. The existing model M is first used to generate a dataset D_0 of input–output data samples through generating input vectors (the input variables take values in their respective range of the problem space where the model performs well) and calculating their corresponding output values using M (Fig. 7.5b). The dataset D_0 is then used to evolve an initial connectionist model M_0 with the use of an evolving connectionist system (ECOS) and to extract rules from it where each rule represents a local model (a prototype).

The model M_0 can be made equivalent to the model M to any degree of accuracy through tuning the parameters of ECOS. The initial ECOS model M_0 is then further trained on the new data D thus tuning the initial rules from M_0 and evolving new rules applicable to the dataset D. The trained ECOS constitutes an integrated new model M_{new} that consists of local adaptable rules. To compare the generalisation ability of M and M_{new}, the datasets D_0 and D are split randomly into training and testing sets: D_{0tr}, D_{0tst}, D_{tr}, D_{tst}. The training sets are used to evolve the new model M_{new} and the test sets are used to validate M_{new} and compare it with the existing model M. The new model M_{new} can be incrementally trained on future incoming

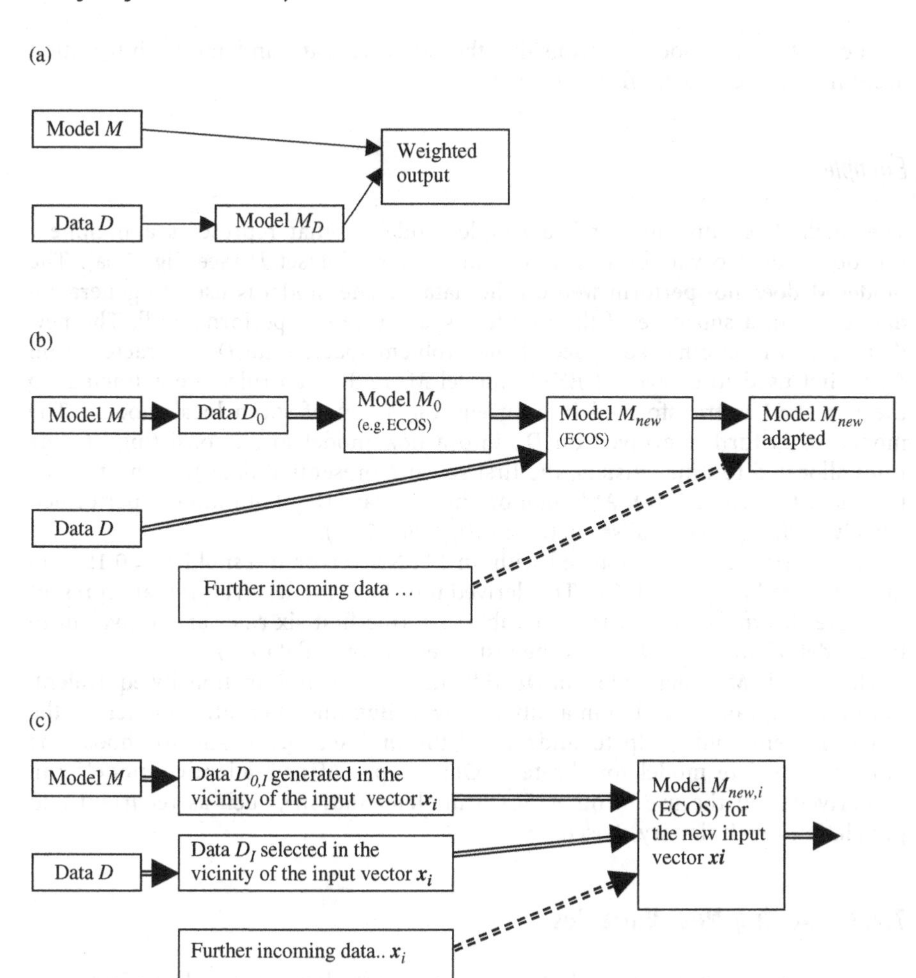

Fig. 7.5 (a) The 'mixture of experts' approach for model M and data D integration combines outputs from different models M and M_D (derived from D); (b) the proposed inductive, local learning method generates data D_0 from an existing model M, creates an ECOS model M_0 from D_0, and further evolves M_0 on the new data D thus creating an integrated, adaptive new model M_{new}; (c) in a transductive approach, for every new input vector x_i, a new model M_{new}, i is created based on generated data from the old model M and selected data from the new dataset D, all of them being in the vicinity of the new input vector (Kasabov, 2006).

data and the changes can be traced over time. New data may contain new variables and have missing values as explained later in the book. The method utilizes the adaptive local training and rule extraction characteristics of ECOS.

In a slightly different scenario (Fig. 7.5c), for any new input vector x_i that needs to be processed by a model M_{new} so that its corresponding output value is calculated, data samples, similar to the new input vector x_i are generated from both dataset D, samples Di, and from the model M, samples $D_{0,i}$, and used to evolve a model $M_{new,i}$ that is tuned to generalise well on the input vector x_i. This approach can be seen as a partial case of the approach from above and in the rest of the book we consider the adaptive data and model integration according to the scheme from Fig. 7.5b.

Example

The method is illustrated with a simple model M that represents a nonlinear function y of two variables x_1 and x_2 and a new dataset D (see Fig. 7.6a). The model M does not perform well on the data D. The model is used to generate a dataset D_0 in a subspace of the problem space where it performs well. The new dataset D is in another subspace of the problem space. Data D_{0tr} extracted from D_0 is first used to evolve a DENFIS model M_0 and seven rules are extracted, so the model M is transformed into an equivalent set of seven local models. The model M_0 is further evolved on D_{tr} into a new model M_{new}, consisting of nine rules allocated to nine clusters, the first seven representing data D_{0tr} and the last two, data D_{tr} (Table 7.1a). Although on the test data D_{0tst} both models performed equally well, M_{new} generalises better on D_{tst} (Fig. 7.6c).

An experiment was conducted with an EFuNN (error threshold $E = 0.15$, and maximum radius $R_{max} = 0.25$). The derived nine local models (rules) that represent M_{new} are shown for comparison in Table 7.1b (the first six rules are equivalent to the model M and data D_{0tr}, and the last three to cover data D_{tr}).

The models M_{new} derived from DENFIS and EFuNN are functionally equivalent, but they integrate M and D in a different way. Building alternative models of the same problem could help to understand the problem better and to choose the most appropriate model for the task. Other types of new adaptive models can be derived with the use of other ECOS implementations, such as recurrent and population (evolutionary) based.

7.2.3 Adding New Variables

The method above is applicable to large-scale multidimensional data where new variables may be added at a later stage. This is possible as partial Euclidean distance between samples and cluster centres can be measured based on a different number of variables. If a current sample S_j contains a new variable x_{new}, having a value x_{newj} and the sample falls into an existing cluster N_c based on the common variables, this cluster centre N is updated so that it takes a coordinate value x_{newj} for the new variable x_{new}, or the new value may be calculated as weighted k-nearest values derived from k new samples allocated to the same cluster. Dealing with new variables in a new model M_{new} may help distinguish samples that have very similar input vectors but different output values and therefore are difficult to deal with in an existing model M.

Example

Samples $S_1 = [x_1 = 0.75, x_2 = 0.824, y = 0.2]$ and $S_2 = [x_1 = 0.75, x_2 = 0.823, y = 0.8]$ are easily learned in a new ECOS model M_{new} when a new variable x_3 is added that has, for example, values of 0.75 and 0.3, respectively, for the samples S_1 and S_2.

Partial Euclidean distance can be used not only to deal with missing values, but also to fill in these values in the input vectors. As every new input vector $\boldsymbol{x}_i$ is mapped into the input cluster (rule node) of the model M_{new} based on the partial Euclidean distance of the existing variable values, the missing value in $\boldsymbol{x}_i$, for an input variable, can be substituted with the weighted average value for this variable across all data samples that fall in this cluster.

(a)

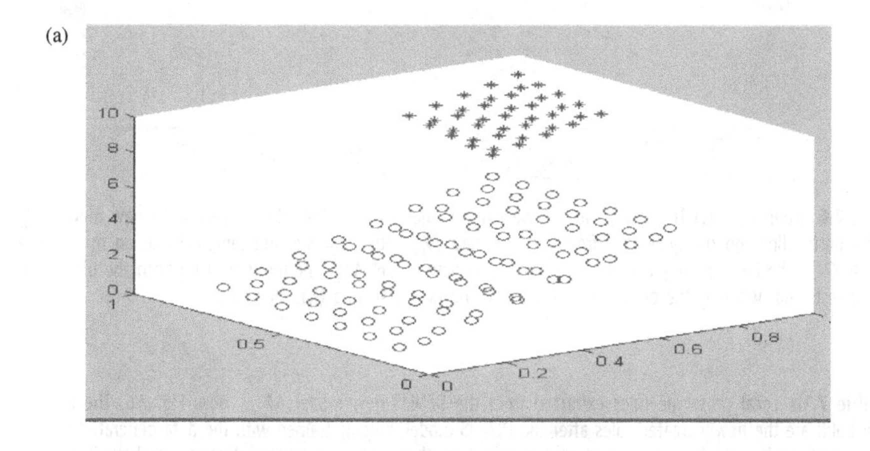

(b)

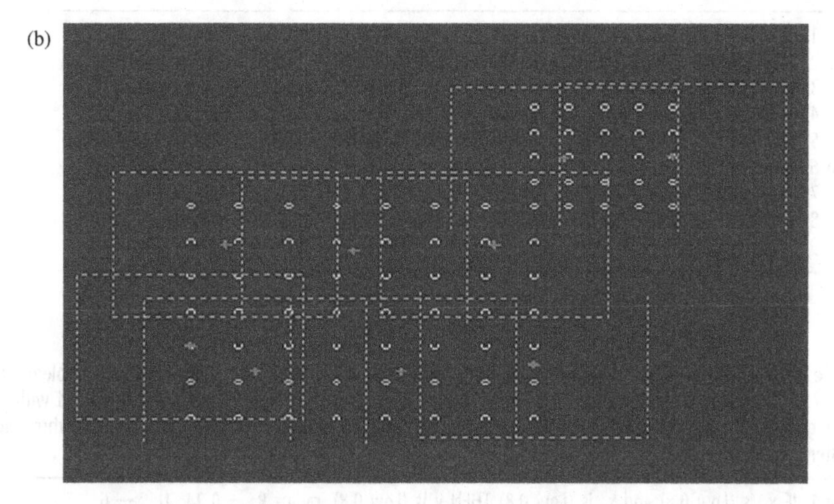

Fig. 7.6 A case study of a model M (formula) and a data set D integration through an inductive, local, integrative learning in ECOS: (a) a 3D plot of data D_0 (data samples denoted o) generated from a model M (formula) $y = 5.1x_1 + 0.345x_1^2 - 0.83x_1 \log_{10} x_2 + 0.45x_2 + 0.57 \exp(x_2^{0.2})$ in the subspace of the problem space defined by x_1 and x_2 both having values between 0 and 0.7, and new data D (samples denoted *) defined by x_1 and x_2 having values between 0.7 and 1. (b) The data clusters of D_0 (the seven clusters on the left, each defined as a cluster centre denoted + and a cluster area) and of the data D (the two upper right clusters) in the 2D input space of x_1 and x_2 input variables from Fig. 7.2a , are formed in a DENFIS ECOS trained with the data D_{0tr} (randomly selected 56 data samples from D_0) and then further trained with the data D_{tr} (randomly selected 25 samples from D)(*Continued overleaf*).

(c)

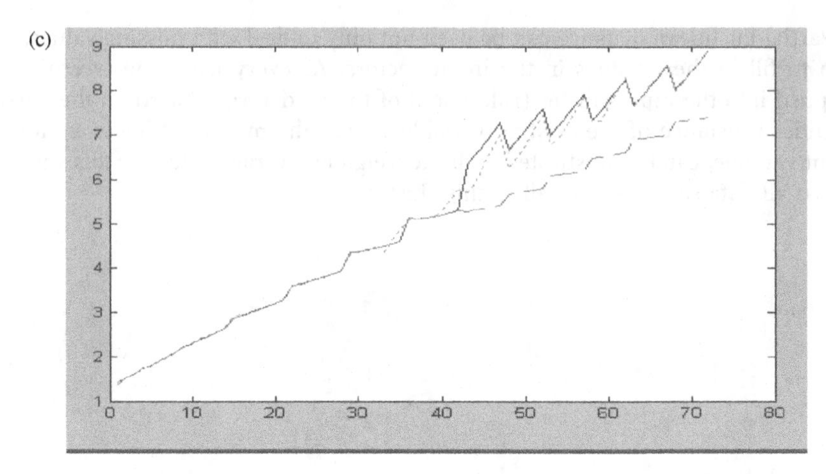

Fig. 7.6 (*continued*) (c) The test results of the initial model M (the dashed line) versus the new model M_{new} (the dotted line) on the generated from M test data D_{0tst} (the first 42 data samples) and on the new test data D_{tst} (the last 30 samples) (the solid line). The new model M_{new} performs well on both the old and the new test data, whereas the old model M fails on the new test data (Kasabov, 2006).

Table 7.1a Local prototype rules extracted from the DENFIS new model M_{new} from Fig. 7.6. The last rules (in bold) are the newly created rules after the DENFIS model, initially trained with the data generated from the existing formula, was further adapted on the new data, thus creating two new clusters (Kasabov, 2006).

Rule 1: IF x_1 is (-0.05, 0.05, 0.14) and x_2 is (0.15,0.25,0.35) THEN $y = 0.01 + 0.7x_1 + 0.12x_2$
Rule 2: IF x_1 is (0.02, 0.11, 0.21) and x_2 is (0.45,0.55, 0.65) THEN $y = 0.03 + 0.67x_1 + 0.09x_2$
Rule 3: IF x_1 is (0.07, 0.17, 0.27) and x_2 is (0.08,0.18,0.28) THEN $y = 0.01 + 0.71x_1 + 0.11x_2$
Rule 4: IF $x_1$is (0.26, 0.36, 0.46) and x_2 is (0.44,0.53,0.63) THEN $y = 0.03 + 0.68x_1 + 0.07x_2$
Rule 5: IF x_1 is (0.35, 0.45, 0.55) and x_2 is (0.08,0.18,0.28) THEN $y = 0.02 + 0.73x_1 + 0.06x_2$
Rule 6: IF x_1 is (0.52, 0.62, 0.72) and x_2 is (0.45,0.55,0.65) THEN $y = -0.21 + 0.95x_1 + 0.28x_2$
Rule 7: IF $x_1$is (0.60, 0.69,0.79) and x_2 is (0.10,0.20,0.30) THEN $y = 0.01 + 0.75x_1 + 0.03x_2$
Rule 8: IF $x_1$is (0.65,0.75,0.85) and x_2 is (0.70,0.80,0.90) THEN $y = -0.22 + 0.75x_1 + 0.51x_2$
Rule 9: IF $x_1$is (0.86,0.95,1.05) and x_2 is (0.71,0.81,0.91) THEN $y = 0.03 + 0.59x_1 + 0.37x_2$

Table 7.1b Local prototype rules extracted from an EFuNN new model M_{new} on the same problem from **Fig.** 7.6. The last rules (in bold) are the newly created rules after the EFuNN model, initially trained with the data generated from the existing formula, was further adapted on the new data, thus creating three new clusters (Kasabov, 2006).

Rule 1: IF x_1 is (Low 0.8) and x_2 is (Low 0.8) THEN y is (Low 0.8), radius $R_1 = 0.24$; $N_{1ex} = 6$
Rule 2: IF x_1 is (Low 0.8) and x_2 is (Medium 0.7) THEN y is (Small 0.7), $R_2 = 0.26$, $N_{2ex} = 9$
Rule 3: IF x_1 is (Medium 0.7) and x_2 is (Medium 0.6) THEN y is (Medium 0.6), $R_3 = 0.17$, $N_{3ex} = 17$
Rule 4: IF x_1 is (Medium 0.9) and x_2 is (Medium 0.7) THEN y is (Medium 0.9), $R_4 = 0.08$, $N_{4ex} = 10$
Rule 5: IF x_1 is (Medium 0.8) and x_2 is (Low 0.6) THEN y is (Medium 0.9), $R_5 = 0.1$, $N_{5ex} = 11$
Rule 6: IF x_1 is (Medium 0.5) and x_2 is (Medium 0.7) THEN y is (Medium 0.7), $R_6 = 0.07$, $N_{6ex} = 5$
Rule 7: IF x1 is (High 0.6) and x2 is (High 0.7) THEN y is (High 0.6), $R_7 = 0.2$, $N_{7ex} = 12$
Rule 8: IF x1 is (High 0.8) and x2 is (Medium 0.6) THEN y is (High 0.6), $R_8 = 0.1$, $N_{8ex} = 5$
Rule 9: IF x1 is (High 0.8) and x2 is (High 0.8) THEN y is (High3 0.8), $R_9 = 0.1$, $N_{9ex} = 6$

7.3 Integrating Kernel Functions and Regression Formulas in Knowledge-Based ANN

7.3.1 Integrating Regression Formulas and Kernel Functions in Locally Adaptive Knowledge-Based Neural Networks

Regression functions are probably the most popular type of prognostic and classification models, especially in medicine. They are derived from data gathered from the whole problem space through inductive learning, and are consequently used to calculate the output value for a new input vector regardless of where it is located in the problem space. For many problems, this can result in different regression formulas for the same problem through the use of different datasets. As a result, such formulas have limited accuracy on new data that are significantly different from those used for the original modelling.

Kernel-based ANNs have radial-based function (RBF) kernels attached to their nodes that are adjusted through learning from data in terms of their centres and radii. They are trained as a set of local models that are integrated at the output. A method for the integration of regression formulas and kernel functions in a knowledge-based neural network (KBNN) model that results in better accuracy and more precise local knowledge is proposed in Song *et al.* 2006). A block diagram of the proposed KBNN structure is given in Fig. 7.7.

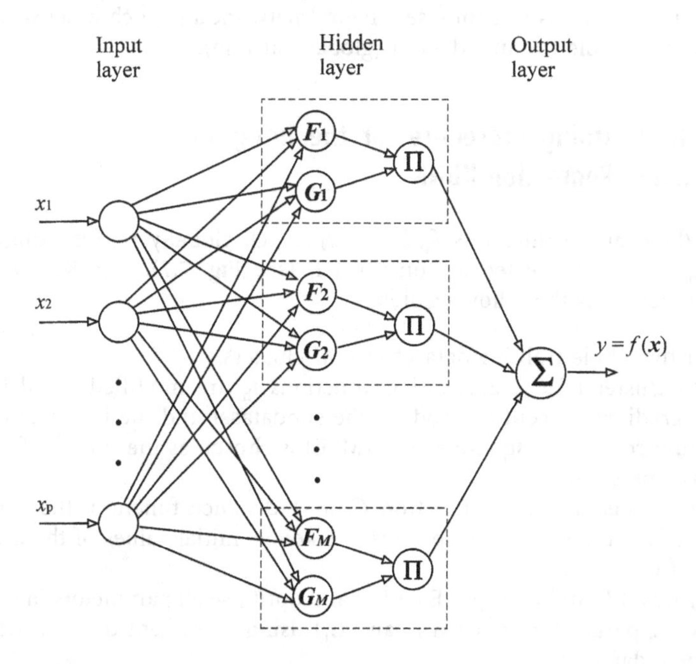

Fig. 7.7 A diagram of a kernel-regression KBNN, combining different kernels Gl with suitable regression functions Fl to approximate data in local clusters $Cl\,(l = 1, 2, \ldots, M)$ (from Song *et al.* (2006)).

The overall functioning of the model can be described by the formula:

$$y(x) = G_1(x)\,F_1(x) + G_2(x)\,F_2(x) + \ldots + G_M(x)\,F_M(x) \tag{7.1}$$

where $x = [x_1, x_2, \ldots, x_P]$ is the input vector; y is the output vector; G_l are kernel functions; and F_l are knowledge-based transfer functions, e.g. regression formulas, $1 = 1, 2, \ldots, M$.

Equation (7.1) can be regarded as a regression function. Using different G_l and F_l, Eq. (7.1) can represent different kinds of neural networks, and describe the different functions associated with neurons in their hidden layer(s). G_l are Gaussian kernel functions and F_l are constants in the case of RBF ANNs. G_l are sigmoid transfer functions and F_l are constants in the case of a generic three-layer multilayer perceptron (MLP) ANN. G_l are fuzzy membership functions and F_l are linear functions in the case of a first-order Takagi–Sugeno–Kang (TSK) fuzzy inference model; and in the simplest case, G_l represents a single input variable and F_l are constants in the case of a linear regression function.

In the KBNN from Fig. 7.7, F_l are nonlinear functions that represent the knowledge in local areas, and G_l are Gaussian kernel functions that control the contribution of each F_l to the system output. The farther an input vector is from the centre of the Gaussian function, the less contribution to the output is produced by the corresponding F_l.

The KBNN model has a cluster-based, multilocal model structure. Every transfer function is selected from existing knowledge (formulas), and it is trained within a cluster (local learning), so that it becomes a modified formula that can optimally represent this area of data. The KBNN aggregates a number of transfer functions and Gaussian functions to compose a neural network and such a network is then trained on the whole training dataset (global learning).

7.3.2 The Learning Procedure for the Integrated Kernel-Regression KBNN

Suppose there are Q functions f_h, $h = 1, 2, \ldots, Q$, globally representing existing knowledge that are selected as functions f_l (see Fig. 7.7). The KBNN learning procedure performs the following steps.

1. Cluster the whole training dataset into M clusters.
2. In each cluster $1, 1 = 1, 2, \ldots, M$, Q functions f_h are modified (local learning) with a gradient descent method on the subdataset and the best one (with the minimum root mean square error, RMSE) is chosen as the transfer function F_1 for this cluster.
3. Create a Gaussian kernel function G_1 as a distance function: the centre and radius of the clusters are, respectively, taken as initial values of the centre and width of G_1.
4. Aggregate all F_1 and G_1 as per Eq. (7.1) and optimise all parameters in the KBNN (including parameters of each F_1 and G_1) using a gradient descent method on the whole dataset.

In the KBNN learning algorithm, the following indexes are used.

- Training data : $i = 1, 2, \ldots, N$.
- Subtraining dataset: $i = 1, 2, \ldots, N_l$.
- Input variables: $j = 1, 2, \ldots, P$.
- Neuron pairs in the hidden layer: $l = 1, 2, \ldots, M$.
- Number of existing functions: $h = 1, 2, \ldots, Q$.
- Number of parameters in F_l $pf = 1, 2, \ldots, L_{pf}$.
- Learning iterations: $k = 1, 2, \ldots$.

The equations for parameter optimisation are described below.

Consider the system having P inputs, one output, and M neuron pairs in the hidden layer; the output value of the system can be calculated for an input vector $x_i = [x_{i1}, x_{i2} \ldots, x_{iP}]$ by Eq. (7.1):

$$y(x_i) = G_1(x_i)\, F_1(x_i) + G_2(x_i) F_2(x_i) + \cdots + G_M(x_i)\, F_M(x_i) \tag{7.2}$$

Here, F_l are transfer functions and each of them has parameters b_{pf}, $pf =$ 1, 2, ..., L_{pf},
and

$$G_l(\mathbf{x}_i) = \alpha_l \prod_{j=1}^{P} \exp[-\frac{(x_{ij} - m_{lj})^2}{2\sigma_{lj}{}^2}] \tag{7.3}$$

are Gaussian kernel functions. Here, α represents a connection vector between the hidden layer and the output layer; m_l is the centre of G_l. σ_l is regarded as the width of G_l, or a 'radius' of the cluster l. If a vector **x** is the same as m_l, the neuron pair $G_l(x)F_l(x)$ has the maximum output, $F_l(x)$; the output will be between $(0.607 \sim 1) \times F_l(x)$ if the distance between x and m_l is smaller than σ_l; the output will be close to 0 if x is far away from m_l.

Suppose the KBNN is given the training input–output data pairs $[\mathrm{x_i}, \mathrm{t_i}]$; the local learning minimizes the following objective function for each transfer function on the corresponding cluster,

$$E_l = \frac{1}{2} \sum_{i=1}^{Nl} [f_h(\mathbf{x}_i) - t_i]^2 \tag{7.4}$$

Here, N_l is the number of data that belong to the lth cluster, and the global learning minimizes the following objective function on the whole training dataset,

$$E = \frac{1}{2} \sum_{i=1}^{N} [y(\mathbf{x}_i) - t_i]^2 \tag{7.5}$$

A gradient descent algorithm (backpropagation algorithm) is used to obtain the recursions for updating the parameters b, α, m, and σ, so that E_l of Eq. (7.4) and E of Eq. (7.5) are minimised. The initial values of these parameters can be obtained from original functions (for b), random values or least-squares method (for α), and the result of clustering (for m and σ):

$$b_{pf}(k+1) = b_{pf}(k) - \eta_b \frac{\partial E_l}{\partial b_{pf}} \quad \text{(for local learning)} \tag{7.6}$$

$$b_{pf}(k+1) = b_{pf}(k) - \eta_b \frac{\partial E}{\partial b_{pf}} \quad \text{(for global learning)} \tag{7.7}$$

$$\alpha_l(k+1) = \alpha_l(k) - \frac{\eta_\alpha}{\alpha_l(k)} \sum_{i=1}^{N} \{G_l(\mathbf{x}_i)F_l(\mathbf{x}_i)[y(\mathbf{x}_i) - t_i]\} \tag{7.8}$$

$$m_{lj}(k+1) = m_{lj}(k) - \eta_m \sum_{i=1}^{N} \left\{ \frac{G_l(\mathbf{x}_i)F_l(\mathbf{x}_i)[y(\mathbf{x}_i) - t_i](x_{ij} - m_{lj})}{\sigma_{lj}^2} \right\} \tag{7.9}$$

$$\sigma_{lj}(k+1) = \sigma_{lj}(k) - \eta_\sigma \sum_{i=1}^{N} \left\{ \frac{G_l(\mathbf{x}_i)F_l(\mathbf{x}_i)[y(\mathbf{x}_i) - t_i](x_{ij} - m_{lj})^2}{\sigma_{lj}^3} \right\} \tag{7.10}$$

Here, η_b, η_α, η_m, and η_σ are learning rates for updating the parameters b, α, m, and σ, respectively;

$$\frac{\partial E_l}{\partial b_{pf}} \text{ and } \frac{\partial E}{\partial b_{pf}}$$

respectively, depend on existing and selected functions, e.g. the MDRD function (this is the GFRR case study introduced in Chapter 5) and the output function can be defined as follows.

$$f(x) = \text{GFR} = b_0 \times x_1^{b1} \times x_2^{b2} \times x_3^{b3} \times x_4^{b4} \times x_5^{b5} \times x_6^{b6} \tag{7.11}$$

In this function, x_1, x_2, x_3, x_4, x_5, and x_6 represent Scr, age, gender, race, BUN, and Alb, respectively. So that, for the local learning:

$$\frac{\partial E_l}{\partial b_0} = \sum_{i=1}^{Nl} [f(x_i) - t_i] \tag{7.12}$$

$$\frac{\partial E_l}{\partial b_p} = x_p^{bp} \ln b_p \sum_{i=1}^{Nl} [f(x_i) - t_i],\ p = 1, 2, \ldots, 6 \tag{7.13}$$

and for the global learning (suppose the MDRD function is selected for the lth cluster):

$$\frac{\partial E}{\partial b_0} = G_l(x_i) \sum_{i=1}^{N} [y(x_i) - t_i] \tag{7.14}$$

$$\frac{\partial E}{\partial b_p} = G_l(x_i) x_p^{bp} \ln b_p \sum_{i=1}^{N} [y(x_i) - t_i], \quad p = 1, 2, \ldots, 6 \tag{7.15}$$

For both local and global learning, the following iterative design method is used:

1. Fix the maximum number of learning iterations (maxK_l for the local learning and max$K$ for the global learning) and the minimum value of the error on training data (minE_l for the local learning and min$E$ for the global learning).
2. Perform Eq. (7.6) repeatedly for the local learning until the number of learning iterations k >maxK_l or the error E_l <= minE_l (E_l is calculated by Eq. (7.4)).
3. Perform Eqs. (7.7)–(7.10) repeatedly for the global learning until the number of learning iterations k >max$K$ or the error E <= min$E$ (E is calculated by Eq. (7.5)).

In this learning procedure, we use the clustering method ECM (evolving clustering method; see Chapter 2) for clustering, and a gradient descent algorithm for parameter optimisation. Although some other clustering methods can be used such as K-means, fuzzy C-means, or the subtractive clustering method, ECM is more appropriate because it is a fast one-pass algorithm and produces well-distributed clusters. The number of clusters M depends on the data distribution in the input space and it can be set up by experience, probing search, or optimisation methods (e.g. the genetic algorithm). In this research, we do not use any optimisation method to adjust M. For generalisation and simplicity, we use in the KBNN learning algorithm a standard general gradient descent method. The Levenberg–Marquardt, one-step second backpropagation algorithm, least squares method, SVD-QR method, or some others can be applied in the KBNN for parameter optimisation instead of a general gradient descent algorithm.

7.3.3 A Case Study Example

The method is applied in Song *et al.* (2006) for the creation of a KBNN model for the prediction of renal functions using the medical dataset described in Chapter 5. Nine existing regression formulas for the prediction of renal function are taken as knowledge-based transfer functions to be used in the KBNN model. Using the proposed model, more accurate results than the existing formulas or other well-known connectionist models have been obtained.

7.4 Ensemble Learning Methods for ECOS

The above two sections presented ECOS-based methods for integrating existing models and new data. Here some methods for evolving several models, including ECOS models, for solving a common problem, are presented. Using several interacting models, instead of one, may improve the performance of the final system as different models may represent different aspects of the problem and the data available (Abbass, 2004; Potter and De Jong, 2000; Kidera *et al.*, 2006; Duell *et al.*, 2006; Liu and Yao, 1999).

7.4.1 Negative Correlation Ensemble Learning

Negative correlation ensemble learning has been developed as an efficient method for improving the accuracy and the speed in ensembles of learning models when compared to single models. Several methods developed by Xin Yao *et al.* have been already published (Liu and Yao, 1999; Duell *et al.*, 2006).

This section presents a cooperative neural network ensemble learning method based on negative correlation learning (from Chan and Kasabov (2005)). It allows integration of different network models and fast implementation on both serial and parallel machines. Results demonstrate competitive performance to the original negative correlation learning method at significantly reduced communication overhead.

Effective use of a neural network ensemble requires three critical factors: first, an efficient ensemble learning scheme that is typically implemented through parallel computing because the training of multiple neural networks is computationally intensive; second, integration of different network models, e.g. MLPs and RBFs to provide a more diversified output; and third, a cooperative learning method to promote interaction between networks.

For cooperative learning, an effective method called negative correlation (NC) learning was proposed by Liu and Yao (1999) and has been shown both theoretically and empirically to improve ensemble generalisation performance. The error functions of the networks are modified to promote negatively correlated prediction errors, which in effect cause the networks to diversify in their outputs and each network to specialise in a particular aspect of the data. However, despite its effectiveness, Liu and Yao's method requires high communication overhead between the networks and it is applicable only to the ensemble of backpropagation-type networks, which hinder parallel speedup and integration of different network models, respectively. Its practicality is therefore diminished.

The novel NC (negative correlation) learning method (Chan and Kasabov, 2005) presented here alleviates the drawbacks of Liu and Yao's method. We generate new sets of data called correlation-corrected data by 'correcting' the original training data with the error correlation information. Now, instead of using penalty functions, NC learning is achieved by simply training the networks with these correlation-corrected data. This method offers two advantages: first, no error function recoding is required and second, updating correlation-corrected data requires much less communication overhead. It is therefore very suitable for parallel execution of NC learning even with an ensemble of different models in a distributed computing environment.

Liu and Yao's NC learning requires: (a) introduction of a correlation penalty function into the error function of each network, and (b) communication between networks on a pattern-by-pattern basis.

Let $T = \{x, d\} = \{(x(1), d(1)), (x(2), d(2)), \ldots, (x(N), d(N))\}$ represents the training data where N is the number of patterns and $\{x, d\}$ are the input and output (target) vectors, respectively. We form an ensemble of M networks whose joint output F is the average of all network outputs F_i, $I = [1, 2, \ldots, M]$.

Consequently, the error E is also the average of all network errors E_i:

$$\mathbf{F}(n) = \frac{1}{M}\sum_{i=1}^{M}\mathbf{F}_i(n) \qquad \mathbf{E}(n) = \frac{1}{M}\sum_{i=1}^{M}\mathbf{E}_i(n) \tag{7.16}$$

The correlation penalty P_i measures the error correlation between the ith network and the rest of the ensemble and it is formulated as follows. Recall that the goal of generalisation is to learn the generating function of the output and not the target data themselves. We use $F(n)$ to approximate the generating function such that $(F_i(n)–F(n))$ approximates the error of the ith network and $\sum_{\forall j\neq i}(F_j(n)–F(n))$ the joint error of the rest of the ensemble from the generating function. The error correlation P_i is then obtained as their product

$$\mathbf{P}_i(n) = (\mathbf{F}_i(n) - \mathbf{F}(n))\sum\nolimits_{\forall j\neq i}(\mathbf{F}_j(n) - \mathbf{F}(n)) \tag{7.17}$$

The new error function E_i is a weighted sum of the original error function and the penalty function P_i, given by

$$\mathbf{E}_i(n) = \frac{1}{2}\,\|\mathbf{F}_i(n) - \mathbf{d}(n)\|^2 + \lambda\mathbf{P}_i(n) \tag{7.18}$$

where $0 \leq \lambda \leq 1$ is the hyperparameter (a term used to describe a similar instance in network regularisation) that adjusts the strength of the correlation penalty. For adjusting the weights of the ith network through standard backpropagation, the derivative of the ensemble error E with respect to F_i is obtained using (7.16), (7.17), and (7.18)

$$\frac{\partial\mathbf{E}(n)}{\partial\mathbf{F}_i(n)} = (\mathbf{F}_i(n) - \mathbf{d}(n)) - \frac{2\lambda(M-1)}{M}(\mathbf{F}_i(n) - \mathbf{F}(n)) \tag{7.19}$$

The computation of the derivative in (7.19) requires periodic updating of the ensemble output $F(n)$ and it is on a pattern-by-pattern basis in Liu's method. The communication overhead is therefore very high.

Correlation-corrected data are transformed target data to which ordinary training of the networks in the ensemble will automate NC learning. Let c_i denote the correlation-corrected data for the ith network; it is derived as the desired network output F_i that minimizes the ensemble error E, i.e. when the derivative of E in (7.19) is set to zero

$$\text{At } \frac{\partial\mathbf{E}(n)}{\partial\mathbf{F}_i(n)} = 0, \qquad \mathbf{F}_i(n) = \mathbf{c}_i(n) = \frac{\mathbf{d}(n) - K\mathbf{F}(n)}{1-K} \qquad \text{where } K = \frac{2\lambda(M-1)}{M} \tag{7.20}$$

The generation of c_i in (7.20) requires only a simple linear combination between the original target $d(n)$ and the ensemble output $F(n)$. Like the original method, $F(n)$ must be updated periodically. However, the frequency of updating is significantly reduced because training to correlation-corrected data (7.20) is more stable and robust to error than training with a correlation-corrected gradient (7.19) as in Liu and Yao's method. This hypothesis is empirically verified (shown later) as we find that updating $F(n)$ over a number of training epochs (each epoch denotes one presentation of the whole set of training patterns) rather than over every training pattern as in Liu and Yao's case offers very similar performance. The longer update interval reduces communication overhead and allows effective parallel execution of NC learning in a coarse-grain distributed computing environment. Figure 7.8 shows the distributed computing environment used in the experiment.

In Fig 7.8 each network of the ensemble operates on a different processor node. A control centre is established to centralize all information flow and its tasks are (a) to generate the correlation-corrected data for each network, (b) to send them out, and (c) to collect the trained network outputs. Let g_{update} and g_{max} denote the number of epochs between each c_i update and the maximum number of epochs allowable. The updating of the correlation-corrected data c_i may be implemented synchronously (after all networks have finished trained for g_{update} epochs) or asynchronously (whenever a network has finished training for g_{update} epochs). In this work we implement the latter as both methods perform similarly. The procedures are summarized in the following pseudo-code.

Step 1: Initialise M networks with random weights. Partially train each network to the training data $T = \{x, d\}$ for g_{update} epochs and then obtain network output.
Step 2: Upon receipt of the ith network output F_i at the control centre:
(a) Update the ensemble output F.
(b) Create the c-corrected target c_i using **(7.20)** and send it to the ith network
(c) Train the ith network to $T_i = \{x, c_i\}$ for g_{update} epochs.
(d) Send network output F_i to control centre.
Step 3: Stop if each network has been trained for a total of g_{max} epochs.

Case Study Experiment

We compare the performance of NC learning using correlation-corrected data with Liu's original method on predicting the Mackey–Glass time series, which is a quasiperiodic and chaotic time series generated by

$$\dot{x}(t) = \beta x(t) + \frac{\alpha x(t-\tau)}{1+x^{10}(t-\tau)}$$

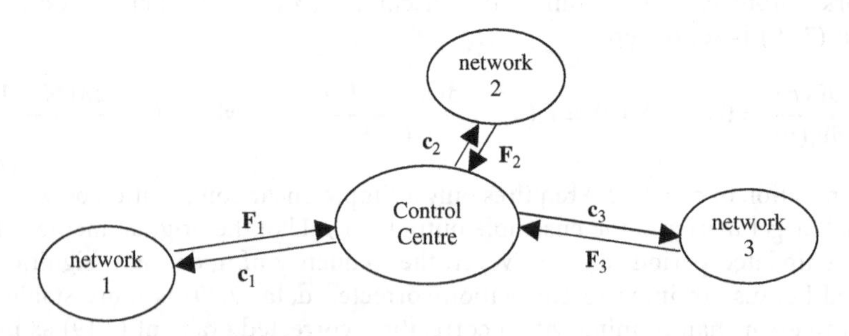

Fig. 7.8 An example of a distributed computing environment, suitable for implementing ensemble negative correlation (NC) learning using correlation-corrected data (Chan and Kasabov, 2005).

with parameters $\alpha = 0.2$, $\beta = 0.1$, and $\tau = 14$ and initial conditions $x(0) = 1.2$, $x(t-\tau) = 0$ for $0 \leq t \leq \tau$ and time-step $= 1$. The input variables are $\{x(t), x(t-6), x(t-12), x(t-18)\}$ and the output variable is $x(t+6)$. Both the training set and test set consist of 500 data points taken from the 118–617th time point and the 618–117th time point, respectively. Most of our ensemble setup follows Liu's setup. The ensemble contains $M = 20$ multilayer perceptron networks (MLPs) in the ensemble. Each network contains one hidden layer of six neurons. The hidden and output activation functions are the hyperbolic tangent and linear function, respectively. The hyperparameter λ is set to 0.5 and the maximum number of training epochs g_{max} is set to 10,000. We experimented with a set of update intervals $g_{update} = [10, 20, 40, \ldots, 2560]$ to investigate their effect on the performance. Each trial was repeated ten times. Performance was assessed by the prediction error on the test set measured in normalised root mean square (NRMS) error, which is simply the root mean square error divided by the standard deviation of the series. The results are plotted in Fig. 7.9 and shown in Table 7.2.

Figure 7.9 shows that NC learning using correlation-corrected data is effective over a range of update intervals g_{update}. The error is highest at 0.018 when $g_{update} = 10$, and decreases gradually with increasing g_{update} until it stabilises to roughly 0.0115 when $g_{update} > 100$. Although this phenomenon is contrary to the intuition that a shorter update interval produces more network interaction and leads to better overall performance, it may be attributed to an inappropriate choice of the hyperparameter λ at different update intervals (Liu and Yao (1999)) show that the ensemble performance is highly sensitive to the value of λ). It causes no problem, as longer update interval g_{update} is actually advantageous in reducing the required communication overhead.

NC learning using correlation-corrected data is clearly more cost-effective in terms of communication overhead when comparing with Liu and Yao's method. At $g_{update} = 100$, it scores slightly higher error (0.0115 cf 0.0100), yet it requires

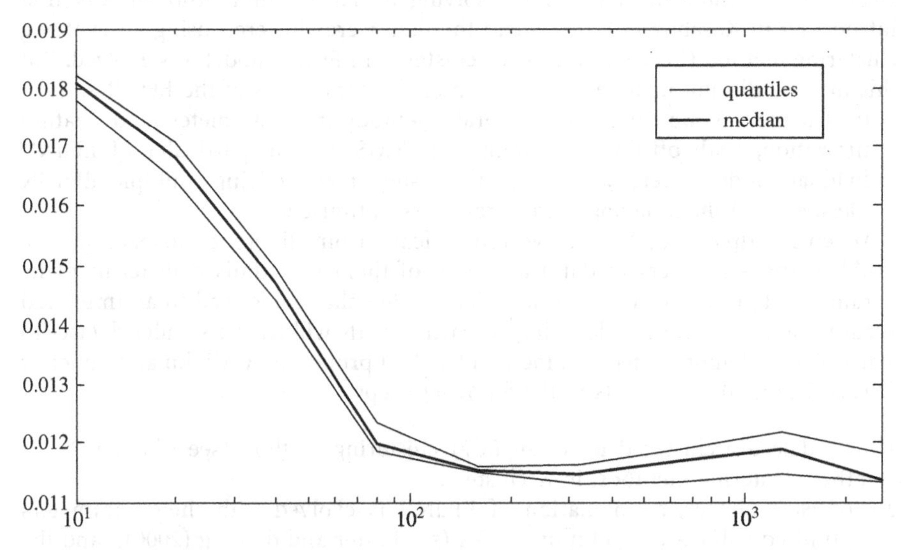

Fig. 7.9 Plot of test error versus update interval g_{update} for the NC-ensemble learning (Chan and Kasabov, 2005).

Table 7.2 Comparison of test error obtained with the use of different methods for ensemble learning (Chan and Kasabov, 2005).

Method	*No. network communication*	*NRMSE*
Cooperative ensemble learning system (CELS)	5×10^6	0.0100
Negative correlation (NC) learning using correlation corrected data ($g_{update} = 100$)	100	0.0115
EPNet	N/A	0.02
Ensemble learning with independent network training	N/A	0.02
Cascade-correlation (CC) learning	N/A	0.06

network communications of only (20 networks $\times$ (10,000 epochs/100 epochs)) = 2000 rather than (500 training patterns $\times 10,000$ epochs) = 5×10^6, which is 2.5×10^3 times smaller. Its error (0.0115) is by far lower than that of other works such as EPNet (0.02), Ensemble learning with independent network training (0.02), and cascade-correlation (CC) learning (0.06) (see Table 7.2).

The use of correlation-corrected data provides a simple and practical way to implement negative correlation ensemble learning. It allows easy integration of models from different sources and it facilitates effective parallel speedup in a coarse-grain distributed environment due to its low communication overhead requirement. Experimental results on Mackey–Glass series show that its generalisation performance is comparable to the original NC method, yet it requires a significantly smaller (2.5×10^3 times) number of network communications.

7.4.2 EFuNN Ensemble Construction Using a Clustering Method and a Co-Evolutionary Genetic Algorithm

Using an ensemble of EFuNNs for solving a classification problem was first introduced in Kasabov (2001c), where the data were clustered using an evolving clustering method (ECM) and for each cluster, an EFuNN model was evolved. But this method did not include optimization of the parameters of the EFuNNs.

In Chapter 6 we presented several methods for parameter and feature optimization, both off-line and online, of ECOS, and in particular of EFuNN individual models. Here we discuss the issue of co-evolving multiple EFuNN models, each of them having their parameters optimised.

When multiple EFuNNs are evolved to learn from different subspaces of the problem space (clusters of data) and each of them is optimised in terms of its parameters relevant to the corresponding cluster, that could lead to an improved accuracy and a speedup in learning, as each EFuNN will have a smaller dataset to learn. This is demonstrated in a method (CONE) proposed by Minku and Ludemir (2006). The method consists of the following steps.

1. The data are clustered using an ECM clustering method (see Chapter 2), or other clustering methods, in K clusters.
2. For each cluster, a population of EFuNN is evolved with their parameters optimised using a co-evolutionary GA (see Potter and de Jong (2000)), and the best one is selected after a certain number of iterations.

3. The output is calculated as the sum of weighted outputs from all the best models for each cluster.

The above procedure is applied on several benchmark datasets, such as Iris, Wine, Glass, and Cancer from the UCI Machine Learning Repository (http://www.ics.uci.edu/~mlearn/MLRepository.html). The classification accuracy of the ensembles of EFuNN is about 20% better than using a single optimised EFuNN. The following ranges for the EFuNN parameters are used: m-of-n [1,15], error threshold [0.01,0.6], maximum radius [0.01, 0.8], initial sensitivity threshold [0.4,0.99], and number of membership functions [2,8]. The maximum radius (*Dthr*) for the clustering method ECF was selected differently for each of the datasets.

7.5 Integrating ECOS and Evolving Ontologies

Ontology is a structured representation of concepts, knowledge, information, and data on a particular problem. Evolving ontologies describe a process rather than a static model (Gottgtroy *et al.*, 2006). The evolving characteristic of an ontology is achieved through the use of learning techniques inherited from data mining and through its meta-knowledge representation. For example, the hierarchical structure of an evolving ontology can be determined using the evolving clustering method ECM (Chapter 2), which permits an instance to be part of more than one cluster. At the same time, evolving ontology uses its meta-knowledge representation to cope with this multiple clustering requirement.

Figure 7.10 gives an example of linking an ECOS-based software environment NeuCom (http://www.theneucom.com) with an ontology system for bioinformatics and biomedical applications, so that the learning process is enacted via ECOS and the resulting knowledge is represented in the ontology formalism (from Gottgtroy *et al.* (2006)).

In recent years ontology structures have been increasingly used to provide a common framework across disparate systems, especially in bioinformatics, medical decision support systems, and knowledge management. The use of ontology is a key towards structuring biological data in a way that helps scientists to understand the relationships that exist between terms in a specialized area of interest, as well as to help them understand the nomenclature in areas with which they are unfamiliar.

For example, gene ontology (GO; http://www.geneontology.org), has been widely used in interdisciplinary research to analyse relationships between genes and proteins across species, including data, literature, and conceptual structured and unstructured information.

In addition to research-based literature, the amount of data produced daily by medical information systems and medical decision support systems is growing at a staggering rate. We must consider that scientific biomedical information can include information stored in the genetic code, but also can include experimental results from various experiments and databases, including patient statistics and clinical data. Large amounts of information and knowledge are available in medicine. Making medical knowledge and medical concepts shared over applications and reusable for different purposes is crucial.

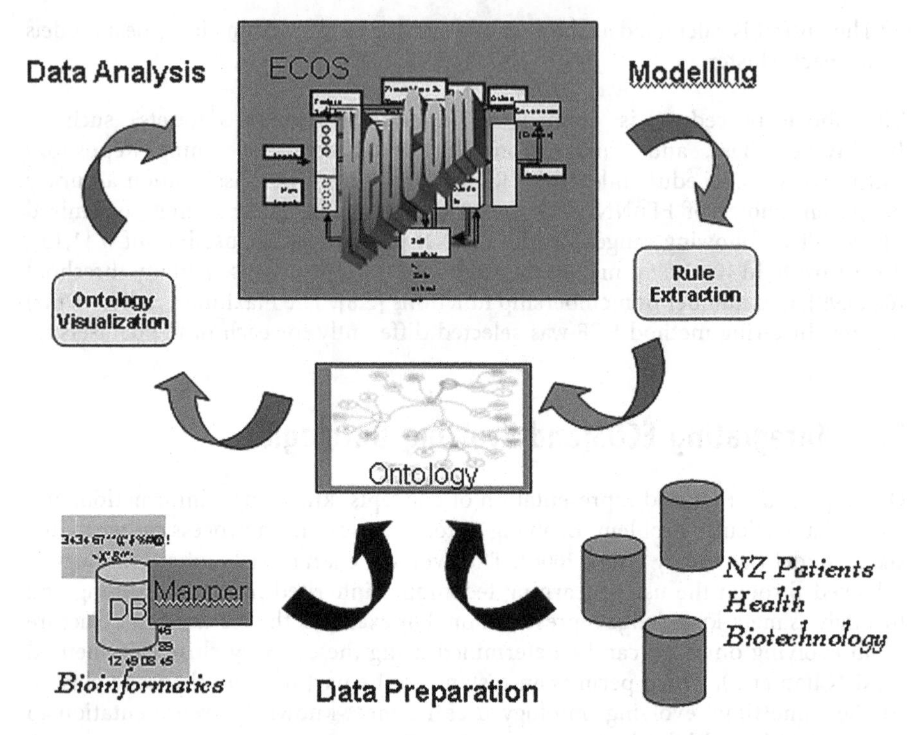

Fig. 7.10 Integrated ECOS and ontology system for applications in bioinformatics and medical decision support (from Gottroy *et al*., 2006).

A biomedical ontology is an organizational framework of the concepts involved in biological entities and processes as well as medical knowledge in a system of hierarchical and associative relations that allows reasoning about biomedical knowledge. A biomedical ontology should provide conceptual links between data from seemingly disparate fields. This might include, for example, the information collected in clinical patient data for clinical trial design, geographical and demographic data, epidemiological data, drugs, and therapeutic data, as well as from different perspectives as those collected by nurses, doctors, laboratory experts, research experiments, and so on.

Figure 7.11 shows a general ontology scheme for bioinformatics and medical decision support.

There are many software environments for building domain-oriented ontology systems, one of them being Protégé, developed at Stanford (http://protege.stanford.edu/).

7.6 Conclusion and Open Questions

This chapter presents general frameworks for building multimodel evolving machines that make use of the methods and the systems presented in Chapters 2 through 6. Issues such as the biological plausibility and complexity of an evolving

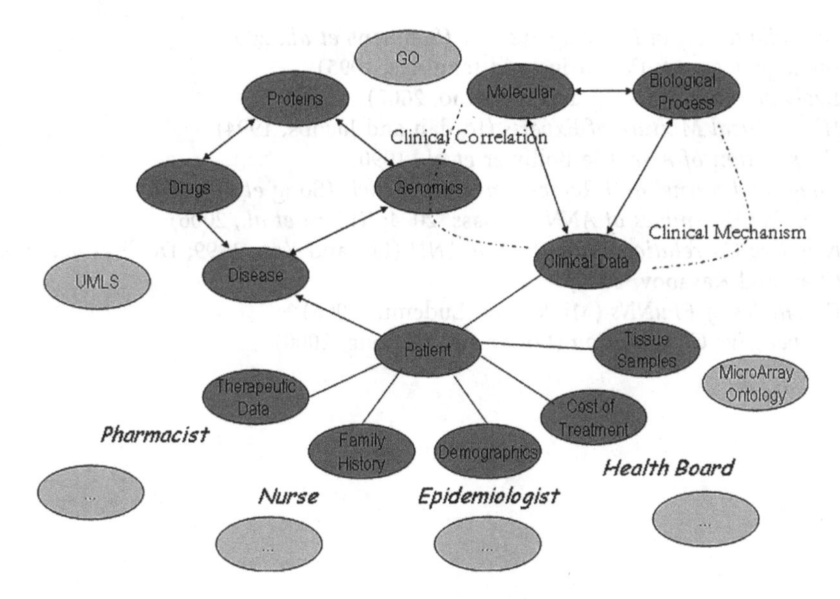

Fig. 7.11 A general ontology scheme for bioinformatics and medical decision support.

system, online parameter analysis and feature selection, and hardware implementation, are difficult and need rigid methodologies that would help the future development and numerous applications of ECOS.

Some open problems raised in the chapter are:

1. How do we build evolving machines that learn the rules that govern the evolving of both their structure and function in an interactive way?
2. How can different ECOS, that are part of an evolving machine, develop links between each other in an unsupervised mode?
3. How can ECOS and modules that are part of an evolving machine learn and improve through communication with each other? They may have a common goal.
4. Can evolving machines evolve their algorithm of operation based on very few prior rules?
5. How can evolving machines create computer programs that are evolving themselves?
6. What do ECOS need in order for them to become reproducible, i.e. new ECOS generated from an existing ECOS?
7. How can we model the instinct for information in an ECOS machine?

7.7 Further Reading

Principles of ECOS and evolving connectionist machines (Kasabov, 1998–2006)

- *Dynamic Statistical Modelling* (West and Harrison, 1989)
- *Artificial Life* (Adami, 1998)

- *Self-adaptation in Evolving Systems* (Stephens *et al.*, 2000)
- *Intelligent Agents* (Woldrige and Jennings, 1995)
- *Evolvable Robots* (Nolfi and Floreano, 2000)
- *Hierarchical Mixture of Experts* (Jordan and Jacobs, 1994)
- *Cooperation of ANN* (de Bollivier *et al.*, 1990)
- *Integrated Kernel and Regression ANN Models* (Song *et al.*, 2006)
- *Evolving Ensembles of ANN* (Abbass, 2004; Kidera *et al.*, 2006)
- *Negative Correlation Ensembles of ANN* (Liu and Yao, 1999; Duell *et al.*, 2006; Chan and Kasabov, 2005)
- *Ensembles of EFuNNs* (Minku and Ludemir, 2006)
- *Cooperative Co-evolution* (Potter and De Jong, 2000)

PART II

Evolving Intelligent Systems

Whereas in Part I of the book generic evolving learning methods are presented, in this part further methods are introduced, along with numerous applications of ECOS to various theoretical and application-oriented problems in:

- Bioinformatics (Chapter 8)
- Brain study (Chapter 9)
- Language modelling (Chapter 10)
- Speech recognition (Chapter 11)
- Image recognition (Chapter 12)
- Multimodal information processing (Chapter 13)
- Robotics and modelling economic and ecological processes (Chapter 14)

All these application-oriented evolving intelligent systems (EIS) are characterised by adaptive, incremental, evolving learning and knowledge discovery. They only illustrate the applicability of ECOS to solving problems and more applications are expected to be developed in the future.

The last chapter, 15, discusses a promising future direction for the development of quantum inspired EIS.

8. *Adaptive Modelling and Knowledge Discovery in Bioinformatics*

Bioinformatics brings together several disciplines: molecular biology, genetics, microbiology, mathematics, chemistry and biochemistry, physics, and, of course, informatics. Many processes in biology, as discussed in the introductory chapter, are dynamically evolving and their modelling requires evolving methods and systems. In bioinformatics new data are being made available with a tremendous speed that would require the models to be continuously adaptive. Knowledge-based modelling, that includes rule and knowledge discovery, is a crucial requirement. All these issues contribute to the evolving connectionist methods and systems needed for problem solving across areas of bioinformatics, from DNA sequence analysis, through gene expression data analysis, through protein analysis, and finally to modelling genetic networks and entire cells as a system biology approach. That will help to discover genetic profiles and to understand better diseases that do not have a cure thus far, and to understand better what the human body is made of and how it works in its complexity at its different levels of organisation (see Fig. 1.1). These topics are presented in the chapter in the following order.

- Bioinformatics: information growth and emergence of knowledge
- DNA and RNA sequence data analysis and knowledge discovery
- Gene expression data analysis, rule extraction, and disease profiling
- Clustering of time-course gene expression data
- Protein structure prediction
- Gene regulatory networks and the system biology approach
- Summary and open problems
- Further reading

8.1 Bioinformatics: Information Growth, and Emergence of Knowledge

8.1.1 The Central Dogma of Molecular Biology: Is That the General Evolving Rule of Life?

With the completion of the first draft of the human genome and the genomes of some other species (see, for example, Macilwain *et al.* (2000) and Friend (2000)) the task is now to be able to process this vast amount of ever-growing information and to create intelligent systems for prediction and knowledge discovery at different levels of life, from cell to whole organisms and species (see Fig. I.1).

The DNA (dioxyribonucleic acid) is a chemical chain, present in the nucleus of each cell of an organism, and it consists of pairs of small chemical molecules (bases, nucleotides) which are: adenine (A), cytosine (C), guanidine (G), and thymidine (T), ordered in a double helix, and linked together by a dioxyribose sugar phosphate nucleic acid backbone.

The central dogma of molecular biology (see Fig. 8.1) states that the DNA is transcribed into RNA, which is translated into proteins, which process is continuous in time as long as the organism is alive (Crick, 1959).

The DNA contains millions of nucleotide base pairs, but only 5% or so is used for the production of proteins, and these are the segments from the DNA that contain genes. Each gene is a sequence of base pairs that is used in the cell to produce proteins. Genes have lengths of hundreds to thousands of bases.

The RNA (ribonucleic acid) has a similar structure as the DNA, but here thymidine (T) is substituted by uridine (U). In the pre-RNA only segments that contain genes are extracted from the DNA. Each gene consists of two types of segments: exons, that are segments translated into proteins, and introns, segments that are considered redundant and do not take part in the protein production. Removing the introns and ordering only the exon parts of the genes in a sequence is called splicing and this process results in the production of messenger RNA (or mRNA) sequences.

mRNAs are directly translated into proteins. Each protein consists of a sequence of amino acids, each of them defined as a base triplet, called a codon. From one DNA sequence there are many copies of mRNA produced; the presence of certain gene in all of them defines the level of the gene expression in the cell and can indicate what and how much of the corresponding protein will be produced in the cell.

The above description of the central dogma of molecular biology is very much a simplified one, but that would help to understand the rationale behind using connectionist and other information models in bioinformatics (Brown, *et al.* 2000).

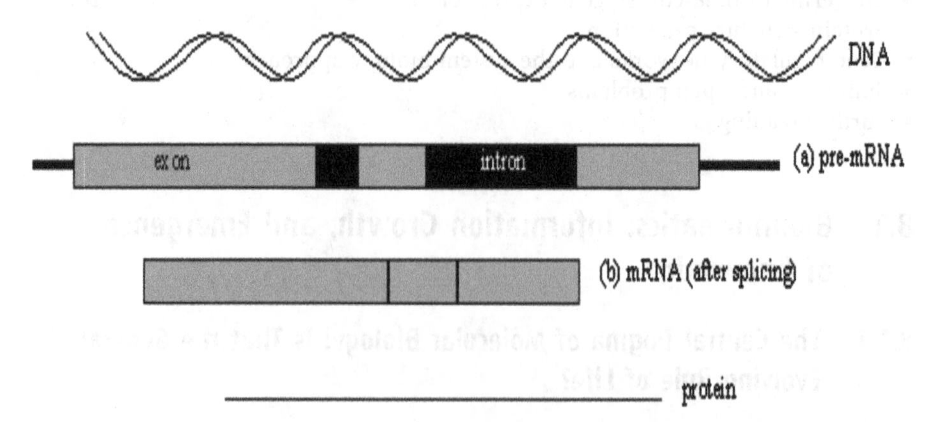

Fig. 8.1 A schematic representation of the central dogma of molecular biology: from DNA to RNA (transcription), and from RNA to proteins (translation).

Genes are complex chemical structures and they cause dynamic transformation of one substance into another during the whole life of an individual, as well as the life of the human population over many generations. When genes are 'in action', the dynamics of the processes in which a single gene is involved are very complex, as this gene interacts with many other genes and proteins, and is influenced by many environmental and developmental factors.

Modelling these interactions, learning about them, and extracting knowledge, is a major goal for bioinformatics. Bioinformatics is concerned with the application of the methods of information sciences for the collection, analysis, and modelling of biological data and the knowledge discovery from biological processes in living organisms (Baldi and Brunak, 1998, 2001; Brown *et al.* 2000).

The whole process of DNA transcription, gene translation, and protein production is continuous and it evolves over time. Proteins have 3D structures that unfold over time and are governed by physical and chemical laws. Proteins make some genes express and may suppress the expression of other genes. The genes in an individual may mutate, slightly change their code, and may therefore express differently at another time. So, genes may change, mutate, and evolve in a lifetime of a living organism.

Modelling these processes is an extremely complex task. The more new information is made available about DNA, gene expression, protein creation, and metabolism, the more accurate the information models will become. They should adapt to the new information in a continuous way. The process of biological knowledge discovery is also evolving in terms of data and information being created continuously.

8.1.2 Life-Long Development and Evolution in Biological Species

Through evolutionary processes (evolution) genes are slowly modified through many generations of populations of individuals and selection processes (e.g. natural selection). Evolutionary processes imply the development of generations of populations of individuals where crossover, mutation, and selection of individuals based on fitness (survival) criteria are applied in addition to the developmental (learning) processes of each individual.

A biological system evolves its structure and functionality through both lifelong learning of an individual and evolution of populations of many such individuals; i.e. an individual is part of a population and is a result of evolution of many generations of populations, as well as a result of its own development, of its lifelong learning process.

The same genes in the genotype of millions of individuals may be expressed differently in different individuals, and within an individual, in different cells of the individual's body. The expression of these genes is a dynamic process depending not only on the types of the genes, but on the interaction between the genes, and the interaction of the individual with the environment (the nurture versus nature issue).

Several principles are useful to take into account from evolutionary biology:

- Evolution preserves or purges genes.
- Evolution is a nonrandom accumulation of random changes.

- New genes cause the creation of new proteins.
- Genes are passed on through evolution: generations of populations and selection processes (e.g. natural selection).

There are different ways of interpreting the DNA information (see Hofstadter (1979)):

- DNA as a source of information and cells as information processing machines (Baldi and Brunak, 2001)
- DNA and the cells as stochastic systems (processes are nonlinear and dynamic, chaotic in a mathematical sense)
- DNA as a source of energy
- DNA as a language
- DNA as music
- DNA as a definition of life

8.1.3 Computational Modelling in Molecular Biology

Following are the main phases of information processing and problem solving in most of the bioinformatics systems (Fig. I.3.)

1. *Data collection:* Collecting biological samples and processing them.
2. *Feature analysis and feature extraction:* Defining which features are more relevant and therefore should be used when creating a model for a particular problem (e.g. classification, prediction, decision making).
3. *Modelling the problem:* Defining inputs, outputs, and type of the model (e.g., probabilistic, rule-based, connectionist), training the model, and statistical verification.
4. *Knowledge discovery in silico*: New knowledge is gained through the analysis of the modelling results and the model itself.
5. *Verifying the discovered knowledge in vitro and in vivo:* Biological experiments in both the laboratory and in real life to confirm the discovered knowledge.

Some tasks in bioinformatics are characterised by:

1. Small datasets, e.g. 100 or fewer samples.
2. Static datasets, i.e. data do not change in time once they are used to create a model.
3. No need for online adaptation and training on new data.

For these tasks the traditional statistical and AI techniques are well suited. The traditional, off-line modelling methods assume that data are static and no new data are going to be added to the model. Before a model is created, data are analysed and relevant features are selected, again in an off-line mode. The off-line mode usually requires many iterations of data propagation for estimating the model parameters. Such methods for data analysis and feature extraction utilise principal component analysis (PCA), correlation analysis, off-line clustering techniques (such as K-means, fuzzy C-means, etc.), self-organising maps (SOMs), and many more. Many modelling techniques are applicable to these tasks, for example: statistical techniques such as regression analysis and support vector machines; AI techniques such as decision trees, hidden Markov models, and finite automata; and neural network techniques such as MLP, LVQ, and fuzzy neural networks.

Some of the modelling techniques allow for extracting knowledge, e.g. rules from the models that can be used for explanation or for knowledge discovery. Such models are the decision trees and the knowledge-based neural networks (KBNN; Cloete and Zurada, (2000)).

Unfortunately, most of the tasks for data analysis and modelling in bioinformatics are characterized by:

1. Large dimensional datasets that are updated regularly.
2. A need for incremental learning and adaptation of the models from input data streams that may change their dynamics in time.
3. Knowledge adaptation based on a continuous stream of new data.

When creating models of complex processes in molecular biology the following issues must be considered.

- How to model complex interactions between genes and proteins, and between the genome and the environment.
- Both stability and repetitiveness are features that need to be modelled, because genes are relatively stable carriers of information.
- Dealing with uncertainty, for example, when modelling gene expressions, there are many sources of uncertainty, such as

  - Alternative splicing (a splicing process of same RNAs resulting in different mRNAs).
  - Mutation in genes caused by: ionizing radiation (e.g. X-rays); chemical contamination, replication errors, viruses that insert genes into host cells, etc. Mutated genes express differently and cause the production of different proteins.

For large datasets and for continuously incoming data streams that require the model and the system to rapidly adapt to new data, it is more appropriate to use online, knowledge-based techniques and ECOS in particular as demonstrated in this chapter.

There are many problems in bioinformatics that require their solutions in the form of a dynamic, learning, knowledge-based system. Typical problems that are also presented in this chapter are:

- Discovering patterns (features) from DNA and RNA sequences (e.g., promoters, RBS binding sites, splice junctions)
- Analysis of gene expression data and gene profiling of diseases
- Protein discovery and protein function analysis
- Modelling the full development (metabolic processes) of a cell (Tomita, 2001)

An ultimate task for bioinformatics would be predicting the development of an organism from its DNA code. We are far from its solution now, but many other tasks on the way can be successfully solved through merging information sciences with biological sciences as demonstrated in this chapter.

8.2 DNA and RNA Sequence Data Analysis and Knowledge Discovery

8.2.1 Problem Definition

Principles of DNA Transcription and RNA Translation and Their Computational Modelling

As mentioned previously, only two to five percent of the human genome (the DNA) contains information that concerns the production of proteins (Brown *et al.*, 2000). The number of genes contained in the human genome is about 40,000 (Friend, 2000). Only the gene segments are transcribed into RNA sequences. The transcription is achieved through special proteins, enzymes called RNA polymerase, that bind to certain parts of the DNA (promoter regions) and start 'reading' and storing in an mRNA sequence each gene code.

Analysis of a DNA sequence and identifying promoter regions is a difficult task. If it is achieved, it may make possible to predict, from DNA information, how this organism will develop or, alternatively, what an organism looked like in retrospect.

In simple organisms, bacteria (prokaryotic organisms), DNA is transcribed directly into mRNA that consists of genes that contain only codons (no intron segments). The translation of the genes into proteins is initiated by proteins called ribosomes, that bind to the beginning of the gene (ribosome binding site) and translate the sequence until reaching the termination area of the gene. Finding ribosome binding sites in bacteria would reveal how the bacteria would act and what proteins would be produced.

In higher organisms (that contain a nucleus in the cell) the DNA is first transcribed into a pre-mRNA that contains all the regions from the DNA that contain genes. The pre-RNA is then transcribed into many sequences of functional mRNAs through a splicing process, so that the intron segments are deleted from the genes and only the exon segments, that account for proteins, are extracted. The functional mRNA is now ready to be translated into proteins.

Finding the splice junctions that separate the introns from the exons in a pre-mRNA structure, is another difficult task for computer modelling and pattern recognition, that once solved would help us understand what proteins would be produced from a certain mRNA sequences. This task is called splice junction recognition.

But even having recognized the splice junctions in a pre-mRNA, it is extremely difficult to predict which genes will really become active, i.e. will be translated into proteins, and how active they will be: how much protein will be produced.That

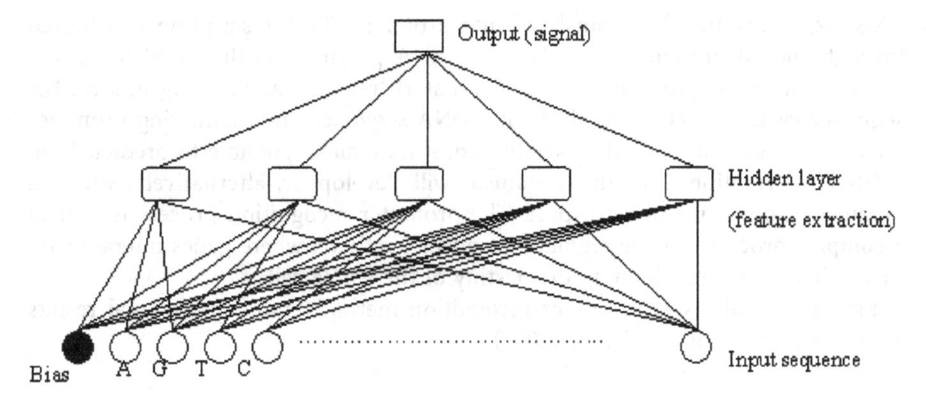

Fig. 8.2 A hypothetical scheme of using neural networks for DNA/RNA sequence analysis.

is why gene expression technologies (e.g. microarrays) have been introduced, to measure the expression of the genes in mRNAs. The level of a gene expression would suggest how much protein of this type would be produced in the cell, but again this would only be an approximation.

Analysis of gene expression data from microarrays is discussed in the next section. Here, some typical tasks of DNA and RNA sequence pattern analysis are presented, namely ribosome binding site identification and splice junction recognition.

Recognizing patterns from DNA or from mRNA sequences is a way of recognizing genes in these sequences and of predicting proteins *in silico* (in a computer). For this purpose, usually a 'window' is moved along the DNA sequence and data from this window are submitted to a classifier (identifier) which identifies if one of the known patterns is contained in this window. A general scheme of using neural networks for sequence pattern identification is given in Fig. 8.2.

Many connectionist models have been developed for identifying patterns in a sequence of RNA or DNA (Fu, 1999; Baldi and Brunak, 2001). Most of them deal with a static dataset and use multilayer perceptrons MLP, or self-organising maps SOMs.

In many cases, however, there is a continuous flow of data that is being made available for a particular pattern recognition task. New labelled data need to be added to existing classifier systems for a better classification performance on future unlabelled data. This can be done with the use of the evolving models and systems.

Several case studies are used here to illustrate the application of evolving systems for sequence DNA and RNA data analysis.

8.2.2 DNA Promoter Recognition

Only two to five percent of the human genome (the DNA) contains useful information that concerns the production of proteins. The number of genes contained in the human genome is about 40,000. Only the gene segments are transcribed into

RNA sequences and then translated into proteins. The transcription is achieved through special proteins, enzymes called RNA polymerase, that bind to certain parts of the DNA (promoter regions) and start 'reading' and storing in a mRNA sequence each gene code. Analysis of a DNA sequence and identifying promoter regions is a difficult task. If it is achieved, it may make possible to predict, from a DNA information, how this organism will develop or, alternatively, what an organism looked like in retrospect. The promoter recognition process is part of a complex process of gene regulatory network activity, where genes interact with each other over time, defining the destiny of the whole cell.

Extensive analysis of promoter recognition methods and experimental results are presented in Bajic and Wee (2005).

Case Study

In Pang and Kasabov (2004) a transductive SVM is compared with SVM methods on a collection of promoter and nonpromoter sequences. The promoter sequences are obtained from the eukaryotic promoter database (EPD) http://www.epd.isb-sib.ch/. There are 793 different vertebrate promoter sequences of length 250 bp. These 250 bp long sequences represent positive training data. We also collected a set of nonoverlapping human exon and intron sequences of length 250 bp each, from the GenBank database: http://www.ncbi.nlm.nih.gov/Genbank/GenbankOverview.html, Rel. 121. For training we used 800 exon and 4000 intron sequences.

8.2.3 Ribosome Binding Site Identification

Case Study. A Ribosome Binding Site Identification in E.coli Bacteria

The following are the premises of the task and the parameters of the developed modelling system. The dataset contains 800 positive (each of them contains a RBS) and 800 negative base sequences (they do not contain a RBS), each of them 33 bases long.

The task is to develop a RBS identification system, so that if a new 33 base long sequence is submitted, the system will identify if there is a RBS within this sequence, or not. This task has been dealt with in several publications (Fu, 1999).

The following encoding is used: A = 1000, T = 0100, G = 0010, C = 0001; binding site = 1, nonbinding site = 0.

An evolving fuzzy neural network EFuNN (Chapter 3) was trained in an online mode for the parameter values: 132 inputs (33×4), 1 output, initial sensitivity threshold $Sthr = 0.9$, error threshold $Erthr = 0.1$, initial learning rate $lr = 0.1$, m-of-n value $m = 3$, evolved rule nodes $Rn = 9$ after the presentation of all 1600 examples; aggregation of rule nodes is performed after every 50 examples – $Nagg = 50$.

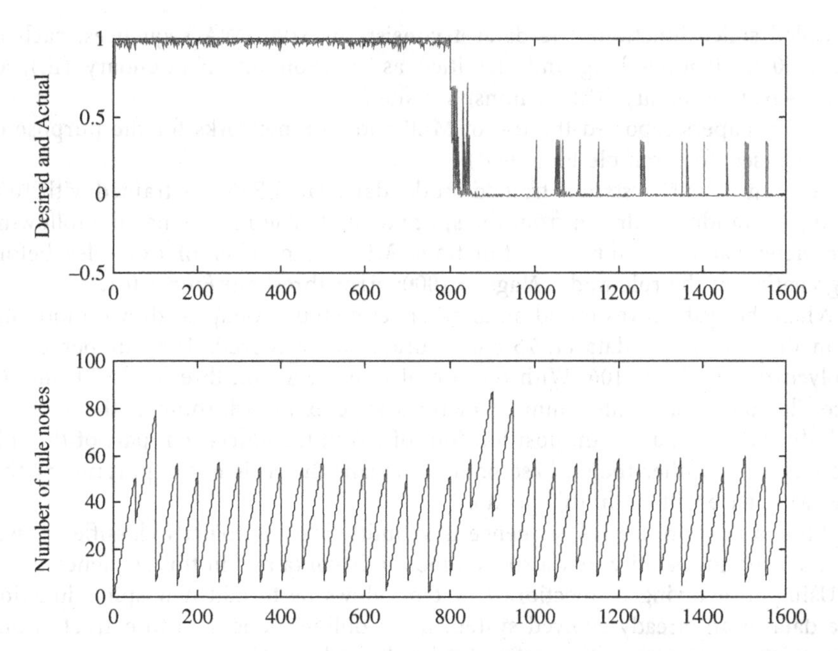

Fig. 8.3 Online learning of ribosome binding site data of two classes (Yes 1, and No 0). EFuNN has learned very quickly to predict new data in an online learning mode. The upper figure gives the desired versus the predicted one-step-ahead existence of RBS in an input vector of 33 bases (nucleotides). The lower figure shows the number of rule nodes created during the learning process. Aggregation is applied after every 50 examples.

Figure 8.3 shows the process of online learning of EFuNN from the ribosome binding site data. It took a very small number of examples for the EFuNN to learn the input patterns of each of the two classes and predict properly the class of the following example from the data stream. Nine rules are extracted from the trained EFuNN that define the knowledge when, in a sequence of 33 bases, one should expect that there would be a RBS, and when a RBS should not be expected. Through aggregation, the number of the rules is kept comparatively small which also improves the generalisation property of the system.

Using an evolving connectionist system allows us to add to the already evolved system new RBS data in an online mode and to extract at any time of the system operation refined rules.

8.2.4 RNA Intron/Exon Splice Junction Identification

Here a benchmark dataset, obtained from the machine learning database repository at the University of California, Irvine (Blake and Merz, 1998) is used. It contains primate splice-junction gene sequences for the identification of splice site boundaries within these sequences. As mentioned before, in eukaryotes the genes that code for proteins are contained in coding regions (exons) that are separated from noncoding regions (introns) of the pre-mRNA at definedboundaries, the so-called splice junction. The dataset consists of 3190 RNA sequences, each of them 60 nucleotides long and classified as an exon–intron boundary (EI), an intron–exon boundary (IE), or nonsplice site (N).

Several papers reported the use of MLP and RBF networks for the purpose of the task (see, for example Fu (1999)).

Here, an EFuNN system is trained on the data. The EFuNN is trained with 1000 examples, randomly drawn from the splice dataset. The EFuNN has the following parameter values: membership functions $MF = 2$; number of examples before aggregation of the rule nodes $Nagg = 1000$; error threshold $Ethr = 0.1$.

When the system was tested on another set of 1000 examples, drawn randomly from the rest of the dataset, 75.7% accuracy was achieved. The number of the evolved rule nodes is 106. With the use of rule extraction thresholds $T1$ and $T2$ (see Chapter 3), a smaller number of rules were extracted, some of them shown in Table 8.1. If on a certain position (out of 33) in the antecedent part of the rule there is a '—' rather than a base name, it means that it is not important for this rule what base will be on this position.

When a new pre-mRNA sequence is submitted to the EFuNN classifier, it will produce the probability that a splice junction is contained in this sequence.

Using an evolving connectionist system allows us to add new splice junction site data to an already evolved system in an online mode, and to extract at any time of the system operation refined splice boundary rules.

Table 8.1 Several rules extracted from a trained EFuNN model on a splice-junction dataset.

Rule4: if ———————————————-GGTGAG—C————————————
then [EI], receptive field = 0.221, max radius = 0.626, examples = 73/1000
Rule20: if —-G————C-G–G———-AGGTG-G-G-G–G—-G–GGG———
then [EI], receptive field = 0.190, max radius = 0.627, examples = 34/1000
Rule25: if ——C–C-C-TCC-G–CTC-GT-C–GGTGAGTG–GGC—C—G-GG-C–CC-
then [EI], receptive field = 0.216, max radius = 0.628, examples = 26/1000
Rule2: if ——-CC————C-TC-CC—CAGG——-——-G————-——-C—-
then [IE], receptive field = 0.223, max radius = 0.358, examples = 64/1000
Rule12: if ——————————-C—C–CAGG———————————————
then [IE], receptive field = 0.318, max radius = 0.359, examples = 83/1000
Rule57: if ————-T——T-C-T-T–T-CAGG————————-C-A———-T-
then [IE], receptive field = 0.242, max radius = 0.371, examples = 21/1000
Rule5: if ———A————T——————————————————————————
then [N], receptive field = 0.266, max radius = 0.391, examples = 50/1000
Rule6: if ——————————————G–G–G–C-G———G————G–GA—-
then [N], receptive field = 0.229, max radius = 0.400, examples = 23/1000
Rule9: if ———G————————————————————————-G————G—-T———
then [N], receptive field = 0.203, max radius = 0.389, examples = 27/1000
Rule10: if ————A-A——A———————A———A—-A———————-A—-
then [N], receptive field = 0.291, max radius = 0.397, examples = 36/1000
Rule14: if —-T—-T—G-TG—-TT-T———————C——C——————C———
then [N], receptive field = 0.223, max radius = 0.397, examples = 21/1000

8.2.5 MicroRNA Data Analysis

RNA molecules are emerging as central 'players' controlling not only the production of proteins from messenger RNAs, but also regulating many essential gene expression and signalling pathways. Mouse cDNA sequencing project FANTOM in Japan showed that noncoding RNAs constitute at least a third of the total number of transcribed mammalian genes. In fact, about 98% of RNA produced in the eukaryotic cell is noncoding, produced from introns of protein-coding genes, non-protein-coding genes, and even from intergenic regions, and it is now estimated that half of the transcripts in human cells are noncoding and functional.

These noncoding transcripts are thus not junk, but could have many crucial roles in the central dogma of molecular biology. The most recently discovered, rapidly expanding group of noncoding RNAs is microRNAs, which are known to have exploded in number during the emergence of vertebrates in evolution. They are already known to function in lower eukaryotes in regulation of cell and tissue development, cell growth and apoptosis, and many metabolic pathways, with similar likely roles in vertebrates (Havukkala *et al.* 2005).

MicroRNAs are encoded by long precursor RNAs, commonly several hundred basepairs long, which typically form foldback structures resembling a straight hairpin with occasional bubbles and short branches. The length and the conservation of these long transcribed RNAs make it possible by a sequence similarity search method to discover and classify many phylogenetically related microRNAs in the Arabidopsis genome (Havukkala *et al.*, 2005). Such analysis has established that most plant microRNA genes have evolved by inverted duplication of target gene sequences. The mechanism of their evolution in mammals is less clear.

Lack of conserved microRNA sequences or microRNA targets between animals and plants suggests that plant microRNAs evolved after the split of the plant lineage from mammalian precursor organisms. This means that the information about plant microRNAs does not help to identify or classify most mammalian microRNAs. Also, in mammalian genomes the foldback structures are much shorter, down to only about 80 basepairs, making a sequence similarity search a less effective method for finding and clustering remotely related microRNA precursors.

Several sequence similarity and RNA-folding-based methods have been developed to find novel microRNAs that include: Simple BLAST similarity search; Screening by RNA-fold prediction algorithms (best known are Mfold and RNAfold) to look for stem-loop structure candidates having a characteristically low deltaG value indicating strong hybridization of the folded molecule, followed by further screening by sequence conservation between genomes of related species; Careful multiple alignment of many different sequences from closely related primate species to find accurate conservation at single nucleotide resolution.

The problem with all these approaches is that they require extensive sequence data and laborious sequence comparisons between many genomes as one key filtering step. Also, findingspecies-specific, recently evolved microRNAs by these methods is difficult, as well as evaluating the phylogenetic distance of remotely related genes which have diverged too much in sequence.

One tenet of this section is that the two-dimensional (2D) structure of many microRNAs (and noncoding RNAs in general) can give additional information which is useful for their discovery and classification, even with data from within only one species. This is analogous to protein three-dimensional (3D) structure analysis showing often functional and/or evolutionary similarities between proteins that cannot easily be seen by sequence similarity methods alone (Havukkala *et al.*, 2006).

Prediction of RNA folding in 2D is more advanced, and reasonably accurate algorithms are available which can simulate the putative most likely and thermodynamically most stable structures of self-hybridizing RNA molecules. Many such structures have been also verified by various experimental methods in the laboratory, corroborating the general accuracy of these folding algorithms.

In Havukkala *et al.* (2005, 2006) we have approached the problem by utilising visual information from images of computer-simulated 2D structures of macromolecules. The innovation is to use suitable artificial intelligence image analysis methods, such as the Gabor filter on bitmap images of the 2D conformation. This is in contrast to the traditional approach of using as a starting point various extracted features, such as the location and size/length of loops/stems/branches etc., which can comprise a preconceived hypothesis of the essential features of the molecule conformation. The procedure is to take a sample of noncoding RNA sequences, calculate their 2D thermodynamically most stable conformation, output the image of the structure to bitmap images, and use a variety of rotation-invariant image analysis methods to cluster and classify the structures without preconceived hypotheses as to what kind of features might be important ones. Although one may lose specific information about the exact location or length of loops/stems or specific sequence motifs, the image analysis could reveal novel relevant features in the image that may not be intuitively obvious to the human eye, e.g. fractal index of the silhouette, ratio of stem/loop areas, handedness of asymmetric configurations, etc.

8.3 Gene Expression Data Analysis, Rule Extraction, and Disease Profiling[1]

8.3.1 Problem Definition

One of the contemporary directions while searching for efficient drugs for many terminal illnesses, such as cancer or HIV, is the creation of gene profiles of these

[1]The gene expression profiling methodology described in this section constitutes intellectual property of the Pacific Edge Biotechnology Ltd. (PEBL) (http://www.peblnz.com). Prior permission from PEBL is needed for any commercial applications of this methodology. The methodology is protected as a PCT patent (Kasabov, 2001b).

diseases and subsequently finding targets for treatment through gene expression regulation. A gene profile is a pattern of expression of a number of genes that is typical for all, or for some of the known samples of a particular disease. A disease profile would look like:

IF (gene g1 is highly expressed) AND (gene g37 is low expressed) AND (gene 134 is very highly expressed) THEN most probably this is cancer type C (123 out of available 130 samples have this profile).

Having such profiles for a particular disease makes it possible to set early diagnostic testing, so a sample can be taken from a patient, the data related to the sample processed, and a profile obtained. This profile can be matched against existing gene profiles and based on similarity, it can be predicted with certain probability if the patient is in an early phase of a disease or he or she is at risk of developing the disease in the future.

A methodology of using DNA or RNA samples, labelled with known diseases, that consists of training an evolving system and extracting rules, that are presented as disease profiles, is illustrated schematically in Fig. 8.4. Each profile is a rule that is extracted from a trained ECOS, which on the figure is visualised through colours: the higher the level of a gene expression, the brighter the colour. Five profiles are visualised in Fig. 8.4. The first three represent a group of samples of class 1 (disease), with the second two representing class two (normal case). Each column in the condition part of the rules (profiles) represents the expression of one gene out of the 100 relevant genes used in this example.

Microarray equipment is used widely at present to evaluate the level of gene expression in a tissue or in a living cell (Schena, 2000). Each point (pixel, cell) in a microarray represents the level of expression of a single gene. Five principal steps in the microarray technology are shown in Fig. 8.5. They are: tissue collection RNA extraction, microarray gene expression recording, scanning, and image processing, and data analysis.

The recent advent of DNA microarray and gene chip technologies means that it is now possible to simultaneously interrogate thousands of genes in tumours. The potential applications of this technology are numerousand include

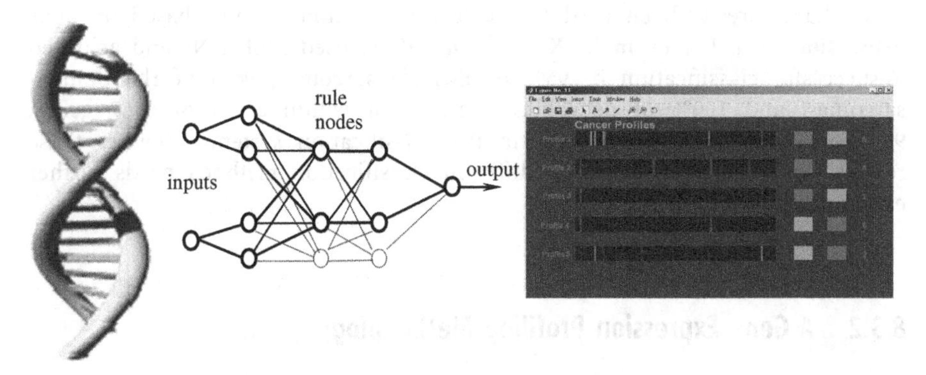

Fig. 8.4 A schematic representation of the idea of using evolving connectionist systems to learn gene profiles of diseases from DNA/RNA data.

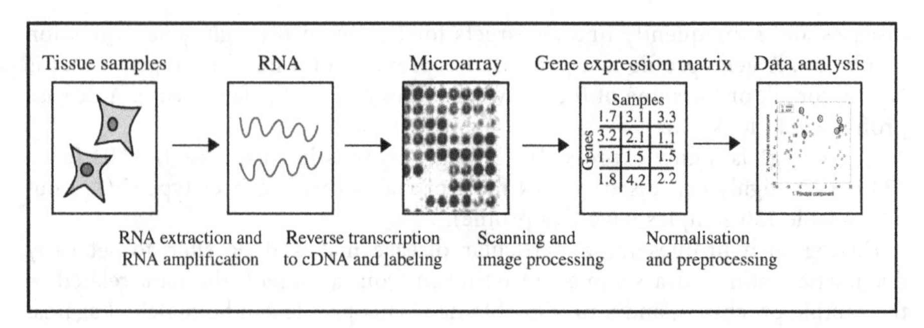

Fig. 8.5 Principal steps in a gene expression microarray experiment with a consecutive data analysis and profile discovery.

identifying markers for classification, diagnosis, disease outcome prediction, therapeutic responsiveness, and target identification. Microarray analysis might not identify unique markers (e.g. a single gene) of clinical utility for a disease because of the heterogeneity of the disease, but a prediction of the biological state of disease is likely to be more sensitive by identifying clusters of gene expressions (profiles; Kasabov (2001a,b)).

For example, gene expression clustering has been used to distinguish normal colon samples from tumours from within a 6500 gene set, although clustering according to clinical parameters has not been undertaken (Alon *et al.*, 1999). Although distinction between normal and tumour tissue can be easily made using microscopy, this analysis represented one of the early attempts to classify biological samples through gene expression clustering. The above dataset is used in this section to extract profiles of colon cancer and normal tissue through using an evolving fuzzy neural network EFuNN (Chapter 3).

Another example of profiling developed in this chapter is for the distinction between two subtypes of leukaemia, namely AML and ALL (Golub *et al.*, 1999).

NN have already been used to create classification systems based on gene expression data. For example, Khan *et al.* (2001) used MLP NNs and achieved a successful classification of 93% of Ewing's sarcomas, 96% of rhabdomyosarcomas, and 100% of neuroblastomas. From within a set of 6567 genes, 96 genes were used as variables in the classification system. Whether these results would be different using different classification methods needs further exploration.

8.3.2 A Gene Expression Profiling Methodology

A comprehensive methodology for profiling of gene expression data from microarrays is described in Futschik *et al.* (2002, 2003a,b). It consists of the following phases.

1. Microarray data preprocessing. This phase aims at eliminating the low expressed genes, or genes that are not expressed sufficiently across the classes(e.g. controlled versus tumour samples, or metastatic versus nonmetastatic tumours). Very often log transformation is applied in order to reduce the range of gene expression data. An example of how this transformation 'squeezes' the gene expression values plotted in the 2D principal components, is given in Fig. 8.6. There are only two samples used (two cell lines) and only 150 genes out of the 4000 on the microarray, that distinguish these samples.
2. Selecting a set of significant differentially expressed genes across the classes. Usually the t-test is applied at this stage with an appropriate threshold used (Metcalfe, 1994). The t-test calculates in principle the difference between the mean expression values for each gene g for each class (e.g. two classes: class 1, normal and class two, tumour)

$$t = (\mu_1 - \mu_2)/\sigma_{12} \tag{8.1}$$

   where μ_1 and μ_2 are the mean expression values of gene g for class 1 and class 2 respectively; σ_{12} is the variance.
3. Finding subsets of (a) underexpressed genes and (b) overexpressed genes from the selected ones in the previous step. Statistical analysis of these subsets is performed.
4. Clustering of the gene sets from phase 3 that would reveal preliminary profiles of jointly overexpressed/underexpressed genes across the classes. An example of hierarchical clustering of 12 microarray vectors (samples), each containing the expression of 50 genes after phases 1 to 3 were applied on the initial 4000 gene expression data from the microarrays, is given in Fig. 8.7. Figure 8.7a plots the samples in a 2D Sammon projection space of the 50D gene expression space. Figure 8.7b presents graphically the similarity between the samples (columns), based on the 50 selected genes, and the similarity between the genes (rows) based on their expression in the 12 samples (see chapter 2).

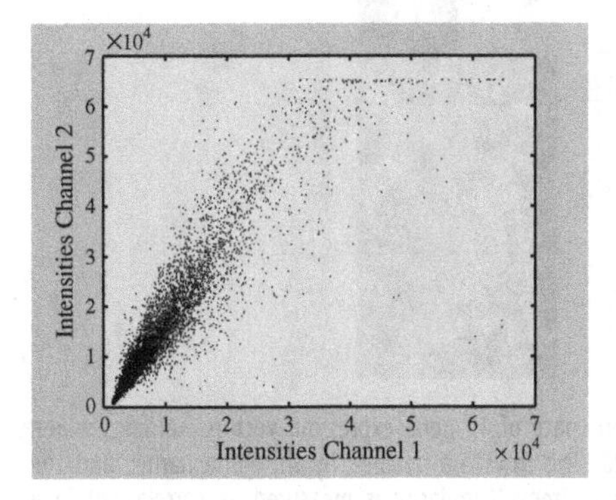

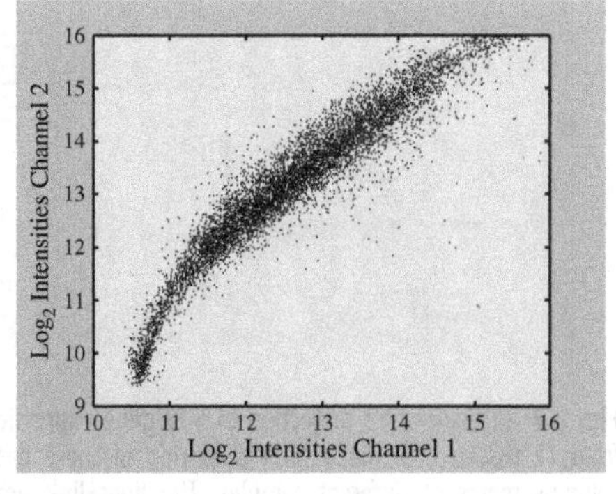

Fig. 8.6 Gene expression data: (a) before log transformation; (b) after log transformation.

(a)

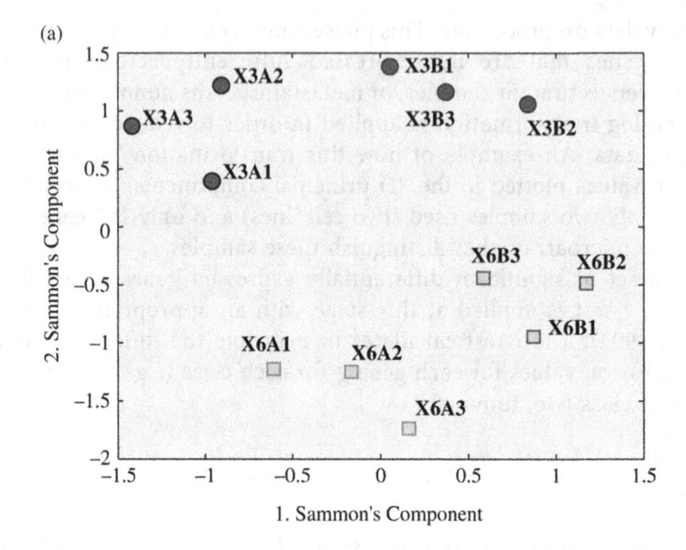

(b)

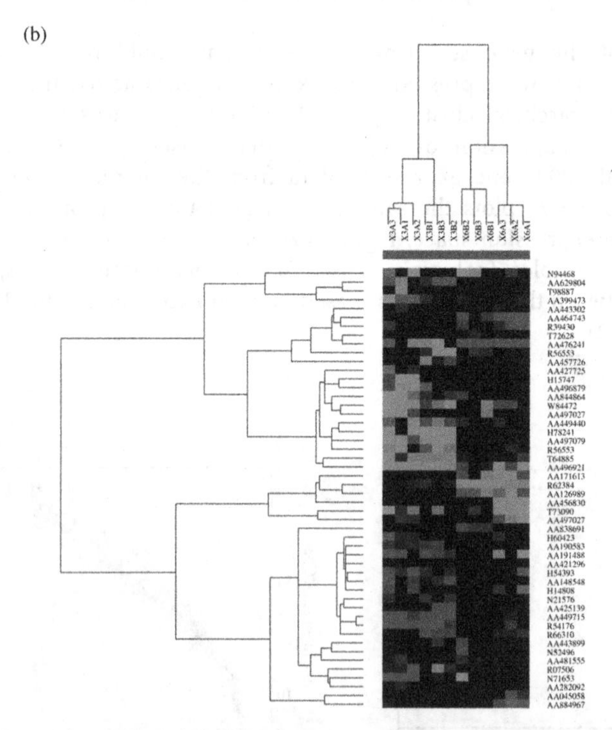

Fig. 8.7 (a) Sammon's projection of 50D gene expression space of 12 gene expression vectors (samples, taken from 12 tissues); (b) hierarchical clustering of these data. The rows are labelled by the gene names and the columns represent different samples. The lines link similar items (similarity is measured as correlation) in a hierarchical fashion.

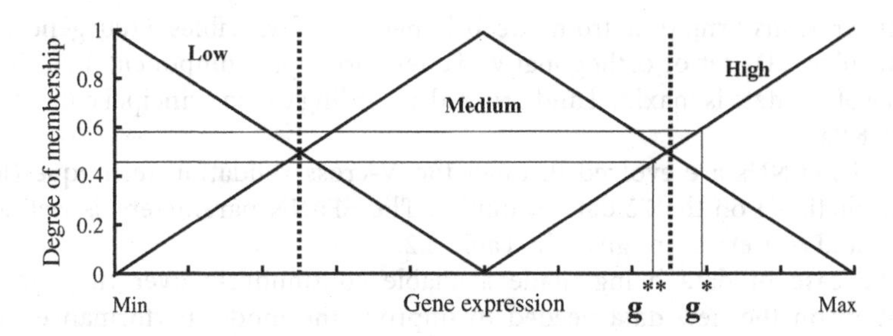

Fig. 8.8 Gene expression values are fuzzified with the use of three triangular membership functions (MF) (Futschik *et al.*, 2002).

5. Building a classification model and extracting rules that define the profiles for each class. The rules would represent the fine grades of the common expression level of groups of genes. Through using thresholds, smaller or larger groups of genes can be selected from the profile. For a better rule representation, gene expression values can be fuzzified as it is illustrated in Fig. 8.8.
6. Further training of the model on new data and updating the profiles. With the arrival of new labelled data (samples) the model needs to be updated, e.g. trained on additional data, and possibly modified rules (profiles) extracted.

Two datasets are used here to illustrate the above methodology that explores evolving systems for microarray data analysis.

8.3.3 Case Study 1: Gene Profiling of Two Classes of Leukaemia with the Use of EFuNN

A dataset of 72 classification examples for leukaemia cancer disease is used, that consists of two classes and a large input space, the expression values of 6817 genes monitored by Affymetrix arrays (Golub et al., 1999). The two types of leukaemia are acute myeloid leukaemia (AML) and acute lymphoblastic leukaemia (ALL). The latter one can be subdivided further into T-cell and B-cell lineage classes. Golub *et al.* split the dataset into 38 cases (27 ALL, 11 AML) for training and 34 cases (20 ALL, 14 AML) for validation of a classifier system. These two sets came from different laboratories. The test set shows a higher heterogeneity with regard to tissue and age of patients making any classification more difficult.

The task is: (1) to find a set of genes distinguishing ALL and AML; (2) to construct a classifier based on these data; and (3) to find a gene profile of each of the classes.

After having applied points 1 and 2 from the methodology above, 100 genes are selected.

A preliminary analysis on the separability of the two classes can be done through plotting the 72 samples in the 2D principal component analysis space.PCA consists of a linear transformation from the original set of variables (100 genes) to a new (smaller, 2D) set of orthogonal variables (principal components) so that the variance of the data is maximal and ordered according to the principal components; see Fig. 8.9a.

Several EFuNNs are evolved through the N-cross-validation technique (leave-one-out method) on the 72 data examples. The EFuNN parameters as well as the training and test error are given in Table 8.2.

In the case of data being made available continuously over time and fast adaptation on the new data needed to improve the model performance, online modelling techniques would be more appropriate, so that any new labelled data

(a)

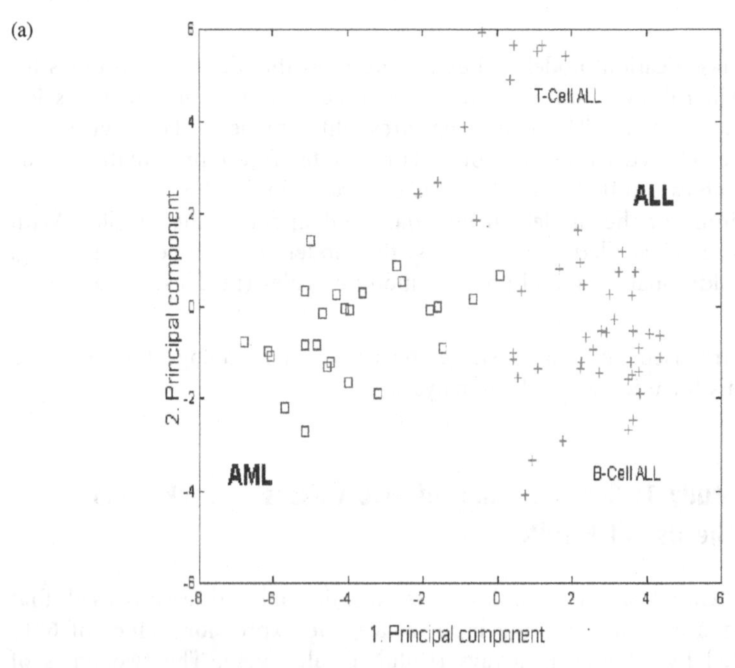

(b)

Fig. 8.9 (a) The first two principal components of the leukaemia 100 genes selected after a t-test is applied: □ AML, + ALL (Futschik *et al.*, 2002); (b) some of the rules extracted from an EFuNN trained on the leukaemia data and visualised as profile patterns (Futschik *et al.*, 2002).

Table 8.2 The parameter values and error results of N-cross-validation EFuNN models for the leukaemia and colon cancer data (Futschik *et al.*, 2002).

Data/Model	*Errthr*	*Rmax*	*Nagg*	*Rule Nodes*	*Classification Accuracy – Training Data*	*Classification Accuracy – Test Data*
Leukaemia-EFuNN	0.9	0.3	20	6.1	97.4	95.8
Leukaemia-EFunN	0.9	0.5	20	2.0	95.2	97.2
Colon cancer-EFuNN	0.9	1.0	40	2.3	88.8	90.3
Colon cancer-EFuNN	0.9	1.0	40	2.3	88.8	91.9

will be added to the EFuNN and the EFuNN will be used to predict the class of any new unlabelled data.

Different EFuNN were evolved with the use of different sets of genes as input variables. The question of which is the optimum number of genes for a particular task is a difficult one to answer. Table 8.3 shows two of the extracted rules after all the examples, each of them having only 11 genes, are learned by the EFuNN. The rules are 'local' and each of them has the meaning of the dominant rule in a particular subcluster of each class from the input space. Each rule covers a cluster of samples that belong to a certain class. These samples are similar to a degree that is defined as the radius of the receptive field of the rule node representing the cluster. For example, Rule 6 from Table 8.3 shows that 12 samples of class 2 (AML) are similar in terms of having genes g2 and g4 overexpressed, and at the same time genes g8 and g9 are underexpressed.

One class may be represented by several rules, profiles, each of them covering a subgroup of similar samples. This can lead to a new investigation on why the subgroups are formed and why they have different profiles (rules), even being part of the same class.

The extracted rules for each class comprise a profile of this class. One way of visually representing these profiles is illustrated in Fig. 8.9b, where rules were extracted from a trained EFuNN with 100 genes.

Table 8.3 Some of the rules extracted from the evolved EFuNN.

Rule 1: if [g1] is (2 0.9) and [g3] is (2 0.9) and [g5] is (2 0.7) and [g6] is (2 0.7) and [g8] is (1 0.8) and [g9] is (2 0.7), receptive field = 0.109 (radius of the cluster), then Class 1, accommodated training examples = 27/72.

Rule 6: if [g2] is (2 0.8) and [g4] is (2 0.874) and [g8] is (1 0.9) and [g9] is (1 0.7), receptive field = 0.100, then Class 2, accommodated training examples =12/72.
....
Denotation: [g1] is (2 0.9) means that the membership degree to which gene 1 expression value belongs to the membership function "High" is 0.9. Alternatively 1 means membership function "Low". There is a membership degree threshold of 0.7 used and values less than this threshold are not shown.

8.3.4 Case Study 2: Gene Profiling of Colon Cancer

The second dataset is two-class gene expression data, the classes being 'colon cancer' and 'normal' (see Alon *et al.* (1999)). The data were collected from 40 tumour and 22 normal colon tissues sampled with the use of the Affymetrix oligonucleotide microarrays. The expression of more than 6500 genes and ESTs is collected for each sample.

After the preprocessing, the normalisation, the log-transformation, and the t-test analysis, only 50 genes are selected for the creation of the classification model and for the knowledge discovery procedure. Figure 8.10 shows the projection of the 62 samples from the 50D gene expression space into the 2D PCA space, and also the ordered gene-samples hierarchical clustering diagram according to a similarity measured through the Pearson correlation coefficients (see Futschik *et al.* (2002, 2003)).

Through N-cross-validation, 62 EfuNNs are evolved on 61 data examples each and tested on the left-out example (the leave-one-out method for cross-validation). The results are shown in Table 8.2. Subgroups of samples are associated with rule nodes, as shown in Fig. 8.10 for the rule nodes 2, 13, and 14. Three membership functions (*MF*) are used in the EFuNN models, representing Low, Medium, and High gene expression values respectively. Figure 8.11 shows the degree to which each input variable, a gene expression value (from the 50 genes selected) belongs to each of the above *MF* for rule node 2.

Table 8.4 shows one of the extracted rules after all the examples are learned by the EFuNN. The rules are 'local' and each of them has its meaning for a particular cluster of the input space. A very preliminary analysis of the rules points to one gene that is highly expressed in a colon cancer tissue (MCM3 – H09351, which gene is already known that it is involved in DNA replication) and several genes that are suppressed in the cancer samples. These are: Caveolin (Z18951), a structural membrane protein involved in the regulation of signalling pathways and also a putative tumour suppressor; and the enzymes carbonic anhydrase I (R93176) and II (J03037), that have already been shown in the literature to be correlated with the aggressiveness of colorectal cancer.

Figure 8.12 visualises some of the rules extracted from an EFuNN model trained on 62 samples from the colon cancer data in the format of profiles.

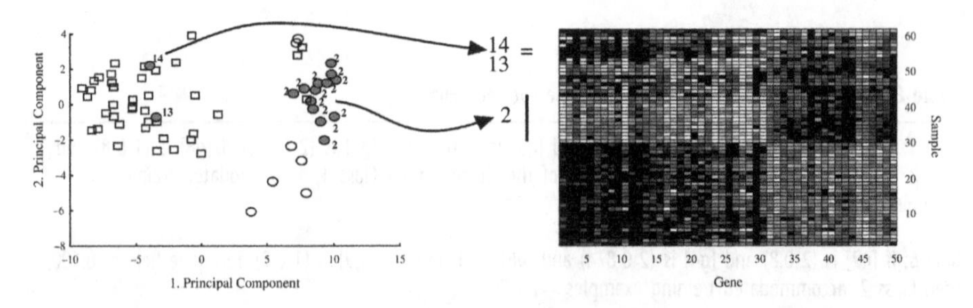

Fig. 8.10 The PCA projection of the 62 samples of colon cancer/normal tissue from 50D gene expression space and the similarity matrix genes/samples calculated based on the Pearson correlation coefficients (Futschik *et al.*, 2002).

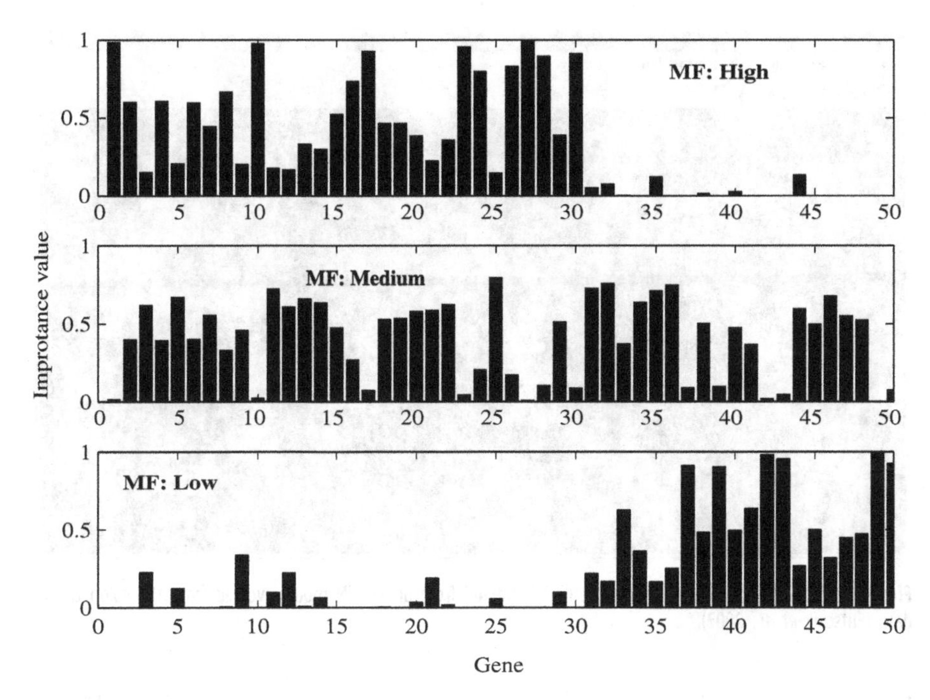

Fig. 8.11 Distribution of fuzzy membership degrees of genes 1 to 50 for rule node 2 from Fig. 8.10 (colon cancer data; Futschik *et al.*, 2002).

Table 8.4 One of the extracted rules that reveal some conditions for a colon cancer against normal tissue (Futschik *et al.*, 2002).

Rule for colon cancer:
IF H57136 is Low (1.0) AND H09351 is High (0.92) AND T46924 is Low (0.9) AND Z18951 is Low (0.97) AND R695523 is Low (0.98) AND J03037 is Low (0.98) AND R93176 is Low (0.97) AND H54425 is Low (0.96) AND T55741 is Low (0.99)
THEN The sample comes from a colon cancer tissue (certainty of 1.0)

8.3.5 How to Choose the Preprocessing Techniques and the Number of Genes for the Profiles

Preprocessing and normalisation affect the performance of the models as illustrated in Fig. 8.13 on the two benchmark data used here. 100 genes are used in the N-cross-validation procedure with the following parameter values for the EFuNN models: $E = 0.9$; $R\text{max} = 0.3$; $N\text{agg} = 20$.

The number of selected genes is another parameter that affects the performance of the classification system. Figure 8.14 shows the N-cross-validation test accuracy of EFuNN models for both the leukaemia and the colon cancer datasets, when the following parameter values are used: $E = 0.1$; $R\text{max} = 0.3$; $N\text{agg} = 20$. For the

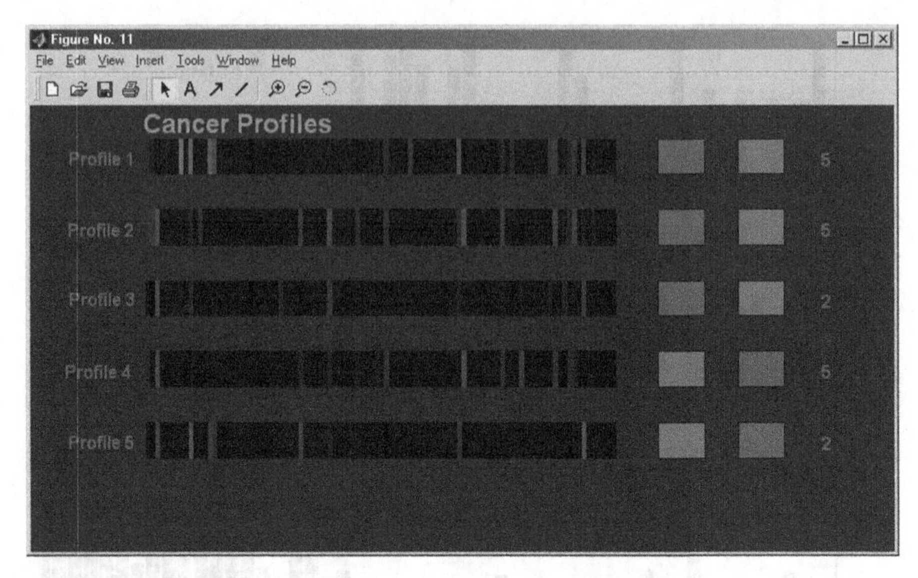

Fig. 8.12 Visualising some the rules (profiles) extracted from an EFuNN model evolved from the colon cancer data (Futschik *et al.*, 2002).

leukaemia dataset the best classification result is achieved for 100 genes, whereas for the colon cancer dataset this number is 300.

8.3.6 SVM and SVM Trees for Gene Expression Classification

In the area of bioinformatics, the identification of gene subsets responsible for classifying available samples to two or more classes (such as 'malignant' or 'benign') is an important task. Most current classifiers are sensitive to disease-marker gene selection. Here we use SVM and SVM-tree (SVMT) on different tasks of the same problem. Whereas the SVM creates a global model and Transductive SVM (TSVM) creates a local model for each sample, the SVMT creates a global model and performs classification in many local subspaces instead in the whole data space as typical classifiers do.

Here we use four different cancer datasets: lymphoma (Ship *et al.*, 2002), leukaemia (Golub *et al.*, 1999), colon (Alon *et al.*, 1999), and leukaemia cell line time-series data (Dimitrov *et al.*, 2004). The lymphoma dataset is a collection of gene expression measurements from 77 malignant lymphocyte samples reported by Shipp *et al* (2002). It contains 58 samples of diffused large B-cell lymphoma (DLBCL) and 19 samples of follicular lymphoma (FL), where DLBCL samples are divided into two groups: those with cured disease ($n = 32$) and those with fatal or refractory disease ($n = 26$). The lymphoma data containing 6817 genes is available at http: //www.genome.wi.mit/MPR/Lymphoma.

The leukaemia data are a collection of gene expression measurements from 72 leukaemia (composed of 62 bone marrow and 10 peripheral blood) samples reported by Golub *et al.* (1999). They contain an initial training set composed

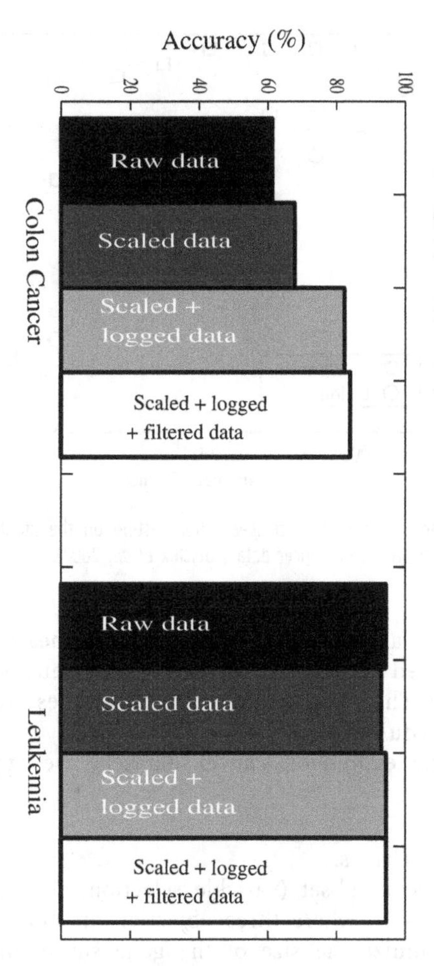

Fig. 8.13 Preprocessing affects the performance of the modelling EFuNN system (Futschik *et al.*, 2002).

of 27 samples of acute lymphoblastic leukaemia (ALL) and 11 samples of acute myeloblastic leukaemia (AML), and an independent test set composed of 20 ALL and 14 AML samples. The gene expression measurements were taken from high-density oligonucleotide microarrays containing 7129 probes for 6817 human genes. These data sets are available at http://www.genome wi.mit.edu/MPR.

The second leukaemia data are a collection of gene expression observations of two cell lines U937 (MINUS, a cancer cell line that is positively affected by retinoic acid and becomes a normal cell after a time interval of 48 hours, and PLUS cell line, that is cancerous and not affected by the drug (Dimitrov *et al.*, 2004). Each of the two time series contains the expression value of 12,000 genes at four time points: CTRL, 6 hours, 24 hours, and 48 hours. We can view this problem also as a classification problem where we have four variables (the time points) and 24,000 examples (the gene expression of a gene over the four time points) classified in two classes, MINUS and PLUS.

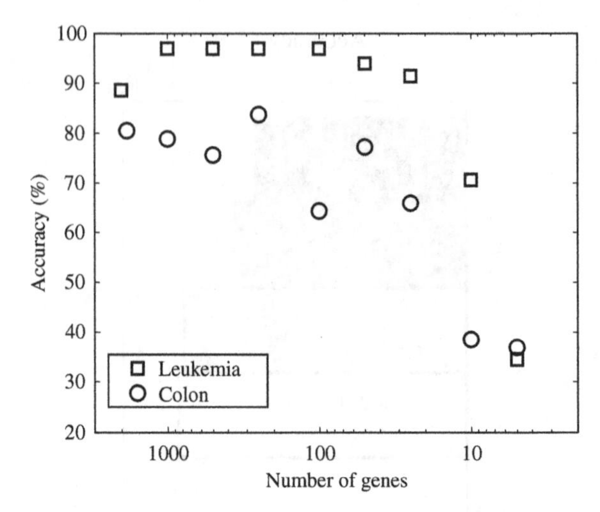

Fig. 8.14 Dependence of the accuracy of N-cross-variation testing on the number of genes in the EFuNN model for both leukaemia data and colon cancer data (Futschik *et al.*, 2002).

The colon dataset is a collection of 62 expression measurements from colon biopsy samples reported by Alon *et al.* (1999). It contains 22 normal and 40 colon cancer samples. The colon data having 2000 genes are available at http://microaaray.princeton.edu/oncology.

On the above gene expression cancer datasets, we applied the following methodology.

Step 1. Define target classes.
Step 2. Identify a gene subset (variable selection). We employed the multi-objective GA (NSGA-II), where three objective functions are used. The first objective is to minimize the size of the gene subset in the classifier. The second objective is to minimize the number of mismatches in the training data samples calculated using the leave-one-out cross-validation procedure. The third objective is to minimize the number of mismatches in the test samples.
Step 3. Filter and normalise data. We eliminate genes with not much variation in the expression values for the two classes to ensure a differentiation of the classes. We normalize data by evaluating the difference of the maximum and minimum gene expression values for every gene, and by measuring its standard deviation.
Step 4. Build a classifier. For each variable set and defined classes we build and test classifiers in a cross-validation mode (leave-one-out) by removing one sample and then using the rest as a training set. Several models are built using different numbers of marker genes and the final chosen model is the one that minimizes the total cross-validation error.
Step 5. Evaluate results. We evaluate prediction results and compute confusion matrices. For the purpose of comparison with past studies, we compare the proposed classifier algorithm with the K-NN model and an inductive global SVM.

Figure 8.15 shows the created SVMT for the lymphoma dataset (Pand et al., 2006; Pang and Kasabov, 2004). Each internal node of the tree identifies an SVM classifier, which is represented as an ellipse with a number as its identity.

When the parent node is labeled i, its two children nodes are identified as $2i$ and $2i+1$, respectively. We also represent the terminal node as a circle or a filled circle, which denotes positive or negative class, respectively.

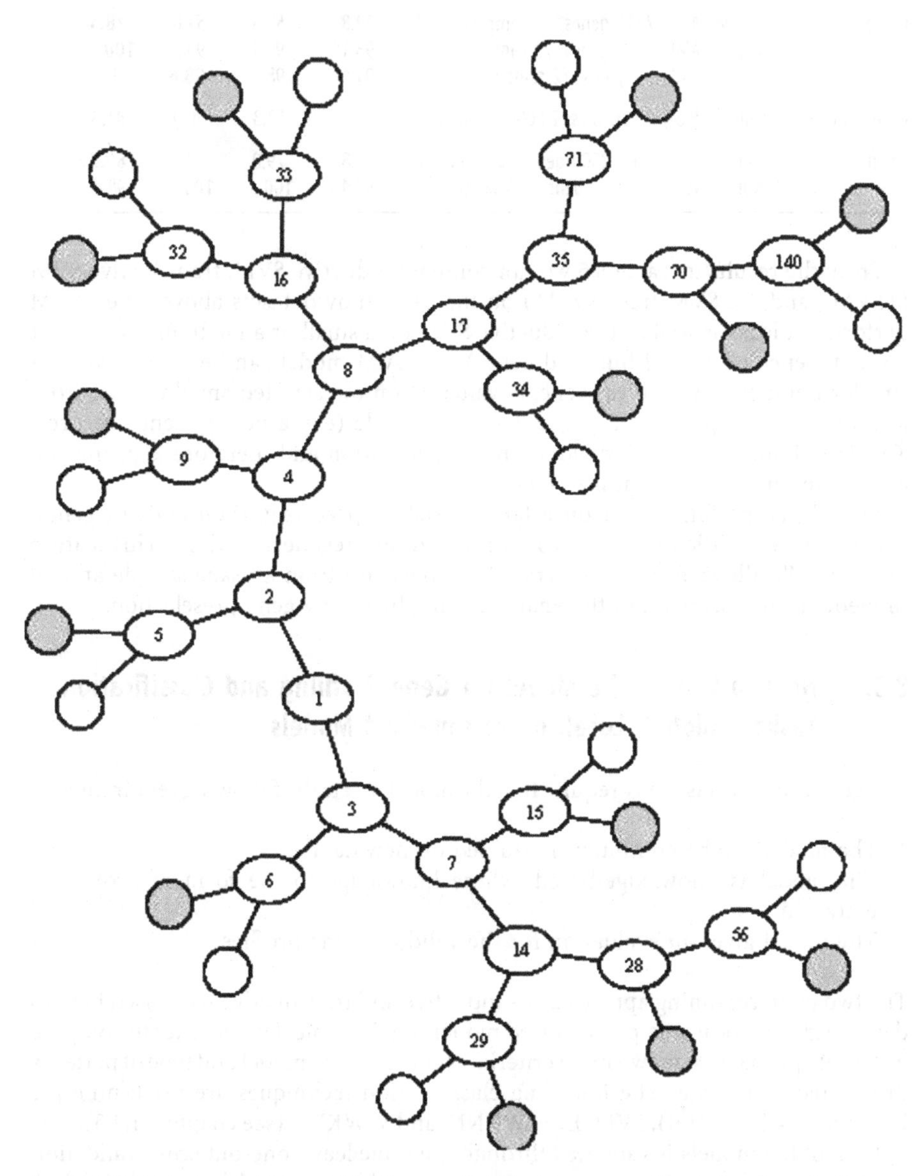

Fig. 8.15 SVMT for the classification of DLBCL versus FL (first class represented as dark nodes as leafs) based on the data from (Ship *et al.*, 2002).

Table 8.5 Results of applying SVM, TSVM, and SVMT on the four gene expression classification problems.

	With/Without Marker Gene Selection	***K-NN*** *(%)*	*SVM (%)*	*TSVM (%)*	*SVM Tree (%)*
Lymphoma	DLBCL vs. FL; 6432 genes, 77 samples	53.2	55.8	57.1	**77.9**
	DLBCL vs. FL; 30 genes, 77 samples	90.9	92.2	**93.5**	92.2
	cured vs. fatal 6432 genes, 58 samples	51.7	53.3	55.1	**72.4**
	cured vs. fatal; 13 genes, 58 samples	70.7	72.4	**79.3**	70.7
Leukaemia	ALL vs. AML; 7219 genes, 72 samples	52.8	55.6	52.8	**78.3**
	ALL vs. AML; 3859 genes, 72 samples	93.1	94.4	95.8	**100**
	ALL vs. AML; 27 genes, 72 samples	91.6	98.6	**98.6**	90.3
L. cell line	Min vs. Plus; 4 variables; 24,000 samples	52.5	53.3	50.0	**81.3**
Colon	Normal vs. Cancer; 2000 genes, 62 samples	75.8	79.0	72.6	**80.7**
	Normal vs. Cancer; 12 genes, 62 samples	98.4	**100**	**100**	98.4

From the results in Table 8.5 we can compare inductive SVM, transductive SVM (TSVM), and the SVM tree (SVMT) on the case study datasets above. The TSVM performs at least as well as the inductive SVM on a small or a medium variable set (several genes or several hundred genes). A TSVM model can be generated on a smaller number of variables (genes) evaluated on the selected small dataset from a local problem space for a particular new sample (e.g. a new patient's record). The TSVM allows for an individual model generation and therefore is promising as a technique for personal medicine.

The SVMT performs best on a large variable space (e.g. thousands of genes, sometimes with little or no preprocessing and no pregene selection). This feature of the SVMT allows for a microarray data collection from a tissue sample and an immediate analysis without the analysis being biased by gene preselection.

8.3.7 How to Choose the Model for Gene Profiling and Classification Tasks – Global, Local, or Personalised Models

The gene profiling task may require that the model meets the following requirements.

1. The model can be continuously trained on new data.
2. The model is knowledge-based, where knowledge in the form of profiles is extracted.
3. The model gives an evaluation for the validity of the profiles.

The two main reasoning approaches – inductive and transductive – are used here to develop global, local, and personalised models on the same data in order to compare different approaches on two main criteria: accuracy of the model and type of patterns discovered from data. The following classification techniques are used: multiple linear regression (MLR), SVM, ECF, WKNN, and WWKNN (see chapters 1,2,3).

Each of the models is validated through the same leave-one-out cross-validation method (Vapnik, 1998). The accuracy of the different models is presented in Table 8.6. It can be seen that the transductive reasoning and personalised modelling

Table 8.6 Experimental results in terms of model accuracy tested through leave-one-out cross-validation method when using different modelling techniques on the DLBCL Lymphoma data for classification of new samples into class 1 – survival, or class 2 – fatal outcome of the disease within five years time (Shipp *et al.*, 2002). The table shows the overall model classification accuracy in % and the specificity and sensitivity values (accuracy for class 1 and class 2, respectively) in brackets.

Model/ Features	*Induct Global MLR [%]*	*Induct Global SVM [%]*	*Induct Local ECF [%]*	*Trans WKNN K = 8 [%],* P_{tthr} *= ..5*	*Trans WKNN K = 26 [%]* $\boldsymbol{P_{thr}}$ *= 0.5*	*Trans WW-KNN K=16*	*Trans MLR K = 8 [%]*	*Trans MLR k = 26 [%]*	*Trans SVM K = 8 [%]*	*Trans SVM k = 26 [%]*	*Trans ECF K = 8 [%]*	*Trans ECF k = 26 [%]*
IPI (one clinical variable)	73 (87,58)	73 (87,58)	46 (0,100)	50 (87,8)	73 (87,56)	68 (63,73)	50 (87,8)	73 (87,58)	46 (100,0)	73 (87,58)	61 (63,58)	46 (0,100)
11 genes	79 (91,65)	83 (88,78)	86 (88,84)	74 (91,54)	73 (93,47)	81 (81,81)	66 (66,65)	78 (81,73)	76 (91,58)	78 (91,62)	78 (81,73)	83 (91,73)
IPI + 11 genes	82 (83,81)	86 (90,81)	**88 (83, 92)**	77 (90,62)	Pthr = . 45 82% (97,65)	80 (80,81)	57 (60,54)	79 (80,77)	77 (93,58)	84 (93,73)	75 (83,65)	77 (87,65)

is sensitive to the selection of the number of the nearest neighbours K. Its optimization is discussed in the next section.

The transductive, pezvnalised WWKNN produces a balanced accuracy of 80 and 81% for each of the two classes (balanced sensitivity and specificity values) along with an individual ranking of the importance of the variables for each individual sample. Having this knowledge, a personalised treatment can be attempted that targets the important genes and clinical variables for each patient.

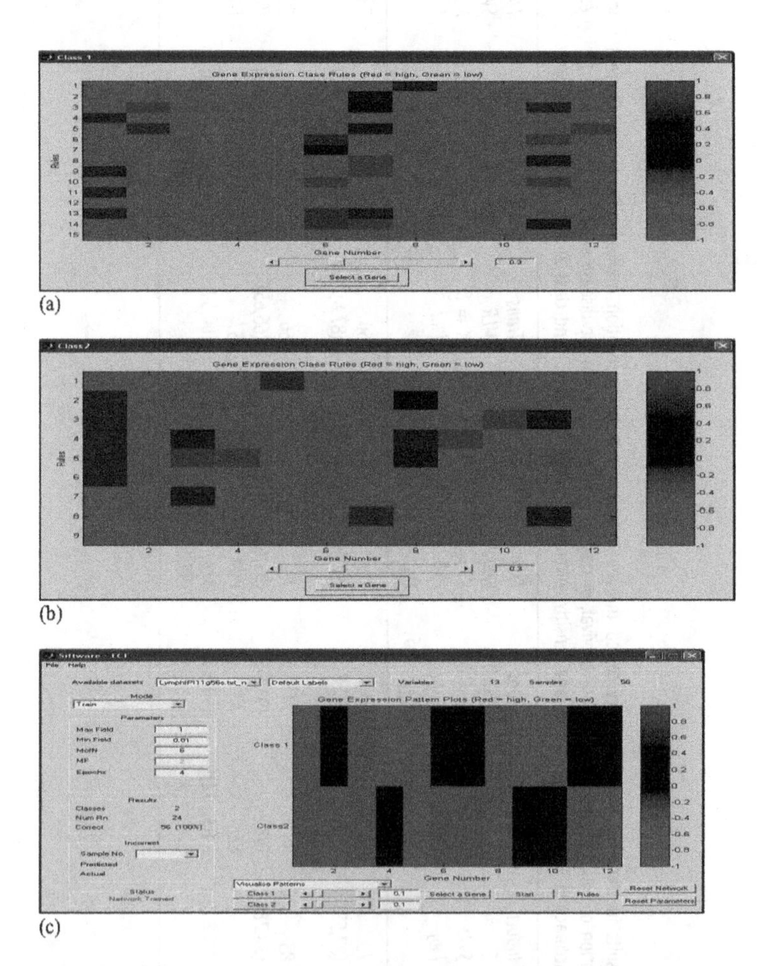

Fig. 8.16 Cluster–based, local patterns (rules) extracted from a trained ECF model from chapter3 (inductive, local training) on 11 gene expression data and clinical data of the lymphoma outcome prediction problem (from M.Slipp et al., 2002). The first variable (first column) is the clinical variable IPI. The accuracy of the model measured through the leave-one-out cross-validation method is 88% (83% class one and 92% class two). The figure shows: (a) 15 local profiles of class 1 (survive), threshold 0.3; (b) 9 local profiles of class 2 (fatal outcome), threshold 0.3; (c) global class profiles (rules) are derived through averaging the variable values (genes or IPI) across all local class profiles from Fig. 8.3 and ignoring low values (below a threshold, e.g. 0.1 as an absolute value). Global profiles for class 1 and class 2 may not be very informative as they may not manifest any variable that is significantly highly expressed in all clusters of any of the two classes if the different class samples are equally scattered in the whole problem space.

The best accuracy is manifested by the local ECF model, trained on a combined feature vector of 11 gene expression variables and the clinical variable IPI. Its prognostic accuracy is 88% (83% for class 1, cured and 92% for class 2, fatal). This compares favourably with the 75% accuracy of the SVM model used in Shipp *et al.* (2002).

In addition, local rules that represent cluster gene profiles of the survival versus the fatal group of patients were extracted as shown graphically in Fig. 8.16. These profiles show that there is no single variable that clearly discriminates the two classes; it is a combination of the variables that discriminates different subgroups of samples within a class and between classes.

The local profiles can be aggregated into global class profiles through averaging the variable values across all local profiles that represent one class; see Fig. 8.16c. Global profiles may not be very informative if data samples are dispersed in the problem space and each class of samples is spread out in the space, but they show the big picture, the common trends across the population of samples.

As each of the global, local, and personalised profiles contains a different level of information, integrating them through the integration of global, local, and personalised models would facilitate a better understanding and better accuracy of the prognosis (chapter 7).

When GA is used to optimise the feature set and the ECF model parameters, a significant improvement of the accuracy is achieved with the use of a smaller number of input variables (features) as a GA optimised ECF model and a feature set on the DLBCL Lymphoma data is shown in Chapter 6, Fig. 6.10. Twenty individual models are used in a population and run for 20 generations with a fitness function, model test accuracy, where the cross-validation method used is fivefold-cross-validation done on every model within a population with 70% of randomly selected data for training and 30% for testing. The same data are used to test all models in a population. The best performing models are used to create a new generation of 20 individual models, etc. The accuracy of the optimal model is now 90.66%, which is higher than the best model from Table 8.6 (no optimization is used there). The best model does not use features 5, 8, and 12 (genes 4, 7, and 11).

8.4 Clustering of Time-Course Gene Expression Data

8.4.1 Problem Definition

Each gene in a cell may express differently over time. And this makes the gene expression analysis based on static data (one shot) not a very reliable mechanism. Measuring the expression rate of each gene over time gives the gene a temporal profile of its expression level. Genes can be grouped together according to their similarity of temporal expression profiles.

This is illustrated here with case study data. For a demonstration of the applicability of our method, we used yeast gene expression data that are available as a public database. We analysed the gene expression during the mitotic cell cycle of different synchronised cultures as reported by Cho *et al.* (1998) and by Spellman *et al.* (1998). The datasets consisted of expression profiles for over 6100 ORFs.

In this study we did not reduce the original dataset by applying a filter in the form of a minimum variance. This leads to a higher number of clusters of weakly cell-regulated genes, however, it diminished the possibility of missing co-regulated genes during the clustering process.

For the search for upstream regulatory sequences we used Hughes' compiled set of upstream regions for the open reading frames (ORFs) in yeast (Church lab: http://atlas.med.harvard.edu/).

One of the main purposes for cluster analysis of time-course gene expression data is to infer the function of novel genes by grouping them with genes of well-known functionality. This is based on the observation that genes which show similar activity patterns over time (co-expressed genes) are often functionally related and are controlled by the same mechanisms of regulation (co-regulated genes). The gene clusters generated by cluster analysis often relate to certain functions, e.g. DNA replication, or protein synthesis. If a novel gene of unknown function falls into such a cluster, it is likely that this gene serves the same function as the other members of this cluster. This 'guilt-by-association' method makes it possible to assign functions to a large number of novel genes by finding groups of co-expressed genes across a microarray experiment (Derisi *et al.*, 1997).

Different clustering algorithms have been applied to the analysis of time-course gene expression data: *k*-means, SOM, and hierarchical clustering, to name just a few (Derisi, 1997. They all assign genes to clusters based on the similarity of their activity patterns. Genes with similar activity patterns should be grouped together, whereas genes with different activation patterns should be placed in distinct clusters. The cluster methods used thus far have been restricted to a one-to-one mapping: one gene belongs to exactly one cluster. Although this principle seems reasonable in many fields of cluster analysis, it might be too limited for the study of microarray time-course gene expression data. Genes can participate in different genetic networks and are frequently coordinated by a variety of regulatory mechanisms. For the analysis of microarray data, we may therefore expect that single genes can belong to several clusters.

8.4.2 Fuzzy Clustering of Time Course Gene Expression Data

Several researchers have noted that genes were frequently highly correlated with multiple classes and that the definition of clear borders between gene expression clusters often seemed arbitrary (Chu *et al.*, 1998). This is a strong motivation to use fuzzy clustering in order to assign single objects to several clusters.

A second reason for applying fuzzy clustering is the large noise component in microarray data due to biological and experimental factors. The activity of genes can show large variations under minor changes of the experimental conditions. Numerous steps in the experimental procedure contribute to additional noise and bias. A usual procedure to reduce the noise in microarray data is setting a threshold for a minimum variance of the abundance of a gene. Genes below this threshold are excluded from further analysis. However, the exact value of the threshold remains arbitrary due to the lack of an established error model and the use of filtering as preprocessing.

Hence we usually have little information about the data structure in advance, a crucial step in cluster analysis is selection of the number of clusters. Finding the 'correct' number of clusters leads to the issue of cluster validity. This has turned out to be a rather difficult problem, as it depends on the definition of a cluster. Without prior information, a common method is the comparison of partitions resulting from different numbers of clusters. For assessing the validity of the partitions, several cluster validity functionals have been introduced (Pal and Bezdek, 1995). These functionals should reach an optimum if the correct number of clusters is chosen. When using evolving clustering techniques the number of the clusters does not need to be defined a priori.

Two fuzzy clustering techniques were applied: the batch mode fuzzy C-means clustering (FCM) and an evolving clustering through evolving self-organised maps (ESOM; see Chapter 2).

In the FCM clustering experiment (for more details see Futschik and Kasabov, (2002)) the fuzzification parameter m (Pal and Bezdek, 1995) turned out to be an important parameter for the cluster analysis. For the randomised dataset, FCM clustering formed clusters only if m was chosen smaller than 1.15. Higher values of m led to uniform membership values in the partition matrix. This can be regarded as an advantage of FCM over exat clustering, which always forms clusters independently of the existence of any structure in the data. An appropriate choice for a lower threshold for m can therefore be set if no cluster artefacts are formed in randomised data. An upper threshold for m is reached if FCM does not indicate any cluster in the original data. This threshold depends mainly on the compactness of the clusters. The cluster analysis with FCM showed that

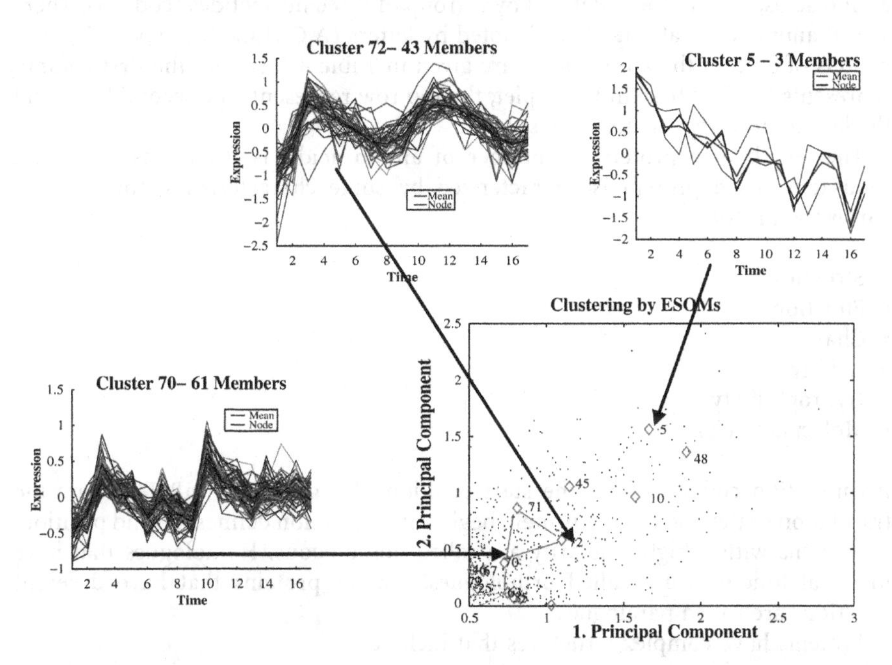

Fig. 8.17 Using evolving self-organised maps (ESOM; see Chapter 2) to cluster temporal profiles of yeast gene expression data (Futschik *et al.*, 1999).

hyperspherical distributions are more stable for increasing m than hyperellipsoid distributions. This may be expected because FCM clustering with Euclidean norm favours spherical clusters.

In another experiment, an evolving self-organising map ESOM was evolved from the yeast gene temporal profiles used as input vectors. The number of clusters did not need to be specified in advance (Fig. 8.17). It can be seen from Fig. 8.17 that clusters 72 and 70 are represented on the ESOM as neighbouring nodes. The ESOM in the figure is plotted as a 2D PCA projection. Cluster 72 has 43 members (genes, that have similar temporal profiles), cluster 70 has 61 members, and cluster 5 has only 3 genes as cluster members.

New cluster vectors will be created in an online mode if the distance between existing clusters and the new data vectors is above a chosen threshold.

8.5 Protein Structure Prediction

8.5.1 Problem Definition

Proteins provide the majority of the structural and functional components of a cell. The area of molecular biology that deals with all aspects of proteins is called proteomics. Thus far about 30,000 proteins have been identified and labelled, but this is considered to be a small part of the total set of proteins that keep our cells alive.

The mRNA is translated by ribosomes into proteins. A protein is a sequence of amino acids, each of them defined by a group of three nucleotides (codons). There are 20 amino acids all together, denoted by letters (A,C-H,I,K-N,P-T,V,W,Y). The codons of each of the amino acids are given in Table 8.7, so that the first column represents the first base in the triplet, the top row represents the second base, and the last column represents the last base.

The length of a protein in number of amino acids, is from tens to several thousands. Each protein is characterized by some characteristics, for example (Brown *et al.*, 1999):

- Structure
- Function
- Charge
- Acidity
- Hydrophilicity
- Molecular weight

An initiation codon defines the start position of a gene in a mRNA where the translation of the mRNA into protein begins. A stop codon defines the end position.

Proteins with a high similarity are called homologous. Homologues that have identical functions are called orthologues. Similar proteins that have different functions are called paralogues.

Proteins have complex structures that include:

- *Primary structure* (a linear sequence of the amino acids): See, for example Fig. 8.18.

Table 8.7 The codons of each of the 20 amino acids. The first column represents the first base in the triplet, the first row represents the second base, and the last column, the last base (Hofstadter, 1979).

	U	*C*	*A*	*G*	
U	Phe	Ser	Tyr	Cys	U
	Phe	Ser	Tyr	Cys	C
	Leu	Ser	–		A
	Leu	Ser	–	Trp	G
C	Leu	Pro	His	Arg	U
	Leu	Pro	His	Arg	C
	Leu	Pro	Gln	Arg	A
	Leu	Pro	Gln	Arg	G
A	Ile	Thr	Asn	Ser	U
	Ile	Thr	Asn	Ser	C
	Ile	Thr	Lys	Arg	A
	Met	Thr	Lys	Arg	G
G	Val	Ala	Asp	Gly	U
	Val	Ala	Asp	Gly	C
	Val	Ala	Glu	Gly	A
	Val	Ala	Glu	Gly	G

- *Secondary structure* (3D, defining functionality): An example of a 3D representation of a protein is given in Fig. 8.19.
- *Tertiary structure* (high-level folding and energy minimisation packing of the protein): Figure 8.20 shows an example of hexokinase (6000 atoms, 48 kD, 457 amino acids). Polypeptides with a tertiary level of structure are usually referred to as globular proteins, because their shape is irregular and globular in form.
- *Quaternary structure* (interaction between two or more protein molecules)

One task that has been explored in the literature is predicting the secondary structure from the primary one. Segments of a protein can have different shapes in their secondary structure, which is defined by many factors, one of them being the amino acid sequence itself. The main types of shape are:

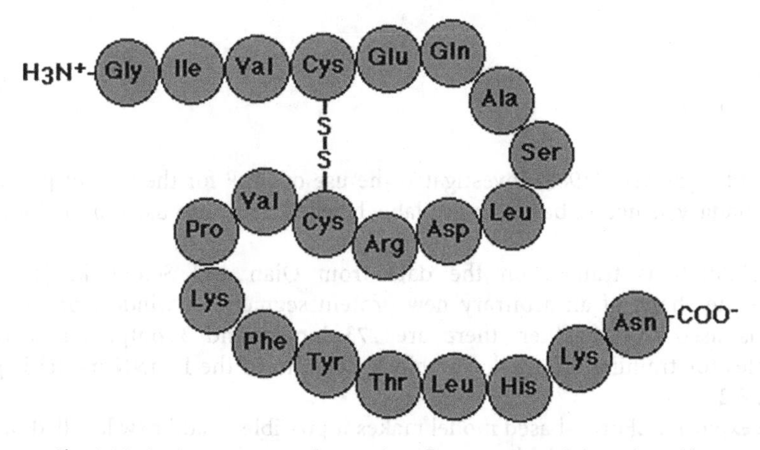

Fig. 8.18 A primary structure of a protein, a linear sequence of the amino acids.

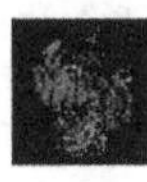

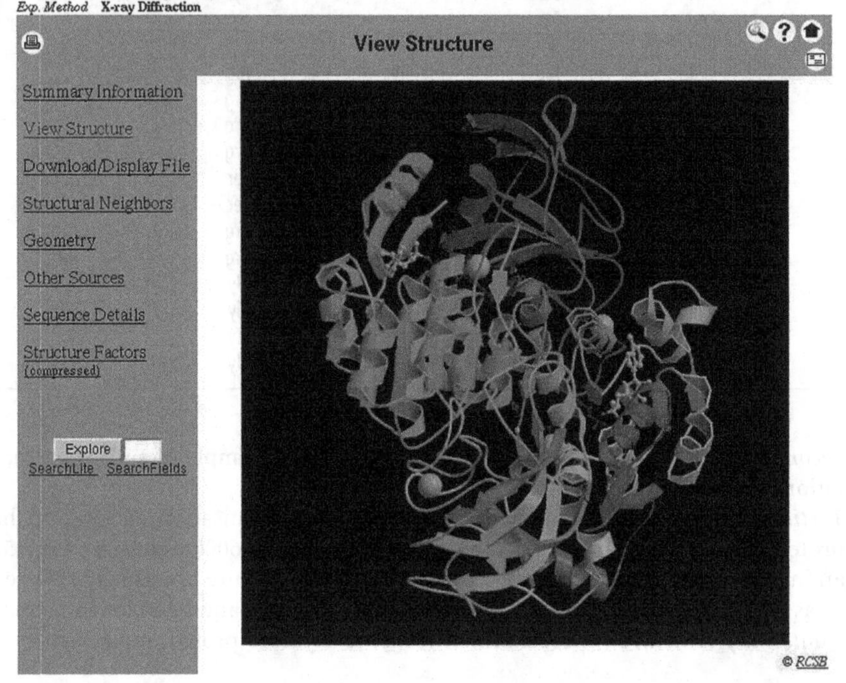

Fig. 8.19 An example of a secondary structure (3D, defining functionality) of a protein obtained with the use of the PDB dataset, maintained by the National Center for Biological Information (NCBI) of the National Institute for Health (NIH) in the United States.

- Helix
- Sheet
- Coil (loop)

Qian and Sejnowski (1988) investigated the use of MLP for the task of predicting the secondary structure based on available labelled data, also used in the following experiment.

An EFuNN is trained on the data from Qian and Sejnowski (1988) to predict the shape of an arbitrary new protein segment. A window of 13 amino acids is used. All together, there are 273 inputs and 3 outputs and 18,000 examples for training are used. The block diagram of the EFuNN model is given in Fig. 8.21.

The explored EFuNN-based model makes it possible to add new labelled protein data as they become available with time.

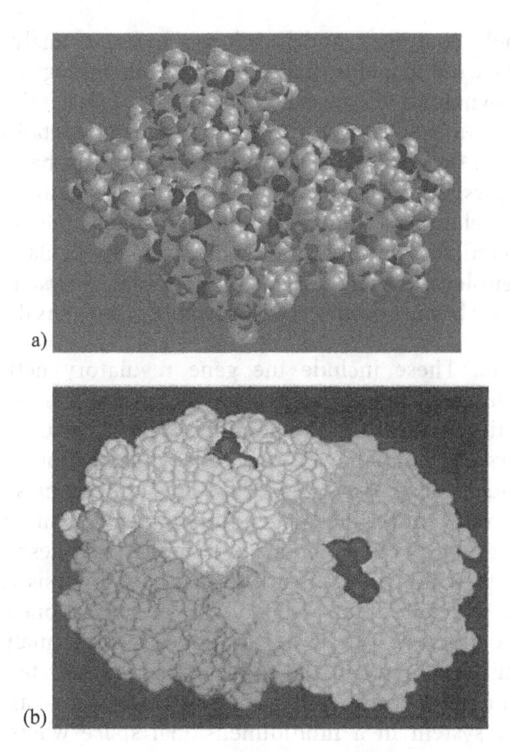

Fig. 8.20 (a) An example of a tertiary structure of a protein (high-level folding and energy minimisation packing); (b) The hexokinase protein (6000 atoms, 48 kD, 457 amino acids; from the PDB database).

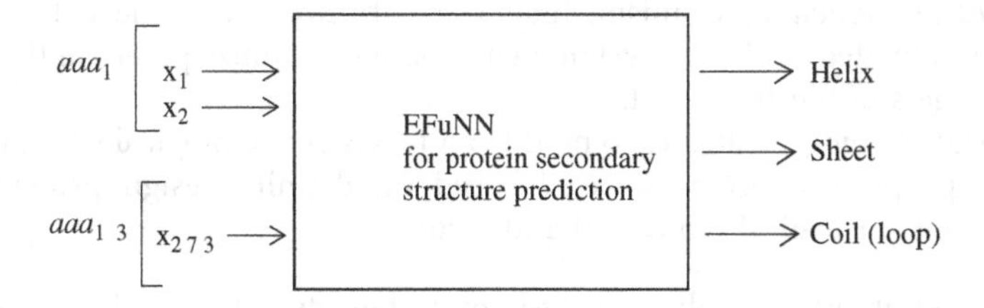

Fig. 8.21 Evolving system for protein secondary structure prediction.

8.6 Gene Regulatory Networks and the System Biology Approach

8.6.1 The System Biology Approach

The aim of computational system biology is to understand complex biological objects in their entirety, i.e. at a system level. It involves the integration of different approaches and tools: computer modeling, large-scale data analysis, and biological experimentation. One of the major challenges of system biology is the identification of the logic and dynamics of gene-regulatory and biochemical networks. The most feasible application of system biology is to create a detailed model of a cell regulation to provide system-level insights into mechanism-based drug discovery.

System-level understanding is a recurrent theme in biology and has a long history. The term 'system-level understanding' is a shift of focus in understanding a system's structure and dynamics as a whole, rather than the particular objects and their interactions. System-level understanding of a biological system can be derived from insight into four key properties (Dimitrov *et al.*, 2004; Kasabov *et al.*, 2005c):

1. *System structures.* These include the gene regulatory network (GRN) and biochemical pathways. They can also include the mechanisms of modulation of the physical properties of intracellular and multicellular structures by interactions.
2. *System dynamics.* System behavior over time under various conditions can be understood by identifying essential mechanisms underlying specific behaviours and through various approaches depending on the system's nature: metabolic analysis (finding a basis of elementary flux modes that describe the dominant reaction pathways within the network), sensitivity analysis (the study of how the variation in the output of a model can be apportioned, qualitatively or quantitatively, to different sources of variation), dynamic analysis methods such as phase portrait (geometry of the trajectories of the system in state space), and bifurcation analysis (bifurcation analysis traces time-varying change(s) in the state of the system in a multidimensional space where each dimension represents a particular system parameter (concentration of the biochemical factor involved, rate of reactions/interactions, etc.). As parameters vary, changes may occur in the qualitative structure of the solutions for certain parameter values. These changes are called bifurcations and the parameter values are called bifurcation values.
3. *The control method.* Mechanisms that systematically control the state of the cell can be modulated to change system behavior and optimize potential therapeutic effect targets of the treatment.
4. *The design method.* Strategies to modify and construct biological systems having desired properties can be devised based on definite design principles and simulations, instead of blind trial and error.

As mentioned above, in reality, analysis of system dynamics and understanding the system structure are overlapping processes. In some cases analysis of the system dynamics can give useful predictions in system structure (new interactions, additional member of system). Different methods can be used to study the dynamical properties of the system:

- Analysis of steady states allows finding the system states when there are no dynamical changes in system components.
- Stability and sensitivity analyses provide insights into how system behaviour changes when stimuli and rate constants are modified to reflect dynamic behaviour.
- Bifurcation analysis, in which a dynamic simulator is coupled with analysis tools, can provide a detailed illustration of dynamic behaviour.

The choice of the analytical methods depends on availability of the data that can be incorporated in the model and the nature of the model. It is important to know the main properties of the complex system under investigation, such as robustness.

Robustness is a central issue in all complex systems and it is very essential for understanding the biological object functioning at the system level. Robust systems exhibit the following phenomenological properties.

- *Adaptation*, which denotes the ability to cope with environmental changes
- *Parameter insensitivity*, which indicates a system's relative insensitivity (to a certain extent) to specific kinetic parameters
- *Graceful degradation*, which reflects the characteristic slow degradation of a system's functions after damage, rather than catastrophic failure

Revealing all these characteristics of a complex living system helps in choosing an appropriate method for their modelling, and also constitutes an inspiration for the development of new CI methods that possess these features.

Modelling living cells *in silico* has many implications; one of them is testing new drugs through simulation rather than on patients. According to recent statistics (Zacks, 2001), human trials fail for 70–75% of the drugs that enter them.

Tomita (2001) stated in his paper, 'The cell is never conquered until its total behaviour is understood, and the total behaviour of the cell is never understood until it is modelled and simulated.'

Computer modelling of processes in living cells is an extremely difficult task for several reasons; among them are that the processes in a cell are dynamic and depend on many variables some of them related to a changing environment, and the processes of DNA transcription and protein translation are not fully understood.

Several cell models have been created and experimented, among them (Bower and Bolouri, 2001):

- The virtual cell model
- The e-cell model and the self-survival model (Tomita *et al.*, 2001)
- A mathematical model of a cell cycle

A starting point to dynamic modelling of a cell would be dynamic modelling of a single gene regulation process. In Gibson and Mjolsness (2001) the following methods for single-gene regulation modelling are discussed, that take into account different aspects of the processes (chemical reactions, physical chemistry, kinetic changes of states, and thermodynamics):

- Boolean models, based on Boolean logic (true/false logic)
- Differential equation models
- Stochastic models
- Hybrid Boolean/differential equation models
- Hybrid differential equations/stochastic models
- Neural network models
- Hybrid connectionist-statistical models

The next step in dynamic cell modelling would be to try to model the regulation of more genes, it is hoped a large set of genes (see Somogyi *et al.* (2001)). Patterns of

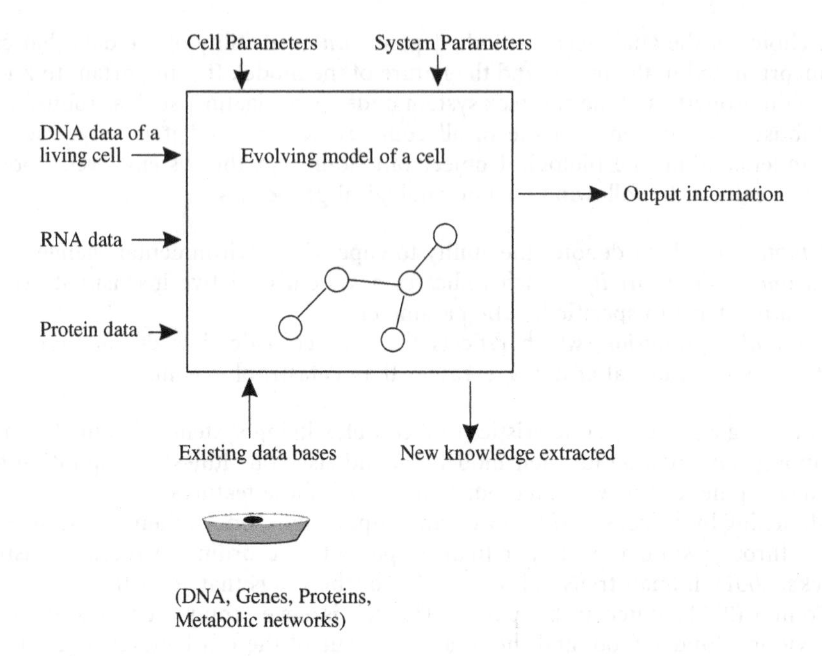

Fig. 8.22 A general, hypothetical evolving model of a cell: the system biology approach.

collective regulation of genes are observed in the above reference, such as chaotic attractors. Mutual information/entropy of clusters of genes can be evaluated.

A general, hypothetical evolving model of a cell is outlined in Fig. 8.22 that encompasses the system biology approach. It is based on the following principles.

1. The model incorporates all the initial information such as analytical formulas, databases, and rules of behaviour.
2. In a dynamic way, the model adjusts and adapts over time during its operation.
3. The model makes use of all current information and knowledge at different stages of its operation (e.g., transcription, translation).
4. The model takes as inputs data from a living cell and models its development over time. New data from the living cell are supplied if such are available over time.
5. The model runs until it is stopped, or the cell has died.

8.6.2 Gene Regulatory Network Modelling

Modelling processes in a cell includes finding the genetic networks (the network of interaction and connections between genes, each connection defining if a gene is causing another one to become active, or to be suppressed). The reverse-engineering approach is used for this task (D'haeseleer *et al.*, 2000). It consists of the following. Gene expression data are taken from a cell (or a cell line) at consecutive time moments. Based on these data a logical gene network is derived.

For example, it is known that clustering of genes with similar expression patterns will suggest that these genes are involved in the same regulatory processes.

Modelling gene regulatory networks (GRN) is the task of creating a dynamic interaction network between genes that defines the next time expression of genes based on their previous levels of expression. A simple GRN of four genes is shown in Fig. 8.23. Each node from Fig. 8.23 represents either a single gene/protein or a cluster of genes that have a similar expression over time, as illustrated in Fig. 8.24.

Models of GRN, derived from gene expression RNA data, have been developed with the use of different mathematical and computational methods, such as statistical correlation techniques; evolutionary computation; ANN; differential equations, both ordinary and partial; Boolean models; kinetic models; state-based models and others.

In Kasabov *et al.* (2004) a simple GRN model of five genes is derived from time course gene expression data of leukaemia cell lines U937 treated with retinoic acid with two phenotype states: positive and negative. The model, derived from time course data, can be used to predict future activity of genes as shown in Fig. 8.25.

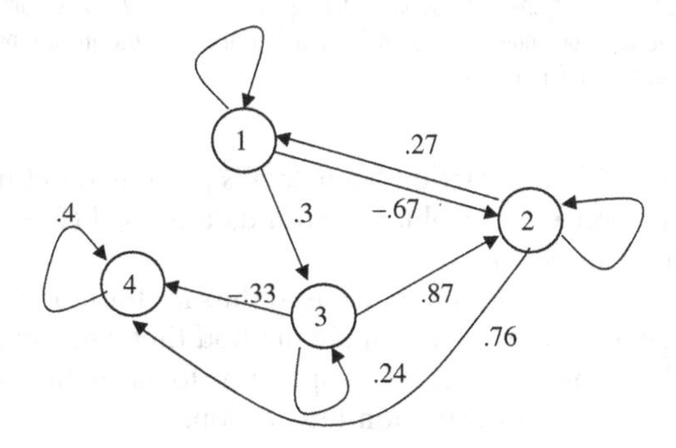

Fig. 8.23 A simplified gene regulatory network where each node represents a gene/protein (or a group of them) and the arcs represent the connection between them, either excitatory (+) or inhibitory (−).

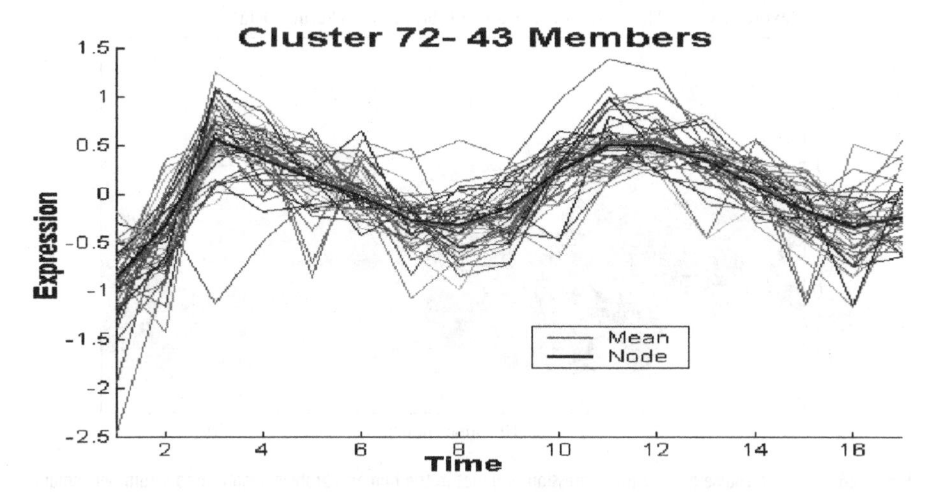

Fig. 8.24 A cluster of genes that are similarly expressed over time (17 hours).

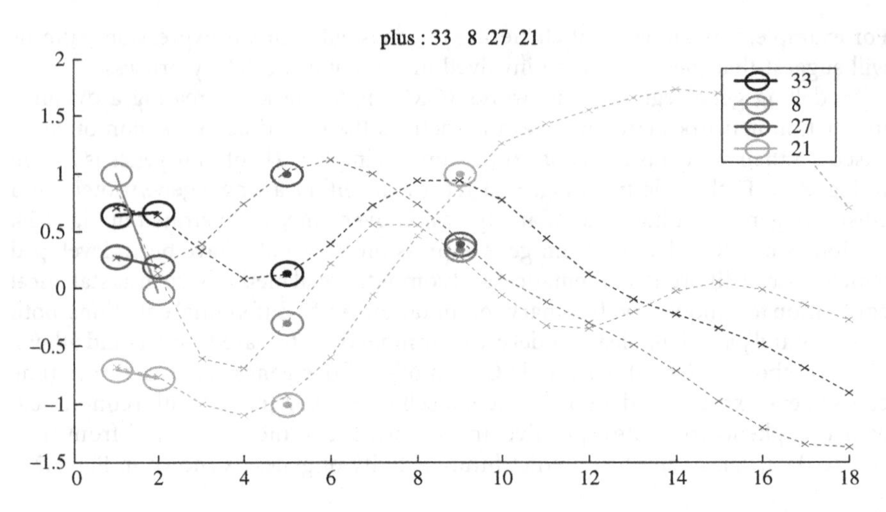

Fig. 8.25 The time course data of the expression of four genes (#33, 8, 27, 21) from the cell line used in (Kasabov *et al.*, 2005). The first four points are used for training and the rest are the predicted by the model expression values of the genes in a future time.

Another example of GRN extraction from data is presented in Chan *et al.* (2006b) where the human response to fibroblast serum data is used (Fig. 8.26) and a GRN is extracted from it (Fig. 8.27).

Despite the variety of different methods used thus far for modelling GRN and for system biology in general, there is no single method that will suit all requirements to model a complex biological system, especially to meet the requirements for adaptation, robustness, and information integration.

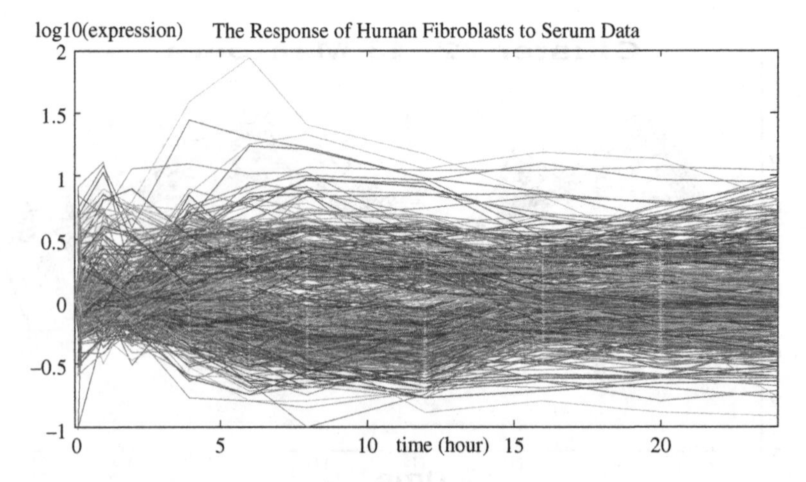

Fig. 8.26 The time course data of the expression of genes in the human fibroblast response to serum benchmark data (Chan *et al.*, 2006b).

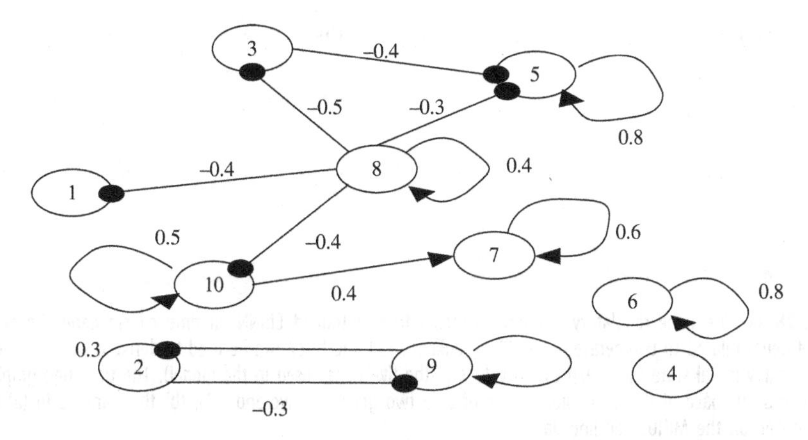

Fig. 8.27 A GRN obtained with the use of the method from Chan *et al.* (2006b) on the data from Fig. 8.26, where ten clusters of gene expression values over time are derived, each cluster represented as a node in the GRN.

8.6.3 Evolving Connectionist Systems for GRN Modelling

Case Study

Here we used the same data of the U937 cell line treated with retinoic acid. The results are taken from Kasabov and Dimitrov (2004). Retinoic acid and other reagents can induce differentiation of cancer cells leading to gradual loss of proliferation activity and in many cases death by apoptosis. Elucidation of the mechanisms of these processes may have important implications not only for our understanding of the fundamental mechanisms of cell differentiation but also for treatment of cancer. We studied differentiation of two subclones of the leukemic cell line U937 induced by retinoic acid. These subclones exhibited highly differential expression of a number of genes including c-Myc, Id1, and Id2 that were correlated with their telomerase activity; the PLUS clones had about 100-fold higher telomerase activity than the MINUS clones. It appears that the MINUS clones are in a more 'differentiated' state. The two subclones were treated with retinoic acid and samples were taken before treatment (time 0) and then at 6 h, 1, 2, 4, 7, and 9 days for the plus clones and until day 2 for the minus clones because of their apoptotic death. The gene expression in these samples was measured by Affymetrix gene chips that contain probes for 12,600 genes. To specifically address the question of telomerase regulation we selected a subset of those genes that were implicated in the telomerase regulation and used ECOS for their analysis.

The task is to find the gene regulatory network $G = \{g1, g2, g3, g_{rest-}, g_{rest+}\}$ of three genes g1 = c-Myc, g2 = Id1, and g3 = Id2 while taking into account the integrated influence of the rest of the changing genes over time denoted g_{rest-} and g_{rest+} representing, respectively, the integrated group of genes which expression level decreases over time (negative correlation with time), and the group of genes which expression increases over time (positive correlation with time).

(a)

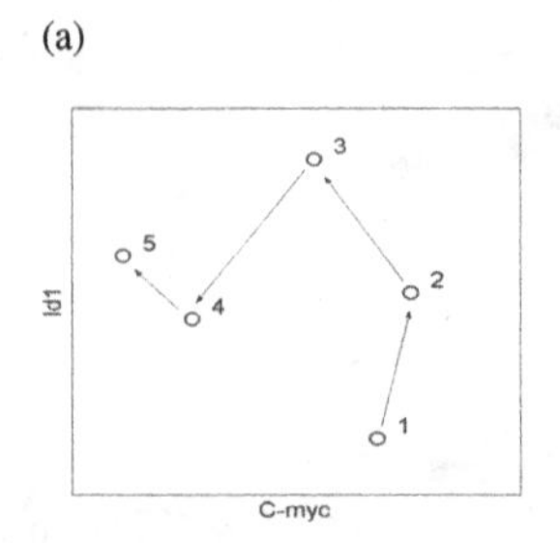

(b)

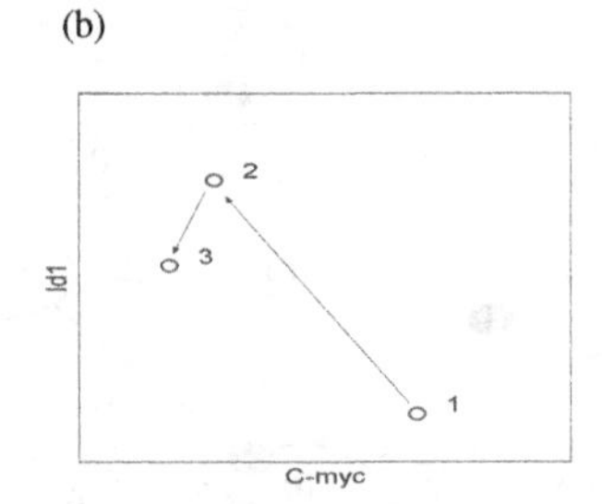

Fig. 8.28 (a) The gene regulatory network extracted from a trained EfuNN on time course gene expression data of genes related to telomerase of the PLUS leukemic cell line U937 can be used to derive a state transition graph for any initial state (gene expression values of the five genes used in the model). The transition graph is shown in a 2D space of the expression values of only two genes (C-myc and Id1); (b) the same as in (a) but here applied on the MINUS cell line data.

Groups of genes g_{rest-}, g_{rest+} were formed for each experiment of PLUS and MINUS cell lines, forming all together four groups of genes. For each group of genes, the average gene expression level of all genes at each time moment was calculated to form a single aggregated variable $g_{rest.}$

Two EFuNN models, one for the PLUS cell, and one for the MINUS cell, were trained on five input vector data, the expression level of the genes G(*t*) at time moment *t*, and five output vectors, the expression level $G(t+1)$ of the same genes recorded at the next time moment. Rules were extracted from the trained structure that describe the transition between the gene states in the problem space. The rules are given as a transition graph in Fig. 8.28.

Using the extracted rules that form a gene regulatory network, one can simulate the development of the cell from initial state $G(t=0)$, through time moments in the future, thus predicting a final state of the cell.

8.7 Summary and Open Problems

Modelling biological processes aims at the creation of models that trace these processes over time. The models should reveal the steps of development, the metamorphoses that occur at different points of time, and the 'trajectories' of the developed patterns.

This chapter demonstrates that biological processes are dynamically evolving and they require appropriate techniques, such as evolving connectionist systems. In Chapter 9 GRN of genes related to brain functions are derived through computational neuro-genetic modelling, which is a step further in this area (Benuskova and Kasabov, 2007).

There are many open problems and questions in bioinformatics that need to be addressed in the future. Some of them are:

1. Finding the gene expression profiles of all possible human diseases, including brain disease. Defining a full set of profiles of all possible diseases *in silico* would allow for early diagnostic tests.
2. Finding the gene expression profiles and the GRN of complex human behaviour, such as the 'instinct for information' speculated in the introduction.

3. Finding genetic networks that describe the gene interaction in a particular diseased tissue, thus suggesting genes that may be targeted for a better treatment of this disease.
4. Linking gene expression profiles with protein data, and then, with DNA data, for a full-circle modelling and complete understanding of the cell processes.

8.8 Further Reading

Further material related to specific sections of this chapter can be found as follows.

- *Computational Molecular Biology* (Pevzner, 2001)
- *Generic Knowledge on Bioinformatics* (Baldi and Brunak, 1998; Brown *et al.*, 2000b; Attwood and Parry-Smith, 1999; Boguski, 1998)
- *Artificial Intelligence and Bioinformatics* (Hofstadter, 1979)
- *Applications of Neural Network Methods, Mainly Multilayer Perceptrons and Self-organising Maps, in the General Area of Genome Informatics* (Wu and McLarty, 2000)
- *A Catalogue of Splice Junction Sequences* (Mount, 1982)
- *Microarray Gene Technologies* (Schena, 2000)
- *Data Mining in Biotechnology* (Persidis, 2000)
- *Application of the Theory of Complex Systems for Dynamic Gene Modelling* (Erdi, 2007)
- *Computational Modelling of Genetic and Biochemical Networks* (Bower and Bolouri, 2001)
- *Dynamic Modelling of the Regulation of a Large Set of Genes* (Somogyi *et al.*, 2001; D'haeseleer *et al.*, 2000)
- *Dynamic Modelling of a Single Gene Regulation Process* (Gibson and Mjolsness, 2001)
- *Methodology for Gene Expression Profiling* (Futschik *et al.*, 2003a; Futschik and Kasabov, 2002)
- *Using Fuzzy Neural Networks and Evolving Fuzzy Neural Networks in Bioinformatics* (Kasabov, 2007b; Kasabov and Dimitrov, 2004)
- *Fuzzy Clustering for Gene Expression Analysis* (Futschik and Kasabov, 2002)
- *Artificial Neural Filters for Pattern Recognition in Protein Sequences* (Schneider and Wrede, 1993)
- *Dynamic Models of the Cell* (Tomita *et al.*, 1999)

9. *Dynamic Modelling of Brain Functions and Cognitive Processes*

The human brain can be viewed as a dynamic, evolving information-processing system, and the most complex one. Processing and analysis of information recorded from brain activity, and modelling of perception, brain functions, and cognitive processes aim at understanding the brain and creating brainlike intelligent systems.

Brain study relies on modelling. This includes modelling of information preprocessing and feature extraction in the brain (e.g. modelling the cochlea), modelling the emergence of elementary concepts (e.g. phonemes and words), modelling complex representation and higher-level functions (e.g. speech and language), and so on. Whatever function or segment of the brain is modelled, the most important requirement is to know or to discover the *evolving rules,* i.e. the rules that allow the brain to learn, to develop in a continuous way. It is demonstrated here that the evolving connectionist systems paradigm can be applied for modelling some brain functions and processes.

This chapter is presented in the following sections.

- Evolving structures and functions in the brain and their modelling
- Auditory, visual, and olfactory information processing and their modelling
- Adaptive modelling of brain states based on EEG and fMRI data
- Computational neuro-genetic modelling: integrating gene and brain information into a single model
- Brain–gene ontology for EIS
- Summary and open problems
- Further reading

9.1 Evolving Structures and Functions in the Brain and Their Modelling

9.1.1 The Brain as an Evolving System

One of the last great frontiers of human knowledge relates to the study of the human brain and human cognition. Models of the cognitive processes are almost without exception qualitative, and are very limited in their applicability. Cognitive science would be greatly advanced by cognitive process models that are both qualitative and quantitative, and which evolve in response to data derived from quantitative measurements of brain activity.

The brain is an evolving system. The brain evolves initially from stem cells (Fig. 9.1). It evolves its structure and functionality from an embryo to a sophisticated biological information processing system (Amit, 1989; Arbib, 1972, 1987, 1998, 1995, 2002; Churchland and Sejnowski, 1992; Deacon, 1988, 1998; Freeman, 2001; Grossberg, 1982; Joseph, 1998; J. G. Taylor, 1998; van Owen, 1994; Wong, 1995). As an embryo, the brain grows and develops mainly based on genetic information. Even at the age of three months, some functional areas are already formed. But identical embryos, with the same genetic information, can develop in different ways to reach the state of an adult brain, and this is because of the environment in which the brain evolves. Both the genetic information (nature) and the environment (nurture) are crucial factors. They determine the evolving rules for the brain. The challenge is how to reveal these rules and eventually use them in brain models. Are they the same for every individual?

The brain evolves its functional modules for vision, for speech and language, for music and logic, and for many cognitive tasks. There are 'predefined' areas of the brain that are 'allocated' for language and visual information processing,

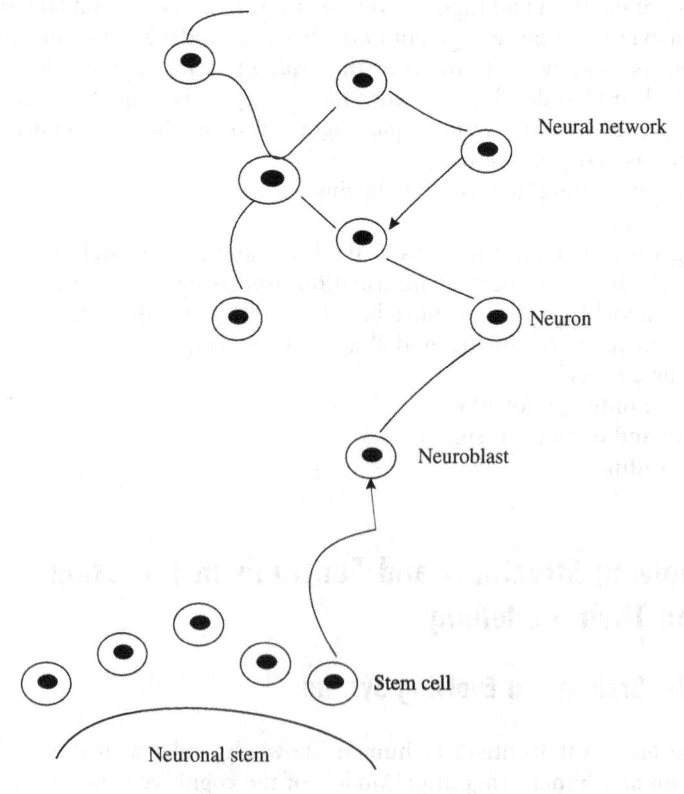

Fig. 9.1 The brain structure evolves from stem cells.

for example, but these areas may change during the neuronal evolving processes. The paths of the signals travelling and the information processes in the brain are complex and different for different types of information. Figure 9.2 shows schematically the pathways for auditory, visual, and sensory motor information processing in the human brain.

The cognitive processes of learning in the brain evolve throughout a lifetime. Intelligence is always evolving. An example is the spoken language learning process. How is this process evolving in the human brain? Can we model it in a computer system, in an evolving system, so that the system learns several languages at a time and adapts all the time to new accents and new dialects? In Kim *et al.* (1997) it is demonstrated that an area of the brain evolves differently when two spoken languages are learned simultaneously, compared with languages that are learned one after another.

Evolution is achieved through both genetically defined information and learning. The evolved neurons have a spatial–temporal representation where similar stimuli activate close neurons. Through dynamic modelling we can trace how musical patterns 'move' from one part of the acoustic space to another in a harmonic and slightly 'chaotic' way. Several principles of the evolving structure, functions, and cognition of the brain are listed below (for details see van Owen, 1994; Wong, 1995; Amit, 1989; Arbib, 1972, 1987, 1998, 1995, 2002; Churchland and Sejnowski, 1992; J. G. Taylor, 1998; Deacon, 1988, 1998; Freeman, 2001; Grossberg, 1982; Joseph, 1998):

- Redundancy, i.e. there are many redundant neurons allocated to a single stimulus or a task; e.g. when a word is heard, there are hundreds of thousands of neurons that are immediately activated.
- Memory-based learning, i.e. the brain stores exemplars of facts that can be recalled at a later stage. Some studies (see Widrow (2006)) suggest that all human actions, including learning and physical actions, are based on the memory.

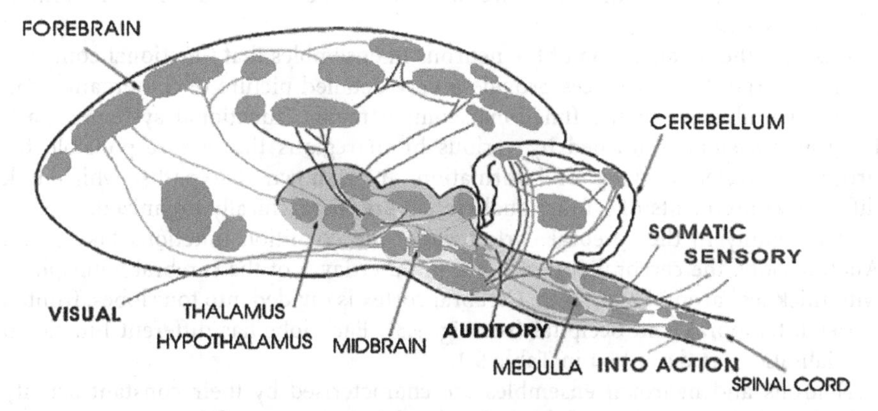

Fig. 9.2 Different areas of the human brain transfer different signals (auditory, visual, somatic-sensory, and control-action) shown as lines (Reproduced with permission from http://brainmuseum.org/circuitry/index.html).

- Evolution is achieved through interaction of an individual with the environment and with other individuals.
- Inner processes take place, e.g. information consolidation through sleep learning.
- The evolving process is continuous and lifelong.
- Through the process of evolving brain structures (neurons, connections) higher-level concepts emerge; they are embodied in the structure and represent a level of abstraction.

It seems that the most appropriate sources of data for brain modelling tasks would come from instrumental measurements of the brain activities. To date, the most effective means available for these types of brain measurement are electroencephalography (EEG), magnetoencephalography (MEG), and functional magnetic resonance imaging (fMRI). Once the data from these measurement protocols have been transformed into an appropriate state space representation, an attempt to model the dynamic cognitive process can be made.

9.1.2 From Neurons to Cognitive Functions

The brain is basically composed of neurons and glial cells. Despite the number of glial cells being 10 to 50 times bigger than the number of neurons, the role of information processing is given exclusively to the neurons (thus far). For this very reason most neural network models do not take into account the glial cells.

Neurons can be of different types according to their main functionality. There are sensory neurons, motor neurons, local interneurons, projection interneurons, and neuroendocrine cells. However, independently of the type, a neuron is basically constituted of four parts: input, trigger, conduction, and output. These parts are commonly represented in neuronal models.

In a very simplified manner, the neurons connect to each other in two basic ways: through divergent and convergent connections. Divergent connection occurs when the output of a neuron is split and is connected to the input of many other neurons. Convergent connections are those where a certain neuron receives input from many neurons.

It is with the organization of the neurons in ensembles that functional compartments emerge. Neurosciences provide a very detailed picture of the organization of the neural units in the functional compartments (functional systems). Each functional system is formed by various brain regions that are responsible for processing different types of information. It is shown that paths, which link different components of a functional system are hierarchically organized.

It is mainly in the cerebral cortex where the cognition functions take place. Anatomically, the cerebral cortex is a thin outer layer of the cerebral hemisphere with thickness around 2 to 4 mm. Cerebral cortex is divided into four lobes: frontal, parietal, temporal, and occipital (see Fig. 9.3). Each lobe has different functional specialisation, as described in Table 9.1.

Neurons and neuronal ensembles are characterised by their constant activity represented as oscillations of wave signals of different main frequencies as shown in Box 9.1.

Box 9.1. Main frequencies of wave signals in ensembles of neurons in the brain

Alpha (8–12 Hz)
Beta (13–28 Hz)
Gamma (28–50 Hz)
Delta (0.5–3.5Hz)
Theta (4–7 Hz)

9.1.3 Modelling Brain Functions

The brain is the most complex information processing machine. It processes data, information, and knowledge at different levels. Modelling the brain as an information processing machine would have different results depending on the goals of the models and the detail with which the models represent the genetic, biological, chemical, physical, physiological, and psychological rules and the laws that govern the functioning and behaviour of the brain.

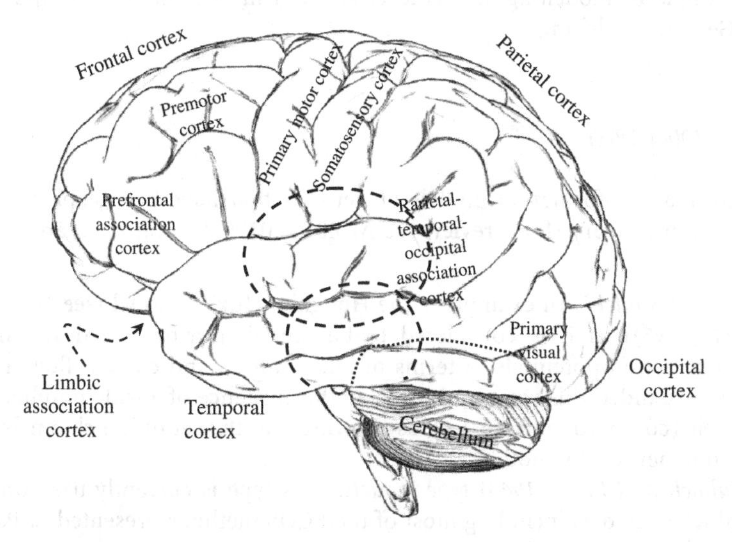

Fig. 9.3 The cerebral cortex and the human brain (from Benuskova and Kasabov (2007)).

Table 9.1 Location of cognitive functions in the cerebral cortex of the brain.

Function	*Cerebral Cortex Location*
Visual perception	Occipital cortex
Auditory perception	Temporal cortex
Multimodal association (visio-spatial location, language)	Parietal-temporal
Multimodal emotions, memory	Temporal, frontal, parietal

Generally speaking there are six levels of information processing in the brain as shown in Fig. I.1. We consider here the following four.

Molecular/Genetic Level

At the genetic level the genome constitutes the input information, whereas the phenotype constitutes the output result, which causes: (1) changes in the neuronal synapses (learning), and (2) changes in the DNA and its gene expression (Marcus, 2004). As pointed out in the introduction, neurons from different parts of the brain, associated with different functions, such as memory, learning, control, hearing, and vision, function in a similar way and their functioning is genetically defined. This principle can be used as a unified approach to building different neuronal models to perform different functions, such as speech recognition, vision, learning, and evolving. The genes relevant to particular functions can be represented as a set of parameters of a neuron. These parameters define the way the neuron is functioning and can be modified through feedback from the output of the neuron. Some genes may get triggered-off, or suppressed, whereas others may get triggered-on, or excited.

An example of modelling at this level is given in Section 9.4, 'Computational Neuro-Genetic Modelling'.

Single Neuronal Level

There are many information models of neurons that have been explored in the neural network theory (for a review see Arbib (2003)). Among them are:

1. *Analytical models.* An example is the Hodgkin–Huxley model (see Nelson and Rinzel (1995)) as it is considered to be the pioneering one describing the neuronal action potentials in terms of ion channels and current flow. Further studies expanded this work and revealed the existence of a wide number of ion channels (compartments) as well as showing that the set of ion channels varies from one neuron to another.
2. *McCulloch and Pitts' (1943) type models.* This type is currently used on traditional ANN models including most of the ECOS methods presented in Part I of this book.
3. *Spiking neuronal models* (see Chapter 4).

According to the neuronal model proposed in Matsumoto (2000) and Shigematsu *et al.*, (1999), a neuron accepts input information through its synapses and, subject to the output value of the neuron, it modifies back some of the synapses, those that, although the feedback signal reaches them, still have a level of information (weights, chemical concentration) above a certain threshold. The weights of the rest of the synapses decrease; see Fig. 9.4. Tsukada et al. (1996) proposed a spatial-temporal learning rule for LTP in hippocampus.

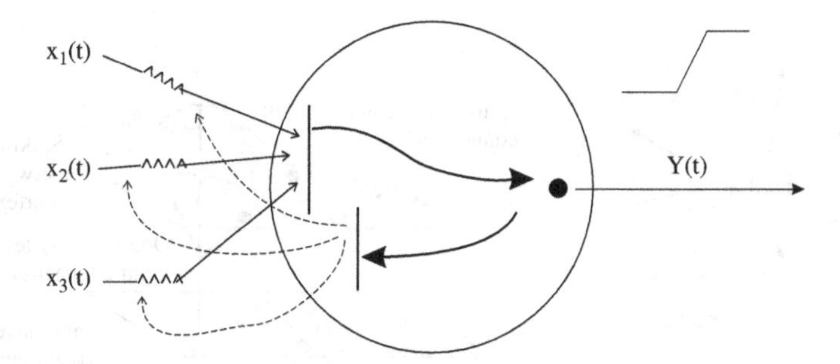

Fig. 9.4 The model of a neuron proposed by Gen Matsumoto (2000). According to the model, each neuron adjusts its synapses through a feedback from its output activation.

Neural Network (Ensemble) Level

Information is processed in ensembles of neurons that form a functionally defined area. A neural network model comprises many neuronal models. The model is an evolving one, and a possible implementation would be with the use of the methods and techniques presented in Part I.

Entire Brain Level

Many neuronal network modules are connected to model a complex brain structure and learning algorithms.

One model is introduced in Matsumoto and Shigematsu (1999). At different levels of information processing, similar, and at the same time, different principles apply. For example, the following common principles of learning across all levels of information processing were used in the model proposed by Matsumoto (2000) and Shigematsu *et al.* (1999):

- Output dependency, i.e. learning is based on both input information and output reaction–action.
- Self-learning, i.e. the brain acquires its function, structure, and algorithm based on both a 'super algorithm' and self-organisation (self-learning).

Modelling the entire brain is far from having been achieved and it will take many years to achieve this goal, but each step in this direction is an useful step towards understanding the brain and towards the creation of intelligent machines that will help people. Single functions and interactions between parts of the brain have been modelled. An illustration of using spiking neural networks (SNN; see Chapter 4) for modelling thalamus–cortical interactions is shown in Fig. 9.5 (Benuskova and Kasabov, 2007) and explained below.

The model from Fig. 9.5 has two layers. The input layer is supposed to represent the thalamus (the main subcortical sensory relay in the brain) and the output layer represents cerebral cortex. Individual model neurons can be based upon the classical spike response model (SRM; Gerstner and Kistler (2002)). The weight of

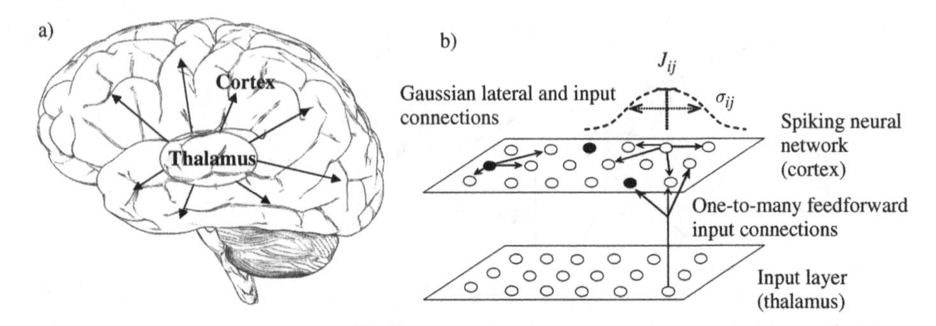

Fig. 9.5 (a) Neural network model represents the thalamocortical (TC) system; (b) the SNN represents cerebral cortex. About 10–20% of the neurons are inhibitory neurons that are randomly positioned on the grid (filled circles). The input layer represents the thalamic input to cortex. The presented model does not have a feedback loop from the cortex to the thalamus (from Benuskova and Kasabov (2007)).

the synaptic connection from neuron j to neuron i is denoted J_{ij}. It takes positive (negative) values for excitatory (inhibitory) connections, respectively. Lateral and input connections have weights that decrease in value with distance from the centre neuron i according to a Gaussian formula whereas the connections themselves can be established at random (for instance with $p = 0.5$).

For example, the asynchronous thalamic activity in the awake state of the brain can be simulated by a series of random input spikes generated in the input layer neurons. For the state of vigilance, a tonic, low-frequency, nonperiodic, and nonbursting firing of thalamocortical input is typical. For simulation of the sleep state we can employ regular oscillatory activity coming out of the input layer, etc. LFP (Local Field Potential) can be defined as an average of all instantaneous membrane potentials; i.e.

$$\Phi(t) = \frac{1}{N} \sum_{i=1}^{N} u_i(t) \tag{9.1}$$

Spiking neurons can be interconnected into neural networks of arbitrary architecture. At the same time it has been shown that SNN have the same computational power as traditional ANNs (Maas, 1996, 1998). With spiking neurons, however, new types of computation can be modelled, such as coincidence detection, synchronization phenomena, etc. Spiking neurons are more easily implemented in hardware than traditional neurons and integrated with neuromorphic systems.

9.2 Auditory, Visual, and Olfactory Information Processing and Their Modelling

The human brain deals mainly with five sensory modalities: vision, hearing, touch, taste, and smell. Each modality has different sensory receptors. After the receptors perform the stimulus transduction, the information is encoded through the excitation of neural action potentials. The information is encoded using pulses and time intervals between pulses. This process seems to follow a common pattern for all sensory modalities, however, there are still many unanswered questions regarding the way the information is encoded in the brain.

9.2.1 Auditory Information Processing

The hearing apparatus of an individual transforms sounds and speech signals into brain signals. These brain signals travel farther to other parts of the brain that model the (meaningful) acoustic space (the space of phones), the space of words, and the space of languages (see Fig. 9.6). The auditory system is adaptive, so new features can be included at a later stage and existing ones can be further tuned.

Precise modelling of hearing functions and the cochlea is an extremely difficult task, but not impossible to achieve (Eriksson and Villa, 2006). A model of the cochlea would be useful for both helping people with disabilities, and for the creation of speech recognition systems. Such systems would be able to learn and adapt as they work.

The ear is the front-end auditory apparatus in mammalians. The task of this hearing apparatus is to transform the environmental sounds into specific features and transmit them to the brain for further processing. The ear consists of three divisions: the outer ear, the middle ear, and the inner ear, as shown in Fig. 9.7.

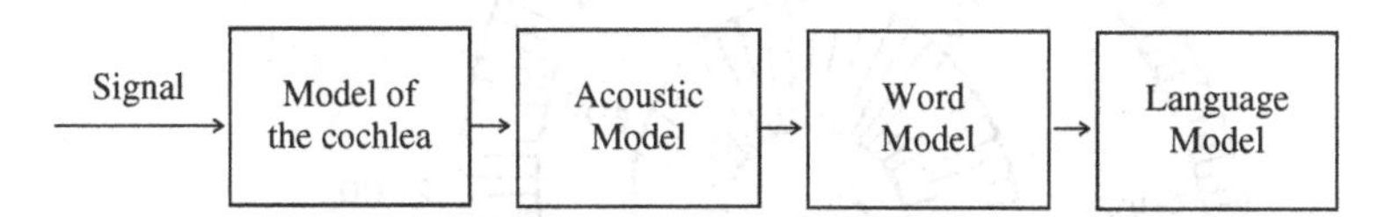

Fig. 9.6 A schematic diagram of a model of the auditory system of the brain.

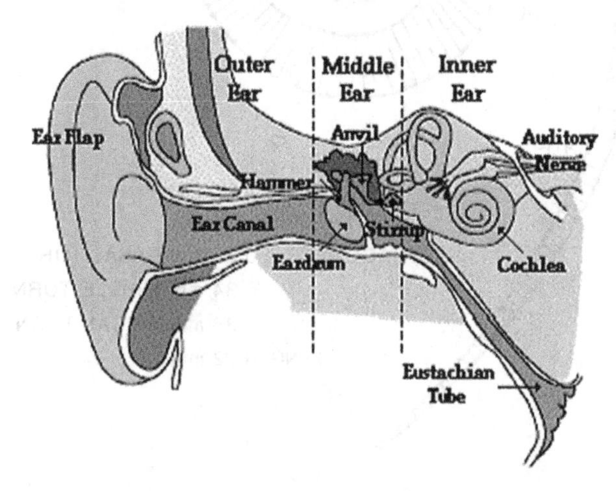

Fig. 9.7 A schematic diagram of the outer ear, the middle ear, and the inner ear. (Reproduced with permission from http://www.glenbrook.k12.il.us/gbssci/phys/Class/sound/u11l2d.html).

Figure 9.8 shows the human basilar membrane and the approximate position of the maximal displacement of tones of different frequencies. This corresponds to a filter bank of several channels, each tuned to a certain band of frequencies.

There are several models that have been developed to model functions of the cochlea (see, e.g. Greenwood (1961, 1990), de-Boer and de Jongh (1978), Allen (1995), Zwicker (1961), Glassberg and Moore (1990), and Eriksson and Villa, 2006). Very common are the Mel filter banks and the Mel scale cepstra coefficients (Cole *et al.*, 1995). For example, the centres of the first 26 Mel filter banks are the following frequencies (in Hertz): 86, 173, 256, 430, 516, 603, 689, 775, 947, 1033, 1120, 1292, 1550, 1723, 1981, 2325, 2670, 3015, 3445, 3962, 4565, 5254, 6029, 6997, 8010, 9216, 11025. The first 20 Mel filter functions are shown in Fig. 9.9.

Other representations use a gammatone function (Aertsen and Johannesma, 1980). It is always challenging to improve the acoustic modelling functions and make them closer to the functioning of the biological organs, which is expected to lead to improved speech recognition systems.

The auditory system is particularly interesting because it allows us not only to recognize sound but also to perform sound source location efficiently. Human ears are able to detect frequencies in the approximate range of 20 to 20,000 Hz.

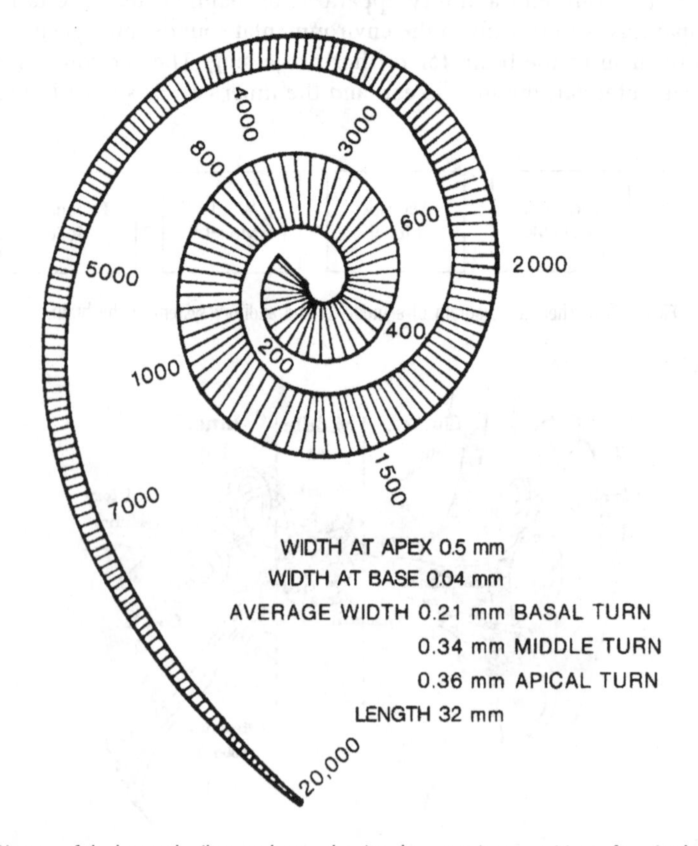

Fig. 9.8 Diagram of the human basilar membrane, showing the approximate positions of maximal displacement to tones of different frequencies.

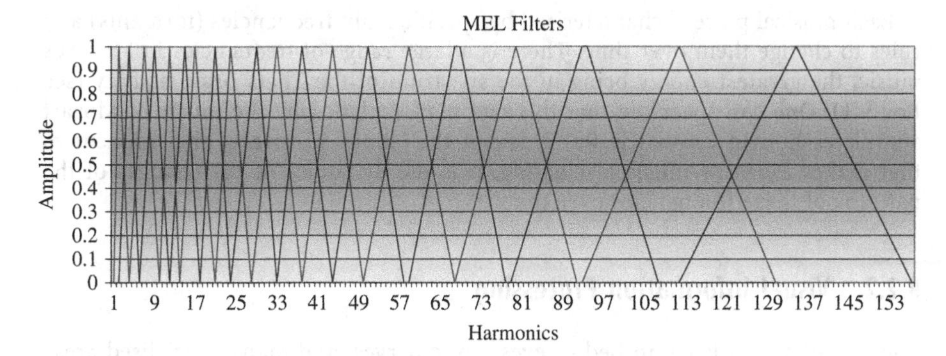

Fig. 9.9 The first 20 Mel filter functions.

Each ear processes the incoming signals independently, which are later integrated considering the signals' timing, amplitudes, and frequencies (see Fig. 9.10). The narrow difference of time between incoming signals from the left and right ear results in a cue to location of signal origin.

How do musical patterns evolve in the human brain? Music causes the emergence of patterns of activities in the human brain. This process is continuous, evolving, although in different pathways depending on the individual.

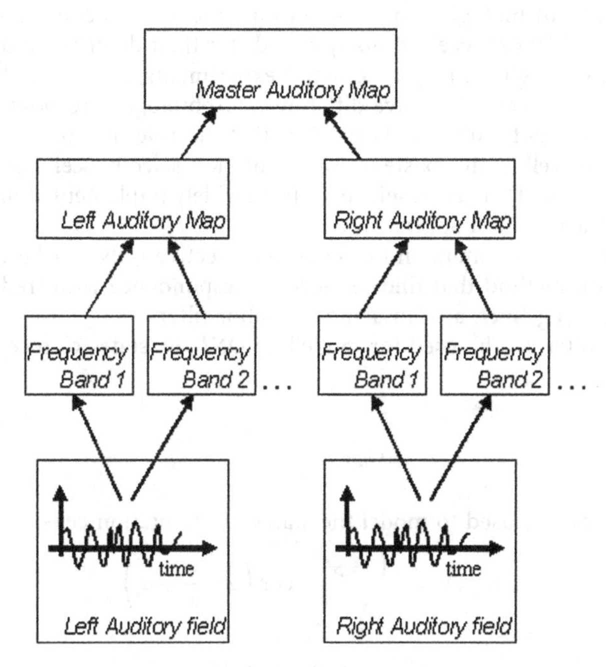

Fig. 9.10 A schematic representation of a model of the auditory system. The left and the right ear information processing are modelled separately and the results are later integrated considering the signal's timing, amplitudes, and frequencies.

Each musical piece is characterised by specific main frequencies (formants) and rules to change them over time. There is a large range of frequencies in Mozart's music, the greatest energy being in the spectrum of the Thetà brain activity (see Box 9.1). One can speculate that this fact may explain why the music of Mozart stimulates human creativity. But it is not the 'static' picture of the frequencies that makes Mozart's music fascinating, it is the dynamics of the changes of the patterns of these frequencies over time.

9.2.2 Visual Information Processing

The visual system is composed of eyes, optic nerves, and many specialised areas of the cortex (the ape for example has more than 30).

The image on the retina is transmitted via the optic nerves to the first visual cortex (V1), which is situated in the posterior lobe of the brain. There the information is divided into two main streams, the 'what' tract and the 'where' tract.

The ventrical ('what') tract separates targets (objects and things) in the field of vision and identifies them. The tract traverses the occipital lobe to the temporal lobe (behind the ears).

The dorsal tract ('where') is specialised in following the location and position of the objects in the surrounding space. The dorsal tract traverses the back of the head to the top of the head.

How and where the information from the two tracts is united to form one complete perception is not completely known.

On the subject of biological approaches for processing incoming information, Hubel and Wiesel (1962) received many awards for their description of the human visual system. Through neuro-physiological experiments, they were able to distinguish some types of cells that have different neurobiological responses according to the pattern of light stimulus. They identified the role that the retina has as a contrast filter as well as the existence of orientation selective cells in the primary visual cortex (Fig. 9.11). Their results have been widely implemented in biologically realistic image acquisition approaches.

The idea of contrast filters and orientation selective cells can be considered a feature selection method that finds a close correspondence with traditional ways of image processing, such as Gaussian and Gabor filters.

A Gaussian filter can be used for modelling ON/OFF states of receptive cells:

$$G(x,y) = e^{\left(\frac{x^2+y^2}{2\sigma^2}\right)} \quad (9.2)$$

A Gabor filter can be used to model the states of orientation cells:

$$G(x,y) = e^{\left(\frac{x'^2+\gamma^2 y'^2}{2\sigma^2}\right)} \cos\left(2\pi\frac{x'}{\lambda}+\varphi\right) \quad (9.3)$$

$$x' = x\cos(\theta)+y\sin(\theta)$$
$$y' = -x\sin(\theta)+y\cos(\theta) \quad (9.4)$$

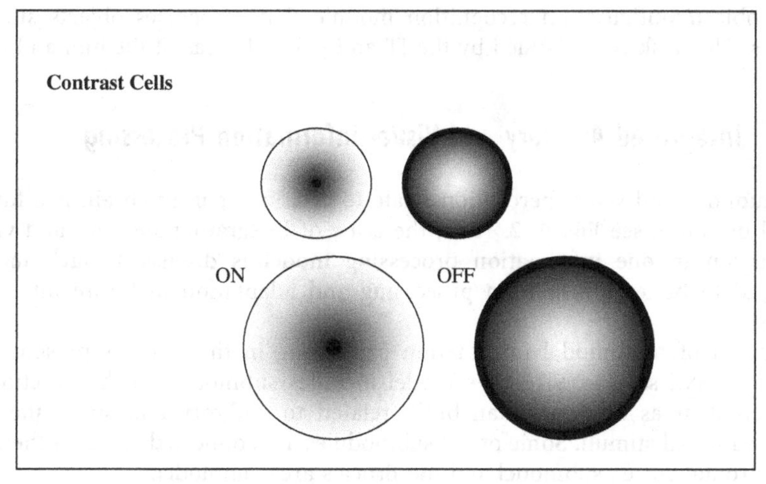

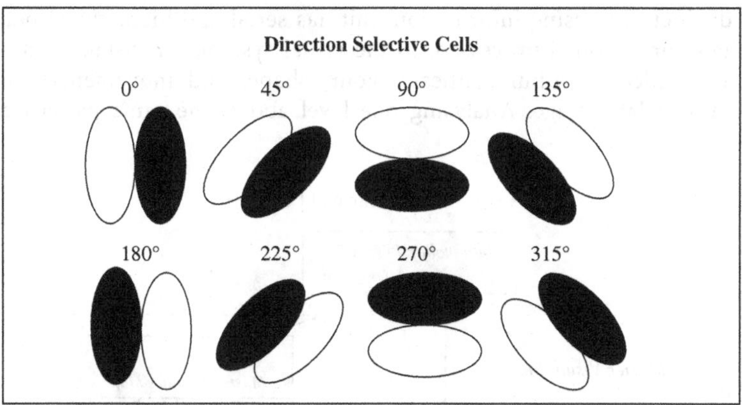

Fig. 9.11 Contrast cells and direction selective visual cells.

where φ = phase offset, θ = orientation (0,360), λ = wavelength, σ = standard deviation of the Gaussian factor of the Gabor function, and γ = aspect ratio (specifies the ellipticity of the support of the Gabor function).

A computational model of the visual subsystem would consist of the following levels.

1. A visual preprocessing module, that mimics the functioning of the retina, the retinal network, and the lateral geniculate nucleus (LGN).
2. An elementary feature recognition module, responsible for the recognition of features such as the curves of lips or the local colour. The peripheral visual areas of the human brain perform a similar task.
3. A dynamic feature recognition module that detects dynamical changes of features in the visual input stream. In the human brain, the processing of visual motion is performed in the V5/MT area of the brain.
4. An object recognition module that recognises elementary shapes and their parts. This task is performed by the inferotemporal (IT) area of the human brain.

5. An object/configuration recognition module that recognises objects such as faces. This task is performed by the IT and parietal areas of the human brain.

9.2.3 Integrated Auditory and Visual Information Processing

How auditory and visual perception relate to each other in the brain is a fundamental question; see Fig. 9.12. Here, the issue of integrating auditory and visual information in one information processing model is discussed. Such models may lead to better information processing and adaptation in future intelligent systems.

A model of multimodal information processing in the brain is presented in Deacon (1988); see Fig. 9.13. The model includes submodels of the functioning of different areas of the human brain related to auditory and simultaneously perceived visual stimuli. Some of the submodules are connected to each other, e.g. the prefrontal cortex submodel and the Broca's area submodel.

Each distinct processing information unit has serial and hierarchical pathways where the information is processed. In the visual system, for instance, the information is divided in submodalities (colour, shape, and movements) that are integrated at a later stage. Analysing one level above, the same pattern can be

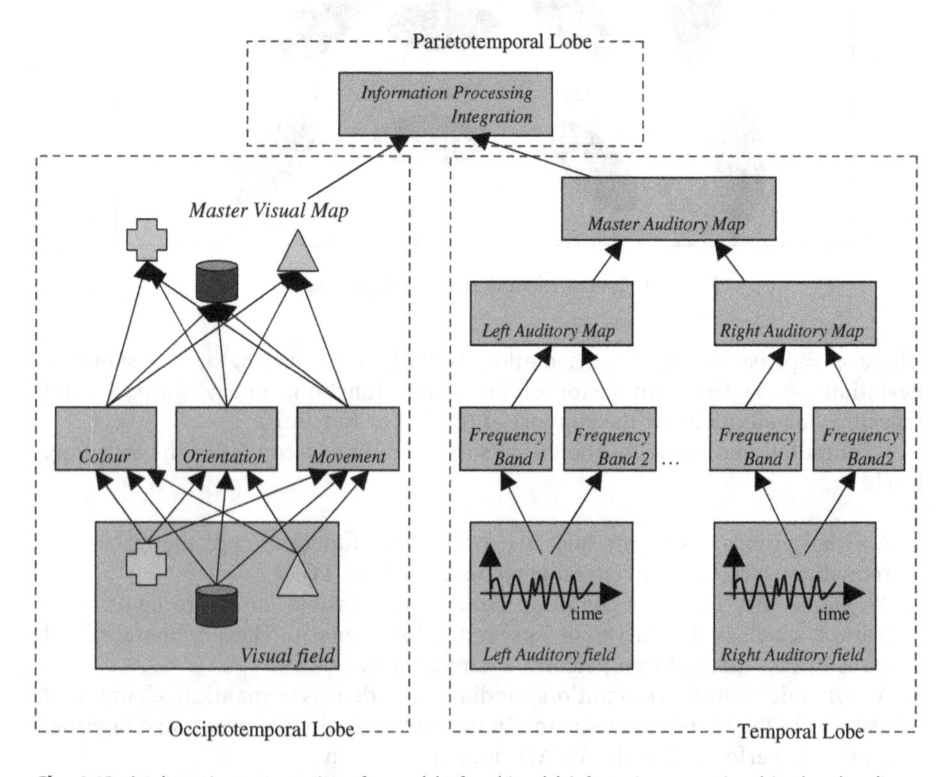

Fig. 9.12 A schematic representation of a model of multimodal information processing (visual and auditory information) in the brain.

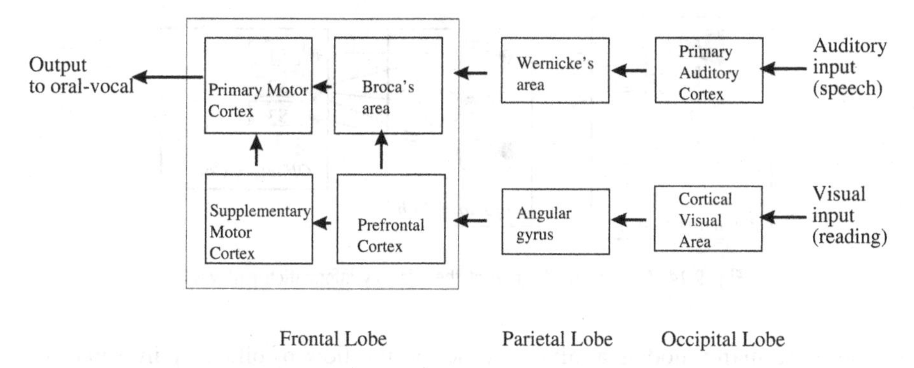

Fig. 9.13 Deacon's model for multi-modal information processing (Deacon, 1988).

noticed. The information from different modules converges to a processing area responsible for the integration. A simple example is the detection of a certain food by the integration of the smell and the visual senses.

9.2.4 Olfactory Information Processing

Smell and taste are chemical senses and the only senses that do not maintain the spatial relations of the input receptors. However, after the transduction of the olfactory stimuli, the encoding is done similarly to the other senses, using pulse rate or pulse time. Furthermore, contrast analysis is done in the first stage of the pathway and parallel processing of olfactory submodalities has been proven to happen in the brain.

There are different types of olfactory sensory neurons that are stimulated by different odorants. Thus, having a large number of different receptor types allows many odorants to be discriminated. The olfactory discrimination capacity in humans varies highly and can reach 5000 different odorants in trained people.

Chemical stimuli are acquired by millions of olfactory sensory neurons that can be of as many as 1000 types. In the olfactory bulb, there is the convergence of sensory neurons to units called glomeruli (approx. 25,000 to 1), that are organized in such a way that information from different receptors is placed in different glomeruli. Each odorant (smell) is recognized using several glomeruli and each glomerula can take part to recognize many odorants. Thus, glomeruli are not odour-specific, but a specific odour is described by a unique set of glomeruli. How the encoding is done is still unknown. The glomeruli can be roughly considered to represent the neural 'image' of the odour stimuli. The information is then sent to different parts of the olfactory cortex for odour recognition; see Fig. 9.14.

Artificial sensors for odour detection are widely available and include metal oxide sensors, polymer resonating sensors, and optical bead sensors. The second step, after data acquisition, is to process the olfactory information. An artificial system that processes olfactory information is required to have mainly three properties: process many sensor inputs; discern a large number of different odours; and handle noisy acquired data.

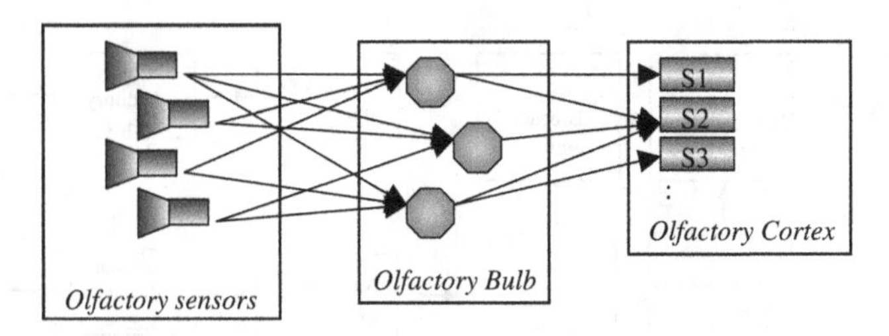

Fig. 9.14 A schematic diagram of the olfactory information pathway.

There are many models aiming to describe the flow of olfactory information. Some of them are oversimplified, tending only to perform pattern recognition without a long biological description and some of them very complex and detailed. A spiking neural network is used in the system described in Allen *et al.* (2002) where odour acquisition is done through a two-dimensional array of sensors (more than 1000). The temporal binary output of the sensor is then passed to a spiking neural network for classification of different scents. The system is then embedded in an FPGA chip.

In Zanchettin and Ludermir (2004) an artificial nose model and a system are experimented on for the recognition of gasses emitted at petrol stations, such as ethane, methane, butane, propane, and carbon monoxide. The model consists of:

- Sensory elements: eight polypyrrol-based gas sensors
- EFuNN for classification of the sensory input vectors into one or several of the output classes (gasses)

The EFuNN model performs at a 99% recognition rate whereas a time-delay ANN performs at the rate of 89%. In addition, the EFuNN model can be further trained on new gasses, new sensors, and new data, also allowing the insertion of some initial rules into the EFuNN structure as initialisation, that are well-known rules from the theory of the gas compounds.

Other olfactory models and 'artificial nose' systems have been developed and implemented in practice (Valova *et al.*, 2004).

9.3 Adaptive Modelling of Brain States Based on EEG and fMRI Data

9.3.1 EEG Measurements

The moving electrical charges associated with a neuronal action potential emit a minute, time-varying electromagnetic field. The complex coordinated activities associated with cognitive processes require the cooperation of millions of neurons, all of which will emit this electrical energy. The electromagnetic fields generated by these actively cooperating neurons linearly sum together via superposition. These summed fields propagate through the various tissues of the cranium to the scalp surface, where EEGs can detect and record this neural activity by means of measuring electrodes placed on the scalp.

Historically, expert analysis via visual inspection of the EEG has tended to focus on the activity in specific wavebands, such as delta, theta, alpha, and beta. As far back as 1969, attempts were made to use computerised analysis of EEG data in order to determine the subject's state of consciousness. Figure 9.15 shows an example of EEG data collected from two states of the same subject: the normal state and an epileptic state.

Noise contaminants in an EEG are called 'artefacts'. For example, the physical movement of the test subject can contaminate an EEG with the noise generated by the action potentials of the skeletal muscles. Even if the skeletal muscle action potentials do not register on the EEG, the shifting mechanical stress on the electrodes can alter the contact with the subject, thereby affecting the measuring electrodes' conductivity. These variations in electrode contact conductivity will also result in the recording of a movement artefact.

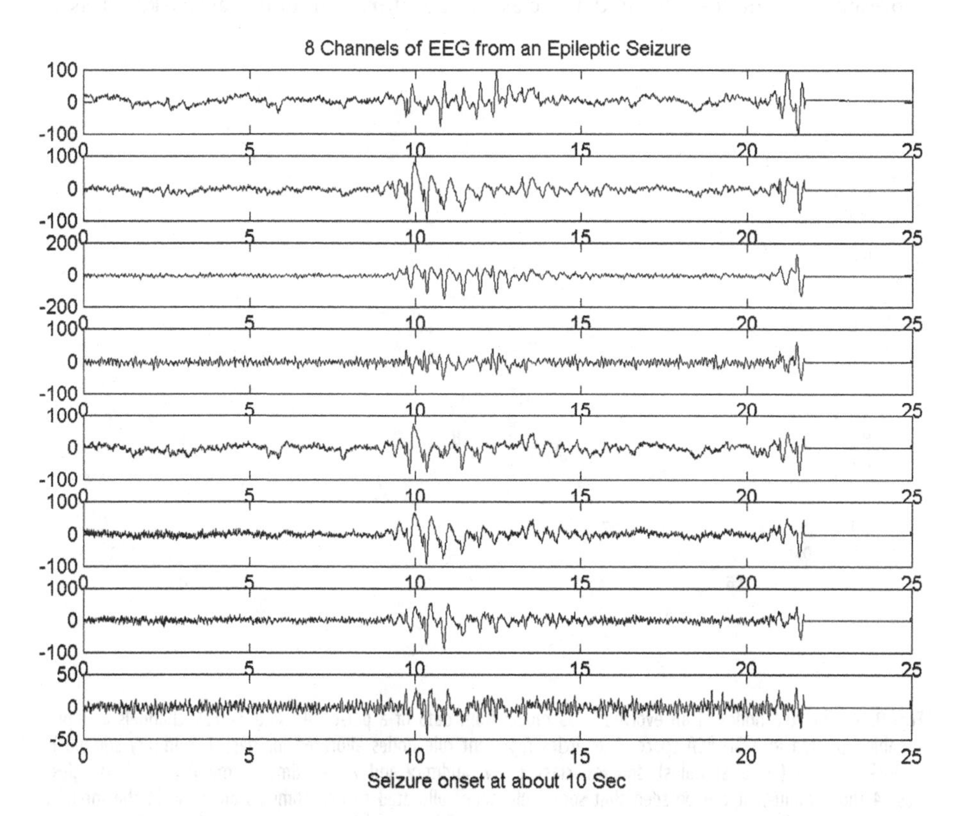

Fig. 9.15 EEG signals recorded in eight channels from a person in a normal state and in an epileptic state (the onset of epilepsy is manifested after the time unit 10).

9.3.2 ECOS for Brain EEG Data Modeling, Classification, and Brain Signal Transition Rule Extraction

In Kasabov *et al.* (2007) a methodology for continuous adaptive learning and classification of human scalp electroencephalographic (EEG) data in response to multiple stimuli is introduced based on ECOS. The methodology is illustrated on a case study of human EEG data, recorded at resting, auditory, visual, and mixed audio-visual stimulation conditions. It allows for incremental continuous adaptation and for the discovery of brain signal transition rules, such as: IF segments S1,S2, . . . , Sn of the brain are active at a time moment *t* THEN segments R1,R2, . . . , Rm will become active at the next time moment (*t+1*). The method results in a good classification accuracy of EEG signals of a single individual, thus suggesting that ECOS could be successfully used in the future for the creation of intelligent personalized human–computer interaction models, continuously adaptable over time, as well as for the adaptive learning and classification of other EEG data, representing different human conditions. The method could help better understand hidden signal transitions in the brain under certain stimuli when EEG measurement is used.

Figure 9.16 shows the rule nodes of an evolved ECOS model from data of a person A using 37 EEG channels as input variables, plotted in a 3D PCA space. The circles represent rule nodes allocated for class 1 (auditory stimulus); asterisks, class 2

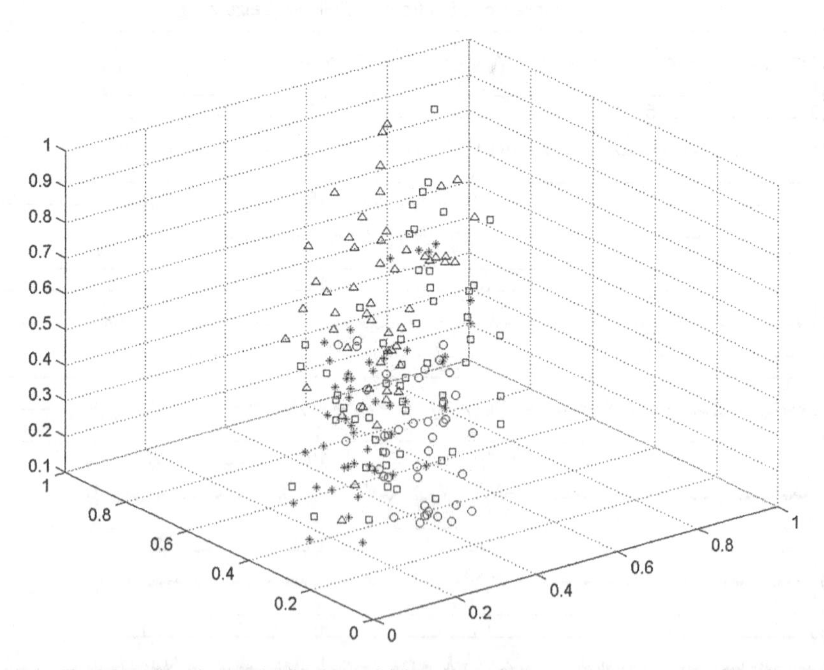

Fig. 9.16 The rule nodes of an evolved ECOS model from data of a person A using 37 EEG channels as input variables, plotted in a 3D PCA space. The circles represent rule nodes allocated for class 1 (auditory stimulus); asterisks, class 2 (visual stimulus); squares, class 3 (AV, auditory and visual stimuli combined); and triangles, class 4 (no stimulus). It can be seen that some rule nodes allocated to one stimulus are close in the model's space, meaning that they represent close location on the EEG surface. At the same time, there are nodes that represent each of the stimuli and are spread all over the whole space, meaning that for a single stimulus the brain activates many areas at a different time of the presentation of the stimulus.

(visual stimulus); squares, class 3 (AV, auditory and visual stimulus combined); and triangles, class 4 (no stimulus). It can be seen that rule nodes allocated to one stimulus are close in the space, which means that their input vectors are similar.

The allocation of the above nodes (cluster centres) back to the EEG channels for each stimulus is shown in Fig. 9.17 and Fig. 9.18 shows the original EEG electrodes allocation on the human scalp.

9.3.3 Computational Modelling Based on fMRI Brain Images

Neural activity is a metabolic process that requires oxygen. Active neurons require more oxygen than quiescent neurons, so they extract more oxygen from the blood haemoglobin.

Functional magnetic resonance imaging (fMRI) makes use of this fact by using deoxygenated haemoglobin as an MRI contrast agent. The entire theory of MRI is based on the fact that different neurons have different relaxation times. In MRI, the nuclear magnetic moments of the neurons to be imaged are aligned with a powerful

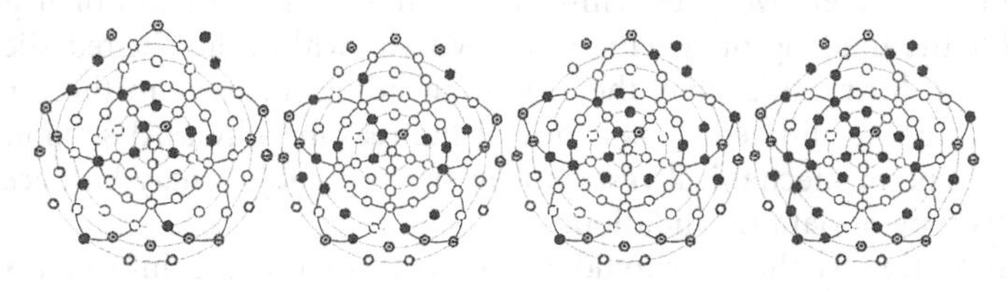

Fig. 9.17 The location of the selected, significantly activated electrodes, from the ECOS model in Fig. 9.16 for each of the stimuli of classes from 1 to 4 (A, V, AV, No, from left to right, respectively).

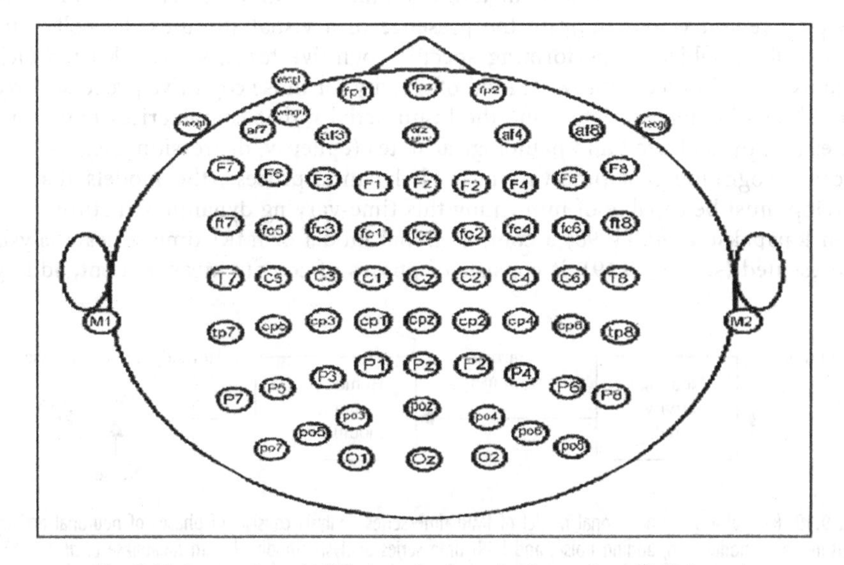

Fig. 9.18 Layout of the 64 EEG electrodes (extended International 10-10 System).

magnetic field. Once the nuclei are aligned, their magnetic moment is excited with a tuned pulse of resonance frequency energy. As these excited nuclear magnetic moments decay back to their rest state, they emit the resonance frequency energy that they have absorbed. The amount of time that is taken for a given neuron to return, or decay, to the rest state depends upon that neuron's histological type. This decay time is referred to as the relaxation time, and nerve tissues can be differentiated from each other on the basis of their varying relaxation times.

In this manner, oxygenated haemoglobin can be differentiated from deoxygenated haemoglobin because of the divergent relaxation times. fMRI seeks to identify the active regions of the brain by locating regions that have increased proportions of deoxygenated haemoglobin.

EEG and fMRI have their own strengths and weaknesses when used to measure the activity of the brain. EEGs are prone to various types of noise contamination. Also, there is nothing intuitive or easy to understand an EEG recording. In principle, fMRI is much easier to interpret. One only has to look for the contrasting regions contained in the image. In fMRI, the resonance frequency pulse is tuned to excite a specific slice of tissue. This localisation is enabled by a small gradient magnetic field imposed along the axis of the imaging chamber. After the 'slice select' resonance frequency excitation pulse, two other magnetic gradients are imposed on the other two axes within the chamber. These additional gradients are used for the imaging of specific tissue 'voxels' within the excited slice. If the subject moves during this time, the spatial encoding inherent in these magnetic gradients can be invalidated. This is one of the reasons why MRI sessions last so long. Some type of mechanical restraint on the test subject may be necessary to prevent this type of data invalidation.

The assumption is that it should be possible to use the information about specific cortical activity in order to make a determination about the underlying cognitive processes. For example, if the temporal regions of the cortex are quite active while the occipital region is relatively quiescent, we can determine that the subject has been presented with an auditory stimulus. On the other hand, an active occipital region would indicate the presence of a visual stimulus. By collecting data while a subject is performing specific cognitive tasks, we can learn which regions of the brain exhibit what kind of activity for those cognitive processes. We should also be able to determine the brain activity that characterises emotional states (happy, sad, etc.) and pathological states (epilepsy, depression, etc.) as well. Because cognition is a time-varying and dynamic process, the models that we develop must be capable of mimicking this time-varying dynamic structure.

In Rajapakse *et al.* (1998) a computational model of fMRI time series analysis is presented (see Fig. 9.19). It consists of phases of activity measurement, adding

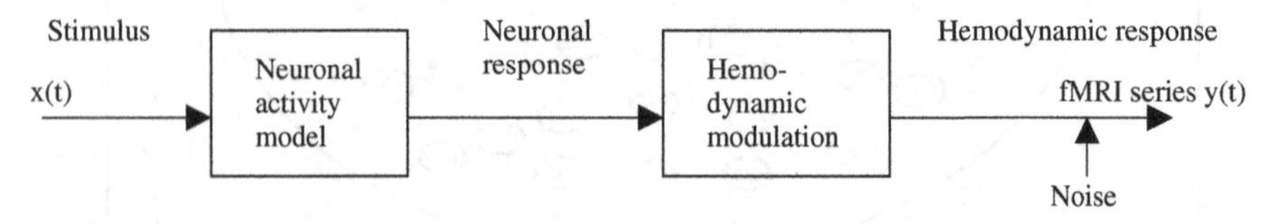

Fig. 9.19 Rajapakse's computational model of fMRI time-series analysis consists of phases of neuronal activity measurement, modulation, adding noise, and fMRI time series analysis (modified from Rajapakse *et al.* (1998)).

noise, modulation, and fMRI time-series analysis. The time series of fMRI are recorded from subjects performing information retrieval tasks.

Online brain image analysis, where brain images are added to the model and the model is updated in a continuous way, is explored in Bruske *et al.* (1998), where a system for online clustering of fMRI data is proposed.

A comprehensive study of brain imaging and brain image analysis related to cognitive processes is presented in J. G. Taylor (1999). Brain imaging has started to be used for the creation of models of consciousness. A three-stage hypothetical model of consciousness, for example is presented in J. G. Taylor (1998).

Using ECOS for online brain image analysis and modelling is a promising area for further research, as ECOS allow for online model creation, model adaptation, and model explanation.

9.4 Computational Neuro-Genetic Modelling (CNGM)

9.4.1 Principles of CNGM

A CNGM integrates genetic, proteomic, and brain activity data and performs data analysis, modelling, prognosis, and knowledge extraction that reveals relationships between brain functions and genetic information (see Fig. I.1).

A future state of a molecule M' or a group of molecules (e.g. genes, proteins) depends on its current state M, and on an external signal Em:

$$M' = Fm\ (M, Em) \tag{9.5}$$

A future state N' of a neuron, or an ensemble of neurons, will depend on its current state N and on the state of the molecules M (e.g. genes) and on external signals En:

$$N' = Fn\ (N, M, En) \tag{9.6}$$

And finally, a future cognitive state C' of the brain will depend on its current state C and also on the neuronal N and the molecular M state and on the external stimuli Ec:

$$C' = Fc\ (C, N, M, Ec) \tag{9.7}$$

The above set of equations (or algorithms) is a general one and in different cases it can be implemented differently as shown in Benuskova and Kasabov (2007) and illustrated in the next section.

9.4.2 Integrating GRN and SNN in CNGM

In Kasabov and Benuskova (2004) and Benuskova and Kasabov (2007) we have introduced a novel computational approach to brain neural network modelling that integrates ANN with an internal dynamic GRN (see Fig. 9.20. Interaction of genes in model neurons affects the dynamics of the whole ANN through neuronal parameters, which are no longer constant, but change as a function of gene expression. Through optimisation of the GRN, initial gene/protein expression values, and ANN parameters, particular target states of the neural network operation can be achieved.

It is illustrated by means of a simple neuro-genetic model of a spiking neural network (SNN). The behaviour of SNN is evaluated by means of the local field potential (LFP), thus making it possible to attempt modelling the role of genes in different brain states, where EEG data are available to test the model. We use the standard FFT signal-processing technique to evaluate the SNN output and compare with real human EEG data. For the objective of this work, we consider the time-frequency resolution reached with the FFT to be sufficient. However, should higher accuracy be critical, wavelet transform, which considers both time and frequency resolution, could be used instead. Broader theoretical and biological background of CNGM construction is given in Kasabov and Benuskova (2004) and Benuskova and Kasabov (2007).

In general, we consider two sets of genes: a set G_{gen} that relates to general cell functions and a set G_{spec} that defines specific neuronal information-processing functions (receptors, ion channels, etc.). The two sets form together a set $\mathbf{G} = \{G_1, G_2, \ldots, G_n\}$. We assume that the expression level of each gene $g_j(t+\Delta t')$ is a nonlinear function of expression levels of all the genes in $\mathbf{G}$,

$$g_j(t+\Delta t') = \sigma\left(\sum_{k=1}^{n} w_{jk} g_k(t)\right) \tag{9.8}$$

We work with normalized gene expression values in the interval (0,1). The coefficients $w_{ij} \in (-5, 5)$ are the elements of the square matrix $\mathbf{W}$ of gene interaction weights. Initial values of gene expressions are small random values; i.e. $g_j(0) \in (0, 0.1)$.

In the current model we assume that: (1) one protein is coded by one gene; (2) the relationship between the protein level and the gene expression level is

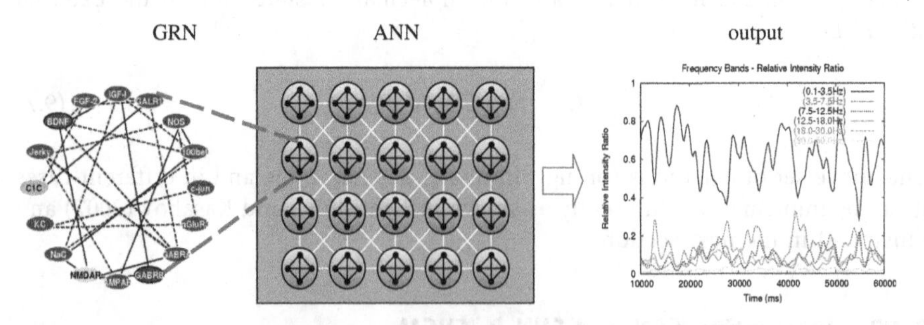

Fig. 9.20 A more complex case of CNGM, where a GRN of many genes is used to represent the interaction of genes, and an ANN is employed to model a brain function. The model spectral output is compared against real brain data for validation of the model and for verifying the derived gene interaction GRN after a GA model optimization is applied (see Chapters 6 and 8) (Kasabov *et al.*, 2005; Benuskova and Kasabov, 2007).

nonlinear; and (3) protein levels lie between the minimal and maximal values. Thus, the protein level $p_j(t+\Delta t)$ is expressed by

$$p_j(t+\Delta t) = (p_j^{\max} - p_j^{\min})\sigma\left(\sum_{k=1}^{n} w_{jk} g_k(t)\right) + p_j^{\min} \tag{9.9}$$

The delay $\Delta t < \Delta t'$ corresponds to the delay caused by the gene transcription, mRNA translation into proteins, and posttranslational protein modifications (Abraham *et al.*, 1993). Delay Δt includes also the delay caused by gene transcription regulation by transcription factors.

The GRN model from Eqs. (9.8) and (9.9) is a general one and can be integrated with any ANN model into a CNGM. Unfortunately the model requires many parameters to be either known in advance or optimized during a model simulation. In the presented experiments we have made several simplifying assumptions:

1. Each neuron has the same GRN, i.e. the same genes and the same interaction gene matrix **W**.
2. Each GRN starts from the same initial values of gene expressions.
3. There is no feedback from neuronal activity or any other external factors to gene expression levels or protein levels.
4. Delays Δt are the same for all proteins and reflect equal time points of gathering protein expression data.

We have integrated the above GRN model with the SNN illustrated in Fig. 9.20. Our spiking neuron model is based on the spike response model, with excitation and inhibition having both fast and slow components, both expressed as double exponentials with amplitudes and the rise and decay time constants (see chapter 4).

Neuronal parameters and their correspondence to particular proteins are summarized in Table 9.2. Several parameters (amplitude, time constants) are linked

Table 9.2 Neuronal parameters and their corresponding proteins (receptors/ion channels).

Neuron's parameter P_j	*Relevant protein p_j*
Amplitude and time constants of:	
Fast excitation	AMPAR
Slow excitation	NMDAR
Fast inhibition	GABRA
Slow inhibition	GABRB
Firing threshold and its decay time constant	SCN and/or KCN and/or CLC

AMPAR = (amino-methylisoxazole- propionic acid) AMPA receptor; NMDAR = (*N*-methyl-D-aspartate acid) NMDA receptor; GABRA = (gamma-aminobutyric acid) GABA receptor A; GABRB = GABA receptor B; SCN = sodium voltage-gated channel; KCN = kalium (potassium) voltage-gated channel; CLC = chloride channel.

to one protein. However their initial values in Eq. (9.3) will be different. Relevant protein levels are directly related to neuronal parameter values P_J such that

$$P_j(t) = P_j(0)p_j(t) \tag{9.10}$$

where $P_j(0)$ is the initial value of the neuronal parameter at time $t = 0$. Moreover, in addition to the gene coding for the proteins mentioned above, we include in our GRN nine more genes that are not directly linked to neuronal information-processing parameters. These genes are: c-jun, mGLuR3, Jerky, BDNF, FGF-2, IGF-I, GALR1, NOS, and S100beta. We have included them for later modelling of some diseases.

We want to achieve a desired SNN output through optimisation of the model 294 parameters (we are optimising also the connectivity and input frequency to the SNN). We evaluate the LFP of the SNN, defined as $\text{LFP} = (1/N)\Sigma u_i(t)$, by means of a FFT in order to compare the SNN output with the EEG signal analysed in the same way. It has been shown that brain LFPs in principle have the same spectral characteristics as EEG (Quiroga, 1998). Because the updating time for SNN dynamics is inherently 1 ms, just for computational reasons, we employ the delays Δt in Eq. (9.9) being equal to just 1 s instead of minutes or tens of minutes. In order to find an optimal GRN within the SNN model so that the frequency characteristics of the LFP of the SNN model are similar to the brain EEG characteristics, we use the following procedure.

1. Generate a population of CNGMs, each with randomly generated values of coefficients for the GRN matrix **W**, initial gene expression values $g(0)$, initial values of SNN parameters $P(0)$, and different connectivity.
2. Run each SNN over a period of time T and record the LFP.
3. Calculate the spectral characteristics of the LFP using the FFT.
4. Compare the spectral characteristics of SNN LFP to the characteristics of the target EEG signal. Evaluate the closeness of the LFP signal for each SNN to the target EEG signal characteristics. Proceed further according to the standard GA algorithm to possibly find a SNN model that matches the EEG spectral characteristics better than previous solutions.
5. Repeat steps 1 to 4 until the desired GRN and SNN model behaviour is obtained.
6. Analyse the GRN and the SNN parameters for significant gene patterns that cause the SNN model behaviour.

Simulation Results

In Benuskova and Kasabov (2007) experimental results were presented on real human interictal EEG data for different clinically relevant subbands over time. These subbands are: delta (0.5–3.5 Hz), theta (3.5–7.5 Hz), alpha (7.5–12.5 Hz), beta 1 (12.5–18 Hz), beta 2 (18–30 Hz), and gamma (above 30 Hz). The average RIRs over the whole time of simulation (i.e., $T = 1$ min) was calculated and used as a fitness function for a GA optimisation. After 50 generations with six solutions in each population we obtained the best solution. Solutions for reproduction were being chosen according to the roulette rule and the crossover between parameter values was performed as an arithmetic average of the parent values. We performed the same FFT analysis as for the real EEG data with the min–max frequency = 0.1/50 Hz. This particular SNN had an evolved GRN with only 5 genes out of 16 periodically changing their expression values (s100beta, GABRB, GABRA, mGLuR3, c-jun) and all other genes having constant expression values.

The preliminary results show that the same signal-processing techniques can be used for the analysis of both the simulated LFP of the SNN CNGM and the real EEG data to yield conclusions about the SNN behaviour and to evaluate the CNGM at a gross level. With respect to our neuro-genetic approach we must emphasize that it is still in an early developmental stage and the experiments assume many simplifications. In particular, we would have to deal with the delays in Eq. (9.9) more realistically to be able to draw any conclusions about real data and real GRNs. The LFP obtained from our simplified model SNN is of course not exactly the same as the real EEG, which is a sum of many LFPs. However LFP's spectral characteristics are very similar to the real EEG data, even in this preliminary example.

Based on our preliminary experimentation, we have come to the conclusion that many gene dynamics, i.e. many interaction matrices **W**s that produce various gene dynamics (e.g., constant, periodic, quasi-periodic, chaotic) can lead to very similar SNN LFPs. In our future work, we want to explore statistics of plausible **W**s more thoroughly and compare them with biological data to draw any conclusions about underlying GRNs. Further research questions are: how many GRNs would lead to similar LFPs and what do they have in common? How can we use CNGM to model gene mutation effects? How can we use CNGM to predict drug effects? And finally, how can we use CNGM for the improvement of individual brain functions, such as memory and learning?

9.5 Brain–Gene Ontology

In Chapter 7 we presented a framework for integrating ECOS and ontology, where interaction between ECOS modelling and an evolving repository of data and knowledge is facilitated. Here we apply this framework to a particular ontology, brain–gene ontology (BGO; *http://www.kedri.info/*) (Kasabov *et al.*, 2006b).

Gene Ontology (GO; *http://www.geneontology.org/*) is a general repository that contains a large amount of information about genes across species, and their relation to each other and to some diseases. The BGO contains specific information about brain structures, brain functions, brain diseases, and also genes and proteins that are related to specific brain-related disorders such as epilepsy and schizophrenia, as well as to general functions, such as learning and memory. Here in this ongoing research we basically focus on the crucial proteins such as AMPA, GABA, NMDA, SCN, KCN, and CLC that are in some way controlling certain brain functions through their direct or indirect interactions with other genes/proteins. The BGO provides a conceptual framework and factual knowledge that is necessary to understand more on the relationship between genes involved during brain disorders and is the best way to provide a semantic repository of systematically ordered concerned molecules.

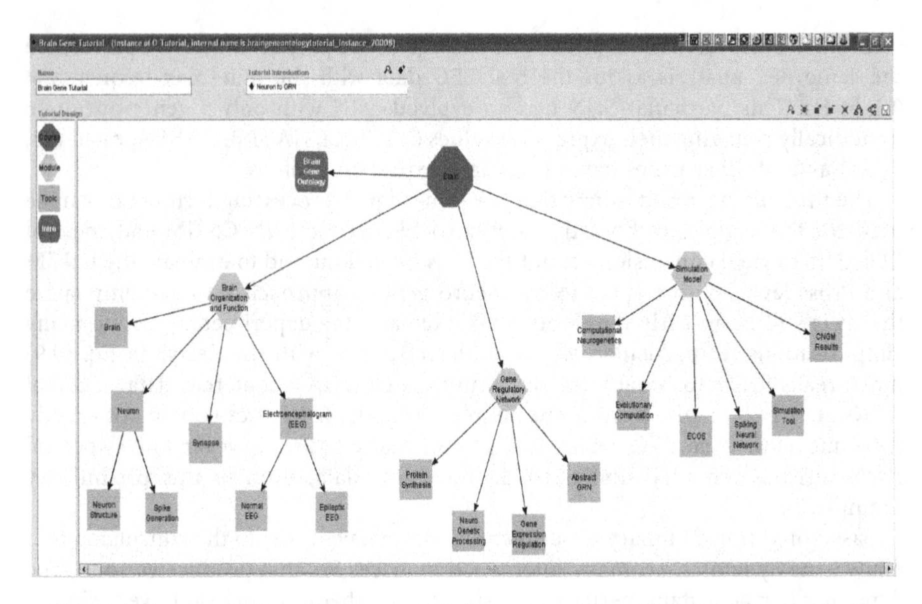

Fig. 9.21 The general information structure of the brain–gene ontology (BGO) (http://www.kedri.info).

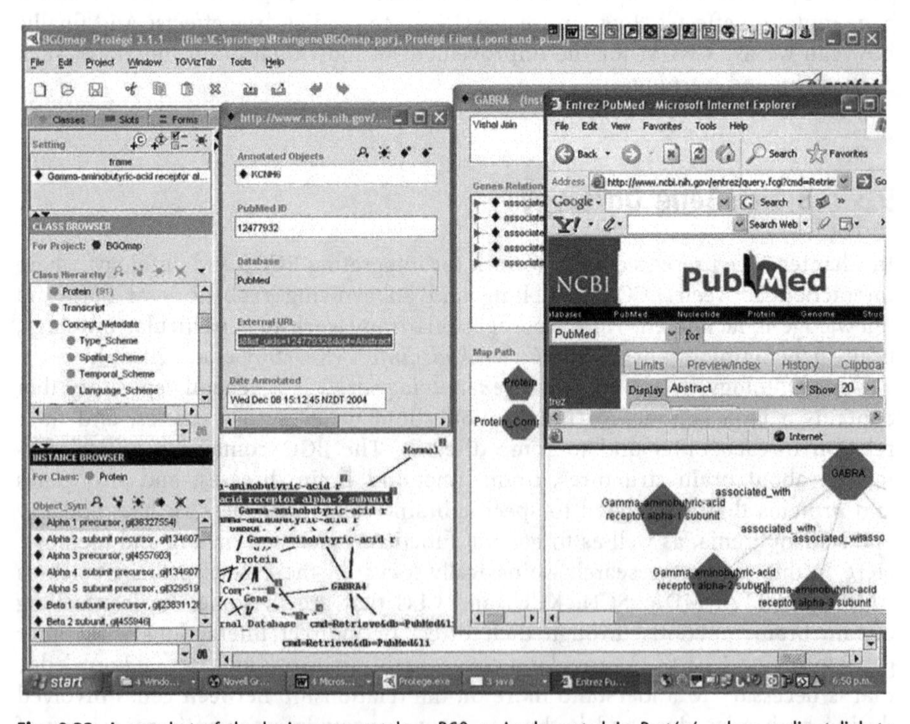

Fig. 9.22 A snapshot of the brain–gene ontology BGO as implemented in Protégé, where a direct link to PubMed and to another general database or an ontology is facilitated.

Ontological representation can be used to bridge the different notions in various databases by explicitly specifying the meaning of and relation between fundamental concepts. In the BGO this relation can be represented graphically, which enables visualisation and creation of new relationships. Each instance in this ontology map is traceable through a query language that allows us, for example, to answer questions, such as, 'Which genes are related to epilepsy?'

The general information structure of the BGO is given in Fig. 9.21. Figure 9.22 presents a snapshot of the BGO as implemented in Protégé, where a direct link to PubMed and to another general database or an ontology is facilitated. The BGO allows for both numerical and graphical information to be derived and presented, such as shown in the Fig. 9.23 histogram of the expression of a gene GRIA1 related to the AMPA receptor (see Table 9.2).

The BGO contains information and data that can be used for computational neuro-genetic modelling (see the previous section). Results from CNGM experiments can be deposited into the BGO using some tags to indicate how the results were obtained and how they have been validated (e.g. *in silico*, in vitro, or in vivo).

9.6 Summary and Open Problems

This chapter discusses issues of modelling dynamic processes in the human brain. The processes are very complex and their modelling requires dynamic adaptive

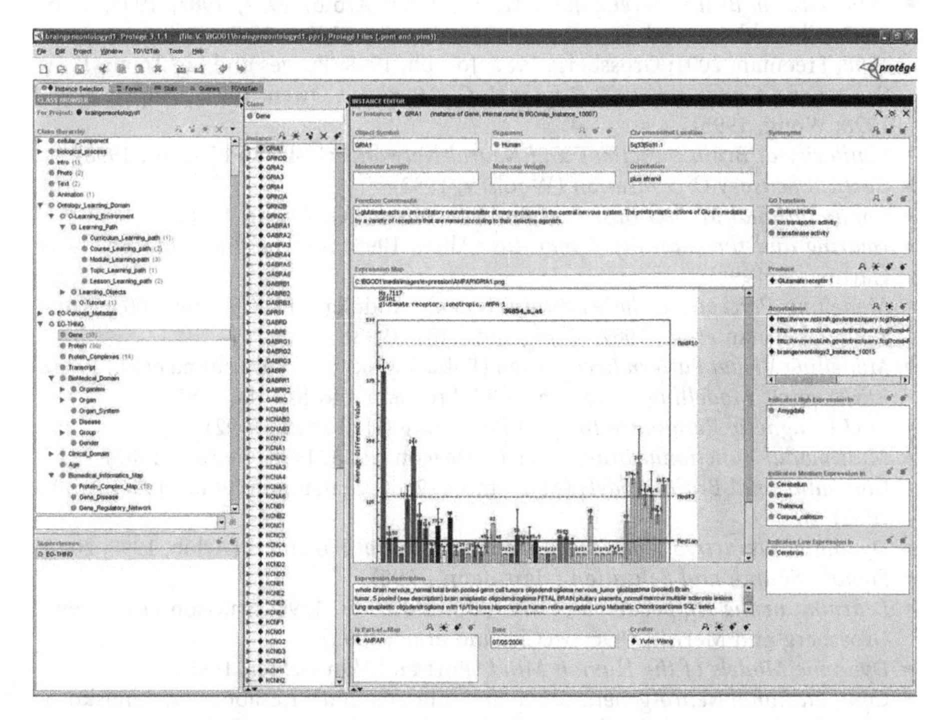

Fig. 9.23 A histogram of the expression of a gene GRIA1 related to the AMPA receptor (see Table 9.2) obtained from the brain–gene ontology BGO.

techniques. This chapter raises many questions and open problems that need to be solved in the future; among them are:

1. How can neural network learning and cell development be combined in one integrated model? Would it be possible to combine fMRI images with gene expression data to create the full picture of the processes in the human brain?
2. How does the generic neuro-genetic principle (see the Introduction) relate to different brain functions and human cognition?
3. Is it possible to create a truly adequate model of the human brain?
4. How can dynamic modelling help trace and understand the development of brain diseases such as epilepsy and Parkinson's disease?
5. How can dynamic modelling of brain activities help understand the instinct for information as speculated in the Introduction?
6. How could precise modelling of the human hearing apparatus help to achieve progress in the area of speech recognition systems?
7. How can we build brain–computer interfaces (see Coyle and McGinnity (2006)?

All these are difficult problems that can be attempted by using different computational methods. Evolving connectionist systems can also be used in this respect.

9.7 Further Reading

- *Principles of Brain Development* (Amit, 1989; Arbib, 1972, 1987, 1998, 1995, 2002; Churchland and Sejnowski, 1992; Deacon, 1988, 1998; Eriksson *et al.*, 1998; Freeman, 2001; Grossberg, 1982; Joseph, 1998; Purves and Lichtman, 1985; Quartz and Sejnowski, 1997; Taylor, J. G., 1998; van Owen, 1994; Wolpert *et al.*, 1998; Wong, 1995)
- *Similarity of Brain Functions and Neural Networks* (Rolls and Treves, 1998)
- *Cortical Sensory Organisation* (Woolsey, 1982)
- *Computational Models Based on Brain-imaging* (J.G. Taylor, 1998)
- *Hearing and the Auditory Apparatus* (Allen, 1995; Glassberg and Moore, 1990; Hartmann, 1998)
- *Modelling Perception, the Auditory System* (Abdulla and Kasabov, 2003; Kuhl, 1994; Liberman *et al.*, 1967; Wang and Jabri, 1998)
- *Modelling Visual Pattern Recognition* (Fukushima, 1987; Fukushima *et al.*, 1983); *EEG Signals Modelling* (Freeman, 1987; Freeman and Skarda, 1985)
- *MRI (Magnetic Resonance Images) Processing* (Hall *et al.*, 1992)
- *Multimodal Functional Brain Models* (Deacon, 1988, 1998; Neisser, 1987)
- *Computational Brain Models* (Matsumoto, 2000; Matsumoto *et al.*, 1996; Arbib, 2002)
- *Dynamic Interactive Models of Vision and Control Functions* (Arbib, 1998; 2002)
- *Signals, Sound, and Sensation* (Hartmann, 1998)
- *Learning in the Hippocampus Brain* (Durand *et al.*, 1996; Eriksson *et al.*, 1998; Grossberg and Merrill, 1996; McClelland *et al.*, 1995)
- *Dynamic Models of the Human Mind* (Port and Van Gelder, 1995)
- *Computational Neuro-genetic Modelling* (Marcus, 2004; Kasabov and Benuskova, 2004; Benuskova and Kasabov, 2007; Howell, 2006)

10. Modelling the Emergence of Acoustic Segments in Spoken Languages

Spoken languages evolve in the human brain through incremental learning and this process can be modelled to a certain degree with the use of evolving connectionist systems. Several assumptions have been hypothesised and proven through simulation in this chapter:

1. The learning system evolves its own representation of spoken language categories (phonemes) in an unsupervised mode through adjusting its structure to continuously flowing examples of spoken words (a learner does not know in advance which phonemes are going to be in a language, nor, for any given word, how many phoneme segments it has).
2. Learning words and phrases is associated with supervised presentation of meaning.
3. It is possible to build a 'lifelong' learning system that acquires spoken languages in an effective way, possibly faster than humans, provided there are fast machines to implement the evolving learning models.

The chapter is presented in the following sections.

- Introduction to the issues of learning spoken languages
- The dilemma 'innateness versus learning', or 'nature versus nurture', revisited
- ECOS for modelling the emergence of acoustic segments (phonemes)
- Modelling evolving bilingual systems

The chapter uses some material published by Taylor and Kasabov (2000).

10.1 Introduction to the Issues of Learning Spoken Languages

The task here is concerned with the process of learning in humans and how this process can be modelled in a program. The following questions are attempted.

- How can continuous learning in humans be modelled?
- What conclusions can be drawn in respect to improved learning and teaching processes, especially learning languages?
- How is learning a second language related to learning a first language?

The aim is computational modelling of processes of phoneme category acquisition, using natural spoken language as training input to an evolving connectionist system. A particular research question concerns the characteristics of 'optimal' input and optimal system parameters that are needed for the phoneme categories of the input language to emerge in the system in the least time. Also, by tracing in a machine how the target behaviour actually emerges, hypotheses about what learning parameters might be critical to the language-acquiring child could be made.

By attempting to simulate the emergence of phoneme categories in the language learner, the chapter addresses some fundamental issues in language acquisition and inputs into linguistic and psycholinguistic theories of acquisition. It has a special bearing on the question of whether language acquisition is driven by general learning mechanisms, or by innate knowledge of the nature of language (Chomsky's Universal Grammar).

The basic methodology consists in the training of an evolving connectionist structure (a modular system of neural networks) with Mel-scale transformations of natural language utterances. The basic research question is whether, and to what extent, the network will organize the input in clusters corresponding to the phoneme categories of the input language. We will be able to trace the emergence of the categories over time, and compare the emergent patterns with those that are known to occur in child language acquisition.

In preliminary experiments, it may be advisable to study circumscribed aspects of a language's phoneme system, such as consonant–vowel syllables. Once the system has proved viable, it will be a relatively simple matter to proceed to more complex inputs, involving the full range of sounds in a natural language, bearing in mind that some languages (such as English) have a relatively large phoneme system compared to other languages (such as Maori) whose phoneme inventory is more limited (see Laws *et al.* (2003)).

Moreover, it will be possible to simulate acquisition under a number of input conditions:

- Input from one or many speakers
- Small input vocabulary versus large input vocabulary
- Simplified input first (e.g. consonant–vowel syllables) followed by phonologically more complex input
- Different sequences of input data

The research presented here is at its initial phase, but the results are expected to contribute to a general theory of human/machine cognition. Technological applications of the research concern the development of self-adaptive systems. These are likely to substantially increase the power of automatic speech recognition systems.

10.2 The Dilemma 'Innateness Versus Learning' or 'Nature Versus Nurture' Revisited

10.2.1 A General Discussion

A major issue in contemporary linguistic theory concerns the extent to which human beings are genetically programmed, not merely to acquire language, but to acquire languages with just the kinds of properties that they have (Pinker, 1994; Taylor and Kasabov, 2000). For the last half century, the dominant view has been that the general architecture of language is innate; the learner only requires minimal exposure to actual language data in order to set the open parameters given by Universal Grammar as hypothesized by Noam Chomsky (1995). Arguments for the innateness position include the rapidity with which all children (barring cases of gross mental deficiency or environmental deprivation) acquire a language, the fact that explicit instruction has little effect on acquisition, and the similarity (at a deep structural level) of all human languages. A negative argument is also invoked: the complexity of natural languages is such that they could not, in principle, be learned by normal learning mechanisms of induction and abstraction.

Recently, this view has been challenged. Even from within the linguistic mainstream, it has been pointed out that natural languages display so much irregularity and idiosyncrasy, that a general learning mechanism has got to be invoked; the parameters of Universal Grammar would be of little use in these cases (Culicover *et al.*, 1999). Moreover, linguists outside the mainstream have proposed theoretical models which do emphasize the role of input data in language learning. In this view, language knowledge resides in abstractions (possibly, rather low-level abstractions) made over rich arrays of input data.

In computational terms, the contrast is between systems with a rich in-built structure, and self-organising systems that learn from data (Elman *et al.*, 1997). Not surprisingly, systems that have been preprogrammed with a good deal of language structure vastly outperform systems which learn the structure from input data. Research on the latter is still in its infancy, and has been largely restricted to modelling circumscribed aspects of a language, most notably, the acquisition of irregular verb morphology (Plunkett, 1996). A major challenge for future research will be to create self-organising systems to model the acquisition of more complex configurations, especially the interaction of phonological, morphological, syntactic, and semantic knowledge.

In the introductory chapter it was mentioned that learning is genetically defined; i.e. there are genes that are associated with long-term potentiation (LTP), learning, and memory (Abraham *et al.*, 1993), but it is unlikely that there are genes associated with learning languages and even less likely that there are genes associated with particular languages, e.g. Italian, English, Bulgarian, Maori, and so on.

The focus of this chapter is the acquisition of phonology, more specifically, the acquisition of phoneme categories. All languages exhibit structuring at the phoneme level. We may, to be sure, attribute this fact to some aspect of the genetically determined language faculty. Alternatively, and perhaps more plausibly, we can regard the existence of phoneme inventories as a converging solution to two different engineering problems. The first problem pertains to a speaker's storage of linguistic units. A speaker of any language has to store a vast (and potentially open-ended) inventory of meaningful units, be they morphemes, words, or fixed phrases. Storage becomes more manageable, to the extent that the meaningful units can be represented as sequences of units selected from a small finite inventory of segments (the phones and the phonemes). The second problem refers to the fact that the acoustic signal contains a vast amount of information. If language learning is based on input, and if language knowledge is a function of heard utterances, a very great deal of the acoustic input has got to be discarded by the language learner. Incoming utterances have got to be stripped down to their linguistically relevant essentials. Reducing incoming utterances to a sequence of discrete phonemes solves this problem, too.

10.2.2 Infant Language Acquisition

Research by Peter Jusczyk (1997) and others has shown that newborn infants are able to discriminate a large number of speech sounds, well in excess of the number of phonetic contrasts that are exploited in the language an infant will subsequently acquire. This is all the more remarkable inasmuch as the infant vocal tract is physically incapable of producing adultlike speech sounds at all (Liberman, 1967). By about six months, perceptual abilities are beginning to adapt to the environmental language, and the ability to discriminate phonetic contrasts that are not utilized in the environmental language declines. At the same time, and especially in the case of vowels, acoustically different sounds begin to cluster around perceptual prototypes, which correspond to the phonemic categories of the target language, a topic researched by Kuhl (1994). Thus, the 'perceptual space' of e.g. the Japanese- or Spanish-learning child becomes increasingly different from the perceptual space of the English- or Swedish-learning child. Japanese, Spanish, English, and Swedish 'cut up' the acoustic vowel space differently, with Japanese and Spanish having far fewer vowel categories than English and Swedish. However, the emergence of phoneme categories is not driven only by acoustic resemblance. Kuhl's research showed that infants are able to filter out speaker-dependent differences, and attend only to the linguistically significant phoneme categories.

It is likely that adults in various cultures, when interacting with infants, modulate their language in ways to optimise the input for learning purposes. This is not just a question of vocabulary selection (although this, no doubt, is important). Features of 'child-directed speech' include exaggerated pitch range and slower articulation rates (Kuhl, 1994). These maximise the acoustic distinctiveness of the different vowels, and therefore reduce the effect of co-articulation and other characteristics of rapid conversational speech.

10.2.3 Phonemes

Although it is plausible to assume that aspects of child-directed speech facilitate the emergence of perceptual prototypes for the different phones and phonemes of a language's sound system, it must be borne in mind that phoneme categories are not established only on the basis of acoustic–phonetic similarity. Phonemes are theoretical entities, at some distance from acoustic events. As long as the child's vocabulary remains very small (up to a maximum of about 40–50 words), it is plausible that each word is represented as a unique pathway through acoustic space, each word being globally distinct from each other word. But with the 'vocabulary spurt', which typically begins around age 16–20 (Bates and Goodman, 1999), this strategy becomes less and less easy to implement. Up to the vocabulary spurt, the child has acquired words slowly and gradually; once the vocabulary spurt begins, the child's vocabulary increases massively, with the child sometimes adding as many as ten words per day to his or her store of words. Under these circumstances, it is highly implausible that the child is associating a unique acoustic pattern with each new word. Limited storage and processing capacity requires that words be broken down into constituent elements, i.e. the phonemes. Rather than learning an open-ended set of distinct acoustic patterns, one for each word (tens of thousands of them!), the words come to be represented as a linear sequence of segments selected from an inventory of a couple of dozen distinct elements.

The above is used as a principle for the experiments conducted later in this chapter.

Phoneme Analysis

Linguists have traditionally appealed to different procedures for identifying the phones and phonemes of a language. One of them is the principle of contrast. The vowels [i:] and [I] are rather close (acoustically, perceptually, and in terms of articulation). In English, however, the distinction is vital, because the sounds differentiate the words *sheep* and *ship* (and countless others). The two sounds are therefore assigned to two different phonemes. The principle of contrast can be used in a modelling system through feedback from a semantic level, back to the acoustic level of modelling.

10.3 ECOS for Modelling the Emergence of Phones and Phonemes

10.3.1 Problem Definition

We can conceptualise the sounding of a word as a path through multidimensional acoustic space. Repeated utterances of the same word will be represented by a bundle of paths that follow rather similar trajectories. As the number of word types is increased, we may assume that the trajectories of different words will overlap in places; these overlaps will correspond to phoneme categories.

It is evident that an infant acquiring a human language does not know, a priori, how many phoneme categories there are going to be in the language that she or he is going to learn, nor, indeed, how many phonemes there are in any given word that the child hears. (We should like to add: the child learner does not know in advance that there are going to be such things as phonemes at all! Each word simply has a different global sound from every other word). A minimum expectation of a learning model is that the language input will be analysed in terms of an appropriate number of phonemelike categories.

The earlier observations on language acquisition and phoneme categories suggest a number of issues that need to be addressed while modelling phoneme acquisition:

1. Does learning require input which approximates the characteristics of 'motherese' with regard to careful exaggerated articulation, also with respect to the frequency of word types in the input language?
2. Does phoneme learning require lexical–semantic information? The English learner will have evidence that sheep and ship are different words, not just variants of one and the same word, because sheep and ship mean different things. Applying this to our learning model, the question becomes: do input utterances need to be classified as tokens of word types?
3. Critical mass: it would be unrealistic to expect stable phoneme categories to emerge after training on only a couple of acoustically nonoverlapping words. We might hypothesize that phonemelike organisation will emerge only when a critical mass of words has been extensively trained, such that each phoneme has been presented in a variety of contexts and word positions. First language acquisition research suggests that the critical mass is around 40–50 words.
4. The speech signal is highly redundant in that it contains vast amounts of acoustic information that is simply not relevant to the linguistically encoded message. We hypothesize that a learning model will need to be trained on input from a variety of speakers, all articulating the 'same' words. The system must be introduced to noise, in order for noise to be ignored during the system's operation.
5. Can the system organize the acoustically defined input without prior knowledge of the characteristics of the input language? If so, this would be a significant finding for language acquisition research! Recall that children learning their mother tongue do not know in advance how many phoneme categories there are going to be in the language, nor even, indeed, that language will have a phoneme level of organization.
6. What is the difference, in terms of acoustic space occupied by a spoken language, between simultaneous acquisition of two languages versus late bilingualism (see Kim *et al.*, 1997)? Will the acquisition of the second language show interference patterns characteristic of human learners?

Underlying our experiments is the basic question of whether the system can organize the acoustic input with minimal specification of the anticipated output. In psycholinguistic terms, this is equivalent to reducing to a minimum the contribution of innate linguistic knowledge. In genetic terms, this means that there are no genes associated with learning languages and learning specific languages in particular. Indeed, our null hypothesis will be that phoneme systems emerge as organizational solutions to massive data input. If it should turn out that learning can be modelled with minimal supervision, this would have very significant consequences for linguistic and psycholinguistic theories of human language learning.

10.3.2 Evolving Clustering for Modelling the Emergence of Phones – A Simple Example

The evolving clustering method ECM from Chapter 2 is used here with inputs that represent features taken from a continuous stream of spoken words. In the experiment shown in Fig. 10.1 frames were extracted from a spoken word 'eight',

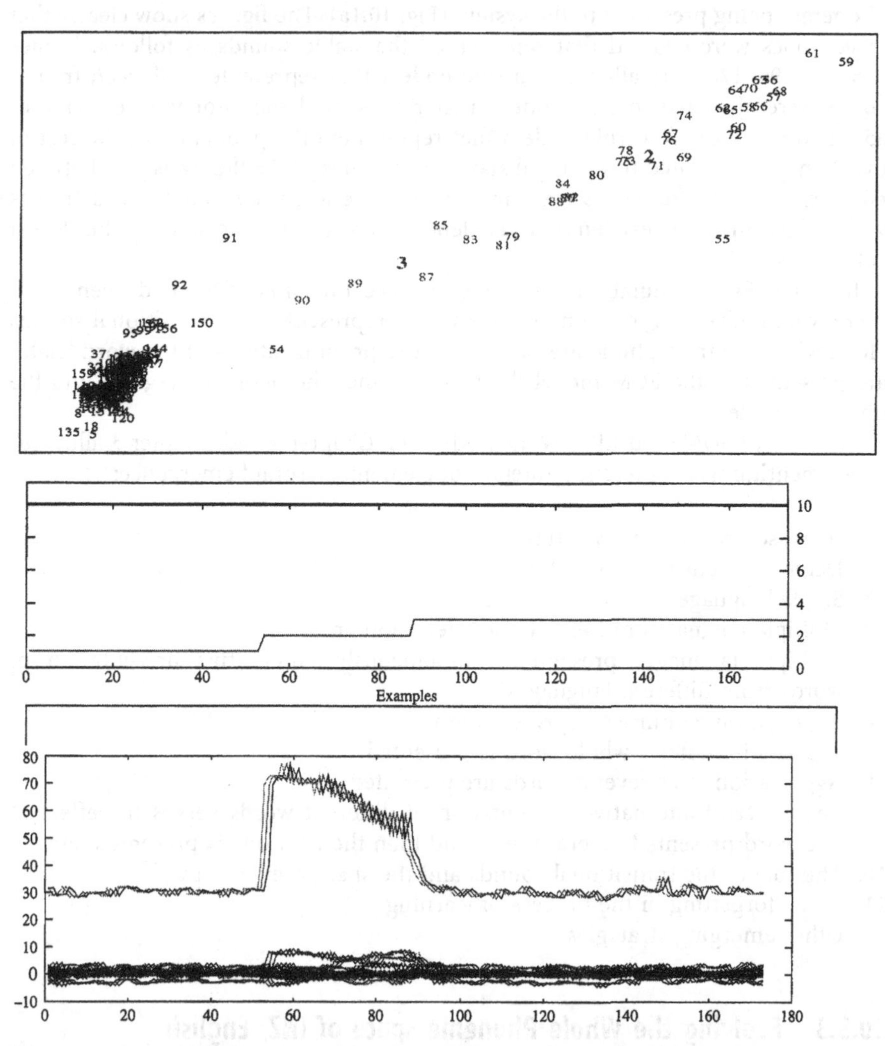

Fig. 10.1 Experimental results with an ECM model for phoneme acquisition, a single pronunciation of the word 'eight'. From top to bottom: (a) the two-dimensional input space of the first two Mel scale coefficients of all frames taken from the speech signal of the pronounced digit 'eight' and numbered with the consecutive time interval, and also the evolved nodes (denoted with larger font) that capture each of the three phonemes of the speech input: /silence/, /ei/, /t/, /silence/; (b) the time of the emergence of the three ECM nodes: cluster centres; (c) all 78 element Mel-vectors of the word 'eight' over time (175 time frames, each 11.6 msec long, with 50% overlap between the frames).

in a phonetic representation it would be: /silence/ /ei/ /t/ /silence/. Three time-lag feature vectors of 26 Mel-scale coefficients each are used, from a window of 11.6 ms, with an overlap of 50% (see Fig. 10.1c).

A cluster structure was evolved with the use of ECM. Each new input vector from the spoken word was either associated with an existing rule node that was modified to accommodate these data, or a new rule node was created. All together, three rule nodes were created (Fig. 10.1b). After the whole word was presented the nodes represented the centres of the phoneme clusters without the concept of phonemes being presented to the system (Fig. 10.1a). The figures show clearly that three nodes were evolved that represented the stable sounds as follows: frames 0–53 and 96–170 were allocated to rule node 1 that represented /silence/; frames 56–78 were allocated to rule node 2 that represented the phoneme /ei/; frames 85–91 were allocated to rule node 3 that represented the phoneme /t/; the rest of the frames represented transitional states, e.g. frames 54–55 the transition between /silence/ and /ei/, frames 79–84, the transition between /ei/ and /t/, and frames 92–96, the transition between /t/ and /silence/, were allocated to some of the closest rule nodes.

If in the ECM simulation a smaller distance threshold *Dthr* had been used, there would have been more nodes evolved to represent short transitional sounds along with the larger phone areas. When more pronunciations of the word 'eight' are presented to the ECM model the model refines the phoneme regions and the phoneme nodes.

The ECM, ESOM, and EFuNN methods from Chapter 2 and Chapter 3 allow for experimenting with different strategies of elementary sound emergence:

1. Increased sensitivity over time
2. Decreased sensitivity over time
3. Single language sound emergence
4. Multiple languages presented one after another
5. Multiple languages presented simultaneously (alternative presentation of words from different languages)
6. Aggregation within word presentation
7. Aggregation after a whole word is presented
8. Aggregation after several words are presented
9. The effect of alternative presentation of different words versus the effect of one word presented several times, and then the next one is presented, etc.
10. The role of the transitional sounds and the space they occupy
11. Using forgetting in the process of learning
12. Other emerging strategies

10.3.3 Evolving the Whole Phoneme space of (NZ) English

To create a clustering model for New Zealand English, data from several speakers from the Otago Speech Corpus (http://translator.kedri.info) were selected to train the model. Here, 18 speakers (9 Male, 9 Female) each spoke 128 words three times. Thus, approximately 6912 utterances were available for training. During the training, a word example was chosen at random from the available words.

The waveform underwent a Mel-scale cepstrum (MSC) transformation to extract 12 frequency coefficients, plus the log energy, from segments of approximately 23.2 ms of data. These segments were overlapped by 50%. Additionally, the delta, and the delta–delta values of the MSC coefficients and log energy were extracted, for an input vector of total dimensionality 39.

The ECM system was trained until the number of cluster nodes became constant for over 100 epochs. A total of 12,000 epochs was performed, each on one of the 12,000 data examples. The distance threshold *Dthr* parameter of the ECM was set to 0.15.

Figure 10.2 shows: (a) the connections of the evolved ECM system (70 nodes are evolved that capture 70 elementary sounds, the columns) from spoken words presented one after another, each frame of the speech signal being represented by 39 features (12 MSC and the power, their delta features, and the delta–delta features, the rows). The darker the colour of a cell, the higher its value is; (b) the evolved cluster nodes and the trajectory of the spoken word 'zero' projected in the MSC1–MSC2 input space; (c) the trajectory of the word 'zero' shown as a sequence of numbered frames, and the labelled cluster nodes projected in the MSC1–MSC2 input space.

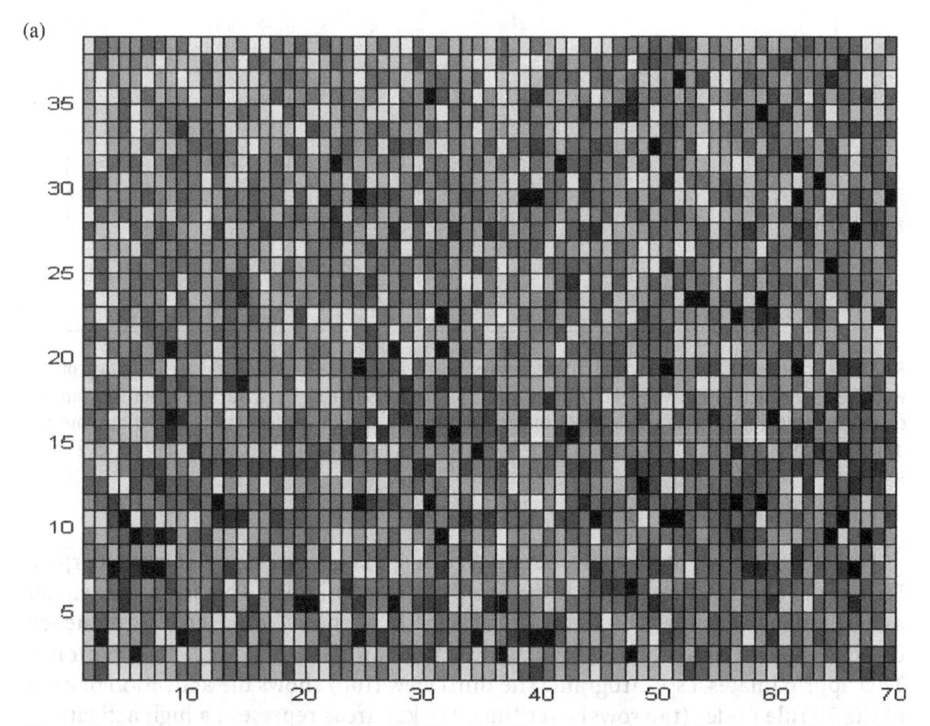

Fig. 10.2 (a) The connections of the evolved ECM model (there are no fuzzified inputs used) from spoken words of NZ English. The 70 cluster nodes are presented on the x-axis and the 39 input variables are presented on the y-axis. The words were presented one after another, each frame of the speech signal being presented by 39 features: 12 MSC and the power, their delta features, and their delta–delta features. The darker the colour is, the higher the value. It can be seen, for example, that node 2 gets activated for high values of MSC 1; (*Continued overleaf*)

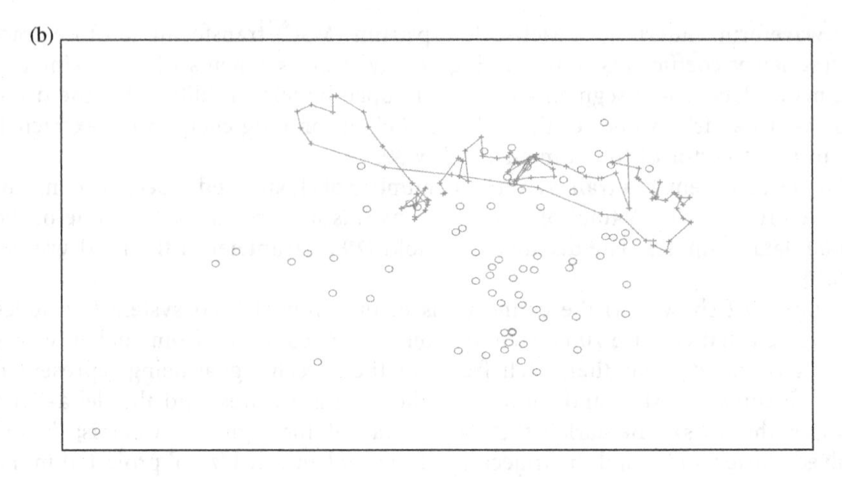

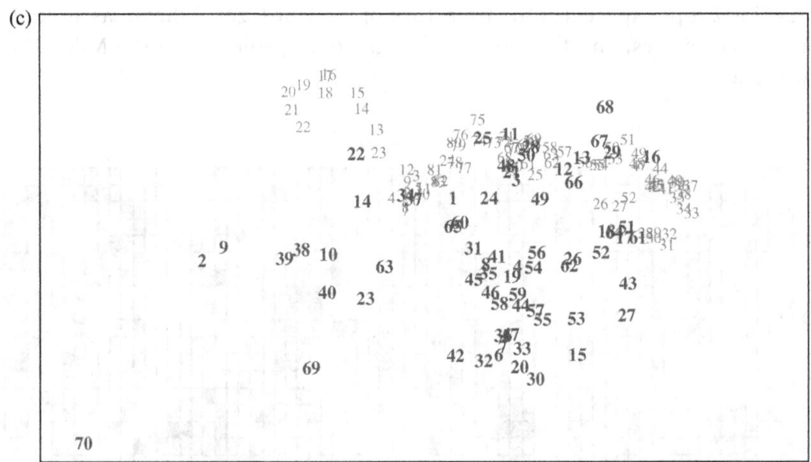

Fig. 10.2 (*continued*) (b) the evolved 70 rule nodes in the ECM model from (a) and the trajectory of the word 'zero' projected in the MSC1–MSC2 input space; (c) the trajectory of the word 'zero' from (b) with the consecutive time frames being labelled with consecutive numbers: the smaller font numbers) and the emerged nodes labelled in a larger font according to their time of emergence, all projected in the MSC1–MSC2 input space.

Figure 10.3 shows three representations of a spoken word 'zero' from the corpus. Firstly, the word is viewed as a waveform (Fig. 10.3, middle). This is the raw signal as amplitude over time. The second view is the MSC space view. Here, 12 frequency components are shown on the y-axis over time on the x-axis (Fig. 10.3, bottom). This approximates a spectrogram. The third view (top) shows the activation of each of the 70 rule nodes (the rows) over time. Darker areas represent a high activation. Additionally, the winning nodes are shown as circles. Numerically, these are: 1, 1, 1, 1, 1, 1, 2, 2,2, 2, 22, 2, 2, 11, 11, 11,11, 11, 24, 11, 19, 19, 19, 19, 15, 15, 16, 5, 5, 16, 5, 15, 16, 2, 2, 2, 11, 2, 2, 1, 1, 1. Some further testing showed that recognition of words depended not only on the winning node, but also on the path of the recognition. Additionally, an n-best selection of nodes may increase discrimination.

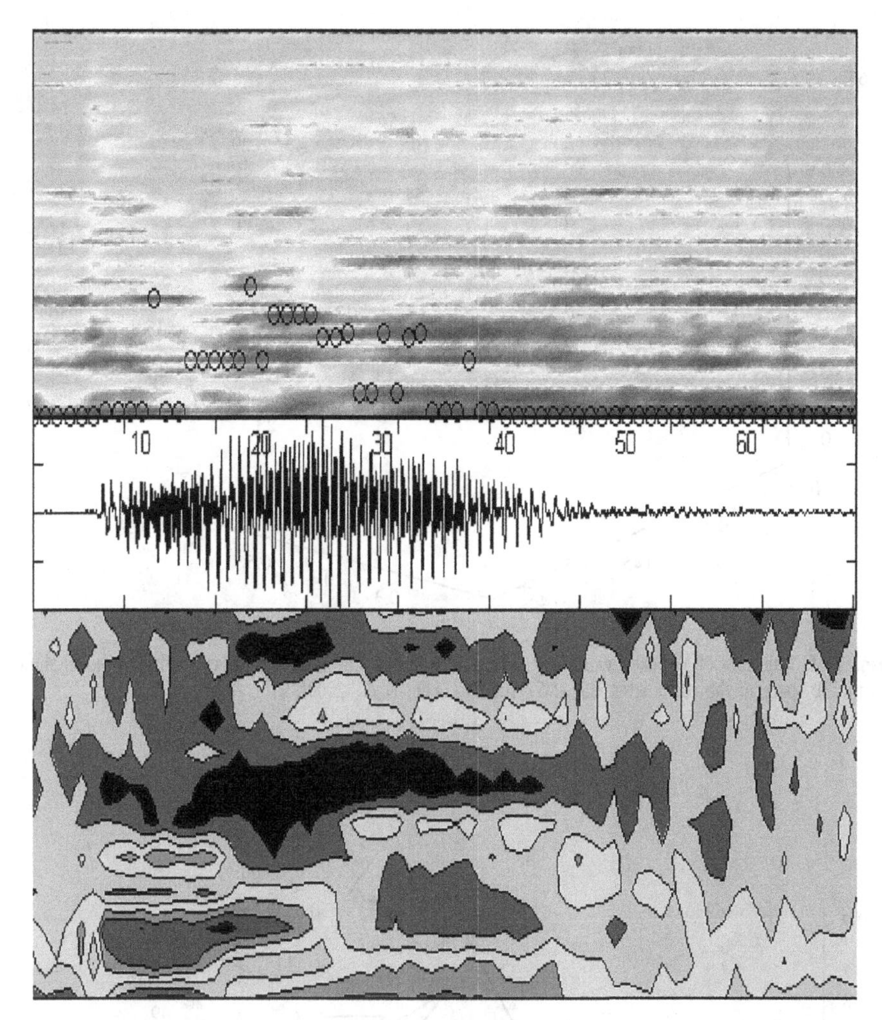

Fig. 10.3 ECM representation of a spoken word 'zero': (upper figure) the activation of the nodes of the evolved ECM model from Fig. 10.2 (nodes from 1 to 70 are shown on the y-axis) when the word was propagated through it (time is presented on the x-axis); (middle figure) the wave form of the signal; (bottom figure) x-axis represents time, and y-axis represents the value of the MSC from 1 to 12 (the darker the colour is, the higher the value).

Trajectory Plots

The trajectory plots, shown in Figs. 10.4 through 10.7, are presented in three of the total 39 dimensions of the input space. Here, the first and seventh MSC are used for the x and y coordinates. The log energy is represented by the z-axis. A single word, 'sue', is shown in Fig. 10.4. The starting point is shown as a square. Several frames represent the hissing sound, which has low log energy. The vowel sound has increased energy, which fades out toward the end of the utterance. Two additional instances of the same word, spoken by the same speaker, are shown in

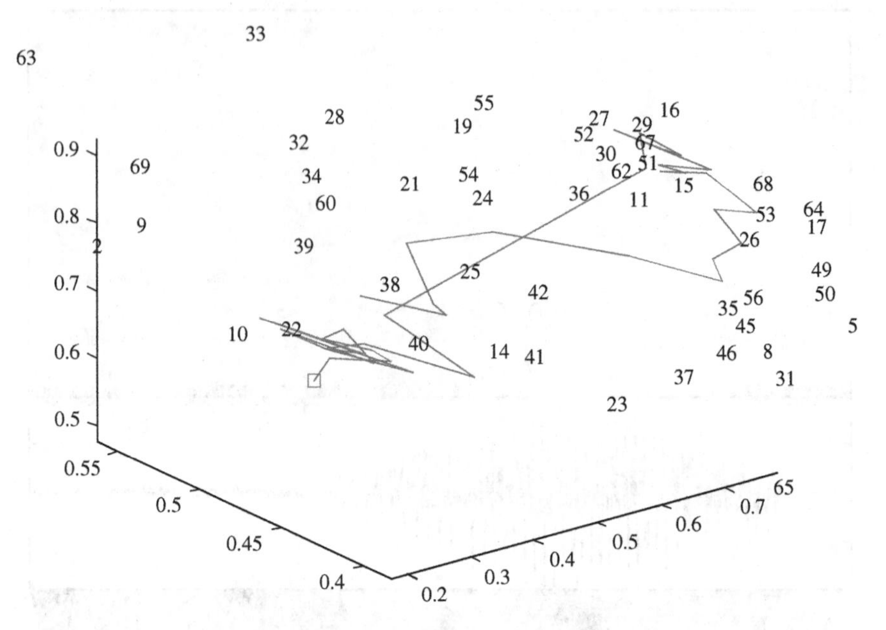

Fig. 10.4 Trajectory of a spoken word 'sue', along with the 70 rule nodes of the evolved ECM model shown in the 3D space of the coordinates MS1–MS7–log E (Taylor *et al.*, 2000).

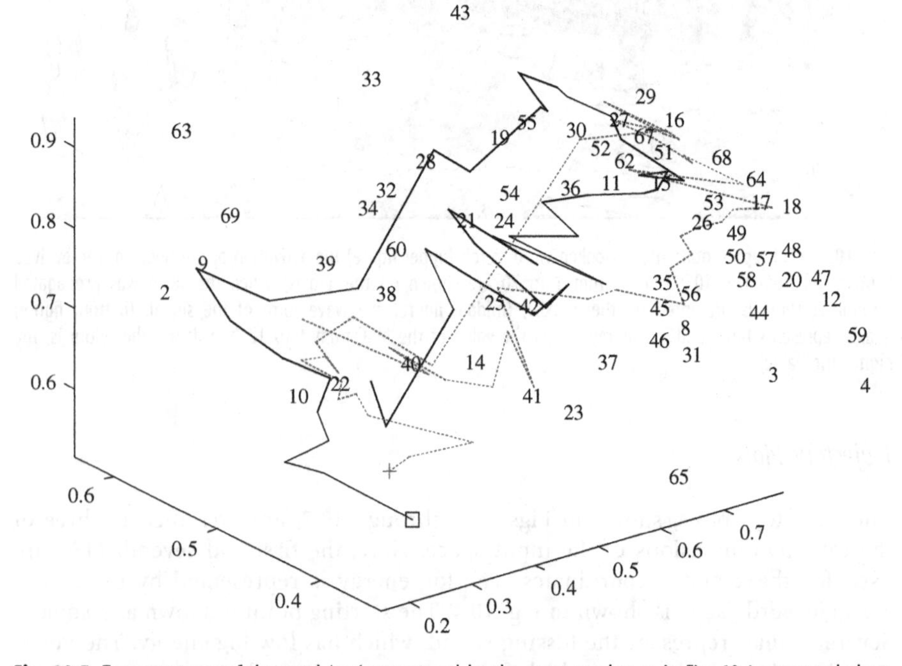

Fig. 10.5 Two utterances of the word 'sue' pronounced by the same speaker as in Fig. 10.4, presented along with the 70 rule nodes of the evolved ECM model in the 3D space of MS1–MS7–log E (see Taylor *et al.* (2000)).

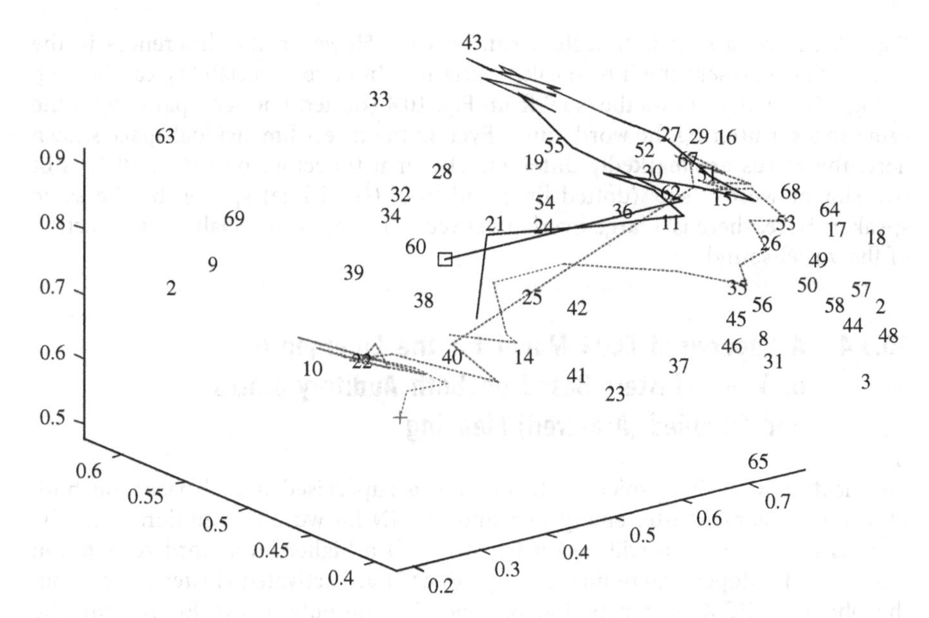

Fig. 10.6 Trajectories of spoken words 'sue' and 'nine' by the same speaker presented in a 3D space MS1–MS7–log E, along with the 70 rule nodes of the evolved ECM model (Taylor *et al.*, 2000).

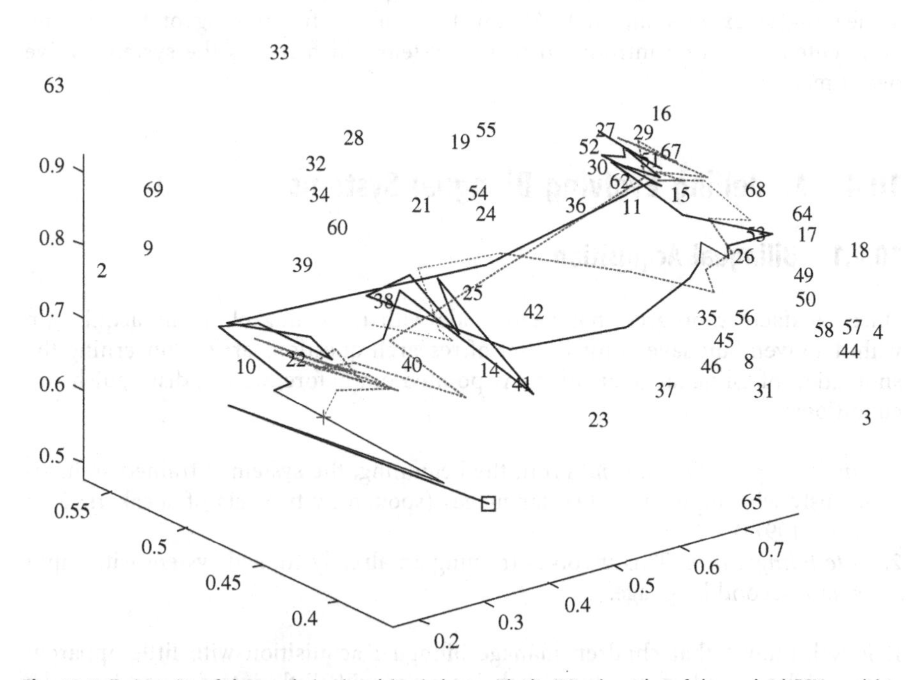

Fig. 10.7 Trajectories of the words 'sue' and 'zoo' along with the 70 rule nodes of the evolved ECM model in the MS1–MS7–log E space (Taylor *et al.*, 2000).

Fig. 10.5. Here, a similar trajectory can be seen. However, the differences in the trajectories represent the intraspeaker variation. Interword variability can be seen in Fig. 10.6, which shows the 'sue' from Fig. 10.4 (dotted line) compared with the same speaker uttering the word 'nine'. Even in the three-dimensional space shown here, the words are markedly different. The final trajectory plot (Fig. 10.7) is of two similar words, 'sue' (dotted line) and 'zoo' (solid line) spoken by the same speaker. Here, there is a large overlap between the words, especially in the section of the vowel sound.

10.3.4 A Supervised ECOS Model for the Emergence of Word Clusters Based on Both Auditory Traces and Supplied (Acquired) Meaning

The next step of this project is to develop a supervised model based on both ECM for phoneme cluster emergence, and EFuNN for word recognition. After the ECM is evolved (it can still be further evolved) a higher-level word recognition module is developed where inputs to the EFuNN are activated cluster nodes from the phoneme ECM over a period of time. The outputs of the EFuNN are the words that are recognized. The number of words can be extended over time thus creating new outputs that are allowable in an EFuNN system (see Chapter 3). A sentence recognition layer can be built on top of this model. This layer will use the input from the previous layer (a sequence of recognized words over time) and will activate an output node that represents a sentence (a command, a meaningful expression, etc.). At any time of the functioning of the system, new sentences can be introduced to the system which makes the system evolve over time.

10.4 Modelling Evolving Bilingual Systems

10.4.1 Bilingual Acquisition

Once satisfactory progress has been made with modelling phoneme acquisition within a given language, a further set of research questions arises concerning the simulation of bilingual acquisition. As pointed out before, we can distinguish two conditions:

1. *Simultaneous bilingualism.* From the beginning, the system is trained simultaneously with input from two languages (spoken by two sets of speakers; Kim *et al.* (1997)).
2. *Late bilingualism.* This involves training an already trained system with input from a second language.

It is well-known that children manage bilingual acquisition with little apparent effort, and succeed in speaking each language with little interference from the other. In terms of an evolving system, this means that the phoneme representations of the two languages are strictly separated. Even though there might be some acoustic similarity between sounds of one language and sounds of the other, the distributions of the sounds (the shape of the trajectories associated with each of the languages) will be quite different.

Late acquisition of a second language, however, is typically characterized by interference from the first language. The foreign language sounds are classified in terms of the categories of the first language. (The late bilingual will typically retain a 'foreign accent', and will 'mishear' second-language utterances.) In terms of our evolving system, there will be considerable overlap between the two languages; the acoustic trajectories of the first language categories are so entrenched, that second language utterances will be forced into the first language categories.

The areas of the human brain that are responsible for the speech and the language abilities of humans evolve through the whole development of an individual (Altman, 1990). Computer modelling of this process, before its biological, physiological, and psychological aspects have been fully discovered, is an extremely difficult task. It requires flexible techniques for adaptive learning through active interaction with a teaching environment.

It can be assumed that in a modular spoken language evolving system, the language modules evolve through using both domain text data and spoken information data fed from the speech recognition part. The language module produces final results as well as a feedback for adaptation in the previous modules. This idea is currently being elaborated with the use of ECOS.

10.4.2 Modelling Evolving Elementary Acoustic Segments (Phones) of Two Spoken Languages

The data comprised 100 words of each Māori and English (see Laws *et al.*, 2003). The same speaker was used for both datasets. One spoken word at a time was presented to the network, frame by frame. In every case, the word was preprocessed in the following manner. A frame of 512 samples was transformed into 26 Mel-scale cepstrum coefficients (MSCC). In addition, the log energy was also calculated. Consecutive frames were overlapped by 50%.

For the English data, a total of 5725 frames was created. For the Māori data, 6832 frames were created. The words used are listed below. The English words are one and two syllables only. The Māori words are up to four syllables, which accounts for the slightly larger number of frames.

English Words

ago, ahead, air, any, are, auto, away, baby, bat, bird, boo, book, boot, buzz, card, carrot, choke, coffee, dart, day, dead, die, dog, dove, each, ear, eight, ether, fashion, fat, five, four, fur, go, guard, gut, hat, hear, how, jacket, joke, joy, judge, lad, ladder, leisure, letter, loyal, mad, nine, nod, one, ooze, other, over, palm, paper, pat, pea, peace, pure, push, rather, recent, reef, riches, river, rod, rouge, rude, school, seven, shoe, shop, sing, singer, six, sue, summer, tan, tart, teeth, teethe, that, thaw, there, three, tour, tragic, tub, two, utter, vat, visit, wad, yard, yellow, zero, zoo

Māori Words

ahakoa, ahau, āhei, ahiahi, āhua, āhuatanga, ake, ako, aku, ākuanei, anake, anei, anō, āpōpō, aroha, ātāhua, atu, atua, aua, auē, āwhina, ēhara, ēnā, ēnei, engari, ērā, ētahi, hanga, heke, hine, hoki, huri, iho, ihu, ika, ināianei, ingoa, inu, iwa, kaha, kaiako, kete, kino, koti, kura, mahi, mea, miraka, motu, muri, nama, nei, ngahere, ngākau, ngaro, ngāwari, ngeru, noa, nōu, nui, ōku, oma, ōna, one, oneone, ono, ora, oti, otirā, pakaru, pekepeke, pikitia, poti, putiputi, rangatira, reri, ringaringa, rongonui, rūma, tahuri, tēnā, tikanga, tokorua, tuna, unu, upoko, uri, uta, utu, waha, waiata, wero, wha, whakahaere, whakapai, whanaunga, whero, whiri, wiki, wūru

Three cases were of interest, to be compared and contrasted. Experiment 1 involved presenting all the English speech samples to the network, followed by all the Māori speech. Experiment 2 was similar, except that the Māori speech was presented first. Both English and Māori speech data were used for Experiment 3, shuffled together.

The evolving clustering method ECM was used for this experiment (Chapter 2). A distance threshold (*Dthr*) of 0.155 was used for all experiments.

Results

Table 10.1 shows the number of clusters resulting from each experiment. The English data alone created 54 clusters and an additional 14 were created by the Māori data used after the English data. The Māori data alone created 49 clusters. The addition of the English data produced 15 more clusters. Both languages presented together and a mixed order produced slightly more clusters than either language presented separately.

The spoken word 'zoo' was presented to the evolved three models. The activation of the nodes is presented as trajectories in the 2D PCA space in Figs. 10.8 through 10.10.

Table 10.1 Number of clusters created for each experiment of the bilingual English and Maori acoustic evolving system based on the ECM evolving clustering method.

	Number of first language clusters	*Total number of clusters*	*Difference*
English then Māori	54	68	14
Māori then English	49	64	15
Both languages	–	70	–

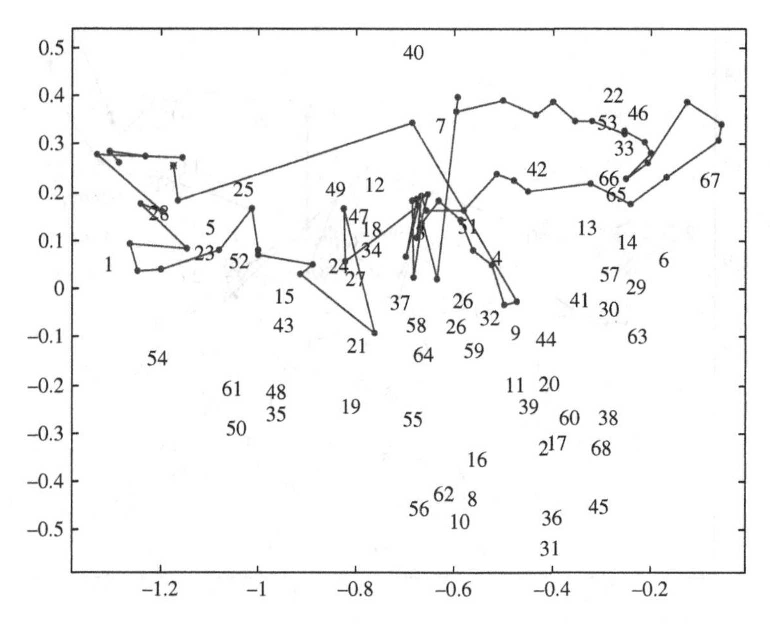

Fig. 10.8 The projection of the spoken word 'zoo' in English and Maori PCA space (see Laws *et al.*, 2003).

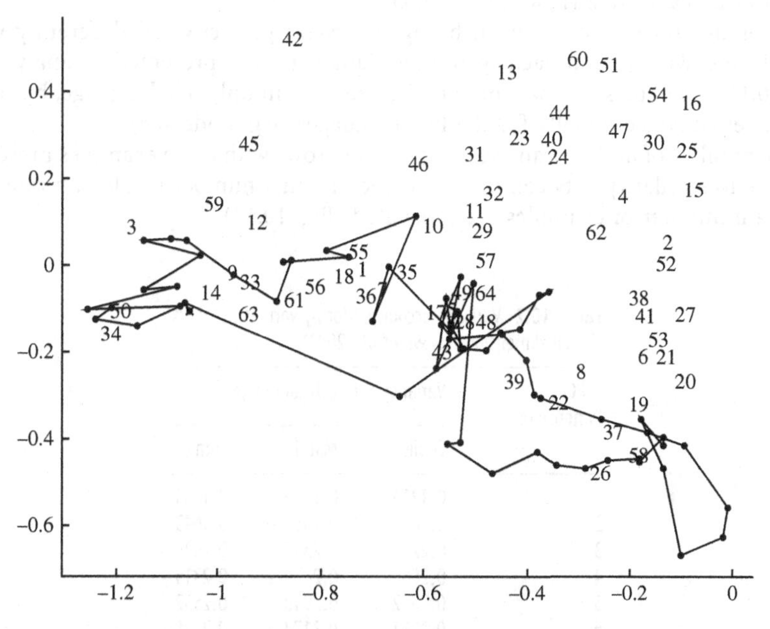

Fig. 10.9 The projection of the spoken word 'zoo' in Maori + English PCA space (Laws *et al.*, 2003).

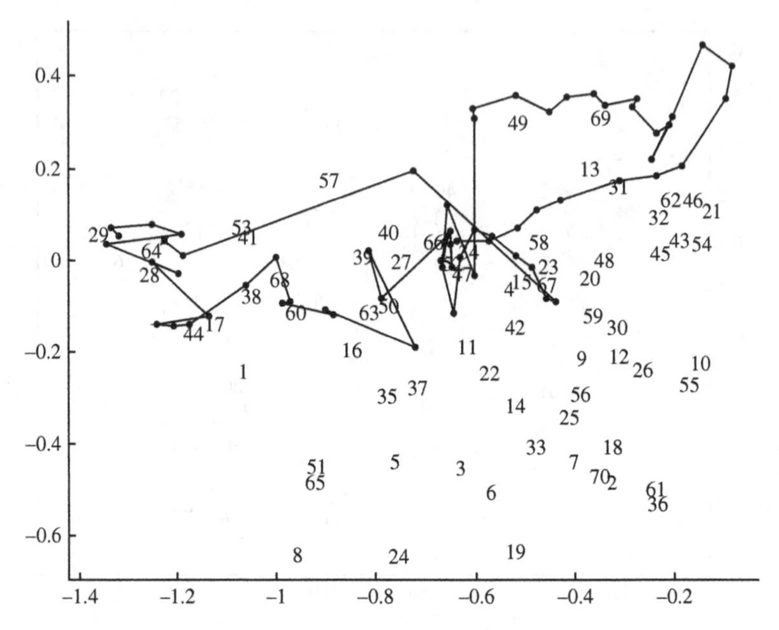

Fig. 10.10 The projection of the spoken word 'zoo' in a mixed English and Maori PCA space (Laws *et al.*, 2003).

As all the above visualisations were done in a PCA 2D space, it is important to know the amount of variance accounted for by the first few PCA dimensions. This is shown in Table 10.2 (Laws *et al.*, 2003).

Different nodes in an evolved bilingual system get activated differently when sounds and words from each of the two languages are presented as analysed in Fig. 10.11. It can be seen that some nodes are used in only one language, but other nodes get activated equally for the two languages (e.g. node #64).

The number of nodes in an evolving system grows with more examples presented but due to similarity between the input vectors, this number would saturate after a certain number of examples are presented (Fig. 10.12).

Table 10.2 Variance accounted for by various numbers of PCA dimensions (Laws *et al.*, 2003).

# PCA Dimensions	Variance for each language		
	English	Māori	Both
1	0.1216	0.1299	0.1084
2	0.1821	0.1956	0.1643
3	0.2229	0.2315	0.2024
4	0.2587	0.2639	0.2315
5	0.2852	0.2940	0.2552
6	0.3081	0.3172	0.2755

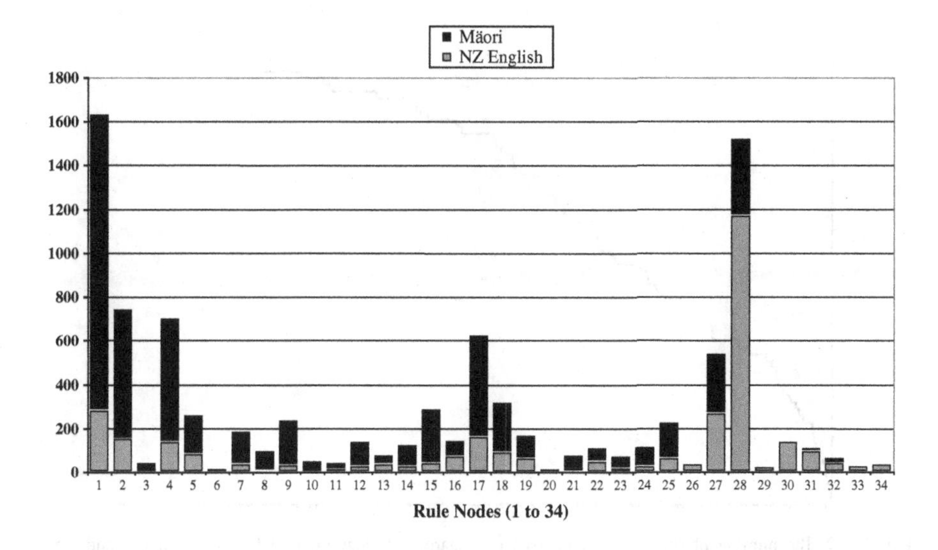

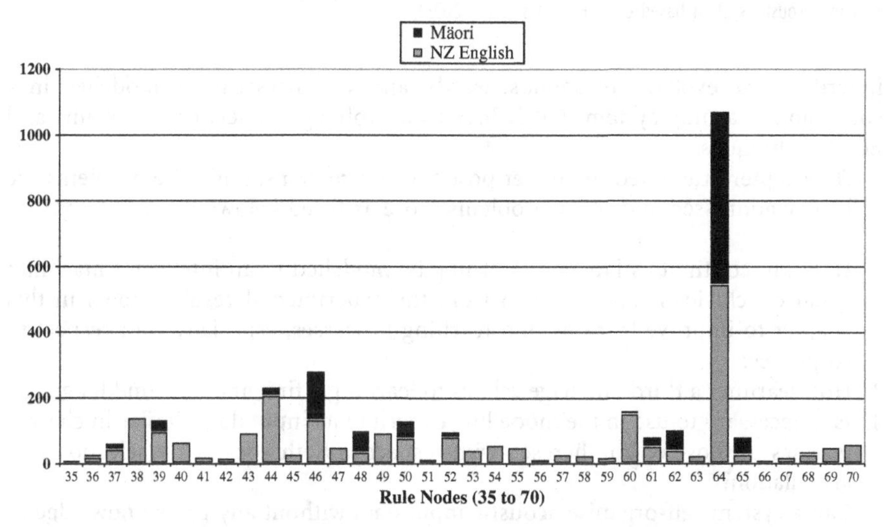

Fig. 10.11 The activation of cluster (rule) nodes of the evolving system evolved through a mixed presentation of both English and Maori, when new words from both languages are presented (Laws *et al.*, 2003).

10.5 Summary and Open Problems

The chapter presents one approach to modelling the emergence of acoustic clusters related to phones in a spoken language, and in multiple spoken languages, namely using ECOS for the purpose of modelling.

Modelling the emergence of spoken languages is an intriguing task that has not been solved thus far despite existing papers and books. The simple evolving model that is presented in this chapter illustrates the main hypothesis raised in this material, that acoustic features such as phones are learned rather than

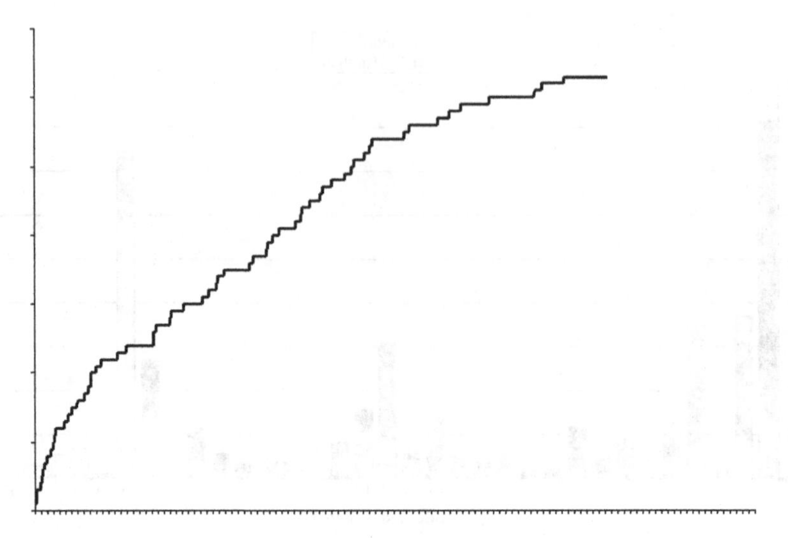

Fig. 10.12 The number of cluster nodes (y-axis) over learning frames (x-axis) for mixed English and Maori evolving acoustic system based on ECM (Laws *et al.*, 2003).

inherited. The evolving of sounds, words, and sentences can be modelled in a continuous learning system that is based on evolving connectionist systems and ECOS techniques.

The chapter attempted to answer posed open problems, but these problems are still to be addressed and other problems arose as listed below:

1. How can continuous language learning be modelled in an intelligent machine?
2. What conclusions can be drawn from the experimental results shown in this chapter to improve learning and teaching processes, especially with respect to languages?
3. How learning a third language relates to learning a first and a second language.
4. Is it necessary to use in the modelling experiments input data similar in characteristics to those of 'motherese', with respect to both vocabulary selection and articulation?
5. Can a system self-organise acoustic input data without any prior knowledge of the characteristics of the input language?
6. Is it possible to answer the dilemma of innateness versus learning in respect to languages through attempting to create a genetic profile of a language (languages) similar to the gene expression profiles of colon cancer and leukaemia presented in Chapter 8?
7. How can we prove or disprove the following hypotheses? There are no specific genes associated with learning the English language, and no genes associated with the Arabic language, and no genes associated with any particular language. Can we use the microarray technology presented in Chapter 8 or a combined microarray technology with fMRI imaging (see Chapter 9)?
8. How does the generic neuro-genetic principle (see Chapter 1 and Chapter 7) relate to learning languages?
9. How is language learning related to the instinct for information (see Chapter 7)?

10.6 Further Reading

For further details of the ideas discussed in this chapter, please refer to Taylor and Kasabov (2000), Laws *et al.*, 2003.

More on the research issues in the chapter can be found in other references, some of them listed below.

- *Generic Readings about Linguistics and Understanding Spoken Languages* (Taylor, 1995, 1999; Segalowitz, 1983; Seidenberg, 1997)
- *The Dilemma 'Innateness Versus Learned' in Learning Languages* (Chomsky, 1995; Lakoff and Johnson, 1999; Elman *et al.*, 1997)
- *Learning Spoken Language by Children* (Snow and Ferguson, 1977; MacWhinney *et al.*, 1996; Juszyk, 1997)
- *The Emergence of Language* (Culicover, 1999; Deacon, 1988; Pinker, 1994; Pinker and Prince, 1988)
- *Models of Language Acquisition* (Parisi, 1997; Regier, 1996)
- *Using Evolving Systems for Modelling the Emergence of Bilingual English–Maori Acoustic Space* (Laws *et al.*, 2003)
- *Modelling the Emergence of New Zealand Spoken English* (Taylor and Kasabov, 2000)

11. Evolving Intelligent Systems for Adaptive Speech Recognition

Speech and signal-processing technologies need new methods that deal with the problems of noise and adaptation in order for these technologies to become common tools for communication and information processing. This chapter is concerned with evolving intelligent systems (EIS) for adaptive speech recognition. An EIS system can learn continuously spoken phonemes, words, and phrases. New words, pronunciations, and languages can be introduced to the system in an incremental adaptive way.

The material is presented in the following sections.

- Introduction to adaptive speech recognition
- Speech signal analysis and feature selection
- A framework of EIS for adaptive speech recognition
- Adaptive phoneme-based speech recognition
- Adaptive whole word and phrase recognition
- Adaptive intelligent human–computer interfaces
- Exercise
- Summary and open problems
- Further reading

11.1 Introduction to Adaptive Speech Recognition

11.1.1 Speech and Speech Recognition

Speech recognition is one of the most challenging applications of signal processing (Cole *et al.*, 1995). Some basic notions about speech and speech recognition systems are given below.

Speech is a sequence of waves that are transmitted over time through a medium and are characterised by some features; among them are intensity and frequency. Speech is perceived by the inner ear in humans (see Chapter 9). It activates oscillations of small elements in the inner ear, which oscillations are transmitted to a specific part of the brain for further processing. The biological background of speech recognition is used by many researchers to develop humanlike automatic speech recognition systems (ASRS), but other researchers take other approaches.

Speech can be represented on different scales:

- Time scale, which representation is called waveform representation.
- Frequency scale, which representation is called spectrum.
- Both time and frequency scale; this is the spectrogram of the speech signal.

The three factors which provide the easiest method of differentiating speech sounds are the perceptual features of loudness, pitch, and quality. *Loudness* is related to the amplitude of the time domain waveform, but it is more correct to say that it is related to the energy of the sound (also known as its intensity). The greater the amplitude of the time domain waveform, the greater is the energy of the sound and the louder the sound appears. *Pitch* is the perceptual correlate of the fundamental frequency of the vocal vibration of the speaker organ.

The *quality* of a sound is the perceptual correlate of its spectral content. The formants of a sound are the frequencies where it has the greatest acoustic energy. The shape of the vocal tract determines which frequency components resonate. The shorthand for the first formant is F1, for the second is F2, and so on. The fundamental frequency is usually indicated by F_0.

A spectrogram of a speech signal shows how the spectrum of speech changes over time. The horizontal axis shows time and the vertical axis shows frequency. The colour scale (the grey scale) shows the energy of the frequency components.

The fundamental difficulty of speech recognition is that the speech signal is highly variable due to different speakers, different speaking rates, different contexts, and different acoustic conditions. The task is to find which of the variations are most relevant for an ASRS (Lee *et al.*, 1993).

There are a great number of factors which cause variability in speech such as the speaker, the context, and the environment. The speech signal is very dependent on the physical characteristics of the vocal tract, which in turn are dependent on age and gender. The country of origin of the speaker and the region in the country the speaker is from can also affect the speech signal. Different accents of English can mean different acoustic realizations of the same phonemes.

There are variations of the main characteristics of speech over time within the same sound, say the sounding of the phoneme /e/. An example is given in Fig. 11.1, where different characteristics of a pronounced phoneme /e/ by a female speaker of New Zealand English are shown.

There can also be different rhythm and intonation due to different accents. If English is the second language of a speaker, there can be an even greater degree of variability in the speech (see Chapter 10).

The same speakers can show variability in the way they speak, depending on whether it is a formal or informal situation. People speak precisely in formal situations and imprecisely in informal situations because they are more relaxed. Therefore the more familiar a speaker is with a computer speech recognition system, the more informal their speech becomes, and the more difficult for the speech recognition systems to recognise the speech. This could pose problems for speech recognition systems if they do not continually adjust.

Co-articulation effects cause phonemes to be pronounced differently depending on the word in which they appear; words are pronounced differently depending on the context; words are pronounced differently depending on where they lie in

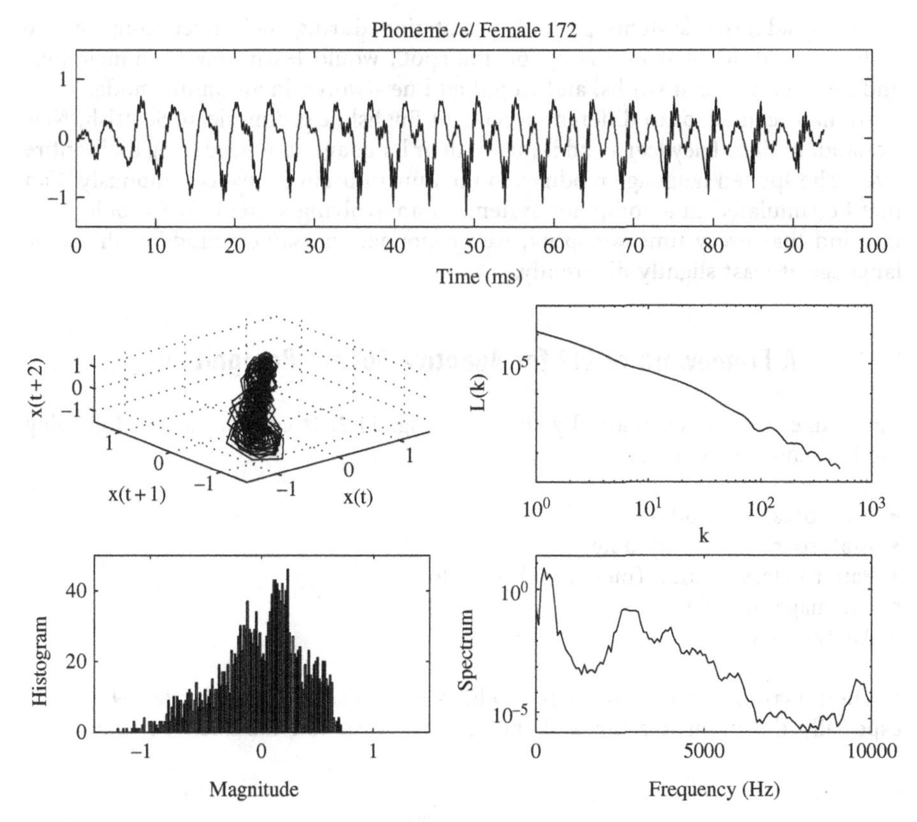

Fig. 11.1 Some of the characteristics of a pronounced English phoneme /e/ by a female speaker (data are taken from the Otago Speech Corpus, sample #172: (http://transaltor.kedri.info/).

a sentence due to the degree of stress placed upon them. In addition, the speaking rate of the speaker can cause variability in speech. The speed of speech varies due to such things as the situation and emotions of the speaker. However, the durations of sounds in fast speech do not reduce proportionally compared to their durations in slow speech.

11.1.2 Adaptive Speech Recognition

The variability of speech explained above requires robust and adaptive systems that would be able to accommodate new variations, new accents, and new pronunciations of speech.

The adaptive speech recognition problem is concerned with the development of methods and systems for speaker-independent recognition with high accuracy, able to adapt quickly to new words, new accents, and new speakers for a small, medium, or large vocabulary of words, phrases, and sentences.

Online adaptive systems perform adaptation during their operation; i.e. the system would adapt if necessary 'on the spot', would learn new pronunciations and new accents as it works, and would add new words in an online mode.

Humans can adapt to different accents of English, e.g. American, Scottish, New Zealand, Indian. They learn and improve their language abilities during their entire lives. The spoken language modules in the human brain evolve continuously. Can that be simulated in a computer system, in an evolving system? We should have in mind that every time we speak, we pronounce the same sounds of the same language at least slightly differently.

11.1.3 A Framework of EIS for Adaptive Speech Recognition

The framework is schematically shown in Fig. 11.2. It consists of the following modules and procedures.

- Preprocessing module
- Feature extraction module
- Pattern classification (modelling) module
- Language module
- Analysis module

The functions of some of these modules were discussed in Chapters 9 and 10, especially the preprocessing and the feature extraction modules. In Chapter 9,

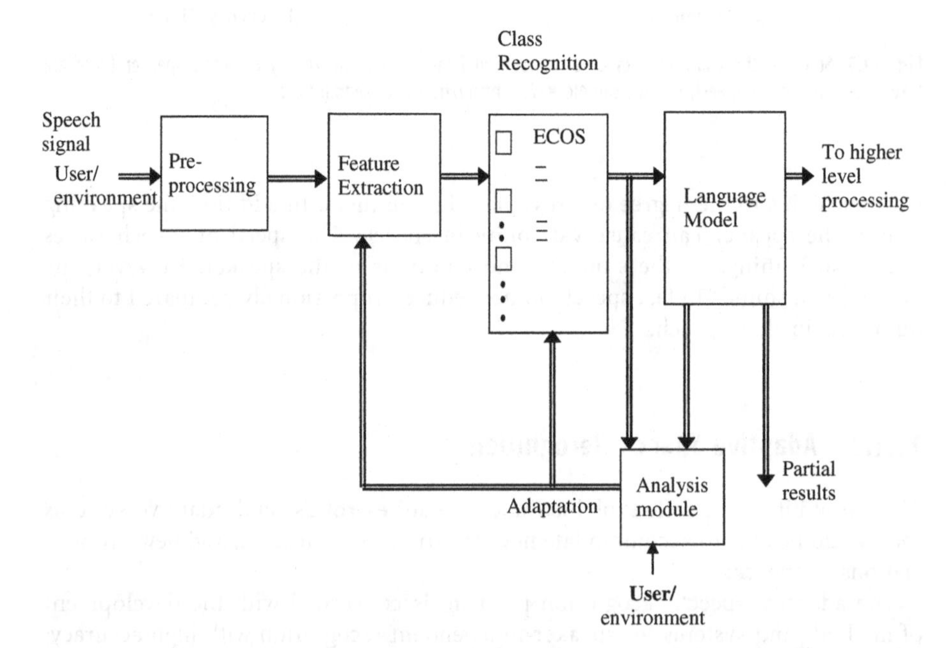

Fig. 11.2 A block diagram of an adaptive speech recognition system framework that utilises ECOS in the recognition part.

Mel-scale coefficients, Mel-scale cepstrum coefficients, gammatone filters, and other acoustic features were discussed.

The set of features selected depends on the organization and on the function of the pattern classifier module (e.g. phoneme recognition, whole word recognition, etc.).

The pattern (class) recognition module can be trained to recognize phonemes, or words, or other elements of a spoken language. The vector that represents the pronounced element is fed into the classifier module, created in advance with the use of the general purpose adaptive learning method, such as ECOS. The ECOS in this case allow for adaptive online learning. New words and phrases can be added to or deleted from the system at any time of its operation, e.g. 'go', 'one', 'connect to the Internet', 'start', 'end', or 'find a parking place'. New speakers can be introduced to the system, new accents, or new languages.

In the recognition mode, when speech is entered to the system, the recognized words and phrases at consecutive time moments are stored in a temporal buffer. The temporal buffer is fed into a sentence recognition module where multiple-word sequences (or sentence) are recognized.

The recognized word, or a sequence of words, can be passed to an action module for an action depending on the application of the system.

11.2 Speech Signal Analysis and Speech Feature Selection

The feature selection process is an extremely important issue for every speech recognition system, regardless of whether it is a phoneme-based or word-based system (see Chapter 1).

Figure 11.1 shows a *histogram* of the speech signal that can be used as a feature vector. Another popular feature set is a vector of FFT (fast Fourier transform) coefficients (or power spectrum as shown in Fig. 11.1). FFT transforms the speech signal from the time domain to the frequency domain. A simple program written in MATLAB that extracts the first 20 FFT coefficients from a spoken signal (e.g. pronounced word) is presented in Appendix B, along with a plot and a print of these coefficients.

Many current approaches towards speech recognition systems use Mel frequency cepstral coefficients (MCCs) vectors to represent, for example each 10–50 ms window of speech samples, taken every 5–25 ms, by a single vector of certain dimension (see Fig. 9.9). The window length and rate as well as the feature vector dimension are decided according to the application task.

For many applications the most effective components of the Mel-scale features are the first 12 coefficients (excluding the zero coefficient). MCCs are considered to be static features, as they do not account for changes in the signal within the speech unit (e.g. the signal sliding window, a phoneme, or a word).

Although MCCs have been very successfully used for off-line learning and static speech recognition systems for online learning adaptive systems that need to adapt to changes in the signal over time, a more appropriate set of features would be a combination of static and dynamic features.

It has been shown (Abdulla and Kasabov, 2002) that the speech recognition rate is noticeably improved when using additional coefficients representing the dynamic behaviour of the signal. These coefficients are the first and second derivatives of the cepstral coefficients of the static feature vectors. The power coefficients, which represent the energy content of the signal, and their first and second derivatives, also have important roles to be included in the representation of the feature vectors. The first and second derivatives are approximated by difference regression equations and accordingly named delta and delta–delta coefficients or first and second deltas, respectively. The power coefficients, which represent the power of the signal within the processed windows, are concatenated with the Mel coefficients. The static coefficients are normally more effective in the clean environment, whereas the dynamic coefficients are more robust in the noisy environment. Concatenating the static coefficients with their first and second derivatives increases the recognition rate and accounts for dynamic changes in the signal.

This approach has some drawbacks as well. Firstly, the static coefficients will dominate the effect of the dynamic coefficients. In this respect, a careful normalisation would be efficient to apply, but not a linear one, in the interval [0,1] for each feature separately. A more appropriate one, for example, is if the delta features are normalised in the range that is 25% of the range of the MCC, and the delta–delta features are normalised in the range that is 50% of the range of the delta features. Using dynamic features also increases the dimensionality of the feature vectors. Figure 11.3 shows the power and Mel coefficients with their derivatives of the phoneme /o/.

Other features that account for dynamic changes in the speech signal are *wavelets* (see Chapter 12) and *gammatone* feature vectors (see Chapter 9).

It is appropriate to use different sets of features in different modules if a modular speech recognition system is built, where a single ANN module is used for one speech class unit (e.g. a phoneme or a word).

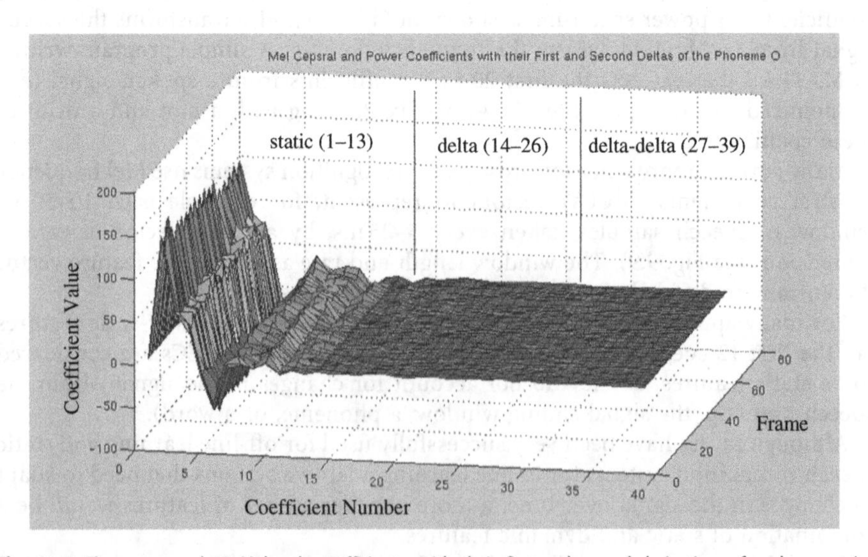

Fig. 11.3 The power and 12 Mel-scale coefficients, with their first and second derivatives of a phoneme /o/ sound signal (Abdulla and Kasabov, 2002).

11.3 Adaptive Phoneme-Based Speech Recognition

11.3.1 Problem Definition

Recognising phonemes from a spoken language is a difficult but important problem. If it is correctly solved, then it would be possible to further recognize the words and the sentences of a spoken language. The pronounced vowels and consonants differ depending on the accent, dialect, health status, and so on of the person.

As an illustration, Fig. 11.4 shows the difference between some vowels in English pronounced by male speakers in the R.P. (received pronunciation) English, Australian English, and New Zealand English, when the first and the second formants are used as a feature space and averaged values are used. A significant difference can be noticed between the same vowels pronounced in different dialects (except the phoneme /I/ for the R.P. and for the Australian English: they coincide on the diagram). In New Zealand English /I/ and /ɜ/ are very close.

There are different artificial neural network (ANN)-based models for speech recognition that utilise MLP, SOM, RBF networks, time-delay NN (Weibel *et al.*, 1989; Picone, 1993), hybrid NN and hidden Markov models (Rabiner, 1989; Trentin, 2001), and so on. All these models usually use one ANN for the classification of all phonemes and they work in an off-line mode. The network has as many outputs as there are phonemes.

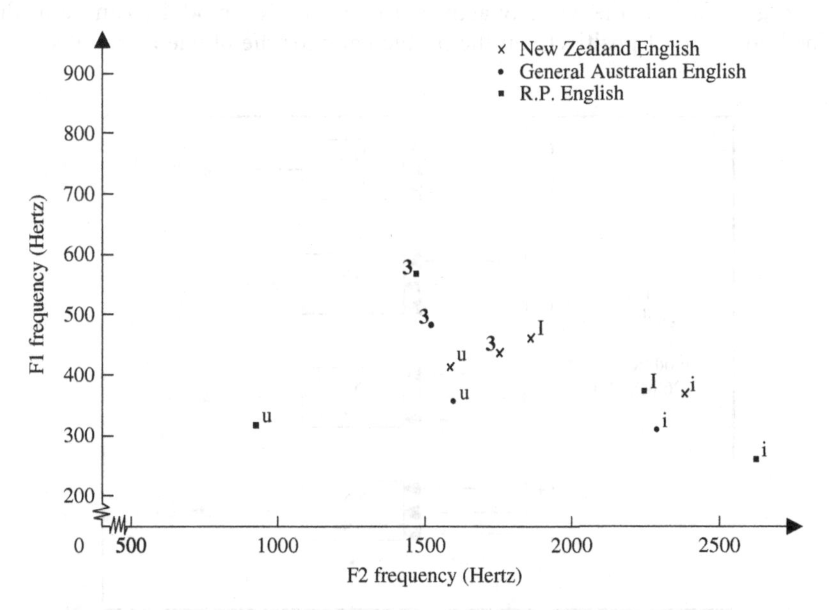

Fig. 11.4 Different phones of received pronunciation English, Australian English, and NZ English presented in the 2D space of the first two formants. Same phonemes are pronounced differently in the different accents, e.g. /I/ (Maclagan, 1982).

11.3.2 Multimodel, Phoneme-Based Adaptive Speech Recognition System

Here an approach is used where each NN module from a multimodular system is trained on a single phoneme data (see Kasabov (1996)) and the training is in an online mode. An illustration of this approach is given in Fig. 11.5 where four phoneme modules are shown, each of them trained on one phoneme data with three time lags of 26 element Mel-scale cepstrum vectors, each vector representing one 11.6 milliseconds timeframe of the speech data, with an overlap of 50% between consecutive timeframes.

The rationale behind this approach is that single phoneme ANN can be adapted to different accents and pronunciations without necessarily retraining the whole system (or the whole ANN in case of a single ANN that recognises all phonemes). Very often, it is just few phonemes that distinguish one accent from another and only these ANN modules need to be adjusted.

Figure 11.6 shows the activation of each of the seven ANN modules trained to recognise different phonemes when a spoken word 'up' is propagated through the whole system over time. Although the /^/ phoneme ANN gets rightly activated when the phoneme /^/ is spoken, and /p/ NN gets rightly activated when the /p/ phoneme is spoken, the /h/ and /n/ phoneme ANNs gets wrongly activated during the silence between /^/ and /p/ in the word 'up', and the /r/ and /d/ phoneme ANNs get wrongly activated when /p/ is spoken.

This phoneme module misactivation problem can be overcome through analysis of the sequence of the recognised phonemes and forming the recognized word through a matching process using a dictionary of words. In order to improve the recognition rate, the wrongly activated phoneme NN modules can be further trained not to react positively on the problematic for the phoneme sounds.

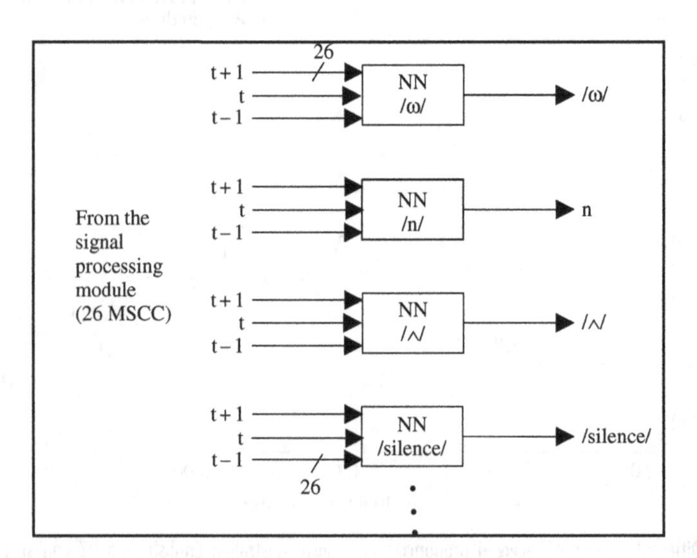

Fig. 11.5 Four ANN modules, each of them trained to recognize one phoneme (from Kasabov (1996), ©MIT Press, reproduced with permission).

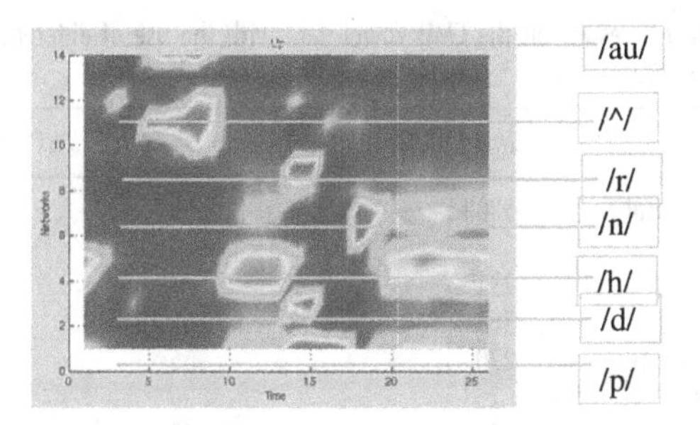

Fig. 11.6 The activation of seven phoneme ANN modules, trained on their corresponding phoneme data, when an input signal of a pronounced word 'up' is submitted. Some of the NN modules are 'wrongly' activated at a time, showing the dynamical features of the phonemes.

Each of the phoneme NN modules, once trained on data of one accent, can be further adapted to a new accent, e.g. Australian English. In order to do that, the NN have to be of a type that allows for such adaptation. Such NN are the evolving connectionist systems ECOS.

In one experiment a single EFuNN is used as a single phoneme recogniser. Each EFuNN from a multimodular ECOS can be further adapted to any new pronunciation of this phoneme (Ghobakhlou *et al.*, 2003). One EFuNN was used for each phoneme. Each EFuNN was further adapted to new pronunciation of this phoneme.

11.3.3 Using Evolving Self-Organising Maps (ESOMs) as Adaptive Phoneme Classifiers

An evolving self-organised map (ESOM; Chapter 2) is used for the classification of phoneme data. The advantage of ESOMs as classifiers is that they can be trained (evolved) in a lifelong mode, thus providing an adaptive, online classification system.

Here, an ESOM is evolved on phoneme frames from the vowel benchmark dataset from the CMU Repository (see also Robinson (1989)). The dataset consists of 990 frames of speech vowels articulated by four male and four female speakers. In traditional experiments 528 frames are used for training and 462 for testing (Robinson, 1989). Here, several models of ESOM are evolved on the training data with the following parameter values: $\varepsilon = 0.5$; $\gamma = 0.05$.

The test results of ESOM and of other classification systems on the same test data are shown in Table 11.1. While using tenfold cross-validation on the whole dataset, much better classification results are obtained in the ESOM models, Table 11.2. Here $\varepsilon = 1.2$.

When online learning is applied on the whole stream of the vowel data, every time testing its classification accuracy on the following incoming data, the error rate decreases with time as can be seen from Fig. 11.7.

Figure 11.7 illustrates that after a certain number of examples drawn from a closed and bounded problem space, the online learning procedure of ECOS can converge to a desired level of accuracy and the error rate decrease (Chapters 2, 3, 5, and 7).

Table 11.1 Classification test results on the CMU vowel data with the use of different classification techniques (Deng and Kasabov, 2000, 2003).

Classifier	*Number of weights*	*% Correct best*	*% Correct average*
5-nearest neighbour with local approximation	5808	53	—
MLP-10	231	—	—
Squared MLP	363	65.1	58.5
5D growing cell structure (Fritzke, 1994) 80 epochs	270	66	
DCS-GCS (Bruske and Sommer, 1995)	216	65	60
ESOM (one epoch only)	207	65.4	61.8

Table 11.2 Tenfold cross-validation classification results on the whole vowel data set (Deng and Kasabov, 2000, 2003).

Classifier	*% Correct (average)*
CART (Classification on a regression tree)	78.2
CART-dB	90
ESOM (average number of nodes is 233)	95.0 +/- 0.5

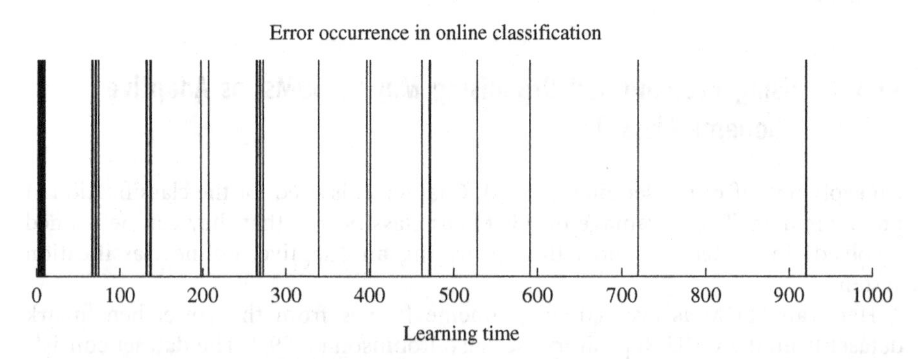

Fig. 11.7 Error rate of an ESOM system trained in an online mode of learning and subsequent classification of frames from the vowel benchmark dataset available from the CMU repository (see explanation in the text). The longer the ESOM is trained on the input stream of data, the less the error rate is. The system is reaching an error convergence (Deng and Kasabov, 2000, 2003).

11.4 Adaptive Whole Word and Phrase Recognition

11.4.1 Problem Definition

In this case, the speech signal is processed so that the segment that represents a spoken word is extracted from the rest of the signal (usually it is separated by silence). Extracting words from a speech signal means identifying the beginning and the end of the spoken word.

There are many problems that need to be addressed while creating a whole word speech recognition system, for example the problems that relate to ambiguity of speech. This ambiguity is resolved by humans through some higher-level processing.

Ambiguity can be caused by:

- Homophones: Words with different spellings and meanings but that sound the same (e.g. 'to, too, two' or 'hear, hair, here'). It is necessary to resort to a higher level of linguistic analysis for distinction.
- Word boundaries: Extracting whole words from a continuous speech signal may lead to ambiguities; for example /greiteip/ could be interpreted as 'grey tape' or 'great ape'. It is necessary to resort to a higher-level linguistic knowledge to properly set the boundaries.
- Syntactic ambiguity: This is the ambiguity arising before all the words of a phrase or a sentence are properly grouped into their appropriate syntactic units. For example, the phrase 'the boy jumped over the stream with the fish' means either the boy with the fish jumped over the stream, or the boy jumped over the stream with a fish in it. The correct interpretation requires more contextual information.

All speech recognition tasks have to be constrained in order to be solved. Through placing constraints on the speech recognition system, the complexity of the speech recognition task can be considerably reduced. The complexity is basically affected by:

1. The vocabulary size and word complexity. Many tasks can be performed with the use of a small vocabulary, although ultimately the most useful systems will have a large vocabulary. In general the vocabulary size of a speech recognition system can vary as follows:.

    - Small, tens of words
    - Medium, hundreds of words
    - Large, thousands of words
    - Very large, tens of thousands of words

2. The format of the input speech data entered to the system, that is: Isolated words (phrases)

    - Connected words; this represents fluent speech but in a highly constrained vocabulary, e.g. digit dialling
    - Continuous speech

3. The degree of speaker dependence of the system:

    - Speaker-dependent (trained to the speech patterns of an individual user)
    - Multiple speakers (trained to the speech patterns of a limited group of people)
    - Speaker-independent (such a system could work reliably with speakers who have never or seldom used the system)

Sometimes a form of task constraint, such as formal syntax and formal semantics, is required to make the task more manageable. This is because if the vocabulary size increases, the possible combinations of words to be recognised grows exponentially.

Figure 11.8 illustrates the idea of using a NN for the recognition of a whole word. As inputs, 26 Mel-scale cepstrum coefficients taken from the whole word signal are used. Each word is an output in the classification system.

11.4.2 A Case Study on Adaptive Spoken Digit Recognition – English Digits

The task is of the recognition of speaker-independent pronunciations of English digits. The English digits are taken from the Otago Corpus database (http://translator.kedri.info). Seventeen speakers (12 males and 5 females) are used for training, and another 17 speakers (12 males and 5 females) are used for testing an EFuNN-based classification system. Each speaker utters 30 instances of English digits during a recording session in a quiet room (clean data) for a total of 510 training and 510 testing utterances (for details see Kasabov and Iliev (2000)).

In order to assess the performance of the evolved EFuNN in this application, a comparison with the linear vector quantization (LVQ) method (Chapter 2, Kohonen (1990, 1997)) is presented. Clean training speech data is used to train both the LVQ and the EFuNN models. Noise is introduced to the clean speech test data to evaluate the behaviour of the recognition systems in a noisy environment. Two different experiments are conducted with the use of the standard EFuNN learning method from Chapter 3. In the first instance, car noise is added to the clean speech. In the second instance office noise is introduced over the clean signal. In both cases, the signal-to-noise ratio SNR ranges from 0 dB to 18 dB.

The results for car noise are shown in Fig. 11.9. The word recognition rate (WRR) ranges from 86.87% at 18 dB to 83.33% at 0 dB. The EFuNN method outperforms the LVQ method, which achieves WRR = 82.16% at 0 dB.

The results for office noise are presented in Fig. 11.10. The WRR of the evolved EFuNN system ranges from 78.63% at 18dB to 71.37% at 0 dB, and is significantly higher than the WRR of LVQ (21.18% at 0 dB).

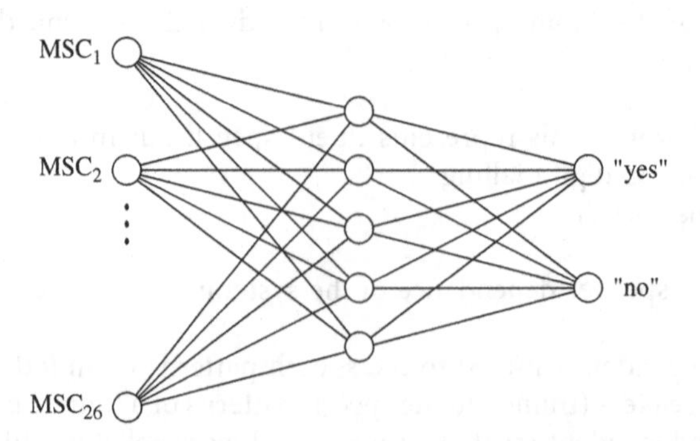

Fig. 11.8 An illustration of an ANN for a whole word recognition problem on the recognition of two words, 'yes' and 'no' (from Kasabov (1996), ©MIT Press, reproduced with permission).

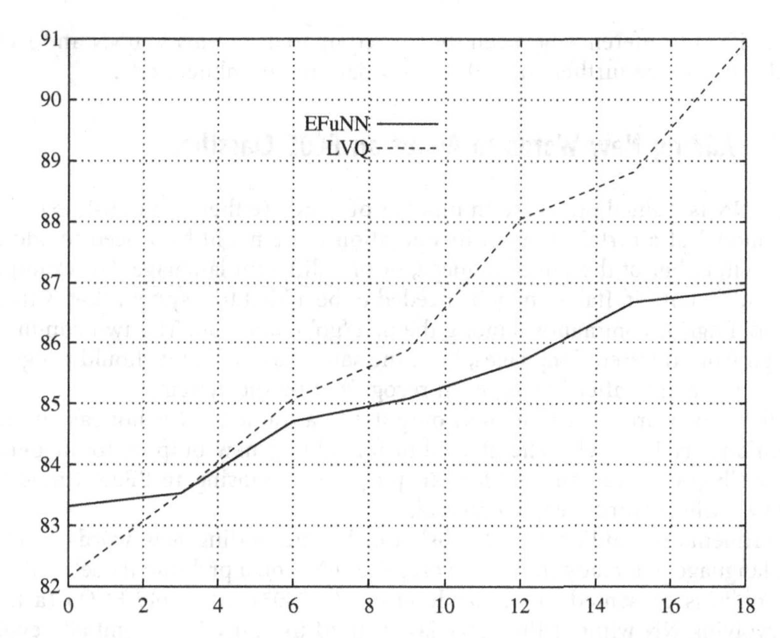

Fig. 11.9 Word recognition rate (WRR) of two speech recognition systems when car noise is added: LVQ, codebook vectors, 396; training iterations, 15,840; EFuNN, 3MF; rule nodes, 157; sensitivity threshold $Sthr = 0.9$; error threshold $Errthr = 0.1$; learning rates $lr1 = 0.01$ and $lr2 = 0.01$; aggregation thresholds $thrw1 = 0.2$, $thrw2 = 0.2$; number of examples for aggregation Nexa = 100; 1 training iteration (Kasabov and Iliev, 2000).

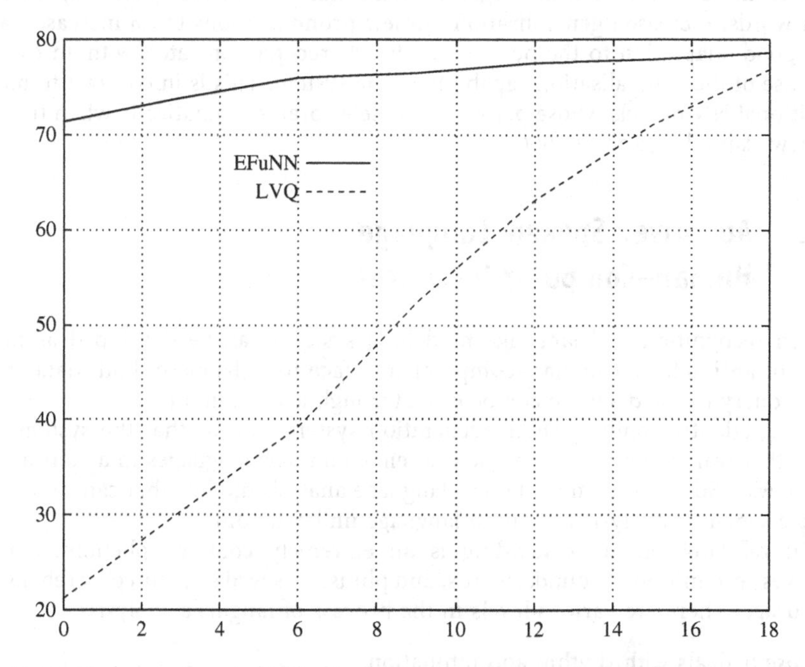

Fig. 11.10 Word recognition rate (WRR) of two speech recognition systems when office noise is added: LVQ, codebook vectors, 396; training iterations, 15,840; EFuNN, 3MF; rule nodes, 157; $Sthr = 0.9$, $Errthr = 0.1$, $lr1 = 0.01$, $lr2 = 0.01$, $thrw1 = 0.2$, $thrw2 = 0.2$, Nexa = 100, 1 training iteration (Kasabov and Iliev, 2000).

A significant difference between the two compared systems EFuNN and LVQ is that EFuNN can be further trained on new data in an online mode.

11.4.3 Adding New Words to Adapt an ECOS Classifier

When a NN is trained on a certain number of words (either in an off-line or in an online mode) at a certain time of its operation there might be a need to add new words to it, either of the same language, or of a different language. For example, a command 'avanti' in Italian may be needed to be added to a system that is trained on many English commands, among them, a 'go' command. The two commands, although from different languages, have the same meaning and should trigger the same output action after having been recognized by the system.

Adding new words (meaning new outputs) to a trained NN is not easy in many conventional NN models. The algorithm for adding new outputs to an EFuNN, given in Chapter 3, can be used for this purpose, supposing an EFuNN module is trained on whole words (e.g. commands).

Experiments on adding new English words, and adding new words from the Maori language to already trained (evolved) EFuNN on a preliminary set of English words only, is presented in Ghobakhlou *et al.* (2003). A simple ECOS (a three-layer evolving NN without the fuzzy layers used in EFuNN) was initially evolved on the digit words (from the Otago Speech Corpus, *http://translator.kedri.info*) and then new words were added in an incremental way to make the system work on both old and new commands. The system was tested on a test set of data. It manifested very little forgetting (less than 2%) of the previously learned digit words, increased generalisation on new pronunciations (10% increase), and very good adaptation to the new words (95.5% recognition rate), with an overall increase of the generalisation capability of the system. This is in contrast to many traditional NN models whose performance deteriorates dramatically when trained on new examples (Robins,1996).

11.5 Adaptive, Spoken Language Human–Computer Interfaces

Speech recognition and language modelling systems can be developed as main parts of an intelligent human–computer interface to a database. Both data entry and a query to the database can be done through a voice input.

Using adaptive online speech recognition systems means that the system can be further trained on new users, new accents, and new languages in a continuous online way. Such a system contains a language analysis module that can vary from simple semantic analysis to natural language understanding.

Natural language understanding is an extremely complex phenomenon. It involves recognition of sounds, words, and phrases, as well as their comprehension and usage. There are various levels in the process of language analysis:

- *Prosody* deals with rhythm and intonation.
- *Phonetics* deals with the main sound units of speech (phonemes) and their correct combination.

- *Lexicology* deals with the lexical content of a language.
- *Semantics* deals with the meaning of words and phrases seen as a function of the meaning of their constituents.
- *Morphology* deals with the semantic components of words (morphemes).
- *Syntax* deals with the rules, which are applied to form sentences.
- *Pragmatics* deals with the language usage and its impact on the listener.

It is the importance of language understanding in communication between humans and computers which was the essence of Alan Turing's test for AI (see Introduction).

Computer systems for language understanding require methods that can represent ambiguity, common-sense knowledge, and hierarchical structures. Humans, when communicating with each other, share a lot of common-sense knowledge which is inherited and learned in a natural way. This is a problem for a computer program. Humans use face expressions, body language, gestures, and eye movement when they communicate with each other. They communicate in a multimodal manner. Computer systems which analyse speech signals, gestures, and face expressions when communicating with users are called multimodal systems. An example of such systems is presented in Chapter 13.

11.6 Exercise

Task: *A small EIS for adaptive signal recognition, written in MATLAB*
Steps:

1. Record or download wave data related to two categories of speech or sound (e.g. male versus female, bird song versus noise, Mozart versus Heavy Metal, 'Yes' versus 'No').
2. Transform the wave data into features.
3. Prepare and label the samples for training an adaptive neural network model (ECOS).
4. Train the model on the data.
5. Test the model/system on new data.
6. Adapt the system (add new data) and test its accuracy on both the new and the old data.
7. Explain what difficulties you have overcome when creating the system.

A simple MATLAB code, that implements only the first part of the task, is given in Appendix B. Screen shots of printed results after a run of the program are included.

11.7 Summary and Open Problems

The applicability of evolving, adaptive speech recognition systems is broad and spans all application areas of computer and information science where systems that communicate with humans in a spoken language ('hands-free and eyes-free environment') are needed. This includes:

- Voice dialling, especially when combined with 'hands-free' operation of a telephone system (e.g. a cell phone) installed in a car. Here a simple vocabulary that includes spoken digits and some other commands would be sufficient.
- Voice control of industrial processes.
- Voice command execution, where the controlled device could be any terminal in an office. This provides a means for people with disabilities to perform simple tasks in an office environment.
- Voice control in an aircraft.

There are several open problems in the area of adaptive speech recognition, some of them discussed in this chapter. They include:

1. Comprehensive speech and language systems that can quickly adapt to every new speaker.
2. Multilingual systems that can learn new languages as they operate. The ultimate speech recognition system would be able to speak any spoken language in the world.
3. Evolving systems that would learn continuously and incrementally spoken languages from all available sources of information (electronic, human voice, text, etc.).

11.8 Further Reading

- *Reviews on Speech Recognition Problems, Methods, and Systems* (Cole *et al.*, 1995; Lippman, 1989; Rabiner, 1989; Kasabov, 1996)
- *Signal Processing* (Owens, 1993; Picone, 1993)
- *Neural Network Models and Systems for Speech Recognition* (Morgan and Scofield, 1991)
- *Phoneme Recognition Using Timedelay Neural Networks* (Waibel *et al.*, 1997)
- *Phoneme Classification Using Radial Basis Functions* (Renals and Rohwer, 1989)
- *Hybrid NN-HMM Models for Speech Recognition* (Trentin, 2001)
- *A Study on Acoustic Difference Between RP English, Australian English, and NZ English* (Maclagan, 1982)
- *Evolving Fuzzy Neural Networks for Phoneme Recognition* (Kasabov, 1998b, 1999)
- *Evolving Fuzzy Neural Networks for Whole Word Recognition, English and Italian Digits* (Kasabov and Iliev, 2000)
- *Evolving Self-organising Maps for Adaptive Online Vowel Classification* (Deng and Kasabov, 2000, 2003)
- *Adaptive Speech and Multimodal Word-Based Speech Recognition Systems* (Ghobaghlou *et al.*, 2003)

12. Evolving Intelligent Systems for Adaptive Image Processing

In adaptive processing of image data it is assumed that a continuous stream of images or videos flows to the system and the system always adapts and improves its ability to classify, recognise, and identify new images. There are many tasks in the image recognition area that require EIS. Some application-oriented models and experimental results of using ECOS, along with other specific or generic methods for image processing, are presented in this chapter. The material here is presented in the following sections.

- Image analysis and feature selection
- Online colour quantisation
- Adaptive image classification
- Online camera operation recognition
- Adaptive face recognition and face membership identification
- Exercise
- Summary and open problems
- Further reading

12.1 Image Analysis and Feature Selection

12.1.1 Image Representation

A 2D image is represented usually as a set of pixels (picture elements), each of them defined by a triplet (x, y, u), where x and y are the coordinates of the pixel and u is its intensity. An image is characterised by spatial and spectral characteristics. The latter represents the colour of a pixel, as identified uniquely through its three components – red, green, and blue (RGB) – each of them reflecting the white light, producing a signal with a different wavelength (in nm; 465 blue; 500 green; 570 red). The RGB model that has 256 levels in each dimension, can represent 16,777,216 colours. The visual spectrum has the range of (400:750) nm wavelengths

Images are represented in computers as numerical objects (rather than perceived objects) and similarity between them is measured as a distance between their corresponding pixels, usually measured as Euclidean distance.

Grey-level images have one number per pixel that represents the parameter u (usually between 0 and 256), rather than three such numbers as it is in the colour images.

12.1.2 Image Analysis and Transformations

Different image analysis and transformation techniques can be applied to an image in order to extract useful information and to process the image in an information system. Some of them are listed below and are illustrated with a MATLAB program in Appendix C, along with prints of resulted images.

- Filtering, using kernels, e.g. Ker = [1 1 1 1 -7 1 1 1 1], where each pixel (or a segment) Im_j and its 8 neighbored pixels of an image Im is convolved (transformed):

$$\mathrm{Conv(Im\_j, Ker) = Sum(Im\_j^{*}Ker)} \tag{12.1}$$

  where"*"means vector multiplication
- Statistical characteristics: e.g. histograms;
- Adding noise:

$$\mathrm{Im\_j = Im\_j + random(Nj)} \tag{12.2}$$

Example

A benchmark image *Lena* is used to illustrate the above techniques; see the MATLAB demo program and the figures in Appendix C.

Different image analysis and image transformation techniques are required for different tasks.

Example

For a task of counting objects (e.g. molecules, atoms, sheep, aircraft, etc.) and their location from a camera image, the following image analysis techniques may be needed as illustrated in Fig. 12.1 in the case of counting sheep.

- Texture analysis
- Finding boundaries of objects (areas of contrasts)
- Finding spatial object location

12.1.3 Image Feature Extraction and Selection

An image can be represented as a set of features forming a feature vector as follows. Some of the most used features are the following ones.

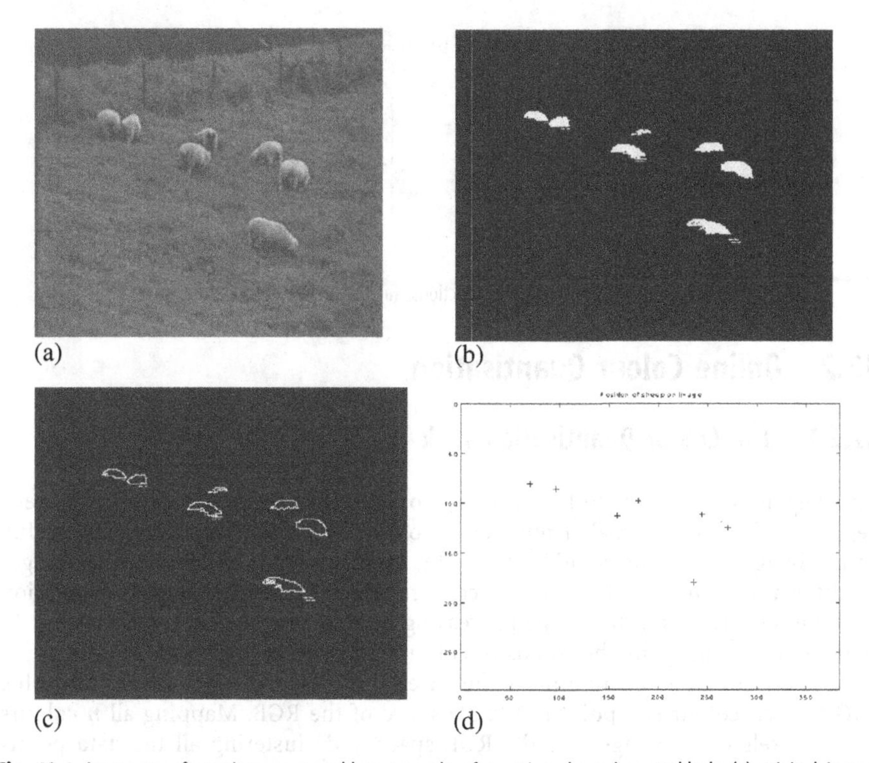

Fig. 12.1 Image transformations on a problem example of counting sheep in a paddock: (a) original image; (b) texture analysis; (c) contour detection; (d) object location identification.

- Raw pixels: Each pixel intensity is a feature (input variable) .
- Horizontal profile: The sum (or average) intensity of each row of the image, in the same order for all images.
- Vertical profile: The sum (or average) intensity of each column of the image, in the same order for all images.
- Composite profile: The sum (or average) intensity of each column plus each row of the image, in the same order for all images.
- Grey histogram.
- Colours as features and/or object shapes.
- Specialised features, e.g. for face recognition.
- FFT frequency coefficients.
- Wavelength features: see Fig. 12.2 for an example and a comparison between a wavelet function and a periodic function (sine).

Example

A small program in MATLAB for extracting a composite profile from a raw image is given in Appendix C.

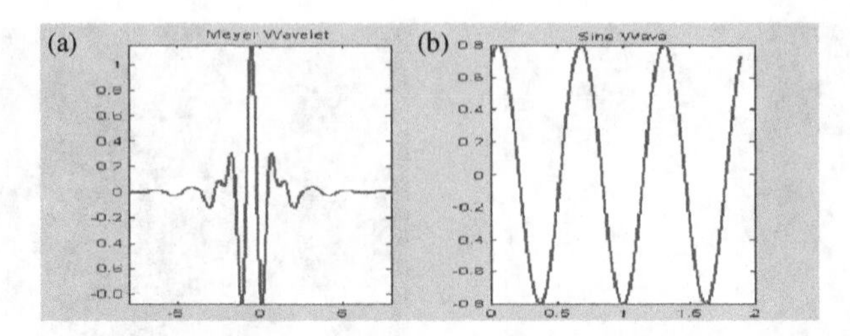

Fig. 12.2 Image transformation functions: (a) Meyer wavelet; (b) sine wave.

12.2 Online Colour Quantisation

12.2.1 The Colour Quantisation Task

This task is concerned with the reduction of the number of colours n an image is represented in, into a smaller number of colours m without degrading the quality of the image (Chen and Smith, 1977). This is necessary in many cases as images might be represented in thousands of colours, which makes the image transmission on long distances and the image processing in a computer unacceptably slow. In many cases keeping the thousands of colours may not be necessary at all.

As each colour is a mixture of the three main colours, red, green, and blue (RGB), each colour is a point in the 3D space of the RGB. Mapping all n colours of the pixels of an image into the RGB space and clustering all the data points into a smaller number of clusters (e.g. 256; colour prototypes) that best represent each colour of the original picture, is the first step of the colour quantisation.

The second step is substituting the original colour for each pixel with its closest prototype. That gives the quantised image.

Many methods are known for colour quantisation (Chen and Smith, 1977; Chaudhuri *et al.*, 1992). Most of them perform in many iterations. In the next section, online colour clustering and quantisation is achieved through applying ESOM from Chapter2.

12.2.2 Online Colour Quantisation Using Evolving Self-Organising Maps (ESOM)

Here, the ESOM algorithm from Chapter 2 is applied to the problem of online colour quantisation. Results are compared with those achieved by applying other methods, including median-cut, octree, and Wu's method. Three test images are chosen: Pool Balls, Mandrill, and Lena, as shown in Figs. 12.3 through 12.5.

The Pool Balls image is artificial and contains smooth colour tones and shades. The Mandrill image is of 262,144 (512×512) pixels but has a very large number of colours (230,427). The Lena image is widely used in the image processing literature and contains both smooth areas and fine details.

Test images are quantised to 256 colours. For the different images, different ESOM parameters are used as follows: (a) Pool Balls, $e = 18.6$, (b) Mandrill, $e = 20.4$, (c) Lena, $e = 31.9$. In all three cases $Tp = 2000$ and $\gamma = 0.05$.

Fig. 12.3 Pool Balls benchmark colour image.

Fig. 12.4 Mandrill benchmark colour image.

Fig. 12.5 Lena colour benchmark image.

Table 12.1 Quantization performance of different methods over the three benchmark images from Figs. 12.1, 12.2, and 12.3, where the quantisation error and the quantisation variance are shown (Deng and Kasabov, 2003).

Methods	*Pool Balls*	*Mandrill*	*Lena*
Median-cut	2.58 / 8.28	11.32 / 5.59	6.03 / 3.50
Octree	4.15 / 3.55	13.17 / 4.98	7.56 / 3.83
Wu's	2.22 / 2.19	9.89 / 4.56	5.52 / 2.94
ESOM	2.43 / 2.56	9.47 / 3.86	5.28 / 2.36

Online clustering is applied directly to the RGB colour space. Here we denote the image as I, with a pixel number of N. The input vector to the ESOM algorithm is now a three-dimensional one: $Ii = (\mathrm{R}i, \mathrm{G}i, \mathrm{B}i)$.

The online clustering process of ESOM will construct a colour map $C = \{cj|j = 1, \ldots, 256\}$. Each image pixel is then quantised to the best-matching palette colour c_m, a process denoted Q: Ii —-> c_m. To speed up the calculation process, the L-norm (see Chaudhuri *et al.* (1992) is adopted as an approximation of the Euclidean metric used in ESOM. The quantisation root mean square error between the original and the quantised images is calculated pixelwise.

Apart from the quantisation error, quantisation error variance is another factor which influences the visual quality of the quantised image.

Quantisation performance of different methods is compared in Table 12.1, where the quantisation error and the quantisation variance are shown.

Generally speaking, ESOM not only achieves a very small value of average quantisation error; its error variance is also the smallest. This explains why images quantised by ESOM have better visual quality than those done by other methods.

Figures 12.6 through 12.9 show the quantized Lena image with the use of the median cut method (Heckbert, 1982), octree (Gervautz and Purghatofer, 1990), Wu's method (Wu, 1992), and ESOM, respectively (Deng and Kasabov, 2003). The

Fig. 12.6 Off-line quantized Lena image with the use of the median cut method.

Fig. 12.7 Off-line quantized Lena image with the use of the octree.

Fig. 12.8 Off-line quantized Lena image with the use of Wu's method.

accuracy of the ESOM model is comparable with the other methods, and in several cases, e.g., the Lena image, the best one achieved.

Using ESOM takes only one epoch of propagating pixels through the ESOM structure, whereas the other methods require many iterations. With the 512×480-sized Lena image, it took two seconds for the ESOM method to construct the quantisation palette on a Pentium-II system running Linux 2.2. By using an evolving model, the time searching for best matching colours is much less than using a model with a fixed number of prototypes. In addition to that, there is a potential of hardware parallel implementation of ESOM, which will increase greatly the speed of colour quantisation and will make it applicable for online realtime applications to video streams.

Fig. 12.9 Online quantized Lena image with the use of ESOM (Deng and Kasabov, 2003).

As ESOM can be trained in an incremental online mode, the already evolved ESOM on a set of images can be further tuned and modified according to new images.

12.3 Adaptive Image Classification

12.3.1 Problem Definition

Some connectionist and hybrid neuro-fuzzy connectionist methods for image classification have been presented in Pal *et al.* (2000) and Hirota (1984). The classification procedure consists of the following steps.

1. *Feature extraction from images.* Different sets of features are used depending on the classification task. Filtering, fast Fourier transformation (FFT), and wavelet transformations (Wang *et al.*, 1996; Szu and Hsu, 1999) are among the most popular ones.
2. *Pattern matching of the feature vectors to a trained model.* For pattern matching, different NN and hybrid neuro-fuzzy-chaos techniques have been used (see e.g. Bezdek (1993) and Szu and Hsu (1999)).
3. *Output formation.* After the pattern matching is achieved, it may be necessary to combine the calculated output values from the pattern classifier with other sources of information to form the final output results. A simple technique is to take the max value of several NN modules, each of them trained on a particular class datum, as is the case in the experiment below.

EIS for adaptive image classification are concerned with the process of incremental model creation when labelled data are made available in a continuous way. The classifier is updated with each new labelled datum entered and is used to classify new, unlabelled data. And this process is a continuous one.

12.3.2 A Case Study of Pest Identification from Images of Damaged Fruit

The case study here is the analysis of damage to pip fruit in orchards with the goal of identifying what pest caused the damage (Wearing, 1998). An image database is also built, allowing for content-based retrieval of damage images using wavelet features (Wearing, 1998). The problem is normally compounded by the fact that the images representing a damaged fruit vary a lot. Images are taken either from the fruit or from the leaves, and are taken at different orientations and distances as shown in Figure 12.10a,b. As features, wavelets are extracted from the images (see Fig. 12.2) for the difference between wavelets and sin functions used in the FFT.

Using Daubechies wavelets for image analysis and image comparison has already been shown to be a successful technique in the analysis of natural images (Wang *et al.*, 1996; Szu and Hsu, 1999). In the experiment here the coefficients resulting from the wavelet analysis are used as inputs to an EFuNN module (see Chapter 3) for image classification.

This section suggests a methodology for classification of images based on evolving fuzzy neural networks (EFuNNs) and compares the results with the use of a fixed size, off-line fuzzy neural network (FuNN; Kasabov *et al.* (1997)) and with some other techniques.

For the experimental modelling, a set of 67 images is used to train five EFuNNs, one for the identification of each of the five pests in apples, denoted as: *alm* – l; *alm* – *f*; cm; *lr* – l; *lr* – *f*. The initial sensitivity threshold is selected as $Sthr = 0.95$ and the error threshold used is $Errthr = 0.01$. The EFuNNs are trained for one epoch. The number of rule nodes generated (*rn*) after training for each of the EFuNN models is as follows: EFuNN – *alm* – l: $rn = 61$, EFuNN – *alm* – *f* $rn = 61$, EFuNN – *cm*: $rn = 51$, EFuNN – *lr* – l: $rn = 62$, and EFuNN – *lr* – *f*: $rn = 61$. The results of the confusion matrix are presented in Table 12.2.

The evolving EFuNN models are significantly better at identifying pests on new test data (what pest has caused the damage to the fruit) than the FuNNs (not

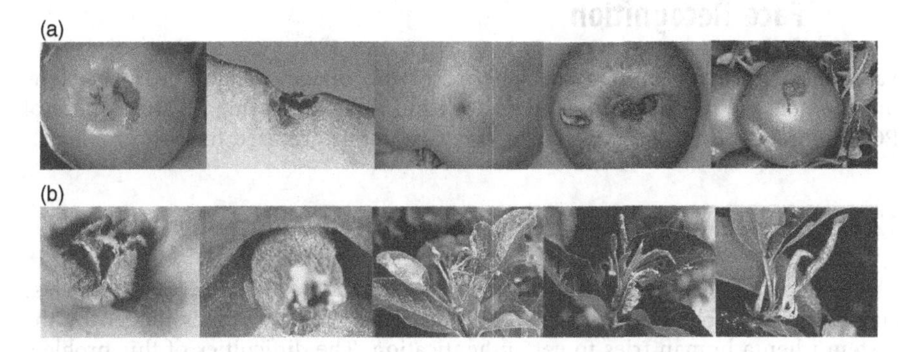

Fig. 12.10 Examples of codling moth damage: (a) on apples; (b) on leaves (Wearing, 1998).

Table 12.2 Test classification results of images of damaged fruit with the use of EFuNN, the confusion matrix over five types of pests (Woodford *et al.*, 1999).

training data							*sums*	*percent*
5 eFunns	lr = 0.0 pune = 0.1 errth = 0.01 sthr = 0.95 fr = 0.							
	alm-l	alm-f	cm	lr-l	lr-f			
am-l	9	0	0	0	0		9	100
alm-f	0	5	0	0	0		5	100
cm	0	0	22	0	0		22	100
lr-l	0	0	0	16	0		16	100
lr-f	0	0	0	0	15		15	100
						67		
sum	9	5	22	16	15		67	
percent	100	100	100	100	100			100.00
rule nodes	*61*	*61*	*61*	*62*	*61*			
test data								
	alm-l	alm-f	cm	lr-l	lr-f		sums	percent
am-l	2	0	1	1	0		4	50
alm-f	0	1	0	0	0		1	100
cm	1	1	7	1	2		12	58
lr-l	0	0	0	2	0		2	100
lr-f	0	0	1	1	2		4	50
						14		
sum	3	2	9	5	4		23	
percent	67	50	78	40	50			60.87

evolving and having a fixed structure; see Kasabov (1996)). Computing the kappa coefficient for both the FuNN and EFuNN confusion matrixes substantiates this with results of 0.10 for the FuNN and 0.45 for the EFuNN.

New images can be added to an EFuNN model in an online mode. Rules can be extracted that represent the relationship between input features, encoding damage on a fruit, and the class of pests that did the damage.

12.4 Incremental Face Membership Authentication and Face Recognition

Face image recognition is a special case of the image recognition task. Here, incremental adaptive learning and recognition in a transformed feature space of PCA and LDA using IPCA and ILDA (see Chapter 1) are used for two problems.

12.4.1 Incremental Face Authentication Based on Incremental PCA

The membership authentication by face classification is considered a two-class classification problem, in which either member or nonmember is judged by the system when a human tries to get authentication. The difficulties of this problem are as follows.

1. The size of the membership/nonmembership group is dynamically changed.
2. In as much as face images have large dimensions, the dimensional reduction must be carried out because of the limitation of the processing time.
3. The size of the membership group is usually smaller than that of nonmembership group.
4. There are few similarities within the same class.

In this research case, the first two difficulties are tackled by using the concept of incremental leaning. In real situations, only a small amount of face data are given to learn a membership authentication system at a time. However, the system must always make a decision as accurately as possible whenever the authentication is needed. To do that, the system must learn given data incrementally so as to improve the performance constantly. In this sense, we can say that membership authentication problems essentially belong to the incremental learning problem as well. However, if large-dimensional data are given to the system as its inputs, it could be faced with the following problems.

1. Face images have large dimensions and the learning may continue for a long time. Therefore, it is unrealistic to keep all or even a part of data in the memory.
2. The system does not know what data will appear in the future. Hence, it is quite difficult to determine appropriate dimensions of feature space in advance.
   The first problem can be solved by introducing one-path incremental learning. On the other hand, for the second problem, we need some method to be able to construct feature space incrementally. If we use principal component analysis (PCA) as a dimensional reduction method, incremental PCA (IPCA; Ozawa *et al.* (2004a,b, 2005a,b)), can be used. This is illustrated by Ozawa *et al.* with a model that consists of three parts.

- Incremental PCA
- ECM (online evolving clustering method; see Chapter 2)
- K-NN classifier (see Chapter 1)

In the first part, the dimensional reduction by IPCA is carried out every time a new face image (or a small batch of face images) is given to the system. In IPCA, depending on given face images, the following two operations can be carried out (see Chapter 1),

- Eigenaxes rotation
- Dimensional augmentation

When only rotation is conducted, the prototypes obtained by ECM can be easily updated; that is, we can calculate the new prototypes by multiplying it and the rotation matrix. On the other hand, if dimensional augmentation is needed, we should note that the dimensions of prototypes in ECM are also increased.

A simple way to cope with this augmentation is to define the following twofold prototypes for ECM: (p_i^N, p_i^D) $(i = 1, \cdots, P_t)$ where p_i^N and p_i^D are, respectively, the ith prototype in the N-dimensional image space and the D-dimensional eigenspace, and P_t is the number of prototypes at time t. Keeping the information on prototypes in the original image space as well as in the eigenspace, it is possible to calculate a new prototype in the augmented $(D+1)$-dimensional eigenspace exactly. Therefore, we do not have to modify the original ECM algorithm at all, except that the projection of all prototypes from the original space to the augmented $(D+1)$-dimensional eigenspace must be carried out before clustering by ECM. In the last part of the k-NN classifier, we do not need any modifications for the classifier even if the rotation and augmentation in the eigenspace are carried out, because they only calculate the distance between a query and each of the prototypes.

To evaluate the performance of the proposed incremental authentication system, we use (see Ozawa *et al.* (2005a,b;2006)) the face dataset that consists of 1355 images (271 persons, 5 images for each person). Here, 4 of 5 images are used for training and the rest are used for test. From this dataset, 5% of persons' images are randomly selected as the initial training (i.e., 56 images in total). The number of incremental stages is 51; hence, a batch of about 20 images is trained at each stage. The original images are preprocessed by wavelets, and transformed into 644-dimensional input vectors. To evaluate the average performance, fivefold cross-validation is carried out. For comparative purposes, the performance is also evaluated for the nonincremental PCA in which the eigenspace is constructed from the initial dataset and the eigenvectors with over 10% power against the cumulative eigenvalue are selected (only one eigenvector is selected in the experiment). The two results are very similar confirming the effectiveness of the incremental algorithm.

12.4.2 Incremental Face Image Learning and Classification Based on Incremental LDA

In Pang *et al.* (2005;2005a,b) a method for incremental LDA (ILDA) is proposed along with its application for incremental image learning and classification (see Chapter 1). It is shown that the performances of ILDA on a database with a large number of classes and high-dimension features are similar to the performance of batch mode learning and classification with the use of LDA. A benchmark MPEG-7 face database is used, which consists of 1355 face images of 271 persons (five different face images per person are taken), where each image has the size of 56×46. The images have been selected from AR(Purdue), AT&T, Yale, UMIST, University of Berne, and some face images obtained from MPEG-7 news videos.

An incremental learning is applied on a database having 271 classes (faces) and 2576 (56×46) dimension features, where the first 30 eigenfeatures of ILDA are taken to perform K-NN leave-one-out classification. The discriminability of ILDA, specifically when bursts of new classes are presented at different times, was evaluated and also the execution time and memory costs of ILDA with the increase of new data addition.

12.5 Online Video-Camera Operation Recognition

12.5.1 The Camera Operation Recognition Task

Advances in multimedia, communications, and computer technologies have led to widespread accessibility of video data. Applications such as digital libraries and digital video broadcast deal with large volumes of video data, which require powerful video indexing and retrieval techniques. One important issue is camera operation recognition.

At the stage of video parsing, it is critical to distinguish gradual shot transitions from the false positives due to camera operation because they both exhibit temporal variances of the same order. Detecting camera operations is also needed at the step of video indexing and retrieval. As camera operations explicitly reflect how the attention of the viewer should be directed, the clues obtained are useful for indexing and summarizing video contents (Koprinska and Carrato, 1998).

This is illustrated on a case study example below, where the EFuNN model used manifested robustness to catastrophic forgetting when new video data were added. Its performance compares favourably with other classifiers in terms of classification accuracy and learning speed.

12.5.2 A Case Study

An evolving fuzzy neural network (EFuNN; Chapter 3) is applied here for camera operation recognition based on motion vector patterns extracted from an MPEG-2 compressed stream (Koprinska and Carrato,1998). The general experimental scheme is shown in Fig. 12.11.

In the presented experiment there are six classes considered.

1. *Static:* Stationary camera and little scene motion
2. *Panning:* Camera rotation around its horizontal axis
3. *Zooming:* Focal length change of a stationary camera
4. *Object motion:* Stationary camera and large scene motion
5. *Tracking:* Moving object being tracked by a camera
6. *Dissolve:* Gradual transition between two sequences where the frames of the first one get dimmer and these of the second one get brighter

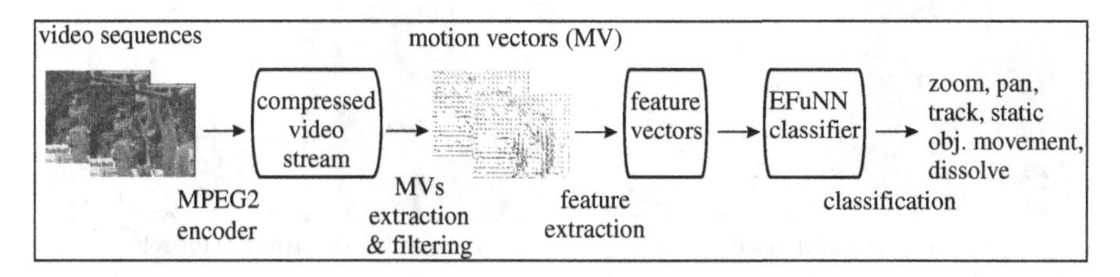

Fig. 12.11 A general framework of a video-camera operation recognition system (Koprinska and Kasabov, 2000).

Although four of these classes are camera operations, object motion and dissolve are added as they introduce false positives. Each of the classes is characterized by specific dynamic patterns represented as motion vectors (MVs) of P and B frames in a MPEG-encoded sequence. The well-known benchmark sequences Akiyo, Salesman, Miss America, Basketball, Football, Tennis, Flower Garden, and Coastguard were used in our experiments; see Fig. 12.12.

It can be seen from Fig. 12.12 that the three static examples have rather different MV fields. The frames of Akiyo can be viewed as ideal static images, however, there are occasionally sharp movements in the Salesman images.

The MV field of Miss America is completely different as the encoder used generates MVs with random orientation for the homogeneous background. Hence, it is advantageous to use a classification system capable to incrementally adapt to new representatives of the static class without the need to retrain the network on the originally used data.

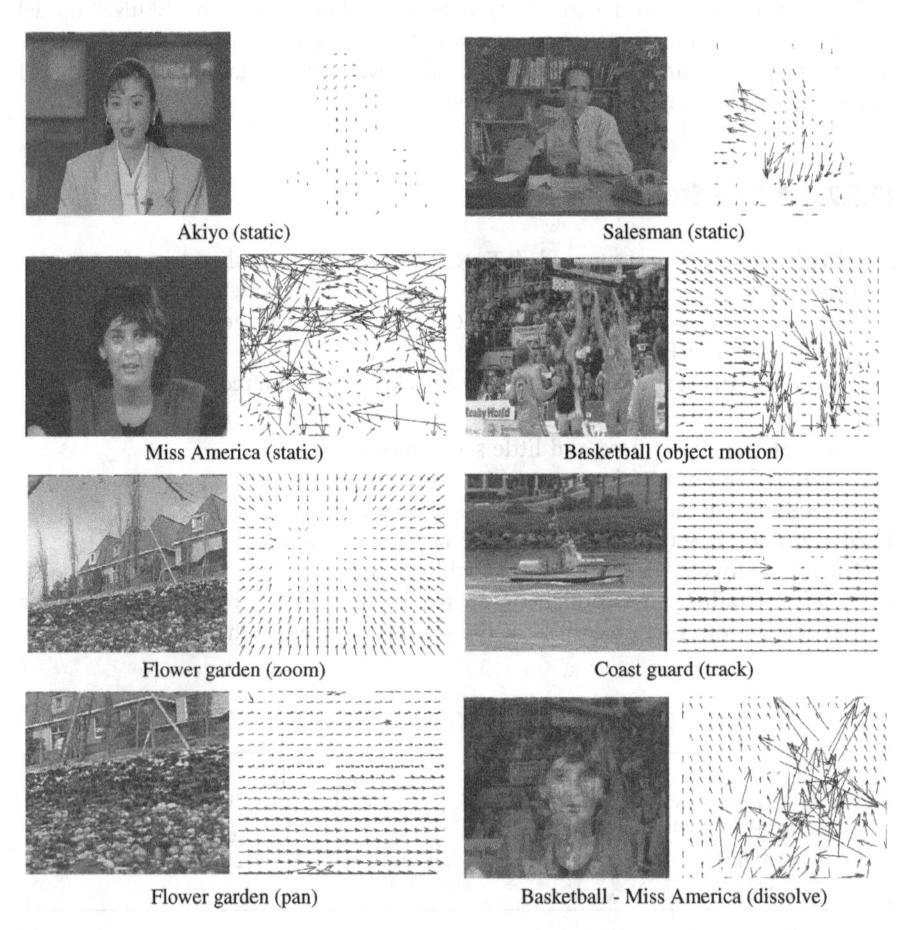

Fig. 12.12 Examples of six classes of camera operation with their motion vectors (MV; Koprinska and Carato (1998)).

Feature Extraction

Data preprocessing and feature extraction are performed as in Koprinska and Carrato (1998). Motion vectors of P and B frames are extracted and smoothed by a vector median filter. Based on them, a 22-dimensional feature vector is created for each frame. The first component is a measure for how static the frame is. It is calculated as the fraction of zero MVs, using both the forward and backward MV components. The forward MV area is then subdivided in seven vertical strips, for which the average and standard deviation of MV directions and the average of MV magnitudes are computed.

In order to build the EFuNN classifier, the MV patterns of 1200 P and B frames (200 for each class), have been visually examined and manually labelled.

Experimental Results and Discussion

The goal of the experiments was fourfold: to test the overall classification performance of EFuNN for online camera operations recognition; to analyse the individual classes detection; to find how the different number of fuzzy membership functions influences the EFuNN performance; and to assess the contribution of rule node aggregation.

For the evaluation of the EFuNN classification results, we used tenfold cross-validation. Apart from the various values for membership function *mf* and thresholds for aggregation *R*max which are discussed below, the EFuNN parameters were set as follows: initial sensitivity threshold $Sthr = 0.92$; error threshold $E = 0.08$; learning rates $lr1 = 0.05$, $lr2 = 0.01$, for the two rule node layers $W1$ and $W2$, respectively; number of examples after which rule node aggregation is applied, $N\text{agg} = 60$.

Table 12.3 shows the classification accuracy of EFuNN with a different number of membership functions when applied to video frame classification. As can be seen from the table, EFuNN achieves the best classification accuracy when four membership functions are used. Further increase in their number almost does not affect the accuracy on the training set but results in worse accuracy on unseen examples due to overtraining.

The respective number of nodes in the EFuNN models is presented in Table 12.4. The table indicates that increasing the number of the membership functions implies

Table 12.3 EFuNN classification accuracy [%] on the training and testing set for different number mf of membership functions used in the EFuNN. The maximum radius of the receptive field is equal to the aggregation thresholds; i.e. $\text{Rmax} = w1Thr = w2Thr = 0.2$ (Koprinska and Kasabov, 2000).

mf	*Acc. [%] on training set*	*Acc. [%] on testing set*
2	85.8 ± 1.5	84.5 ± 2.4
3	91.4 ± 1.1	86.8 ± 4.5
4	95.5 ± 0.6	91.6 ± 2.9
5	95.5 ± 0.4	89.3 ± 4.3
6	95.2 ± 0.9	88.6 ± 4.5

Table 12.4 Number of nodes (inputs, fuzzy input nodes, rule nodes, fuzzy output nodes, and outputs) for various EFuNN architectures for the video camera operation recognition case study. The number of rule nodes increases with the increase of membership functions mf from 2 to 6 (Koprinska and Kasabov, 2000).

Nodes	mf				
	2	3	4	5	6
Input	22	22	22	22	22
Fuzzy inp.	44	66	88	110	132
Rule	30 ± 2	101.3 ± 5.5	183.1 ± 5.5	204.9 ± 9.6	229.5 ± 7.9
Fuzzy out.	12	18	24	30	36
Output	6	6	6	6	6
Total	114	213	323	362	425

considerable growth in the number of rule nodes and, hence, the computational complexity of the EFuNN's training algorithm. As a result, learning speed slows down significantly. However, depending on the specific application, a suitable trade-off between the learning time and the accuracy can be found.

For the sake of comparison, Table 12.5 summarizes the results achieved by the method of learning vector quantisation LVQ, using the public domain package LVQ Pack. The performance of the online EFuNN model compares favourably with LVQ in terms of classification accuracy.

The EFuNN model requires only 1 epoch for training in contrast to LVQ's multipass learning algorithms that need 1520 epochs in our case study.

Table 12.6 summarises the EFuNN classification of the individual classes. It was found that although 'zoom' and 'pan' are easily identified, the recognition of object movement, tracking, and dissolve are more difficult. Despite the fact that the MV

Table 12.5 LVQ performance on the video camera operation recognition case study (Koprinska and Kasabov, 2000).

Accuracy [%] on training set	*Accuracy [%] on testing set*	*Nodes (input & codebook)*	*Training epochs*
85.4 ± 2.5	85.8 ± 2.2	60 (22 inp., 38 cod.)	1520

Table 12.6 EFuNN classification accuracy in [%] of the individual video classes for the video camera operation recognition case study (see Koprinska and Kasabov (2000)). Membership functions are between $mf = 1$ and $mf = 6$. Optimum performance of the EFuNN classifier is achieved when for each individual class a specific number of mf is used (Koprinska and Kasabov, 2000).

mf	*Zoom*	*Pan*	*Object motion.*	*Static*	*Tracking*	*Dissolve*
2	100	97.2 ± 5.4	74.6 ± 12.0	95.0 ± 6.6	75.9 ± 15.6	62.1 ± 14.8
3	100	92.8 ± 4.1	78.1 ± 12.7	97.8 ± 4.4	72.9 ± 20.6	77.3 ± 14.7
4	100	97.4 ± 2.8	88.1 ± 10.7	98.1 ± 2.5	83.0 ± 11.5	84.0 ± 10.8
5	100	99.0 ± 2.1	85.0 ± 9.2	97.7 ± 3.1	69.7 ± 21.5	85.2 ± 9.5
6	100	94.3 ± 5.2	88.4 ± 9.7	97.7 ± 3.3	63.5 ± 18.8	87.6 ± 8.0

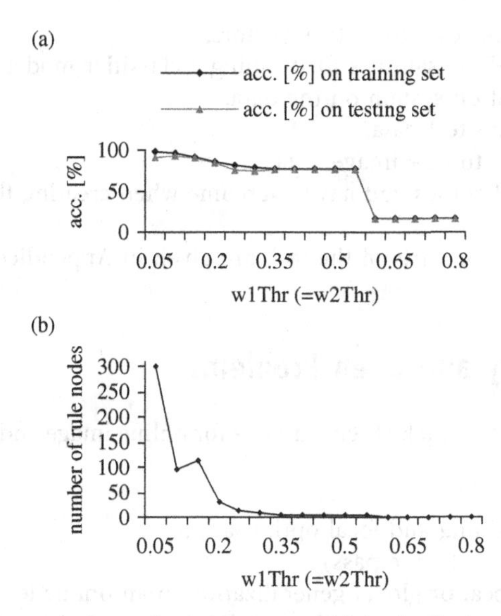

Fig. 12.13 Impact of the aggregation on the evolving systems: (a) the classification accuracy drops significantly if the aggregation threshold is above 0.5; (b) the number of rule nodes drops significantly for an aggregation threshold above 0.15. The number of membership functions is $mf = 2$ (Koprinska and Kasabov, 2000).

fields of Miss America were not typical for static videos and complicated learning, they are learned incrementally and classified correctly in the EFuNN system.

Figures 12.13a,b show the impact of aggregation on the classification accuracy and on the number of rule nodes, respectively. Again, tenfold cross-validation was applied and each bullet represents the mean value for the ten runs. As expected, the results demonstrate that the aggregation is an important factor for achieving good generalization and keeping the EFuNN architecture at a reasonable size. The best performance in terms of a good trade-off between the recognition accuracy and net size was obtained for $R\max = 0.15$, and 0.2. When the aggregation parameter $R\max$ is between 0.25 and 0.55, the accuracy on the testing set drops with about 10% as the number of rule nodes becomes insufficient. Further increase in the values of the aggregation coefficients results in networks with one rule node which obviously cannot be expected to generalise well.

In conclusion of this section, an application of EFuNN for camera operation recognition was presented. EFuNN learns from examples in the form of motion vector patterns extracted from MPEG-2 streams. The success of the EFuNN model can be summarised as high classification accuracy and fast training.

12.6 Exercise

Task specification: *A small EIS for image analysis and adaptive image recognition*

Steps:

1. Record or download images related to two categories (e.g. male versus female, face versus no-face, cancer cell versus noncancer cell, damaged object versus normal, etc.).

2. Transform the images into feature vectors.
3. Prepare and label the samples for training a classifier model /system.
4. Train a classification system on the data.
5. Test the system on test data.
6. Adapt the system to new image data.
7. Explain what difficulties you have overcome when creating the system.

Exemplar programs for parts of the task are given in Appendices B and C.

12.7 Summary and Open Problems

ECOS have features that make them suitable for online image and video processing. These features are:

- Local element training and local optimisation
- Fast learning (possibly one pass)
- Achieving high local or global generalization in an online learning mode
- Memorising exemplars for a further retrieval or for a system's improvement
- Interpretation of the ECOS structure as a set of fuzzy rules
- Dynamic self-organisation achieved through growing and pruning

Some open problems in this research area are:

1. Identifying artists by their paintings (see Herik and Postma (2000))
2. Combining online image analysis with online analysis of other modalities of information
3. Other online image feature selection methods, in addition to incremental LDA (Pang *et al.*, 2005a,b) and incremental PCA (Ozawa *et al.*, 2005a,b, 2006).

12.8 Further Reading

For further details of the material presented in this chapter, the reader may refer to Deng and Kasabov (2003), Kasabov *et al.* (2000a,b), Koprinska and Kasabov (2000), and Ozawa *et al.* (2005a,b, 2006).

Further material on the issues discussed in this chapter can be found as follows.

- *Adaptive Pattern Recognition* (Pao, 1989)
- *Image Colour Quantisation* (Heckbert, 1982; Gervautz and Purghatofer, 1990; Wu, 1992)
- *Neuro-fuzzy Methods for Image Analysis* (Pal *et al.*, 2000; Ray and Ghoshal, 1997; Szu and Hsu, 1999)
- *Image Classification* (Hirota, 1984)
- *Image Segmentation* (Bezdek *et al.*, 1993)
- *Recognition of Facial Expression using Fuzzy Logic* (Ralescu and Iwamoto, 1993)
- *Applying Wavelets in Image Database Retrieval* (Wang *et al.*, 1996)

- *Classification of Satellite Images with the Use of Neuro-fuzzy Systems* (Israel *et al.*, 1996)
- *MPEG Image Transformations* (Koprinska and Carrato, 1998)
- *Image Analysis of Damaged Fruit* (Wearing, 1998; Woodford *et al.*, 1999)
- *Incremental Image Learning and Classification* (Ozawa *et al.*, 2005a,b)

13. *Evolving Intelligent Systems for Adaptive Multimodal Information Processing*

The chapter presents a general framework of an EIS in which auditory and visual information are integrated. The framework allows for learning, adaptation, knowledge discovery, and decision making. Applications of the framework are: a person-identification system in which face and voice recognition are combined in one system, and a person verification system. Experiments are performed using visual and auditory dynamic features which are extracted in a synchronised way from visual and auditory data. The experimental results support the hypothesis that the recognition rate is considerably enhanced by combining visual and auditory features.

The chapter is presented in the following sections.

- Multimodal information processing
- A framework for adaptive integration of auditory and visual information
- Person identification based on auditory and visual information
- Person verification based on auditory and visual information
- Summary and open problems
- Further reading

13.1 Multimodal Information Processing

Many processes of perception and cognition are multimodal, involving auditory, visual, tactile, and other types of information processing. All these processes are extremely difficult to model without having a flexible, multimodular evolving system in place. Some of these modalities are smoothly added at a later stage of the development of a system without the need to 'reset' the whole system.

Information from different modalities can support the performance of a computer system originally designed for a task with a unimodal nature. Thus, a system for speech recognition may benefit from an additional visual information stream. For instance, visual information from the lips and the eyes of a speaker improves the spoken-language recognition rate of a speech-recognition system substantially. The improvement per se is already obvious from the use of two sources of information (i.e. sound and images) in one system.

Research on multimodal speech-recognition systems has already shown promising results. A notable example is the successful recognition of words pronounced in a noisy environment, i.e. the 'cocktail party problem' (also known as the 'source separation problem'; Massaro and Cohen (1983)). The additional visual information can also be used for solving important problems in the area of spoken-language recognition, such as the segmentation of words from continuous speech and the adaptation to new speakers or to new accents.

Conversely, image information, auditory information, and textual input (possibly synchronised with the image and the auditory signal) can be used to enhance the recognition of objects, for instance, the identification of moving objects based on information coming from their blurred images and their sounds. Obviously, the auditory information does not have to be speech or sound within the audible spectrum of human perceivers. It could also be a signal characterised by its frequency, time, and intensity (e.g. the echolocation of dolphins). The main question is how much auditory or textual input information is required in order to support or improve an image-recognition process significantly. A second derived question is how to synchronise two (or more) flows of information in a multimodal computer system.

Integrating auditory and visual information in one system requires the following four questions to be addressed.

- Auditory and visual information processing are both multilevel and hierarchical (ranging from an elementary feature extraction level up to a conceptual level). Therefore, at which level and to what degree should the two information processes be integrated? One model for such integration was discussed in Chapter 9 (see Fig. 9.24).
- How should time be represented in an integrated audiovisual information-processing system? This is a problem related to the synchronisation of two flows of information. There could be different scales of integration, e.g. milliseconds, seconds, minutes, hours.
- How should adaptive learning be realised in an integrated audiovisual information-processing system? Should the system adapt each of its modules dependent on the information processing in the other modalities?
- How should new knowledge (e.g. new rules) be acquired about the auditory and the visual inputs of the real world?

This chapter describes a general framework for integrating auditory and visual information to answer these questions. The application of the framework is illustrated on a person identification task involving audiovisual inputs.

13.2 Adaptive, Integrated, Auditory and Visual Information Processing

13.2.1 A Framework of EIS for Multimodal Auditory and Visual Information Processing

Below we describe a connectionist framework for auditory and visual information-processing systems, abbreviated as AVIS. The architecture of AVIS is illustrated in Fig. 13.1 and consists of three subsystems: an auditory subsystem, a visual

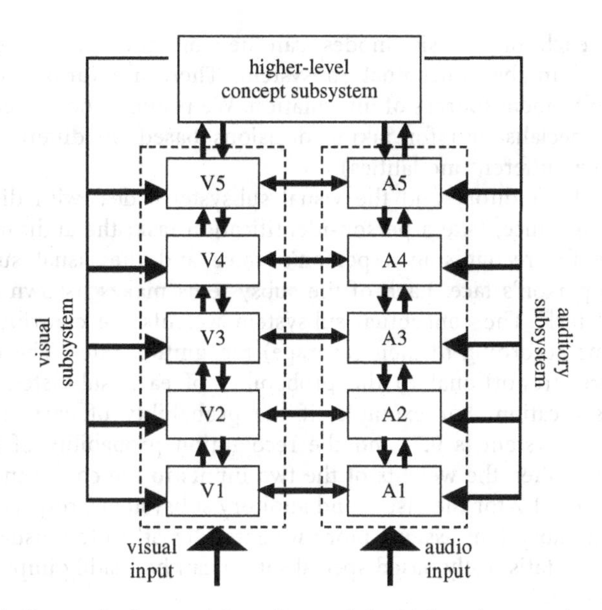

Fig. 13.1 A block diagram of a framework for auditory and visual information processing systems (AVIS; see Kasabov *et al.* (2000c)).

subsystem, and a higher-level conceptual subsystem. Each of them is specified below, followed by a description of the modes of operation.

13.2.2 Modes of Operation

AVIS allows an auditory subsystem, as well as a visual subsystem, to operate either as a separate subsystem, or in concert. Their distinct outputs will be combined in a higher-level subsystem. In addition, each subsystem in isolation is able to accommodate both unimodal and bimodal input streams. Altogether, AVIS can operate in six main modes of operation:

- *Unimodal auditory mode:* The auditory subsystem processes auditory input only (e.g. spoken language recognition from speech).
- *Cross-modal auditory mode:* The auditory subsystem processes visual input only (e.g. speech recognition from lip movements).
- *Bimodal auditory mode:* The auditory subsystem processes both visual and auditory inputs (e.g. spoken language recognition from speech and lip movement).
- *Unimodal visual mode:* The visual subsystem processes visual input only (e.g. face recognition).
- *Cross-modal visual mode:* The visual subsystem processes auditory input only (e.g. an image-recognition system trained on audiovisual inputs recalls images from their associated sounds).
- *Bimodal visual mode:* The visual subsystem processes both visual and auditory inputs (e.g. recognising a speaker by his or her speech and face).

Furthermore, each of the six modes can be combined with the conceptual processing level in the conceptual subsystem. There are various strategies for combining multimodal sources of information. We propose the principle of statistically based specialisation for taking decisions based on different sources of information (i.e. different modalities).

In general, the auditory and the visual subsystems deal with different parts of a task. For instance, take a person-identification task; the auditory subsystem is responsible for recognising a person's voice and the visual subsystem for recognising a person's face. Each of the subsystems makes its own contribution to the overall task. The conceptual subsystem weights the contributions of the two subsystems according to their (average) recognition rates. The weights have values that are proportional to the probability of each subsystem to produce a correct classification. For example, if the probability of correct recognition of the visual subsystem is 0.7, and the recognition probability of the auditory subsystem is 0.5, then the weights of the two inputs to the conceptual subsystem are 0.7/1.2 and 0.5/1.2 for the visual and auditory subsystems, respectively. Hence, the conceptual subsystem assigns more weighted 'trust' to the visual subsystem. The principle of statistically based specialisation can be readily implemented in a connectionist way.

13.3 Adaptive Person Identification Based on Integrated Auditory and Visual Information

The AVIS framework can be applied to tasks involving audiovisual data. As an example of the AVIS framework, here an exemplar application system for person identification based on auditory and visual information is presented, abbreviated as PIAVI. The system identifies moving persons from dynamic audiovisual information.

13.3.1 System Architecture

The global structure of PIAVI resembles the structure of AVIS. However, in PIAVI the auditory and visual subsystems each consist of single modules. Each of the subsystems is responsible for a modality-specific subtask of the person-identification task. The visual subsystem processes visual data associated with the person, i.e. lip-reading, or facial expressions. The inputs to this subsystem are raw visual signals. These signals are preprocessed, e.g. by normalising or edge-enhancing the input image. Further processing subserves the visual identification of the person's face. The auditory subsystem of PIAVI comprises the processing stages required for recognising a person by speech. The inputs of the subsystem are raw auditory signals. These signals are preprocessed, i.e. transformed into frequency features, such as Mel-scale coefficients, and further processed to generate an output suitable for identification of a person by speech. The conceptual subsystem combines inputs from the two subsystems to make a final decision.

13.3.2 Modes of Operation

PIAVI has four modes of operation, briefly described below.

1. *Unimodal visual mode* takes visual information as input (e.g. a face), and classifies it. The classification result is passed to the conceptual subsystem for identification.
2. *Unimodal auditory mode* deals with the task of voice recognition. The classification result is passed to the conceptual subsystem for identification.
3. *Bimodal (or early-integration) mode* combines the bimodal and cross-modal modes of AVIS by merging auditory and visual information into a single (multimodal) subsystem for person identification.
4. *Combined mode* synthesises the results of all three modes. The three classification results are fed into the conceptual subsystem for person identification.

In two case studies we examine the above modes of processing of audiovisual information in PIAVI. The first case study consists of a preliminary investigation using a small dataset with the aim of assessing the beneficial effects of integrating auditory and visual information streams at an early locus of processing. The second case study employs a larger dataset to evaluate the relative efficiencies of unimodal and bimodal processing in solving the person-identification task.

13.3.3 Case Study 1

The first case study aims at evaluating the added value of combining auditory and visual signals in a person-identification task. An additional goal is to assess the complexity of the task of identifying persons from dynamic auditory and visual input.

Given the goals of the study, the dataset has to fulfil two requirements. First, it should contain multiple persons. Second, the persons contained in the dataset should be audible and visible simultaneously. To meet these two requirements, we downloaded a digital video containing small fragments of four American talk-show hosts, from CNN's Web site. The movie contains visual frames accompanied by an audio track. Each frame lasts approximately 125 milliseconds. During most of the frames, the hosts are both visible and audible. The dataset is created as follows. Twenty suitable frames, i.e. frames containing both visual and auditory information, are selected for each of the four persons (hosts). The visual and auditory features are extracted from 2.5-second fragments (20 frames).

Feature Extraction

Person recognition relies on an integration of auditory and visual data. Although static images may suffice for person recognition, in our study we rely on dynamic visual information for two reasons in particular. First, dynamic features avoid recognition on the basis of unreliable properties, such as the accidental colour of the skin or the overall level of lighting. Second, the added value of integrating auditory and visual information at an early level lies in their joint temporal variation.

Our emphasis on dynamical aspects implies that the integration of auditory and visual information requires an extended period of time. The duration required for integration varies depending on the type of audiovisual event. For early integration, a duration of about 100 milliseconds may suffice when short-duration visual events (e.g. the appearance of a light) are to be coupled to short-duration auditory events (e.g. a sound). However, when dynamical visual events such as face and lip movements are to be coupled to speech, a duration of at least half a second is required. To accommodate early integration, we defined aggregate features encompassing the full duration (i.e. 125 milliseconds) of the video segments for both modalities.

Visual Features

The images (i.e. frames of video data) contained in each segment need to be transformed into a representation of the spatiotemporal dynamics of a person's head. It is well known that spatiotemporal features are important for person-identification tasks. Behavioural studies show that facial expressions, potentially person-specific, flicker rapidly across the face within a few hundred milliseconds (in Stork and Hennecke (1996)). Because the head moves in several directions during a segment, a means of compensating for these variations in a three-dimensional pose is required. Moreover, the head should be segmented from the background to remove background noise. To fulfil these requirements, we used a straightforward spatial-selection method. A single initial template was defined for each person in the dataset. The size of the template was set at $M \times N$ pixels, with $M = 15$ and $N = 7$. The templates intended to cover the entire head. The content of each template was cross-correlated with the content of the next video frame. The best-matching $M \times N$ part of the next frame served as a starting point for the extraction of visual features and was defined as the new template.

A commonly used technique for extracting features from images is based on principal component analysis (PCA). For instance, in their lip-reading studies Luettin *et al.* (1996) employed PCA on the visual lip-shape data. However, a comparative study of (dynamic) features for speech-reading showed that a 'delta' representation, based on the differences in grey values between successive frames, works better than a representation based on PCA. For this reason we used the delta representation to generate our visual features.

The visual features were obtained as follows. The absolute values of the changes of subsequent frames yielded the elements of a delta image, defined as

$$\Delta(x, y) = I(t+1, x, y) - I(t, x, y) \tag{13.1}$$

with $I(t, x, y)$ the grey value of the pixel at co-ordinate (x, y) of frame t (t represents the frame number).

Auditory Features

The audio signal is transformed into the frequency domain using standard FFT (256 points; sampling rate 11 kHz; one channel, one byte accuracy) in combination with a Hamming window, yielding a sequence of vectors containing 26 Mel-scale coefficients. Each vector represents an audio segment of 11.6 milliseconds, with 50% overlap between the segments. The Mel-scale vectors averaged over a duration of approximately 125 milliseconds are represented as 'audio frames' (i.e. vectors containing averaged Mel-scale coefficients). By subtracting subsequent frames, a delta representation is obtained that corresponds to the visual delta representation. The auditory features are vectors containing three delta representations obtained at three different time lags, each of 125 msec, with an overlapping of 50%.

Modelling the Subsystems

The subsystems of PIAVI are modelled using evolving fuzzy neural networks (EFuNNs; see Chapter 3). Each of the input and output nodes of an EFuNN has a semantic meaning. EFuNNs are designed to facilitate the use of both data and fuzzy rules in a connectionist framework. They allow for easy adaptation, modification, rule insertion, and rule extraction. The unimodal visual mode of operation is modelled as an EFuNN with 105 input nodes, 315 input membership functions (3 per input node, i.e. representing the fuzzy representations 'small', 'medium', and 'high'), 4 output nodes (for the four hosts), and 8 output membership functions (2 per output, i.e. representing the fuzzy classifications 'unlikely' and 'likely'). The unimodal auditory mode of operation is modelled as an EFuNN with 78 input nodes, 234 input membership functions, 4 output nodes, and 8 output membership functions. The bimodal, early-integration mode of operation is modelled by an EFuNN with the same dimensions except for the input. There are 183 input nodes (105 for visual features plus 78 for auditory features) and 549 membership functions. Finally, in the combined mode of operation the two unimodal modes and the bimodal mode are combined. The higher-level concept subsystem is modelled according to the principle of statistically based specialisation. The criterion for classification is as follows. The output node with the largest activation defines the class assigned to the input pattern.

Experimental Procedure

The EFuNNs assembling PIAVI are trained. To assess the generalisation performance, the dataset is split into a training set and a test set, each containing ten examples (ten frames of 125 msec each). The EFuNNs are trained on the training set. The generalisation performance is defined as the classification performance on the test set.

Results

The overall recognition rate achieved in the combined mode is 22% higher than the recognition rate in the unimodal visual mode, 17% higher than the recognition rate in the unimodal auditory mode, and 4% higher than the recognition rate in the early-integration mode of operation.

Figure 13.2 shows the results obtained in Case Study 1. The test frames are shown on the x-axis (first ten for person one, etc.). The output activation values of the four outputs are shown on the y-axis.

The experimental results confirm the main hypothesis of this research, that the AVIS framework and its realisation PIAVI achieve a better performance when auditory and visual information are integrated and processed together.

13.3.4 Case Study 2

The second case study attempts to improve and extend the results obtained in the first case study by employing a larger (more realistic) dataset and by defining aggregate features representing longer temporal intervals.

The Dataset

We (see Kasabov *et al.* (2000c)) recorded CNN broadcasts of eight fully visible and audibly speaking presenters of sport and news programs. An example is given in Fig. 13.3. All recordings were captured in a digital format. The digital video files so obtained were edited with a standard video editor. This yielded

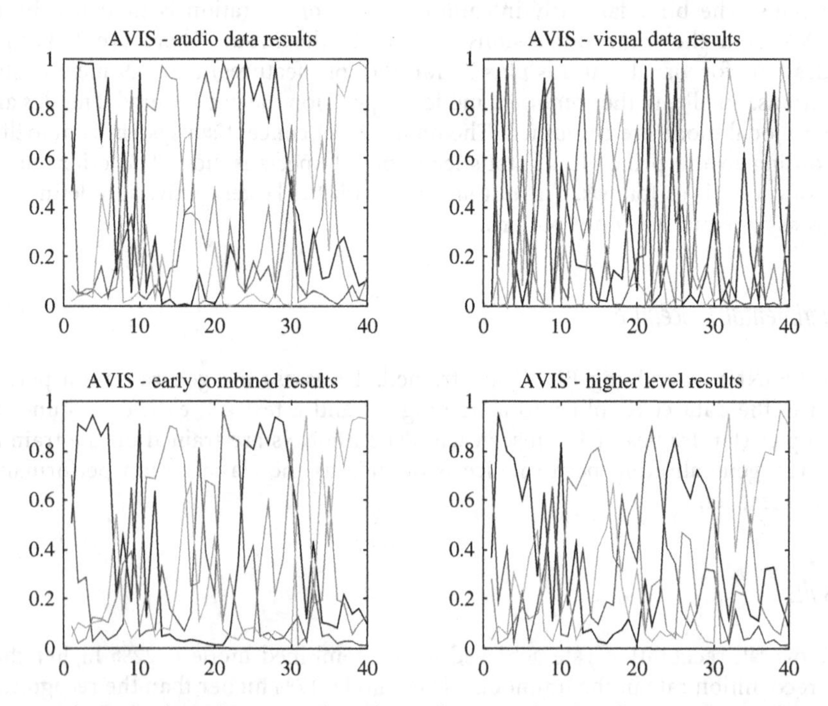

Fig. 13.2 The results obtained in Case Study 1. The test frames are shown on the x-axis (first ten for person one, etc.). The output activation values are shown on the y-axis for the four outputs (one for each person).

Fig. 13.3 An example of a frame used in the dataset (Kasabov *et al.*, 2000c).

video segments of one second length at $F = 15$ frames per second. Each segment contains $F + 1$ frames. The visual and auditory features were extracted from these segments.

Visual Features

As in the first case study, the $F + 1$ images (i.e. the frames of video data) contained in each segment were transformed into a representation of the spatiotemporal dynamics of a person's face. The extraction of visual features in this case study differed in three respects from the extraction in the first case study. First, in segmenting the face from the background, a fixed template was used for each person, instead of redefining the template with each new frame. The size of the template was defined as $M \times N$ pixels, with $M = 40$ and $N = 20$. Figure 13.4 shows the face template used for the person displayed in Figure 13.3.

Second, the temporal extent of the aggregate features is extended over a one second period to accommodate temporal variations over a longer interval. The aggregate features were obtained as follows. The absolute values of the changes

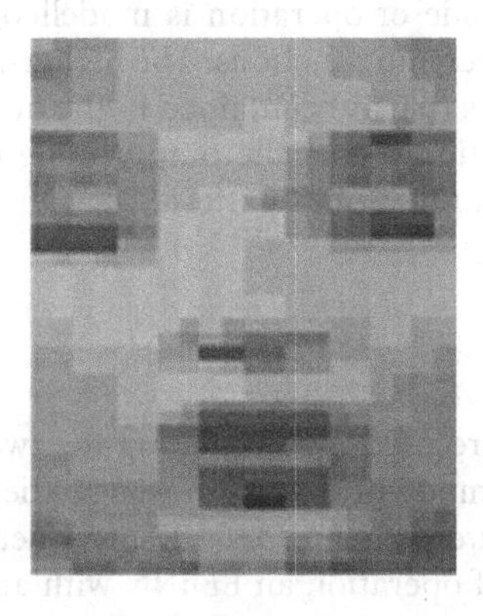

Fig. 13.4 The face template used for video frames, applied on the speaker video data from Fig. 13.3.

for a one second period (i.e. $F+1$ frames) are summed pixelwise, yielding an average-delta image, the elements of which are defined as

$$\Delta(x,y) = \sum_{i=1,F} |I(t+1, x, y) - I(t,x,y)| / F \tag{13.2}$$

with $I(t,x,y)$ the colour value of the pixel at co-ordinate (x,y) of frame t (t represents the frame number).

Third, a compressed representation of the delta image is used instead of a representation based on all pixel values. The final aggregate visual features are contained in a vector v, the elements of which are the summed row values and the summed column values of the average-delta image. Formally, the elements $v(i)$ of v are defined as

$$v(i) = \sum_{j=1,N} \Delta(j,i), \text{ for } 1 <= i =< M \tag{13.3a}$$

and

$$v = \sum_{i=1,M} v(i) \tag{13.3b}$$

Auditory Features

The auditory features are extracted according to the procedure described in the first case study, except for the aggregated features which are obtained by averaging over a one second interval.

Modelling the Subsystems

The unimodal visual mode of operation is modelled in an EFuNN with 60 ($N+M$) input nodes, 180 input membership functions, and 16 output membership functions. The bimodal mode of operation is modelled using an EFuNN with 86 input nodes and 258 membership functions. The higher-level decision subsystem is not modelled explicitly. It is (partly) contained in the output layers of the EFuNNs. The criterion for classification is that the output node with the largest activation defines the class assigned to the input pattern.

Experimental Procedure

The experimental procedure used for simulating the two modes of operation was as follows. The unimodal mode of operation was studied by presenting the visual examples to the appropriately dimensioned EFuNN network (Experiments 1 and 2). In the bimodal mode of operation an EFuNN with an extended input was used to accommodate the audiovisual input pattern (Experiment 3).

Experiment 1: Unimodal Processing

To assess the performance of PIAVI in the unimodal mode of operation the visual subsystem was trained on a training set that contained 385 examples (corresponding to a total of 385 seconds of video playing time). The test set contained 100 examples.

Experiment 2: Unimodal Processing with a Small Training Set

The setting is as in Experiment 1, but here a small training set is used (5 examples per class). The test set contains 25 examples per class, except for the class corresponding to speaker one which contains 5 examples only. An overall generalisation performance of 75% correct classification was obtained.

Experiment 3: Bimodal Processing

To assess the performance of PIAVI in the bimodal mode of operation, 40 early integrated audiovisual examples (5 per class) each of them having 86-feature were presented to an EFuNN. An overall generalisation performance of 91% correct classification on the test set was obtained after 1000 epochs (150 seconds of simulation time).

Considerations are corroborated by the graphs in Fig. 13.5 displaying the values $\nu(i)$ as a function of i for ten examples of a single class. The peaks at $i = 15$ and $i = 30$ correspond to the dynamics of the eyes and mouth during a one second interval.

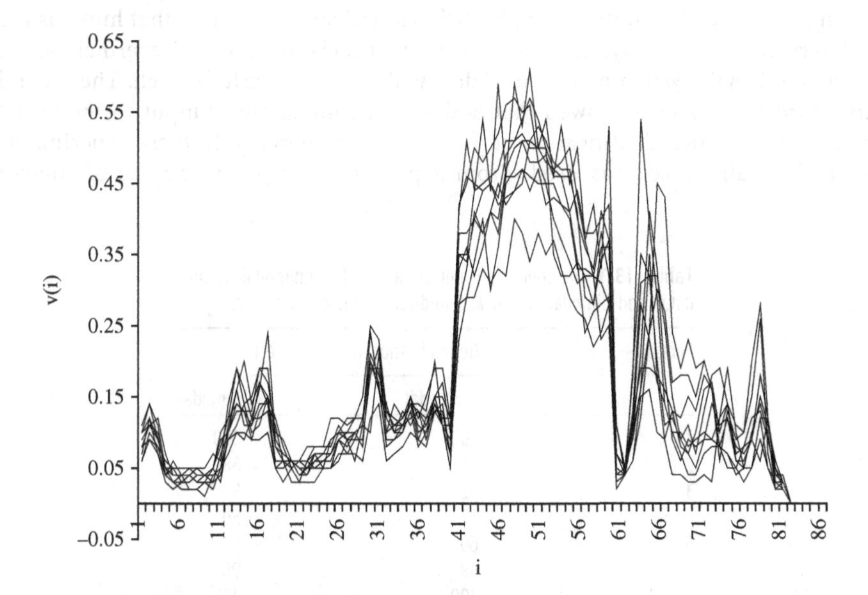

Fig. 13.5 Graphs showing dynamic, integrated visual (first 60) and audio (last 26) features $\nu(i)$ as a function of i for ten examples of one class (Kasabov *et al.*, 2000c).

Combination of the visual and auditory data in Experiment 3 yielded a major improvement in generalisation performance (Table 13.1).

The contribution of the nonlinear EFuNNs to the results becomes evident by considering the generalisation performances obtained with standard linear statistical methods such as linear discriminant functions. Figure 13.6 displays a plot of all auditory examples mapped on the first two discriminant functions. Figure 13.7 shows the same plot for all visual examples. The mapping of the audio-visual examples is displayed in Fig. 13.8. A subset of these examples is plotted in Fig. 13.9 as a two-dimensional configuration obtained with a multidimensional scaling procedure. The low-dimensional configuration represents the high-dimensional configuration in audiovisual feature space by preserving the inter-point distances as much as possible. The generalisation performance obtained in the discriminant-function models (measured using the leaving-one-out procedure) are 37.3%, 64.1%, and 66.4%, for the auditory, visual, and audiovisual examples, respectively. Evidently, the EFuNNs contribute significantly to the generalisation performances obtained in this case study.

13.3.5 Discussion

The results of the two case studies prove the added value of integrating auditory and visual information for person identification. In the first case study, which used a small training set, combining the auditory and visual information enhanced the generalisation performance. In the second case study, the first experiment showed that with a large number of training examples unimodal processing on the basis of dynamic visual features leads to a perfect performance on a large dataset. This finding is interesting in its own right. Behavioural studies suggest that humans are not very good at identifying persons from their facial dynamics. Nevertheless, the unimodal PIAVI system managed to deal with this task perfectly well. The second and third experiments showed that adding dynamic auditory input to the visual input enhances the identification performance considerably. In these experiments a smaller training set was used. From a practical viewpoint, the use of smaller

Table 13.1 Generalisation performances for unimodal (visual data) and bimodal (visual and auditory data) experiments.

Person	Generalisation performance (%)	
	Unimodal	Bimodal
1	100	100
2	28	88
3	92	96
4	72	80
5	60	64
6	48	96
7	100	100
8	96	100

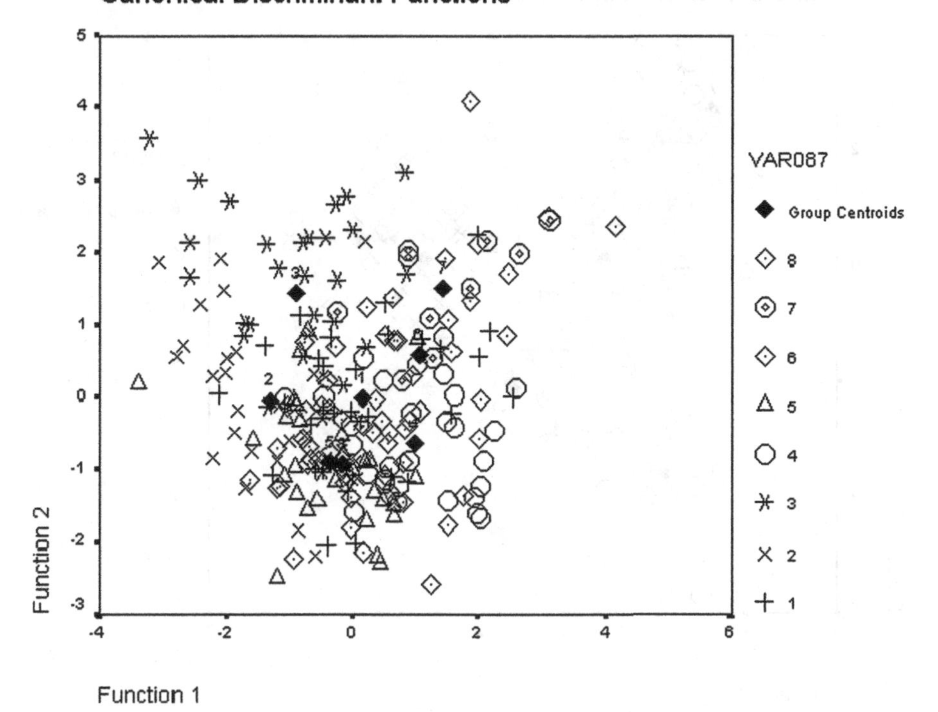

Fig. 13.6 Mapping the auditory dataset on the first two linear discriminant functions (Kasabov *et al.*, 2000c).

training sets facilitates the speed at which the PIAVI system in its bimodal mode of operation learns to classify persons from video data.

13.4 Person Verification Based on Auditory and Visual Information

13.4.1 Problem Definition

Biometric verification can be defined as a process of uniquely identifying a person by evaluating one or more distinguishing biological traits. Unique identifiers include fingerprints, hand geometry, retina, iris patterns, face image, and voice. There are many biometric features that distinguish individuals from each other, thus many different sensing modalities have been developed (Brunelli and Falavigna, 1995). These identifiers may be used individually, as exemplified by the iris scan system deployed in the banking sector and currently being tested for airport security (Luettin *et al.*, 1996).

Over the past few years, interest has been growing in the use of multiple modalities to solve automatic person verification problems. The motivation for using multiple modalities is multifold. In the first instance different modalities

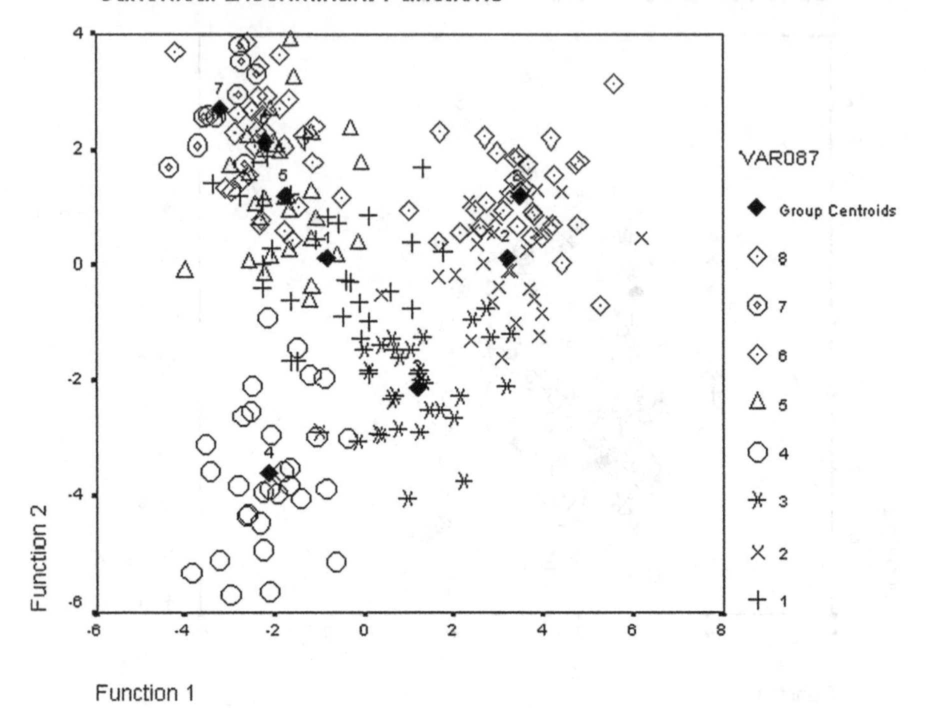

Fig. 13.7 Mapping the visual dataset on the first two linear discriminant functions (Kasabov *et al.*, 2000c).

measure complementary information and by this virtue multimodal systems can achieve better performance than single modalities. Single features may fail to be exact enough for identification of individuals.

A person-verification system is essentially a two-class decision task where the system can make two types of errors. The first error is a false acceptance, where an impostor is accepted. The second error is false rejection, where a true claimant is rejected. False acceptance rate (FAR) and false rejection rate (FRR) are calculated according to the following equations,

$$FAR = \frac{I_A}{I_T} \quad (1) \qquad FRR = \frac{C_R}{C_T} \tag{13.4}$$

where I_A is the number of impostors classified as true claimants, I_T is the total number of impostors presented, C_R is the number of true claimants classified as impostors, and C_T is the total number of true claimants presented. The trade-offs among these errors are adjusted using an acceptance threshold.

In Ghobabklou *et al.* (2004) a multimodal ECOS is proposed for person verification, based on speech and face image integrated features. This system is described below.

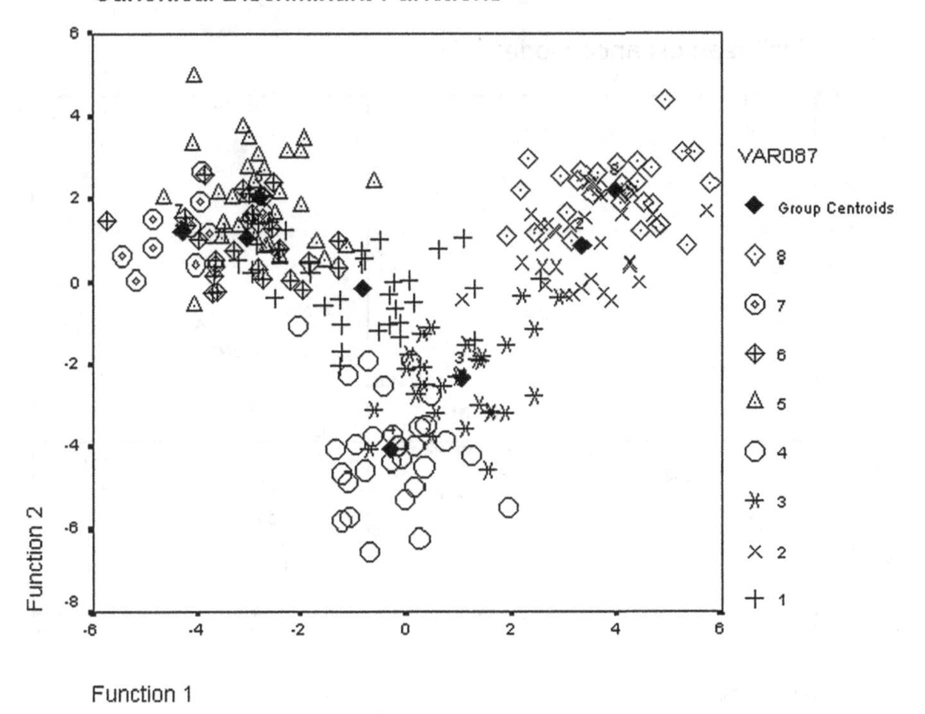

Fig. 13.8 Mapping the integrated visual-auditory dataset on the first two linear discriminant functions (Kasabov *et al.*, 2000c).

13.4.2 Adaptive Multimodal Person Verification

Here we use an implementation of a simple ECOS, called the evolving classifier function (ECF; see Chapter 3). The ECF algorithm classifies input data into a number of classes and finds their class centres in the n-dimensional input space by 'placing' a rule node in the evolving layer. Each rule node is associated with a class and an influence (receptive) field representing a part of the n-dimensional space around the rule node. Generally such an influence field in the n-dimensional space is a hypersphere. Essentially each person is modelled by a number of rule nodes that represent this person. Here the recall algorithm of ECF was modified for the task of person verification. Accordingly we call it the verification algorithm. The verification algorithm consists of the following steps.

- With the trained ECF module, when a new test sample^I is presented, first it is checked to see whether it falls within the influence field of the rule nodes representing the claimed identity of the sample^I. This is achieved by calculating the Euclidean distance between this sample and appropriate rule nodes, then comparing this distance D_i with the corresponding influence field Inf_i. The sample ^{I}is verified as person i if the relation (3.5) is satisfied.

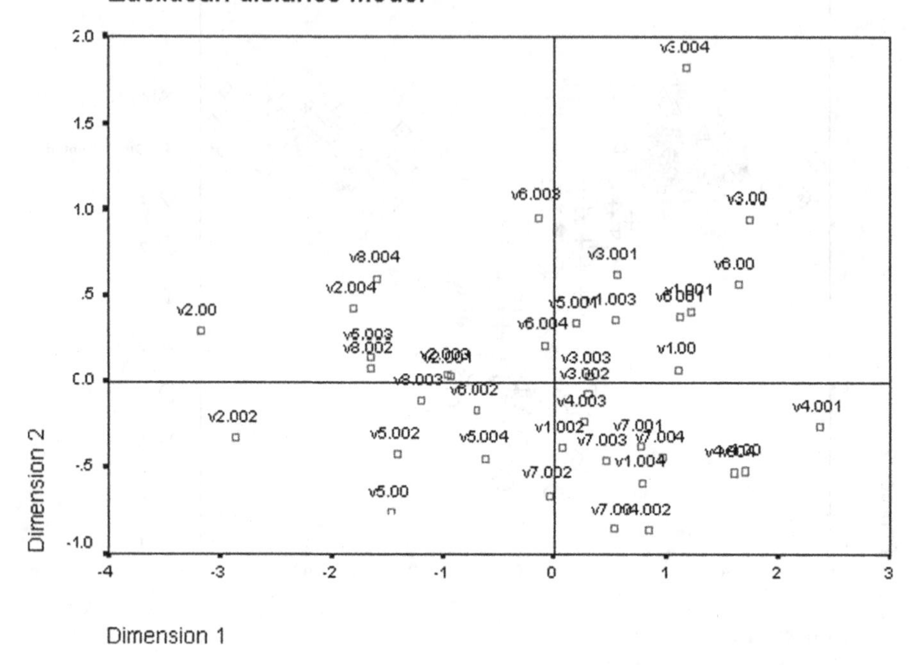

Fig. 13.9 Multidimensional scaling map of the Euclidean distance in the audiovisual space into two dimensions (*vp*.00*n* means the vector of the *n*th example of person *p*; only five examples per person have been mapped) (Kasabov *et al.*, 2000c).

$$D_i <= \mathrm{Inf}_i \tag{13.5}$$

- If the sample I does not fall in the influence field of any existing rule node,
  - Find the rule node which has the shortest distance to this sample; note this distance as $D_{\min}$.
  - If this distance $D_{\min}$ is less than a preset acceptance threshold θ, the sample I is verified as person i. Otherwise, this sample is rejected by this verification module.

This verification algorithm was applied to speaker face image, and integrated verification modules. Figure 13.10 illustrates the overall process of an adaptive person-verification system.

13.4.3 Integrated Speech and Image Feature Selection

Features obtained from speech and face images of persons were merged to form integrated input features. There are various strategies of combining multimodal

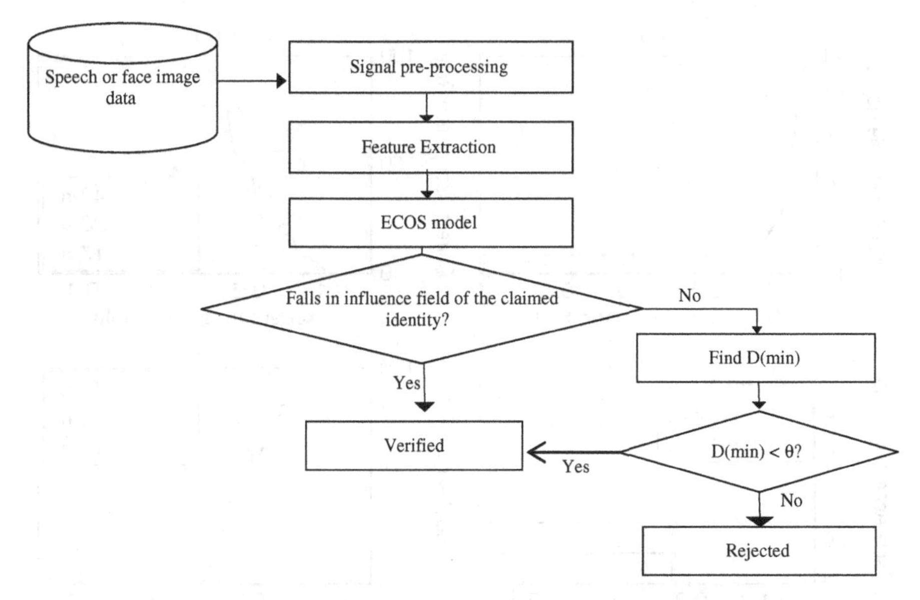

Fig. 13.10 Overall view of an adaptive connectionist person-verification system.

sources of information. In this approach, speech and face image information were integrated at the feature level. In a particular implementation, there were 100 input features in a speech sample and 64 input features in a face image sample. These two sets of features were concatenated to form integrated input features.

In one implementation, the speech data were captured using a close-to-mouth microphone. The speech was sampled at 22.05 kHz and quantized to a 16 bit signed number. In order to extract Mel frequency cepstrum coefficients (MFCC) as acoustic features, spectral analysis of the speech signal was performed over 20 ms with Hamming window and 50% overlap. Discrete cosine transformation (DCT) was applied on the MFCC of the whole word to obtain input feature vectors.

The images were captured using a Web-cam with a resolution of 320×240. Once a new image was captured, features were extracted using the composite profile technique. The composite profile features were composed of the average intensity value of the columns in the image followed by the average value of rows in the image. It is a relevant feature to characterise symmetric and circular patterns, or patterns isolated in a uniform background. This feature can be useful to verify the alignment of objects. In order to reduce the number of features, the interpolation technique was applied to the 560 features.

13.4.4 A Case Study Implementation

In this study, speech data were taken from seven members of the KEDRI Institute (http://www.kedri.info). As speech is text-dependent, all speakers were requested to say the word 'security' for speech-based speaker verification. Five samples from

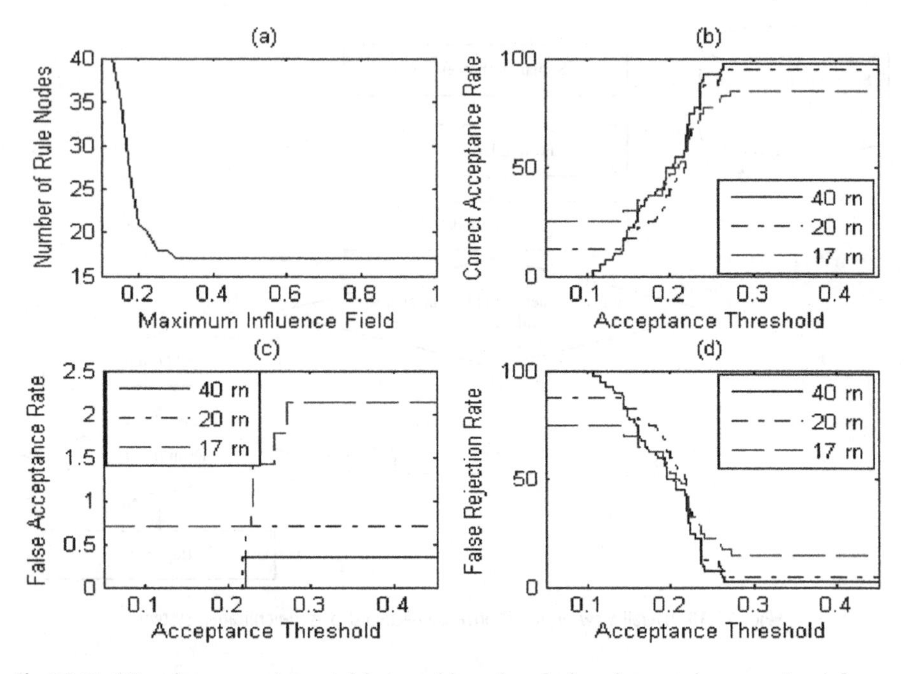

Fig. 13.11 ECF performance on integrated features: (a) number of rule nodes created versus various influence field values; (b) correct acceptance rate versus acceptance thresholds; (c) false acceptance rate (FAR) versus acceptance thresholds; (d) false rejection rate (FRR) versus acceptance thresholds.

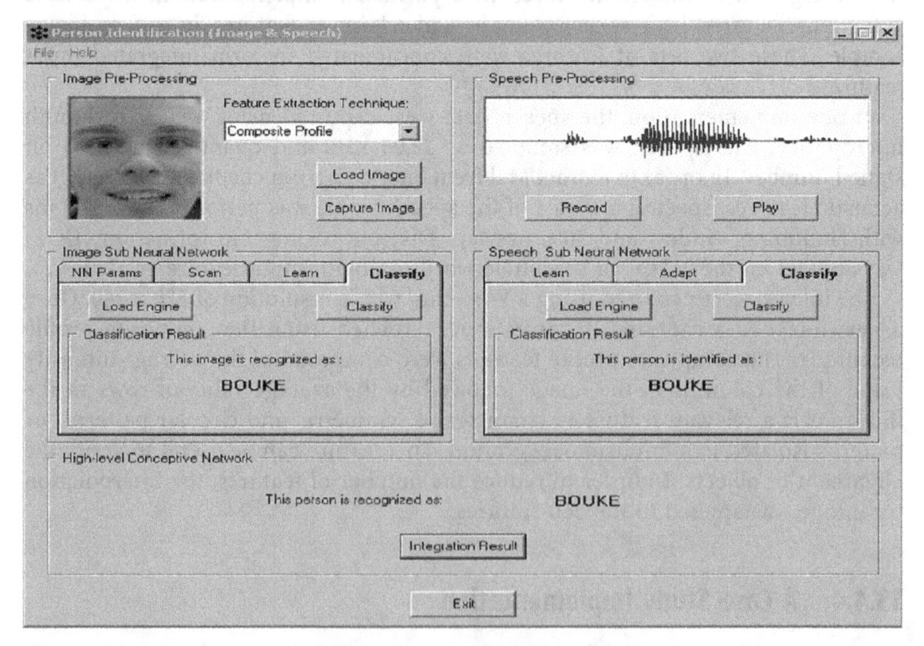

Fig. 13.12 A software implementation of a system that consists of a speech recognition module, image recognition module, and higher-level decision making for a person verification or person identification application (Ghobaklou *et al.*, 2004).

Table 13.2 Performance of the speaker recognition model, face recognition module, and the integration model.

Person	Speaker recognition model (%)	Face recognition model (%)	Integration model (%)
A	90	75	90
B	80	80	90
C	70	65	80
D	85	90	95
E	90	80	95
F	85	85	90
G	95	80	100
Average	85	79.29	91.43

each speaker were collected to form the training dataset. Another five samples from each of these speakers were used to form a testing dataset. In a similar fashion the face images of the same people were captured to prepare training and testing datasets. The input features from speech and face image were integrated according to the method described in Section 13.4.

A person-verification system was built on integrated voice and face features using ECF. Each integrated sample has 164 input features. The test results are shown in Fig. 13.11.

The results in Fig. 13.11 show that the smaller the maximum influence field, the more rule nodes are allocated for each person. The best ECF performance was achieved with correct acceptance rate of 97% and FAR error of just less than 0.5%.

Figure 13.12 shows a software implementation of a system that consists of a speech-recognition module, image-recognition module and higher-level decision making for person-verification or person-identification applications (Ghobaklou *et al.*, 2004).

The system is tested on a small dataset. The results shown in Table 13.2 illustrate the improvement of accuracy when the speech and the image input data are integrated.

13.5 Summary and Open Problems

We have introduced a framework and two systems that integrate auditory and visual information. The framework facilitates the study of:

- Different types of interaction between modules from hierarchically organised subsystems for auditory and visual information processing
- Early and late integration of the auditory and the visual information flows,
- Dynamic auditory and visual features
- Pure connectionist implementations at different levels of information processing
- Evolving fuzzy neural networks that allow for learning, adaptation, and rule extraction.

The integrated processing of auditory and visual information may yield:

- An improved performance on classification tasks involving information from both modalities and
- Reduced recognition latencies on these tasks.

The framework has the potential for many applications for solving difficult AI problems. Examples of such problems are: adaptive speech recognition in a noisy environment, face tracking and face recognition, person identification, person verification tracking dynamic (moving) objects, recognising the mood or emotional state of subjects based on their facial expression and their speech, solving the blind-source separation problem. Through solving these problems, the development of intelligent multimodal information systems can be facilitated.

For the integration of auditory and visual information the following open questions need to be addressed.

1. At which level and to what degree should the auditory and visual information processes be integrated? The AVIS framework accommodates integration at multiple levels and at various degrees. It seems that early integration works fine, but a further fine-tuning is required to obtain a better insight into the problem.
2. How should time be represented in an integrated audiovisual information-processing system? In our two case studies we examined bimodal processing using an aggregate vector representation with time included. This representation turned out to be especially effective when using longer time intervals. For shorter time intervals other ways of representing time should be investigated.
3. How should adaptive learning be realised in an integrated audiovisual information processing system? A deeper investigation of ECOS is realisable and will most probably lead to good results.
4. How should new knowledge be acquired about the auditory and visual inputs of the real world? Translating the hidden representations of ECOS into rules provides a first answer to this question.
5. How does the neuro-genetic principle (see Chapter 9) apply to multimodal information processing?

13.6 Further Reading

For further details of the ideas discussed in this chapter, refer to Kasabov *et al.* (2000c), Ghobakglou and Kasabov (2004), Ghobakglou *et al.* (2004).

Further material on the issues discussed in this chapter can be found as follows.

- *Sources of Neural Structure in Speech and Language Processing* (Stork, 1991)
- *Integrating Audio and Video Information at Different Levels* (Waibel *et al.*, 1997)
- *Using Lip Movement for Speech Recognition* (Stork and Hennecke, 1996; Massaro and Cohen, 1983; Gray *et al.*, 1997; Luettin *et al.*, 1996)
- *Person Identification and Verification* (Brunelli and Falavigna, 1995)

14. *Evolving Intelligent Systems for Robotics and Decision Support*

This chapter presents examples of how the methods of EIS and ECOS in particular can be applied on various applications for robotics and decision support. In the examples here we have used some of the methods presented in Part I of the book, but more sophisticated methods can be further developed and applied in the future on the same problems. The reason for using ECOS is that many social and environmental systems are characterised by a continuous change and by a complex interaction of many variables over time; they are evolving. One of the challenges for information sciences is to be able to represent the dynamic processes, to model them, and to reveal the 'rules' that govern the adaptation and the variable interaction over time. Decision making, related to complex and dynamically changing processes, requires sophisticated decision support systems (DSS) that are able to:

- Learn and adapt quickly to new data in an online mode.
- Continuously learn patterns of variable relationship from data streams.
- Deal with vague, fuzzy, and incomplete information, as well as with crisp information.

Such systems can be realised as DSS or as intelligent robots. The chapter includes the following topics.

- Adaptive learning robots
- Modelling of evolving financial and socioeconomic processes
- Adaptive Environment risk evaluation
- Summary and open problems
- Further reading

14.1 Adaptive Learning Robots

14.1.1 Multimodal Interactive Robots

The robotics field is growing with more robots being created continuously for different purposes (see Fig. 14.1). Intelligent robots should be able to communicate through spoken commands and images. They should be able to recognise

Fig. 14.1 The robotics field is growing with more robots being created continuously for different purposes.

a command (e.g. 'bring an orange'), analyse the meaning of the command (e.g. to find and bring an orange), recognise the object (e.g. image recognition of an orange in a scene), and act appropriately.

A small example of this scenario is shown in Fig. 14.2a (a functional block diagram), Fig. 14.2b (examples of recognisable objects), and Fig. 14.2c (a block diagram of a system realisation).

Furthermore, an intelligent robot should be able to learn new commands and recognise new objects on the fly, without forgetting previously learned objects. Such an experiment with the use of ECOS is presented in Zhang *et al.* (2004).

14.1.2 Adaptive Multiple Robot Control

When several robots take part in a common task, e.g. playing football at a robocup competition, the robots should adapt their strategy on the fly depending on the current circumstances, e.g. the way the opponent plays.

In Huang *et al.* (2005) an adaptive control of the positioning of the robot players on the soccer field is implemented with the use of the ECOS ECF (Fig. 14.3. The positioning rules can evolve and change during the match, rather than the robot team playing with fixed positioning rules.

14.2 Modelling of Evolving Financial and Socioeconomic Processes

14.2.1 Adaptive Prediction of Financial Indexes

Financial data may change dynamically where many variables are involved. For the prediction of financial indexes, adaptive incremental learning models are needed. ECOS would be suitable techniques for the purpose as illustrated below.

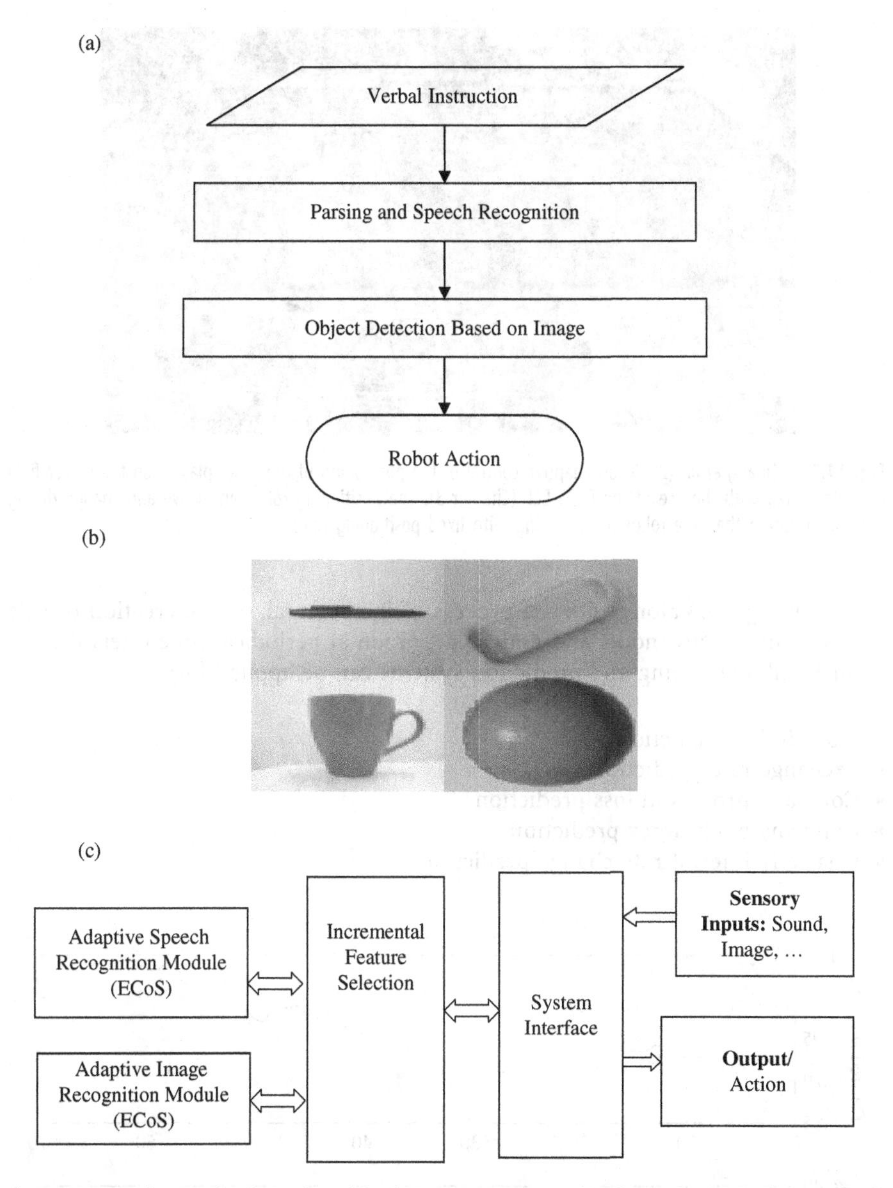

Fig. 14.2 A small example of a multimodal information processing robot for recognising spoken commands and image objects: (a) a functional block diagram; (b) examples of recognisable objects; (c) a block diagram of a system realisation for adaptive multimodal speech command recognition and object/image recognition and action robot.

Example

Prediction of the exchange rate Euro/US$ (1, 2, 3, and 4 weeks ahead) using three input variables: Exchange Rate, Euro/Yen, Stock-E/US, with four time lags each, is

Fig. 14.3 In Huang *et al.* (2005) an adaptive control of the positioning of the robot players on the soccer field is implemented with the use of the ECOS ECF (Chapter 3). The positioning rules can evolve and change during the match, rather than the robot team playing with fixed positioning rules.

shown in Fig. 14.4 along with the process of incremental, online creation of rule nodes in an EFuNN model and node aggregation at periodical time intervals.

Such online learning and prediction systems can be applied for:

- Stock index prediction
- Exchange rate prediction
- Company profit and loss prediction
- Company bankruptcy prediction
- Quarterly interest rate change prediction

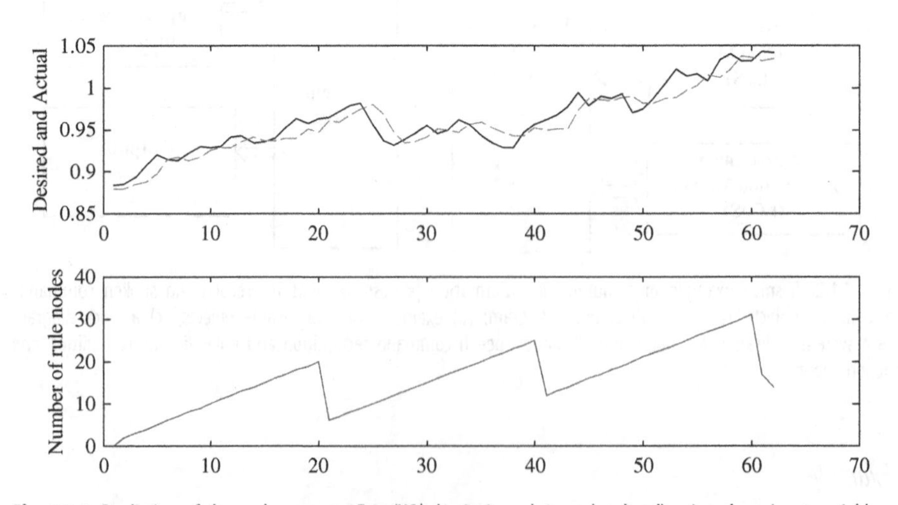

Fig. 14.4 Prediction of the exchange rate Euro/US$ (1, 2, 3, and 4 weeks ahead) using three input variables: Exchange Rate, Euro/Yen, Stock-E/US, with four time lags each, along with the process of incremental, online creation of rule nodes in an EFuNN model and node aggregation at periodical time intervals.

14.2.2 Modelling of Evolving Macroeconomic Clusters

Countries and regions can be grouped (clustered) together based on similarity, measured on the basis of several macroeconomic variables, such as GDP, inflation rate, unemployment rate, index of goods, and so on. With a quarterly flow of information on these variables, the clusters may change, indicating a trend in the development of some countries or group of countries.

This can be traced and analysed with the use of evolving clustering methods, such as ECM and ESOM from Chapter 2, as shown in the example below (see Kasabov (2006), Rizzi *et al.*, 2003). The data used in the example are given in Appendix D and some clusters are shown in Fig. 14.5a,b.

Example: Tracing Evolving Macroeconomic Clusters in Europe/United States

In Fig. 14.5a the evolved clusters in an EFuNN model, when data from 1994 until 1998 are used, are shown. The upper figure shows a plot of the rule nodes: their cluster centres and receptive fields in the input space 'x = Unemployment $(t-1)$, y = GDP$(t-1)$'. The lower figure shows the same nodes in the input space '$x = CPI(t-1)$ and y = Interest rate$(t-1)$'. The data examples are represented as 'o'. The rule nodes are numbered in a larger font with the consecutive numbers of their evolution. Data examples are numbered from 1 to 45 meaning the consecutive input vectors used for the evolution of the EFuNN in the shown order. BE45 for example means the four parameter values for Belgium for the year 1994 and 1995 as $(t = 1)$ and (t) input values to the EFuNN model.

Figure 14.5b shows the changes of the clusters from Fig. 14.5a when data for 1999 are added. That will help to trace the evolving macroeconomic clusters in Europe/United States over time. The figure shows the evolved clusters, when the EFuNN model was updated on the 1999 data in the input space 'x = Unemployment rate $(t-1)$, y = GDP per capita$(t-1)$' (upper figure), and in the input space '$x = CPI(t-1)$ and y = Interest rate $(t-1)$' (lower figure). The data examples and the cluster centres (rule nodes) are represented in the same way as in Fig. 14.5a.

14.3 Adaptive Environmental Risk of Event Evaluation

14.3.1 Problem Definition

We are given a domain dataset, or a stream of data: $D = \{X_1, X_2, \cdots, X_i, \ldots\}$, where X_i are vectors, each of them consisting of k input variables $Xi = (x_1, x_2, \cdots, x_k)$and an output variable y_i that takes a value of 1 (an event has occurred) and 0 (event has not occurred), for a new vector $X^{(\text{new})}$ to predict the risk, between 0 and 1, for the event to happen.

This problem is very common in environmental modeling, e.g. based on geographical locations and environmental variables, to evaluate the risk of:

- Establishment of invasive species at different locations in the world (Worner, 1988, 2002; Gevrey *et al.*, 2006)

(a)

1	2	3	4	5	6	7	8	9	10	11	12	13	14	15	16	17	18
BE45	DK45	DE45	EL45	ES45	FR45	IR45	IT45	NL45	AS45	IT45	IR45	SW45	UK45	US45	BE56	DK56	DE56
19	20	21	22	23	24	25	26	27	28	29	30	31	32	33	34	35	36
EL56	ES56	FR56	IR56	IT56	NL56	AS56	PT56	FI56	SW56	UK56	US56	BE67	DK67	DE67	EL67	ES67	FR67
37	38	39	40	41	42	43	44	45									
IR67	IT67	NL67	AS67	PT67	FI67	SW67	UK67	US67.									

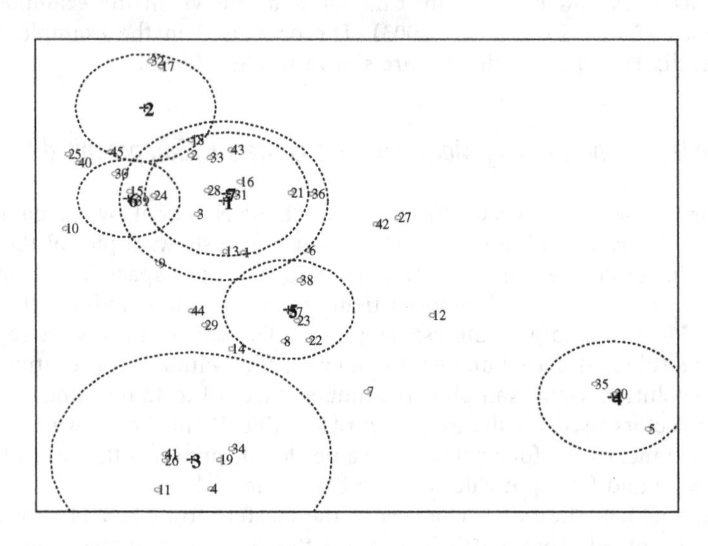

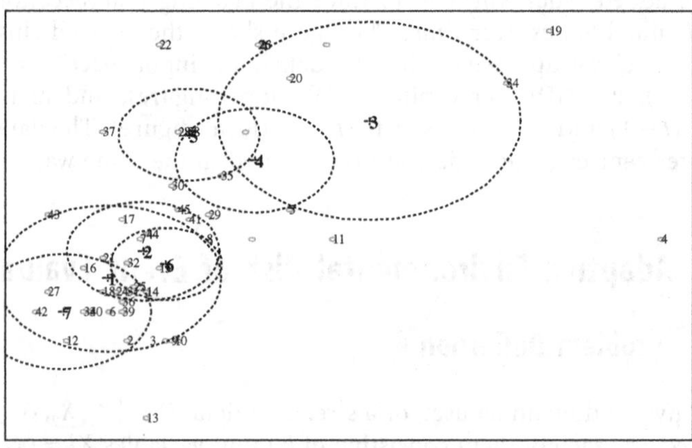

Fig. 14.5 (a) The evolved clusters in an EFuNN model when macroeconomic data are used. The upper figure shows a plot of the rule nodes, their cluster centres and receptive fields in the input space '$x=$ Unemployment $(t-1)$, $y=$ GDP$(t-1)$'. The lower figure shows the same nodes in the input space 'x CPI$(t-1)$ and $y=$ Interest rate$(t-1)$'. The data examples are represented as o. The rule nodes are numbered in a larger font with the consecutive numbers of their evolvement. Data examples are numbered from 1 to 45 meaning the consecutive input vectors used for the evolvement of the EFuNN in the shown order. BE45 for example means the four parameter values for Belgium for the year 1994 and 1995 as $(t=1)$ and (t) input values to the EFuNN model. *(Continued overleaf)*

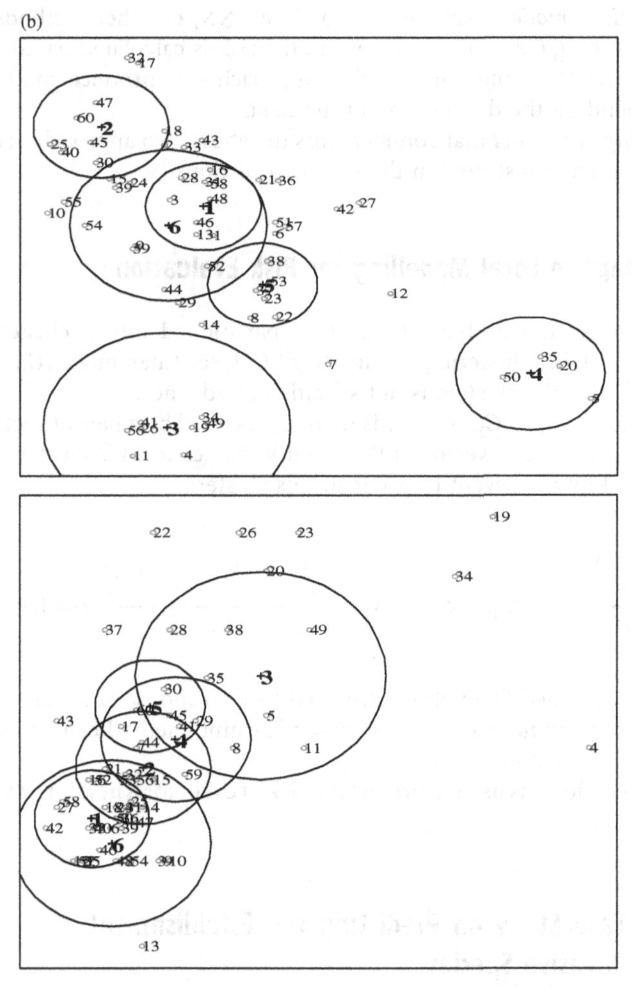

Fig. 14.5 (*continued*) (b) The changes of the clusters from (a) when new data for 1999 are added. That will help to trace the evolving macroeconomic clusters in Europe/U.S. over time. The figure shows the evolved clusters when the EFuNN model was updated on the 1999 data in the input space '$x =$ Unemployment rate $(t-1)$, $y =$ GDP per capita$(t-1)$' (upper figure), and in the input space '$x =$ CPI$(t-1)$ and $y =$ Interest rate $(t-1)$'(lower figure). The data examples and the cluster centres (rule nodes) are represented in the same way as in (a).

- Earthquakes
- Floods and other disasters
- Spread of infectious disease
- Global Climate Change
- And others

Several approaches can be used for solving the above problem:

- *A global regression model.* This model is difficult to adapt on new data without using the previous data.

- *A personalised model*, using WKNN or WWKNN, or other methods for transducive reasoning (see Chapter 1), where the risk is calculated based on distance to nearest neighbouring samples. This approach may produce good results, but it will depend on the distribution of the data.
- *A local adaptive model* that complements the above two approaches: an example is presented and illustrated in the next two sections.

14.3.2 Adaptive Local Modelling for Risk Evaluation

1. The data from the problem space D is partitioned into n clusters using an adaptive, evolving clustering method (e.g.ECM, see Chapter 2): $\{C_1, C_2, \cdots, C_n\}$. The number of the clusters is not specified in advance.
2. Each cluster $C_i \in \{C_1, C_2, \cdots, C_n\}$ is then represented by a pair of vectors (X_i^c, p_i^c), where X_i^c is the mean vector of the input variables for this cluster and $p_i^c$is the risk (probability) of event to occur in this cluster:

$$X_i^c = \frac{\sum_{j=1}^{|C_i|} X}{|C_i|}, \quad p_i^c(Y|x_1, x_2, \cdots, x_k) = \frac{\sum_{j=1}^{|C_i|} p(y|x_1, x_2, \cdots, x_k)}{|C_i|}, i = 1, \cdots n. \quad (14.1)$$

3. $\mathbf{P}^c$ and $\mathbf{X}^c$ obtained from above are used to fit response surfaces as a function of the predictors; here a local supervised learning method can be used, such as DENFIS (see Chapter 3).
4. The evolved model is used to predict the risk of event for a new input vector$X^{(\text{new})}$.

14.3.3 A Case Study on Predicting the Establishment of Invasive Species

This section describes a model based on ECM evolving clustering and DENFIS for predicting the establishment potential of a pest insect into new locations, using as a case study *Planocuccus citri* (Risso), the citrus mealybug (Worner 1988, 2002; Soltic *et al.*, 2004; Gevrey *et al.*, 2006). The model is based on the relationship between climate at locations from the world map and *P. citri* presence or absence.

If new yet unseen data become available the model will adapt its structure and produce output to accommodate the new data. The model can be trained incrementally and produce rules that have the potential to explain the relationship between the climate factors and the potential of pest establishment. The evolving model proposed here can complement qualitative assessments and the knowledge of expert advisers, and can be used to assess the possibility of establishment before an actual introduction of the pest insect. Finally, predictive maps are produced for visual inspection, showing general trends and possible risk spots.

The meteorological data for more than 7000 worldwide locations, where the *Planocuccus citri* (Risso) has been recorded as either present (223 locations) or considered absent, was assembled from published sources (Lora Peacock, unpublished data; PhD). It is important to note that in this dataset, locations where the species is recorded as absent may be either areas where *P. citri* has never had the opportunity to establish, or has not been found, rather than the fact that the environment is not suitable for its establishment. Each location was described by a selection of temperature and moisture attributes (predictor variables) known to indicate the climatic suitability of the habitat for insect establishment: maximum summer temperature (T_{max}), minimum winter temperature (T_{min}), mean total rainfall (R_{mean}), annual actual evapotranspiration (AET), and an aridity index represented by the ratio of precipitation/ potential evapotranspiration (P/PE) with low P/PE index indicating dryness.

After clustering the data with the use of ECM, 20 clusters were created; for each of them a risk (probability of event) was calculated as in Eq. (14.1). As an example, the samples clustered in cluster number 4 with a probability of event 0.73 are mapped back on the geographical map and shown in Fig. 14.6.

Then cluster information is used to train a DENFIS model. The DENFIS model predicts better than a global regression model the risk of the pest establishment as tested on real data.

The DENFIS prediction model is then used to calculate the risk at any location of the world map as shown as bar graphs in Fig. 14.7.

Fig. 14.6 Spatially distributed 454 locations are grouped by the ECM clustering method into 20 clusters. The number associated with each cluster indicates the ratio between the number of locations occupied by the insect *P. citri* and the total number of locations. The risk of establishment of the insect in this cluster is measured as probability. Only cluster 4 is shown in the figure (the biggest cluster, containing 100 locations) with a risk of insect establishment of 0.73.

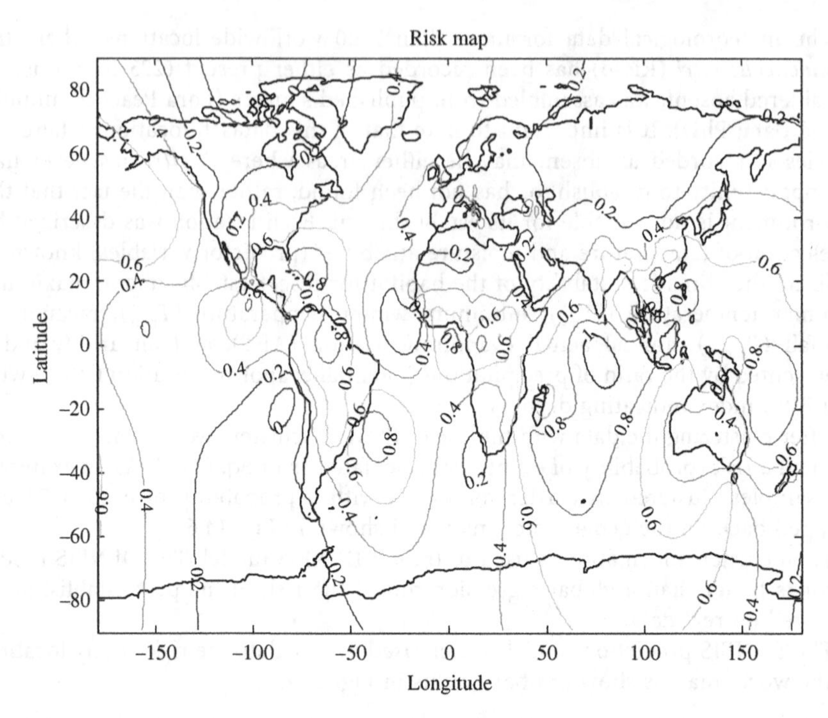

Fig. 14.7 A world risk map for the establishment of the insect *P. citri* from Fig. 14.6 is developed with the use of a trained ECOS DENFIS prediction model to calculate the risk at any location of the world map. If new data become available, e.g. the pest is established at new locations, or the climate is changing, this information can be added to the ECM clusters and to the DENFIS model to update this map dynamically.

If new data become available, e.g. the pest is established at new locations, this information can be added to the ECM clusters and to the DENFIS model to update the model and the map from Fig. 14.7 dynamically.

14.4 Summary and Open Questions

This chapter only illustrates the large number of various application problems that can be solved with the use of EIS and ECOS in particular. New evolving learning methods can be developed in the future in the search for more efficient solutions for problems such as

- Spread of viruses
- Spread of pandemic diseases
- Predicting earthquakes
- Predicting tsunami events
- Predicting solar eruptions and cosmic events
- Searching for signals from extraterrestrial intelligence (see WWW of the Institute 'Search for Extraterrestrial Intelligence', SETI: http:// www.seti.com)
- Predicting global Climate changes

14.5 Further Reading

- *Adaptive Robots* (Fukuda *et al.*, 1997)
- *Applications of Evolving Fuzzy Rule-based System for Control* (Angelov, 2002; Angelov and Filev, 2004; Angelov *et al.*, 2004)
- *Evolving Fuzzy Systems for Real-time Landmark Recognition* (Zhou and Angelov, 2006)
- *Adaptive, Evolving Robots* (Nolfi and Floreano, 2000)
- *EFuNN for Tennis Coach Decision Support* (Bacic, 2004)
- *Adaptive Fuzzy Decision Support Systems* (Furuhashi *et al.*, 1993, 1994)
- *Macroeconomic Decision Support Systems* (Rizzi *et al.*, 2003)
- *Ecological Modelling* (Worner, 1988, 2002; Gevrey *et al.*, 2006, Soltic *et al.*, 2004)
- *Searching for Signals from Extraterrestrial Intelligence* (www.seti.com)

15. What Is Next: Quantum Inspired Evolving Intelligent Systems?

This chapter presents some promising future directions for the development of methods and EIS inspired by the principles of quantum processes, as part of the whole process of information processing (see Fig. ??). Many open questions and challenges are presented here, to be addressed in future research. The chapter discusses the following issues.

- Why quantum inspired EIS?
- Quantum information processing
- Quantum inspired optimisation techniques
- Quantum inspired connectionist systems
- Linking quantum to neuro-genetic information processing: is this the challenge for the future?
- Summary and open problems
- Further reading

15.1 Why Quantum Inspired EIS?

Quantum computation is based upon physical principles from the theory of quantum mechanics (Feynman *et al.*, 1965). One of the basic principles that is likely to trigger the development of new methods and EIS is the linear superposition of states.

At a macroscopic or classical level a system exists only in a single basis state as energy, momentum, position, spin, and so on. However, at a microscopic or quantum level, the system at any time represents a superposition of all possible basis states. At the microscopic level any particle can assume different positions at the same time, can have different values of energy, can have two values of the spins, and so on. This superposition phenomenon is counterintuitive because in classical physics one particle has only one position, energy, spin, and so on.

If the system interacts in any way with its environment, the superposition is destroyed and the system collapses into one single real state as in classical physics. This process is governed by a probability amplitude (Feynman *et al.*, 1965). The square of the intensity for the probability amplitude is the quantum probability to observe the state. Quantum mechanical computers and quantum algorithms try to exploit the massive quantum parallelism which is expressed in the principle of superposition.

The principle of superposition can be applied to many existing methods of EIS, where instead of a single state (e.g. a parameter value, a finite automata state, or a connection weight, etc.) a superposition of states will be used described by a wave probability function, so that all these states will be computed in parallel increasing the speed of computation exponentially And not only the speed can be increased, the accuracy of the solutions would also increase. New problems, that have not been possible to solve thus far, may be possible to solve with the use of quantum inspired EIS (QEIS).

There are already examples of how much more efficient a quantum inspired algorithm can be when compared to classical algorithms for some specific tasks. Quantum mechanical computers were proposed in the early 1980s and a description was formalised in the late 1980s (Benioff, 1980). This kind of computer proved to be superior to classical computers in various specialized problems. Many efforts were undertaken to extend the principal ideas of quantum mechanics to other fields of interest. There are well-known quantum algorithms such as Shor's quantum factoring algorithm (Shor, 1997) and Grover's database search algorithm (Grover, 1996). Hogg extended the work of Grover in order to demonstrate the application of quantum algorithms in the context of combinatorial search (Hogg and Portnov, 2000).

15.2 Quantum Information Processing

The concept of quantum computing utilises the special nonlocal properties of the quantum phenomena. A quantum atomic or subatomic particle (e.g. atoms, electrons, protons, neutrons, bosons, fermions, photons) exists in a probabilistic superposition of states rather than in a single definite state. For example, an electron circling around a nucleus jumps to different orbits, states, due to absorbing or releasing energy (Fig. 15.1). Particles in general are characterized by: charge, spin, position, velocity, and energy.

Some principles, assumptions, and facts in quantum information processing are listed below:

- *Heisenberg's uncertainty principle:* Both the position and the momentum of an electron, or generally, of a particle, cannot be known, because to know it means to measure it, but measuring causes interfering and change of both the position and the momentum. Making an observation of the system 'collapses' the system to one possible state, or universe.
- *The superposition principle:* A particle can be in several states at the same time, with certain probabilities. It is illustrated by Schroedinger by his famous thought experiment of seeing with one eye open, a creature (a cat) in both alive and dead states with certain probabilities (see also Koch and Hepp, 2006).
- *The entanglement principle:* Two or more particles, regardless of their location, are in the same state with the same probability. The two particles can be viewed as 'correlated', undistinguishable, or 'synchronised', coherent. An example is a laser beam consisting of millions of photons having the same characteristics and states.

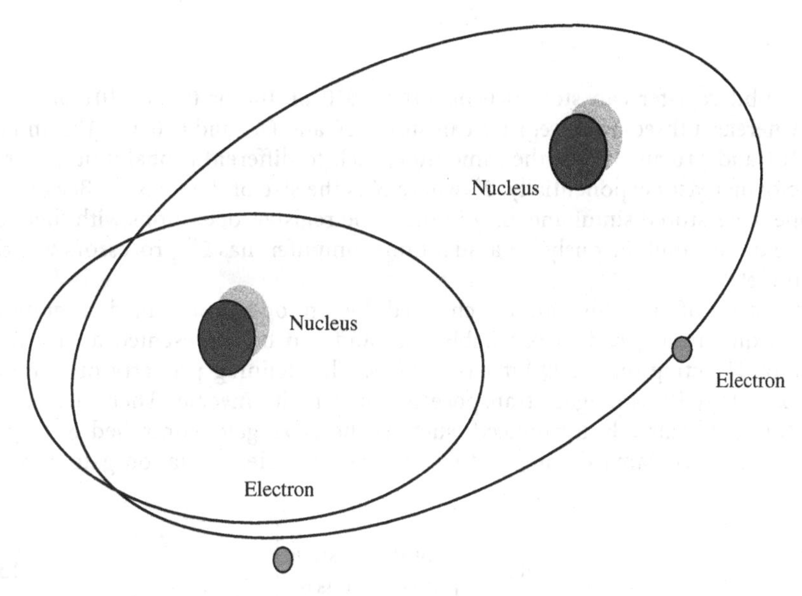

Fig. 15.1 Heisenberg's uncertainty principle: both the position and the momentum of an electron, or generally of a particle, cannot be known, because to know it, means to measure it, but measuring causes interfering and change of both the position and the momentum.

- *Electromagnetic radiation:* Emitted in discrete quanta whose energy E is proportional to the frequency:

$$E = h.f, \tag{15.1}$$

where h is the Max Planck constant (approx. $6.62608 \,.\, 10^{-34}$) and f is the frequency.

The advantage of quantum computing is that, while a system is uncollapsed, it can carry out more computing than a collapsed system, because, in a sense, it is computing in an infinite number of universes at once.

Ordinary computers are based on bits, which always take one of the two values 0 or 1. Quantum computers are based instead on what are called *Q-bits* (or qubits). A Q-bit may be simply considered as the spin state of an electron. An electron can have spin Up or spin Down; or three quarters Up and one quarter Down. A Q-bit contains more information than a bit, but in a strange sense, not in the same sense in which two bits contain more information than a bit.

The state of a Q-bit can be represented as below, where α and β are complex numbers that specify the probability amplitudes of the corresponding states 0 and 1.

$$|\psi\rangle = \alpha|0\rangle + \beta|1\rangle \tag{15.2}$$

Because the Q-bit can only be in these two states, it should satisfy the condition:

$$|\alpha|^2 + |\beta|^2 = 1 \tag{15.3}$$

Example

A three-bit register can store 000 or 001 or 010 or 100 or 011 or 101 or 110 or 111, whereas a three-qubit register can store 000 and 001 and 010 and 100 and 011 and 101 and 110 and 111 at the same time, each to different probabilities. Storage capacity increases exponentially, 2^N where N is the size of the register. Because the numbers are stored simultaneously in the same register, operations with them can also be done simultaneously, so a quantum 'computer' has 2^N processors working in parallel.

The state of a Q-bit can be changed by an operation called a *quantum gate*. A quantum gate is a reversible gate and can be represented as a unitary operator U acting on the Q-bit basis states. The defining property of a unitary matrix is that its conjugate transpose is equal to its inverse. There are several quantum gates already introduced, such as the NOT gate, controlled NOT gate, rotation gate, Hadamard gate, and so on. For example, a rotation gate is represented as

$$U(\theta) = \begin{bmatrix} \cos\theta & -\sin\theta \\ \sin\theta & \cos\theta \end{bmatrix} \tag{15.4}$$

15.3 Quantum Inspired Evolutionary Optimisation Techniques

15.3.1 General Principles

Quantum inspired methods of evolutionary computation (QIEC) have already been discussed in Han and Kim (2002) and Jang *et al.* (2003), that include: genetic programming (Spector, 2004), particle swarm optimizers (Liu *et al.*, 2005), and finite automata and Turing machines (Benioff, 1980).

The quantum inspired evolutionary computation (QIEC) methods for optimisation are based on the following principles.

- Whereas in EC the representation of individuals ('chromosomes') is usually made in the form of bit strings, real-valued vectors, symbols, and the like, the QIEC uses a q-bit representation based on the concept of q-bits. Each q-bit is defined as a pair of numbers (α, β). A Q-bit individual is a string of m q-bits and is represented as

$$\left[\begin{array}{c|c|c|c} \alpha_1 & \alpha_2 & \dots & \alpha_m \\ \beta_2 & \beta_2 & \dots & \beta_m \end{array}\right] \tag{15.5}$$

where the following holds for $i = 1, 2, \dots, m$,

$$|\alpha_i|^2 + |\beta_i|^2 = 1 \tag{15.6}$$

- Evolutionary computing with Q-bit representation has a better characteristic of population diversity than other representations, because it can represent linear superposition of states probabilistically. Here, only one Q-bit individual with m q-bits is enough to represent 2^m states whereas in binary representation, 2^m individuals will be required for the same.
- The Q-bit representation leads to a quantum parallelism in the system as it is able to evaluate the function on a superposition of all possible inputs. The output obtained is also in the form of superposition which needs to be collapsed to get the actual solution.
- In QIEC, the population of Q-bit individuals at time t can be represented as

$$Q(t) = \{q_1^t, q_2^t, \ldots, q_n^t\} \tag{15.7}$$

where n is the size of the population.

15.3.2 Quantum Inspired Evolutionary Particle Swarm Optimisation Algorithms

The pseudocode for an algorithm, called QEA, proposed by Han and Kim is given in Fig. 15.2. It extends the PS algorithm from chapter 6.

Quantum inspired swarm optimization algorithm by Han and Kim
(See the text for an explanation of the notations)

begin
  $t \leftarrow 0$
i) initialize a population of Q-bit individuals, $Q(t)$
ii) make $P(t)$ by observing the states of $Q(t)$
iii) evaluate $P(t)$
iv) store the best solutions among $P(t)$ into $B(t)$
  while (**not** termination condition) **do**
  begin
  $t \leftarrow t+1$
v) make $P(t)$ by observing the states of $Q(t-1)$
vi) evaluate $P(t)$
vii) update $Q(t)$ using Q-gates
viii) store the best solutions among $B(t-1)$ and $P(t)$ into $B(t)$
ix) store the best solution b among $B(t)$
x) **if** (migration condition)
  then migrate b or b_j^t to $B(t)$ globally or locally, respectively
  end

end

Fig. 15.2 The pseudocode for a quantum inspired evolutionary algorithm, called QEA, proposed by Han and Kim (2002).

In the above QEA algorithm the population of Q-bit individuals at time t can be represented as $Q(t) = \{q_1^t, q_2^t, \ldots, q_n^t\}$ where n is the size of the population. The rotation gate which is used as the Q-gate is represented as

$$U(\theta) = \begin{bmatrix} \cos\theta & -\sin\theta \\ \sin\theta & \cos\theta \end{bmatrix} \tag{15.8}$$

15.4 Quantum Inspired Connectionist Systems

15.4.1 Motivation Behind Quantum Inspired Neural Networks

Recent research activities focus on a combination of quantum computation and artificial neural networks. Neural networks are biological inspired information-processing systems that were shown to be powerful in solving classification problems, predicting developments, and controlling robots because of their fault tolerance, robustness, and their ability of massive parallel processing. Considering quantum neural networks seems to be important for two reasons.

First, there is evidence for the essential role that quantum processes may play in realizing information processing in the living brain. Roger Penrose argued that a new physics binding quantum phenomena with general relativity can explain such mental abilities as *understanding, awareness,* and *consciousness* (Penrose, 1994).

The second motivation is the possibility that the field of classical neural networks could be generalized to the promising new field of quantum computation (Brooks, 1999). Both considerations suggest a new understanding of mind and brain function as well as new unprecedented abilities in information processing. Ezhov and Ventura (2000) are considering the quantum neural networks as the next natural step in the evolution of neurocomputing systems.

Naraymen and Meneer (2000) simulated classical and various types of quantum inspired neural networks and compared their performance. Their work suggests that there are indeed certain types of problems for which quantum neural networks will prove superior to classical ones.

Other relevant work includes quantum decision making (Brooks, 1999), quantum learning models (Kouda *et al.*, 2005), quantum networks for signal recognition (Tsai *et al.*, 2005) and quantum associative memory (Trugenberger, 2002; Ventura and Martinez, 2000). There are also recent approaches to quantum competitive learning where the quantum system's potential for excellent performance is demonstrated on real-world datasets (Ventura, 1999; Xie and Zhuang, 2003).

The quantum inspired neural network (QUINN), proposed by Narayanan and Meneer (2000), interprets each input pattern Sp $(p = 1, 2, \ldots, k)$ as a particle, being learned in a separate ANNp model in a separate universe Up, the superposition of all NN constituting the whole NN model. The structure of all NN is the same, so that a connection weight between neuron Ni and neuron Nj in the total model is a superposition of all connection weights Wij (k) of all k NNs. When an input pattern S is presented for recognition, the NN model 'collapses' into a particular NN-S that recognises this pattern. Each pattern needs to be presented only once in order for a NN model to be created for this pattern and become part of the superposition of all NN models.

QUINNs are in general evolving systems. Presenting a new pattern S_{k+1} (a new particle) to the evolving QUINN model means creating a new NN model that becomes part of the superposition of connection weights and states.

15.4.2 Neural Network Models Trained with Quantum Inspired Evolutionary Algorithms

The Han and Kim QEA algorithm from Fig. 15.2 was modified in Venayagamoorthy and Singhal (2005) and implemented for a neural network training. The pseudocode for the modified QEA is given in Fig. 15.3.

In the modified QEA, the local best solutions of the individuals are not stored in memory. The updates of the parameters α and β are done by comparing the individual solutions $P(t)$ with the global best solution $Best(t-1)$. This makes the modified QEA a greedy algorithm. But this issue is taken care of by introducing a variable *threshold* in step (viii) of the algorithm. This parameter allows for the solution to come out of the local optimum by introducing some 'noise'.

The parameter theta (θ) has been made adaptive in the modified QEA as opposed to keeping it fixed throughout the simulation. The value of theta is made smaller after the error reaches a certain acceptable value. This parameter is similar to the learning rate used in backpropagation and a few other neural network training

The modified by (Venayagamoorthy *et al.*, 2005) quantum inspired swarm optimisation algorithm by Han and Kim (2002) applied to train a neural network model (see the text for an explanation of the notations):

begin
$t \leftarrow 0$
i) initialize $Q(t)$
ii) make $P(t)$ by observing the states of $Q(t)$
iii) evaluate $P(t)$
iv) store the best solution among $P(t)$ into *Best(t)*
while (**not** termination condition) **do**
begin
$t \leftarrow t+1$
v) make $P(t)$ by observing the states of $Q(t-1)$
vi) evaluate $P(t)$
vii) update $Q(t)$ using Q-gates
viii) store the best solution among *(Best(t-1) + threshold) and P(t)* into *Best(t)*
ix) update *threshold* (make it smaller) if best solution among $P(t)$ was stored in *Best(t)* in the step viii).
x) **if** (*Best(t)* error < limit)
update *theta*
end
end
end

Fig. 15.3 Han and Kim's QEA algorithm from Fig. 15.2 modified in Venayagamoorthy and Singhal (2005) and implemented for neural network training.

algorithms. Just as making the learning rate smaller helps to exploit the solution space better, similarly, reducing theta helps in exploiting the search space.

15.5 Linking Quantum to Neuro-Genetic Information Processing: Is This The Challenge For the Future?

In the section on computational neuro-genetic modelling (Chapter 9) we presented a model that links the level of expression of genes and proteins in a neuron to the neuronal spiking activity, and then to the information processing of a neuronal ensemble that is measured as local field potentials (LFP).

But how do quantum information processes in the atoms and particles (ions, electrons, etc.), that make the large protein molecules, relate to the spiking activity of a neuron and to the activity of a neuronal ensemble? This is a challenging question that is not possible to answer now, but here we can make some speculative steps, we hope in the right direction.

The spiking activity of a neuron relates to the transmission of thousands of ions and neurotransmitter molecules across the synaptic cleft and to the emission of spikes. Spikes, as carriers of information, are electrical signals made of ions and electrons that are emitted in one neuron and transmitted along the nerves to many other neurons. But ions and electrons are characterised by their quantum properties as discussed in a previous section of this chapter. Therefore, quantum properties would influence the spiking activity of neurons and the whole brain and therefore brains obey the laws of quantum mechanics.

Similarly to a chemical effect of a drug to protein and gene expression levels in the brain that may affect the spiking activity and the functioning of the whole brain (modelling of these effects is the subject of the computational neuro-genetic modelling CNGM; see Chapter 9), external quantum factors such as radiation, high-frequency signals, and the like can influence the quantum properties of the particles in the brain through gate operators. According to Penrose (1989), microtubules in the neurons are associated with quantum gates.

Therefore, the challenge is, similar to the CNGM, to create quantum CNGM, that also take into account quantum properties of the particles in a neuron and in the brain as a whole.

In the first instance, is it possible at all? At this stage the answer is not known, but we describe the above relationships in an abstract theoretical way, hoping to be able to refine this framework, modify it, proof it, and use it in the future, at least partially.

Figure 15.4 shows hypothetical and abstract links between the different levels of information processing, following the opening picture in Fig. I.1. Here the interaction at different levels is shown as hypothetical aggregated functions:

$$Q' = Fq(Q, Eq) \tag{15.11}$$

Quantum information processes:
$Q = Fq(Q, Eq)$

Molecular/gene information processes in a neuron
$M = Fm(Q, M, Em)$

Neuronal ensemble (e.g. of spiking neurons)
$N = Fn\ (N,M,Q,En)$

Brain cognitive processing
$C = Fc\ (C,N,M,Q,Ec)$

Fig. 15.4 Hypothetical and abstract links between the different levels of information processing, following the opening picture in Fig. I.1. Here the interaction at different levels is shown in terms of hypothetical aggregated functions.

a future state Q' of a particle of group of particles (e.g. ions, electrons, etc.) depends on the current state Q and on the frequency spectrum Eq of an external signal

$$M' = Fm\,(Q, M, Em) \tag{15.12}$$

A future state of a molecule M' or a group of molecules (e.g. genes, proteins) depends on its current state M, on the quantum state Q of the particles, and on an external signal Em.

$$N' = Fn(N, M, Q, En) \tag{15.13}$$

A future state N' of a spiking neuron or an ensemble of neurons will depend on its current state N, on the state of the molecules M, on the state of the particles Q and on external signals En.

$$C' = Fc(C, N, M, Q, Ec), \tag{15.14}$$

a future cognitive state C' of the brain will depend on its current state C and also on the neuronal N, on the molecular M, and on the quantum Q states of the brain.

We can support the above hypothetical model of integrated representation, by stating the following assumptions, some of them already supported by experimental results (Penrose, 1989).

1. A large amount of atoms are characterised by the same quantum properties, possibly related to the same gene/protein expression profile of a large amount of neurons characterised by spiking activity.
2. A large neuronal ensemble can be represented by a single LFP.
3. A cognitive process can be represented perhaps as a complex but single function Fc that depends on all previous levels.

The model above is too simplistic, and at the same time, too complex to implement at this stage, but even linking two levels of information processing in a computational model will be useful for the further understanding of complex information processes and for modelling complex brain functions.

15.6 Summary and Open Questions

We will be aiming in the future at creating quantum inspired models, with the intention of:

- Using quantum principles to create more powerful information-processing methods and systems
- Understanding the quantum level information processing in nature as a promising direction for science in the future
  - Understanding molecular and quantum information processing important as Knowledge for all areas of science.
  - Modelling molecular processes, needed for biology, chemistry, and physics.
  - Using quantum processes as inspiration for new computer devices – exponential times faster and more accurate quantum computers are predicted to be produced in 20 years' time – not so far away.

New theories (speculative at this stage) have been already formulated, for example:

- Deutsch (see Brooks, 1999) argues that NP-hard problems (e.g. time complexity grows exponentially with the size of the problem) can be solved by a quantum computer.
- Penrose (1994) argues that solving the quantum measurement problem is prerequisite for understanding the mind.
- Hameroff (see Brooks, 1999) argues that consciousness emerges as a macroscopic quantum state due to a coherence of quantum-level events within neurons.

Many open questions need to be answered in this respect. Some of them are listed below:

- How do quantum processes affect the functioning of a living system in general?
- How do quantum processes affect cognitive and mental functions?
- Is the brain a quantum machine, working in a probabilistic space with many states (e.g. thoughts) being in a superposition all the time and only when we formulate our thought through speech or writing does the brain then 'collapse' in a single state?
- Is the fast pattern recognition process in the brain, involving far away segments, a result of both parallel spike transmissions and particle *entanglement* across areas of the brain?
- Is communication between people, and living organisms in general, also a result of *entanglement* processes? What about "connecting" with 'ghosts', or with extraterrestrial intelligence?
- How does the energy in the atoms relate to the energy of the proteins, the cells, and the whole living system?
- How does energy relate to information?
- Would it be beneficial to develop different quantum inspired (QI) computational intelligence techniques, such as QI-SVM, QI-GA, QI-decision trees, QI-logistic regression, and QI-cellular automata?

- How do we implement the QI computational intelligence algorithms on existing computer platforms in order to benefit from their high potential speed and accuracy? Should we wait for the quantum computers to be realised many years from now, or we can implement them efficiently on specialised computing devices based on classical principles of physics?

15.7 Further Reading

- *Quantum Information Processing* (Feynman *et al.*, 1965; Resconi *et al.*, 1999)
- *Quantum Computation* (Brooks, 1999; Benioff, 1980; Hey, 1999)
- *Quantum Processes and the Brain* (Penrose, 1989, 1994; Pribram, 1993; Koch and Hepp, 2006)
- *Quantum Neural Networks* (Ezhov and Ventura, 2000; Narayanan and Meneer, 2000)
- *Quantum Search and Optimisation Algorithms* (Grover, 1996; Han and Kim, 2002)
- *Feedback Quantum Neuron* (Li *et al.*, 2006)

Appendix A. A Sample Program in MATLAB for Time-Series Analysis

```
% Time series analysis
%
xx=fname;
% the time interval for the phase plot
[d1, d2]=size(xx);
%plot raw data time series
fprintf('Plotting raw data time series\n');
tt=(1:d1);
subplot(221)
plot(tt,xx)
xlabel('Time t in days')
ylabel('The time series')
title('Time series analysis')
pause
%%plot 2d phase plot analysis
%fprintf('Plotting 2d phase plot analysis\n');
%dd=d1-4;
%tau=1;
%subplot(222)
%plot(xx(1:dd),xx(1+tau:dd+tau),'b.')
%xlabel('x(t)')
%ylabel('x(t+1)')
%title('The 2D phase space')
%pause;
%plot 3d phase plot analysis
fprintf('Plotting 3d phase plot analysis\n');
tau=1;
dd=d1-4;
subplot(222)
plot3(xx(1:dd),xx(1+tau:dd+tau),xx(1+2*tau:dd+2*tau),'b.')
grid
xlabel('x(t)')
ylabel('x(t+1)')
```

```
zlabel('x(t+2)')
title('3D phase space')
pause;
% Histograms
fprintf('Plotting probability distribution of the series');
subplot (223)
hist(xx)
xlabel('x values')
ylabel('Repetitive occurance')
title('Histogram')
pause;
% Power spectrum
yy=psd(xx,512);
subplot(224)
semilogy(yy)
xlabel('Frequency')
ylabel('PSD')
title('Power spectrum')
```

Appendix B. A Sample MATLAB Program to Record Speech and to Transform It into FFT Coefficients as Features

The feature vectors extracted in the program are ready to be used to train a classifier. A trained ECOS classifier can be further adapted on new speech samples of the same categories (see Fig. B.1 and Table B.1).

```
% Part A: Recording speech samples
Fs=22050; %sampling frequency
display('Press a button and say in 1 sec a word, e.g. Ja, or Yes, or ...');
pause;
Ja = wavrecord(1*Fs,Fs);
      wavplay (Ja,Fs); % listen to the recording
      figure (1)
      plot(Ja);
      save Ja;
 display ('Press a button and say in 1 sec another word, e.g. Nein, or No, or ...');
 pause;
 Nein = wavrecord(1*Fs,Fs);
      wavplay (Nein,Fs); % listen to the recording
      figure (2)
      plot(Nein);
      save Nein;
        pause; %Press a button to continue
      % Part B: Feature extraction: The first 20 FFT coefficients are extracted as
 % a feature vector that represents a speech sample.
 y = fft(Ja); % Compute DFT of x
 m = abs(y); % Magnitude
 % Plot the magnitude
figure (3); plot(m); title('Magnitude Ja');
FJa=m(1:20);
save FJa;
pause;
```

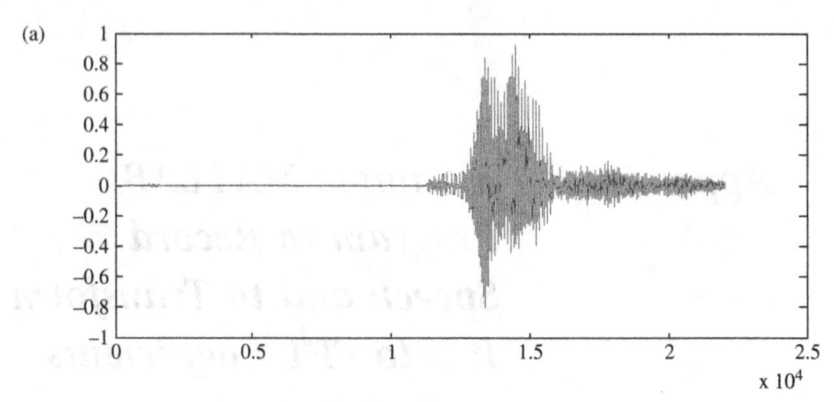

(a)
x 10^4

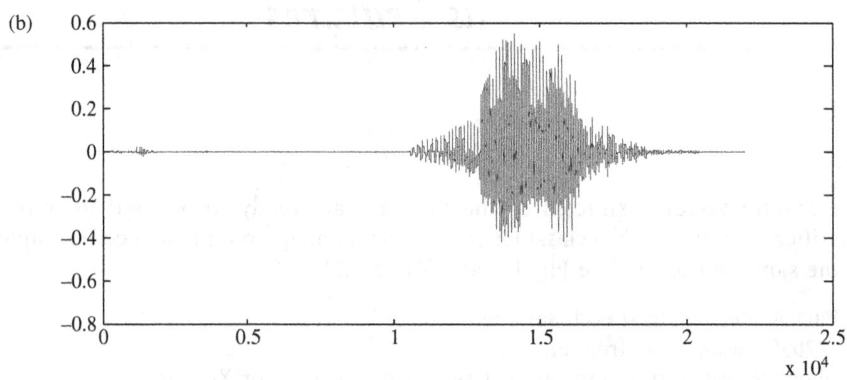

(b)
x 10^4

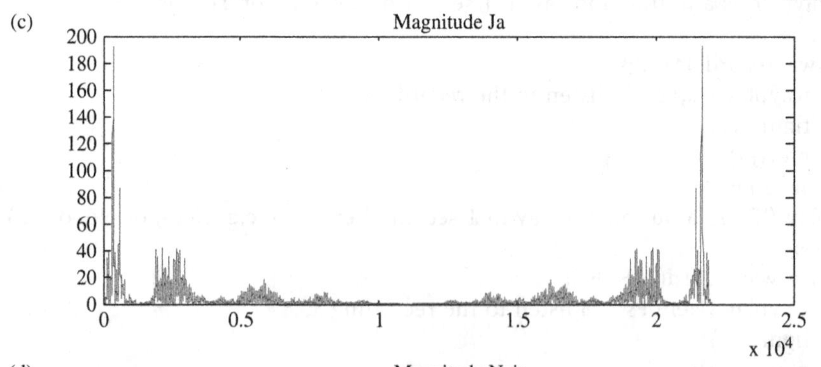

(c)
Magnitude Ja
x 10^4

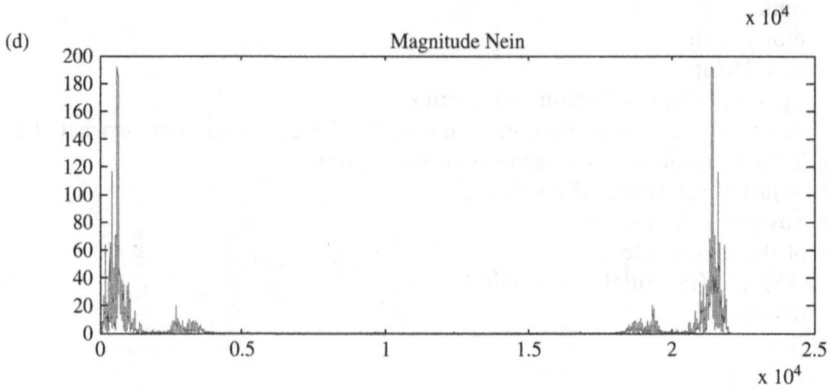

(d)
Magnitude Nein
x 10^4

Table B.1 Two exemplar feature vectors of 20 FFT coefficients each obtained from a run of the program in Appendix B .

	A Feature Vector of 'Ja' – the First 20 FFT Coefficients	*A Feature Vector of 'Nein' – the First 20 FFT Coefficients*
FFT1	0.0463	0.1806
FFT2	0.1269	0.2580
FFT3	0.1477	0.2839
......	0.2374	0.2693
	0.2201	0.3294
	0.3701	0.0596
	0.3343	0.1675
	0.3993	0.1247
	0.2334	0.3774
	0.4385	0.1717
	0.6971	0.5194
	1.2087	0.2292
	1.1161	0.4361
	0.9162	0.4029
	0.7397	0.1440
	0.4951	0.1322
	0.5241	0.4435
	0.2495	0.4441
	0.5335	0.1277
FFT20	1.0511	0.2506

```
 y = fft(Nein); % Compute DFT of x
m = abs(y); % Magnitute
% Plot the magnitude
figure (4); plot(m); title('Magnitude Nein');
FNein=m(1:20);
save FNein;
 % The next parts are to be implemented as an individual work:
% .......................................................
% Part C Train an ECOS classifier (e.g. EFuNN, ECF, etc)
% Part D Recall the model on new (unlabelled) data
% Part E Adapt the model on new (labelled) data through additional training
```

Fig. B.1 Screen shots of printed results of a sample run of the program: (a) the waveform of a pronounced word 'Ja' within 1 sec (sampling frequency of 22,050 Hz) in a noisy environment; (b) the wave form of a pronounced word 'Nein' within 1 sec (sampling frequency of 22,050 Hz) in a noisy environment; (c)the magnitude of the FFT for the pronounced word 'Ja'; (d) magnitude of the FFT for the pronounced word 'Nein'.

Appendix C. A Sample MATLAB Program for Image Analysis and Feature Extraction

% Image Filtering and Analysis Exercise

See Figs. C.1a–f.

```
clc
%Read Lena.tif image
I=imread('lena.tif');
%Display the image
figure; imshow(I); title('Original Image');
pause;
%IMAGE FILTERING AND CONVOLUTION
%Define filter
kernel=[+1 +1 +1; +1 -7 +1; +1 +1 +1];
%Convolve image with kernel
convolved=filter2(kernel,I);
%Display filtered image
figure; imshow(mat2gray(convolved)); title('Filtered Image');
pause;
%Change the kernel to blur the image using a Gaussian kernel
kernel=fspecial('gaussian',5,4);
%Convolve the image
convolved=filter2(kernel,I);
%Display the image
figure; imshow(mat2gray(convolved)); title('Blurred Image');
pause;
% IMAGE AND NOISE
noisyimage = imnoise(I,'salt & pepper');title('Image with Noise')
figure; imshow(noisyimage);
pause;
% IMAGE STATISTICAL ANALYSIS
% surface plot
m=mat2gray(double(I));
```

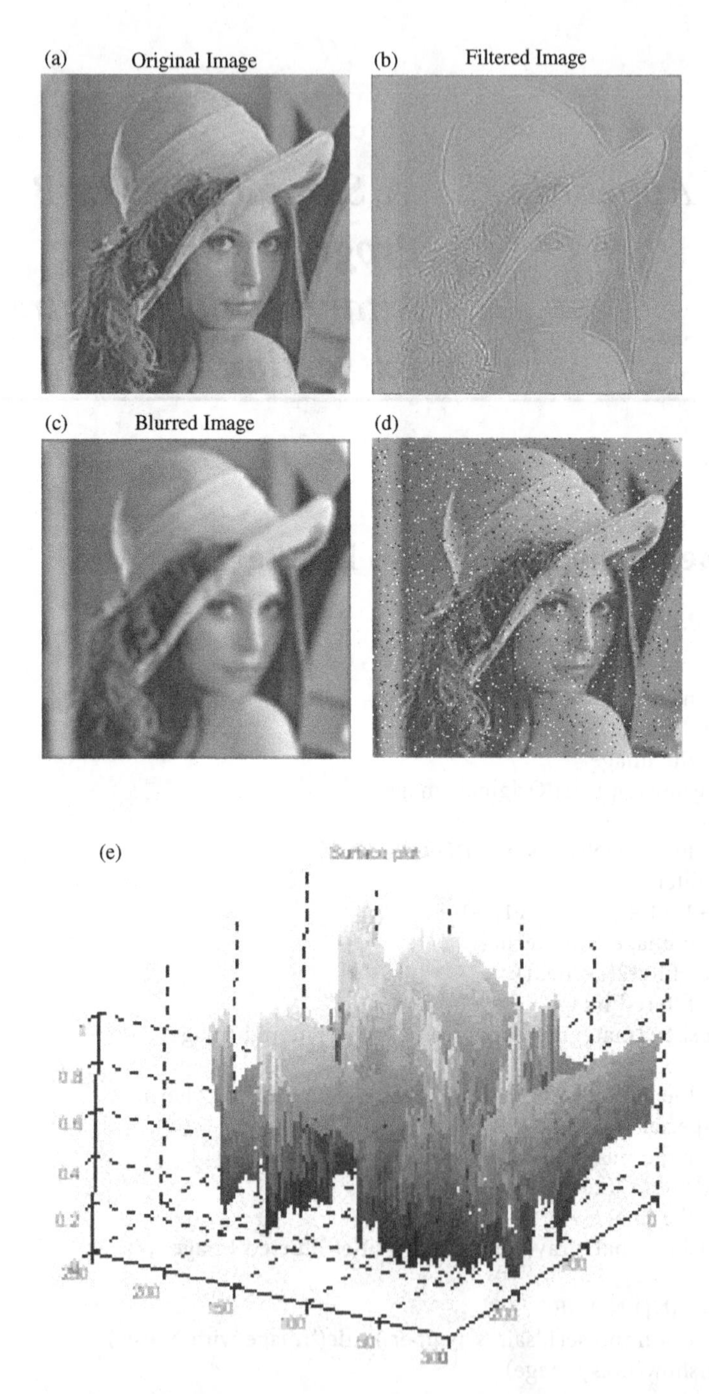

Fig. C.1 Image analysis of Lena image performed in the sample MATLAB program: (a) original image; (b) convoluted image; (c) blurred image; (d) added noise; (e) 3D image of Lena, where the third dimension is a bar graph of the intensity of each pixel; (*Continued overleaf*)

(f)

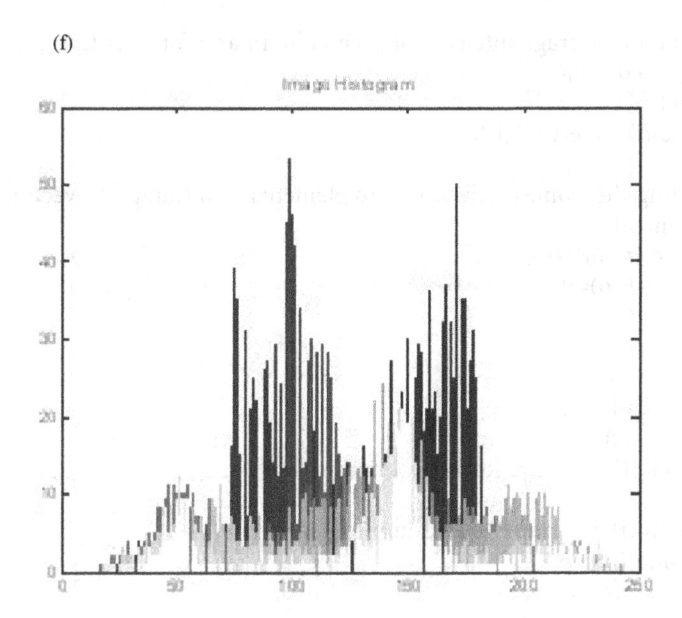

Fig. C.1 (*continued*) (f) histogram of the intensity of the Lena image.

```
figure; mesh(m), rotate3d on; colormap(gray); title ('Surface plot');
pause
% obtain histogram of the image
figure; hist(double(I),256); colormap(gray);title('Image Histogram')
pause;
```

Image Feature Extraction MATLAB Program: Composite Profile Feature Vector Extraction

```
clc
%Read Lena.tif image
I=imread('lena.tif');
%Display the image
figure; imshow(I); title('Original Image');
pause;
```

%Image Feature Extraction - Composite Profile Vector of 16+16=32 elements

```
NFro=16; %number of composite row features
NFcol=16; %number of composite column features
[Nro,Ncol]=size(I);
%calculating the average intensity of each row and forming CompRo vector
CompRo(1:Nro)=0;
for i=1:Nro
    CompRo(i)= mean(I(i,:));
end;
```

```
%calculating the average intensity of each column and forming CompCol vector
CompCol(1:Ncol)=0;
for j=1:Ncol
   CompCol(j)= mean(I(:,j));
end;
% aggregating the ComRo vector of Nro elements into CompFRo vector of
% NFro elements
CompRoStep=floor(Nro/NFro);
CompFRo(1:NFro)=0;
i=1;
k=1;
while i <= NFro
   l=k + CompRoStep -1;
        for j= k :l
     CompFRo(i)= CompFRo(i) + CompRo(j);
   end;
   CompFRo (i) = CompFRo(i)/CompRoStep;
 k=k + CompRoStep;
 i=i+1;
end; %while
% aggregating the CompCol vector of NCol elements into CompFCol vector of
% NFcol elements
CompColStep=floor(Ncol/NFcol);
d=NFcol;
CompFCol(1:d)=0;
i=1;
k=1;
while i <= d
   l=k + CompColStep -1;
     for j= k :l
     CompFCol(i)= CompFCol(i) + CompCol(j);
   end;
   CompFCol (i) = CompFCol(i)/CompColStep;
 k=k + CompColStep;
 i=i+1;
end; %while
% Composite profile feature vector
CompFVec= [CompFRo CompFCol]
save CompFVec;
```

Appendix D. Macroeconomic Data Used in Section 14.2 (Chapter 14)

EU Member Countr.	CPI	Int. Rates	Une-mpl.	GDP Cap.	EU Candid. Countr.	CPI	Int. Rates	Une-mpl.	GDP Cap.	Asia-Pacific and USA	CPI	Int. Rates	Une-mpl.	GDP Cap.
BE94	2.4	6.6	10.0	23501.88	BG94	96.0	102.5	12.8	1070.584	AU94	1.9	5.4	5.6	18864.80
DK94	2.1	5.6	8.2	29203.53	CZ94	10.0	15.0	3.2	3977.484	CA94	0.2	5.8	10.4	19339.92
DE94	2.7	5.6	8.4	25703.15	EE94	47.6	20.0	7.6	1551.433	JP94	0.7	4.5	2.9	37523.78
EL94	10.7	7.7	8.9	9493.891	HU94	18.8	27.3	11.4	4088.381	US94	2.6	7.1	6.1	27064.55
ES94	4.7	8.3	24.1	13069.69	LV94	35.8	35.3	20.0	1387.861	AU95	4.6	7.5	8.5	20011.71
FR94	1.8	6.2	12.3	23603.64	LT94	72.1	100.0	17.3	1128.341	CA95	2.1	7.3	9.4	20022.78
IR94	2.3	7.7	14.3	15249.09	PL94	32.2	42.2	16.0	2552.231	JP95	−0.1	3.4	3.2	41016.32
IT94	4.1	7.7	11.4	18223.49	RO94	137.0	93.1	10.9	1321.433	US95	2.8	8.8	5.6	28159.58
NL94	2.8	5.6	7.1	22839.21	SK94	13.3	17.6	14.8	2575.209	AU96	2.6	7.1	8.6	22125.82
AS94	2.9	5.6	3.8	24893.14	SI94	21.0	37.7	14.2	7228.965	CA96	1.6	4.5	9.6	20393.80
PT94	5.4	7.7	7.0	9406.548	TR94	106.3	104.0	8.1	2136.126	JP96	0.1	3.1	3.4	36635.78
FI94	1.1	5.6	16.6	19814.19	BG95	62.1	79.8	11.1	1450.371	US96	2.9	8.3	5.4	29447.22
SW94	2.4	4.0	9.4	23522.05	CZ95	9.2	14.3	2.9	5042.864	AU97	0.3	5.4	8.6	21893.79
UK94	2.4	6.6	9.6	17748.93	EE95	29.0	15.9	9.7	2323.289	CA97	1.6	3.5	9.1	20823.87
BE95	1.4	7.1	9.9	27688.33	HU95	28.4	32.5	11.3	4371.519	JP97	1.7	2.6	3.4	33470.18
DK95	2.0	8.1	7.2	34468.86	LV95	25.0	28.3	18.9	1760.944	US97	2.3	8.4	4.9	30978.79
DE95	1.7	6.6	8.2	30118.61	LT95	39.7	91.8	17.5	1603.128	AU98	0.8	5.0	8.0	19296.98
EL95	8.9	12.0	9.2	11268.59	PL95	27.9	36.7	15.2	3268.609	CA98	1.0	5.1	8.3	19913.63
ES95	4.7	11.0	22.9	15116.65	RO95	32.3	45.1	9.5	1562.484	JP98	0.7	2.4	4.1	30177.30
FR95	1.7	7.3	11.7	27027.11	SK95	9.9	18.3	13.1	3249.357	US98	1.6	8.4	4.5	32371.24
IR95	2.6	11.7	12.3	18313.38	SI95	13.5	20.7	14.5	9418.932	AU99	1.5	4.8	7.2	20695.62
IT95	5.3	11.7	11.9	19465.62	TR95	93.2	91.5	6.9	2793.032	CA99	1.8	4.9	7.6	20874.28
NL95	1.9	6.6	6.9	26818.29	BG96	121.6	300.3	12.5	1094.422	JP99	−0.3	2.3	4.7	34402.24
AS95	2.2	6.7	3.9	29274.04	CZ96	8.8	13.9	3.5	5618.062	US99	2.1	8.0	4.2	33933.58
PT95	4.2	11.7	7.3	11150.55	EE96	23.0	13.8	10.0	2835.259	CH94	24.1	15	2.8	453.8093
FI95	0.8	6.6	15.4	25519.55	HU96	23.5	27.8	10.7	4437.277	HK94	8.8	7.3	1.9	21844.09
SW95	2.9	9.9	8.8	27153.07	LV96	17.6	19.1	18.3	1981.451	KR94	6.2	12.5	2.4	9035.619
UK95	3.4	8.2	8.7	19207.55	LT96	24.6	62.3	16.4	2099.703	NZ94	2.8	8.4	8.1	14562.10
BE96	2.1	6.6	9.7	26878.00	PL96	19.9	25.0	13.1	3696.670	SN94	3.1	6.5	2.6	23783.86
DK96	2.1	7.2	6.8	34816.05	RO96	38.8	43.5	6.6	1560.051	CH95	17.1	12	2.9	579.6171
DE96	1.4	6.2	8.9	29112.06	SK96	5.8	16.2	12.8	3505.221	HK95	4.5	8	3.2	22765.22
EL96	8.2	10.9	9.6	11897.31	SI96	9.9	21.5	14.5	9486.336	KR95	4.5	12.5	2	10872.87

(Continued)

EU Member Countr.	*CPI*	*Int. Rates*	*Une-mpl.*	*GDP Cap.*	*EU Candid. Countr.*	*CPI*	*Int. Rates*	*Une-mpl.*	*GDP Cap.*	*Asia-Pacific and USA*	*CPI*	*Int. Rates*	*Une-mpl.*	*GDP Cap.*
ES96	3.6	9.0	22.2	15708.41	TR96	79.4	92.8	6.1	2801.376	NZ95	2.9	10.1	6.3	16818.37
FR96	2.0	6.4	12.4	26941.92	BG97	1061.5	209.8	14.0	1136.308	SN95	1.7	6	2.7	27523.79
IR96	1.7	9.9	11.6	19973.93	CZ97	8.5	13.9	5.2	5165.963	CH96	8.3	10.1	3	671.2046
IT96	4.0	9.9	12.0	21842.17	EE97	11.2	18.4	9.7	3036.381	HK96	6.3	8	2.8	24716.34
NL96	2.0	6.2	6.3	26506.04	HU97	18.3	22.4	10.9	4510.119	KR96	4.9	11.1	2	11446.39
AS96	1.5	6.2	4.3	28758.02	LV97	8.4	15.1	14.4	2228.753	NZ96	2.6	10.3	6.1	18166.88
PT96	3.1	8.1	7.3	11580.36	LT97	8.9	27.1	14.1	2550.053	SN96	1.4	5.5	3	28963.74
FI96	0.6	6.2	14.6	25125.13	PL97	14.8	25.0	10.5	3698.310	CH97	2.8	8.6	3	730.2216
SW96	0.8	8.2	9.6	29575.41	RO97	160.9	56.0	8.9	1556.907	HK97	5.8	8	2.2	26623.61
UK96	2.4	7.8	8.2	20060.55	SK97	6.0	15.9	12.5	3623.796	KR97	4.5	15.3	2.6	10381.88
BE97	1.6	5.8	9.4	24336.20	SI97	8.4	19.1	14.9	9548.725	NZ97	0.8	9.4	6.6	17775.99
DK97	2.3	6.3	5.6	31961.49	TR97	85.3	93.4	6.4	2975.614	SN97	2	5.5	2.4	28970.36
DE97	1.9	5.6	9.9	25780.23	BG98	18.7	14.1	12.2	1377.469	CH98	−0.7	7.1	3.1	772.4022
EL97	5.5	9.9	9.8	11514.43	CZ98	10.7	13.5	7.5	5488.726	HK98	2.8	9.9	4.7	24893.97
ES97	1.9	6.4	20.8	14393.66	EE98	10.5	16.5	9.9	3391.875	KR98	7.5	11.1	6.8	6840.121
FR97	1.2	5.6	12.3	24325.34	HU98	14.1	19.7	9.9	4659.209	NZ98	0.4	8.9	7.5	14427.66
IR97	1.5	7.1	9.8	21535.54	LV98	4.7	13.1	13.8	2513.289	SN98	−0.3	5.9	3.2	24496.42
IT97	2.0	7.1	12.1	20586.60	LT98	5.1	21.6	13.3	2863.991	CH99	0	5	3	791.3046
NL97	2.2	5.6	5.2	24130.14	PL98	11.6	24.5	10.9	4059.731	HK99	−4	8.5	6	23639.57
AS97	1.3	5.6	4.4	25615.78	RO98	59.1	38.8	10.4	1839.840	KR99	0.8	8.5	6.3	8711.929
PT97	2.3	7.1	6.8	11041.68	SK98	6.7	16.5	15.6	3786.201	NZ99	0.5	7.1	6.8	14596.51
FI97	1.2	5.6	12.7	24022.38	SI98	7.9	16.0	14.5	10024.23	SN99	0	5.8	3.3	24807.76
SW97	0.9	6.7	9.9	26786.31	TR98	83.7	93.9	6.4	3087.431					
UK97	3.2	7.2	7.0	22641.23	BG99	2.6	13.6	13.7	1422.346					
BE98	1.0	4.8	9.5	24981.72	CZ99	2.1	9.0	9.4	5180.776					
DK98	1.8	5.0	5.1	32903.07	EE99	3.3	8.6	12.0	3503.575					
DE98	1.0	4.6	9.4	26232.61	HU99	10.0	16.7	9.6	5070.581					
EL98	4.7	8.5	10.7	11535.17	LV99	2.3	13.6	9.1	2622.108					
ES98	1.8	4.9	18.7	14995.52	LT99	0.8	14.4	10.0	2817.907					
FR98	0.8	4.7	11.7	24958.29	PL99	7.3	17.5	13.3	3977.692					
IR98	2.4	5.0	7.8	23025.41	RO99	43.2	35.0	11.5	1507.497					
IT98	2.0	5.0	12.2	21050.44	SK99	10.5	14.9	19.2	3555.579					
NL98	2.0	4.6	4.0	24925.68	SI99	6.2	12.0	13.1	10802.41					
AS98	1.0	4.8	4.7	26109.71	TR99	63.6	79.3	7.3	2889.758					
PT98	2.7	5.0	5.1	11669.40										
FI98	1.5	4.6	11.4	25167.72										
SW98	0.4	5.2	8.3	26818.57										
UK98	3.4	5.7	6.3	24097.07										
BE99	1.1	4.7	9.0	24760.10										
DK99	2.4	5.0	5.2	32727.21										
DE99	0.6	4.5	8.7	25782.08										
EL99	2.7	6.4	10.4	11873.06										
ES99	2.3	4.4	15.9	15368.53										
FR99	0.6	4.9	11.3	24593.61										
IR99	1.6	4.8	5.7	24529.16										
IT99	1.7	4.0	11.4	20734.37										
NL99	2.2	4.6	3.3	24987.81										
AS99	0.6	4.3	3.7	25793.41										
PT99	2.3	4.8	4.5	11823.92										
FI99	1.2	4.7	10.2	25194.63										
SW99	0.3	5.0	7.2	26869.68										
UK99	1.6	5.1	6.2	24632.55										

References

Abbass, H.A. (2004) Evolving neural network ensembles by minimization of mutual information, *Int. J. Hybrid Syst.* **1**(1).

Abdulla, W. and Kasabov, N. (2003) Reduced feature-set based parallel CHMM speech recognition systems, *Inf. Sci.* **156**: 23–38.

Abe, S. and Lan, M.S. (1995) A method for fuzzy rules extraction directly from numerical data and its application to pattern classification. *IEEE Trans.Fuzzy Syst.* **3**: 18–28.

Abraham, W.C. et al. (1993) Correlating between immediate early gene induction and the persistence of long-term potentiation, *Neuroscience* **56**(3): 717–727.

Adami, C. (1998) *Introduction to Artificial Life*, Springer Verlag, New York.

Aertsen, A. and Johannesma, P. (1980) Spectro-temporal receptive fields of auditory neurons in grass frog. I. Categorization of tonal and natural stimuli, *Biol.Cybern.* **38**: 223–234.

Aha, D.W, Kibler, D., and Albert, M.K. (1991) Instance-based learning algorithms, *Mach. Learn.***6**: 37–66.

Ajjanagadde, V. and Shastri, L. (1991) Rules and variables in neural networks. *Neural Comput.* **3**(1): 121–134.

Albus, J.S. (1975) A new approach to manipulator control: The cerebellar model articulation controller (CMAC), *Trans. ASME, Dynamic Syst. Meas. Contr.* **97**(3): 220–227.

Aleksander, I. (Ed.) (1989) *Neural Computing Architectures. The Design of Brain-Like Machines.* MIT Press, Cambridge, MA.

Aleksander, I. and Morton, H. (1990) *An Introduction to Neural Computing*, Chapman & Hall, London.

Allen, J.B. (1995) *Speech and Hearing in Communication*, Acoustic Society of America, ASA ed., New York.

Allen, J.N., Abdel-Aty-Zohdy, H.S., Ewing, R.L., and Chang, T.S. (2002) Spiking networks for biochemical detection, circuits and systems, MWSCAS-2002. The 2002 45th Midwest Symposium vol. 3, 4–7 Aug. pp. III-129 - III-132 vol.3.

Almeida, L., Langlois, T., and Amaral, J. (1997) Online step size adaptation, Technical Report, INESC RT07/97.

Alon, A. et al. (1999) Patterns of gene expression revealed by clustering analysis of tumor and normal colon tissues probed by oligonucleotide arrays, *Proc. Natl. Acad. Sci.USA* **96**: 6745–6750.

Altman, G. (1990) *Cognitive Models of Speech Processing*, MIT Press, Cambridge, MA.

Amari, S. (1967) A theory of adaptive pattern classifiers. *IEEE Trans. Electron. Comput.* **16**: 299–307.

Amari, S. (1990) Mathematical foundations of neurocomputing.*Proc. IEEEE* **78**: 1143–1163.

Amari, S. and Kasabov, N. (Eds.) (1998) *Brain-like Computing and Intelligent Information Systems*, Springer Verlag, Singapore.

Amari, S., Cichocki, A., and Yang, H. (2000) Blind signal separation and extraction: Neural and information-theoretic approach. In: S. Haykin (Ed.) *Unsupervised Adaptive Filtering*, vol.1, John Wiley & Sons, New York, pp. 63–138.

Amit, D. (1989) *Modeling Brain Function: The World of Attractor Neural Networks*, Cambridge University Press, Cambridge, UK.

Anand, R., Mehrotra, K., Mohan, C.K., and Ranka, S. (1995) Efficient classification for multiclass problems using modular neural networks. *IEEE Trans. Neural Netw.* **6**: 117–124.

Anderson, J. (1995) *An Introduction to Neural Networks*, MIT Press, Cambridge, MA.

Anderson, J.A. (1983) Cognitive and psychological computations with neural models, *IEEE Trans. Syst. Man Cybern.* **SMC-13**: 799–815.

Anderson, J.R. (1983) *The Architecture of Cognition*, Harvard University Press, Cambridge, MA.

Anderson, K.M., Odell, P.M., Wilson, P.W.F., and Kannel, W.B. (1991) Cardiovascular disease risk profiles, *Am. Heart J.* **121**(1 Part 2): 293–298.

Andrews, R., Diederich, J., and Tickle, A.B. (1995) A survey and critique of techniques for extracting rules from trained artificial neural networks. Knowl. Based Syst. **8**: 373–389.

Angelov, P. (2002) *Evolving Rule-based Models: A Tool for Design of Flexible Adaptive Systems.* Springer-Verlag, Heidelberg.

Angelov, P. and Filev, D. (2004) An approach to online identification of evolving Takagi-Sugeno models, *IEEE Trans. Syst. Man Cybern. B* **34**(1): 484–498.

Angelov, P. and Kasabov, N. (2005) Evolving computational intelligence systems, In: R. Alcala et al. (Eds.) *Proceedings of the I Workshop on Genetic Fuzzy Systems*, Granada, March 17–19, pp.76–82.

Angelov, P., Victor, J., Dourado, A., and Filev, D. (2004) Online evolution of Takagi-Sugeno fuzzy models. In: *Proceedings of the 2nd IFAC Workshop on Advanced Fuzzy/Neural Control*, 16–17 September 2004, Oulu, Finland, pp. 67–72.

Anthony, M. and Biggs, N. (1992) *Computational Learning Theory: An Introduction*, Cambridge University Press, Cambridge, UK.

Arbib, M. (1972) *The Metaphorical Brain - An Introduction to Cybernetics as Artificial Intelligence and Brain Theory.* Wiley Interscience, New York.

Arbib, M. (1987) *Brains, Machines and Mathematics.* Springer Verlag, Berlin.

Arbib M, (1995, 2002) *The Handbook of Brain Theory and Neural Networks.* MIT Press, Cambridge, MA.

Arbib, M. (1998) From vision to action via distributed computation. In: Amari and Kasabov (Eds.) *Brain-Like Computing and Intelligent Information Systems*, Springer, Singapore and Heidelberg.

Arhns, I., Bruske, J., and Sommer, G. (1995) Online learning with dynamic cell structures. In: *Proceedings of the International Conference on Artificial Neural Networks*, pp. 141–146

Attwood, T. and Parry-Smith, D. (1999) *Introduction to Bioinformatics. Cell and Molecular Biology in Action.* Addison-Wesley Longman, Reading, MA.

Bacic, B. (2004) Towards a neuro fuzzy tennis coach: Automated extraction of the region of interest (ROI). In: *Proceedings of the IEEE International Joint Conference on Neural Networks*, Budapest, July 26–29, IEEE Press, Washington, DC.

Bajic, V.B. and Wee, T.T. (Eds.) (2005) *Information Processing and Living Systems, Series on Advances in Bioinformatics and Computational Biology*, Y. Xu and L. Wong (Series Eds.), Imperial College Press, Singapore.

Baker, P. and Brass, A. (1998) Recent developments in biological sequence databases. *Curr. Opinion Biotechnol.* **9**: 54–58.

Baker, R.H.A. (2002) Predicting the limits to the potential distribution of alien crop pests. In: G. Halman and C.P. Schwalbe (Eds.), *Invasive Arthropods and Agriculture: Problems and Solutions*, Science, Enfield, NH, Chapter 11, pp. 208–241.

Baldi, P. and Brunaks, S. (1998, 2001) *Bioinformatics - A Machine Learning Approach*, MIT Press, Cambridge, MA.

Baldi, P. and Hatfield, G.W. (2002) DNA *Microarrays and Gene Expression*, Cambridge University Press, Cambridge, UK.

Barlett, P. (1993) The sample size necessary for learning in multi-layer networks. In: *Proceedings of Australian Conference on Neural Networks*, pp.14–17.

Barnden, J. and Shrinivas, K. (1991) Encoding techniques for complex information structures in connectionist systems. *Connection Sci.* **3**(3): 269–315.

Barndorff-Nielsen, O., Jensen, J., and Kendall, W. (Eds.) (1993) *Networks and Chaos - Statistical and Probabilistic Aspects*, Chapman and Hall, London.

Bates, E. and Goodman, J. (1999) On the emergence of grammar from the lexicon. In: B. MacWhinney (Ed.), *The Emergence of Language.* Lawrence Erlbaum, Mahwah, NJ, pp. 29–79.

Baxevanis, A. (2000) The molecular biology database collection: An online compilation of relevant database resources. *Nucleic Acids Res.* **28**: 1–7.

Benioff, P. (1980) The computer as a physical system: A microscopic quantum mechanical Hamiltonian model of computers as represented by Turing machines, *J. Statist. Phys.* 22, 563.

Benson, D., Mizrachi-Karsch, I., Lipman, D., Ostell, J., Rapp, B., and Wheeler, D. (2000) Genbank. *Nucleic Acids Res.* **28**(1): 15–18.

Benuskova, L. and Kasabov, N. (2007) *Computational Neurogenetic Modelling*, Springer, New York.

Benuskova, L., Jain, V., Wysoski, S.G., and Kasabov, N. (2006) Computational neurogenetic modeling: a pathway to new discoveries in genetic neuroscience, *Int. J. Neural Syst.* **16**(3): 215–227.

Benuskova, L., Wysoski, S., and Kasabov, N. (2006) Computational neuro-genetic modelling: A methodology to study gene interactions underlying neural oscillations. In: *Proceedings of IJCNN 2006*, IEEE Press, Washington, DC.

Berenji, H.R. (1992) Fuzzy logic controllers. In: R.R. Yager and L.A. Zadeh (Eds.) *An Introduction to Fuzzy Logic Applications in Intelligent Systems.* Kluwer Academic, Higham, MA.

Bezdek, J. (1981) *Pattern Recognition with Fuzzy Objective Function Algorithms.* Plenum Press. New York.

Bezdek, J. (Ed.) (1987) *Analysis of Fuzzy Information*, vols. 1–3, CRC Press, Boca Raton, FL.

Bezdek J. (1993) A review of probabilistic, fuzzy, and neural models for pattern recognition. *J. Intell. Fuzzy Syst.* **1**: 1–25.

Bezdek, J., Hall, L.O., and Clarke, L.P. (1993) Review of MR image segmentation techniques using pattern recognition. *Med. Phys.* **20**:1033–1048.

Biehl, M., Freking, A., Holzer, M., Reents, G., and Schlosser, E. (1998) Online learning of prototypes and principal components. In: D. Saad (Ed.), *online Learning in Neural Networks*, Cambridge University Press, Cambridge, UK, pp 231–249.

Bienenstock, E.L., Cooper, L.N., and Munro, P.W. (1982) Theory for the development of neuron selectivity: Orientation specificity and binocular interaction in visual cortex. *J. Neurosci.* **2**: 32–48.

Bishop, C. (2000) *Neural Networks for Pattern Recognition*, Oxford University Press, New York.

Bishop, C.M., Svensen, M., and Williams, C.K.I. (1998) GTM: The generative topographic mapping. *Neural Comput.* **10**(1): 215–234.

Blake, C. and Merz, C. (1998) Repository of machine learning databases. http://www.ics.uci.edu/~mlearn/MLRepository.html: University of California, Irvine, CA. Department of Information and Computer Science.

Blanzieri, E. and Katenkamp, P. (1996) Learning radial basis function networks online. In: *Proceedings of the International Conference on Machine Learning*, Bari, Italy, Morgan Kaufmann, San Francisco, pp. 37–45.

Boguski, M. (1998) *Bioinformatics – A New Era. Trends Guide to Bioinformatics*, pp. 1–3.

Bohte, S.M., La Poutre, H.A., and Kok, J.N. (2000) Spikeprop: Error backpropagation of network of spiking neurons. In: *Proceedings of ESANN' 2000*, pp. 419–425.

Bottu, J. and Vapnik, V. (1992) Local learning computation. *Neural Comput.* **4**: 888–900.

Boubacar, H.A., Lecoeuche, S., and Maouche, S., (2005) Self-adaptive kernel machine: Online clustering in RKHS. In: *Proceedings of IEEE International Joint Conference on Neural Networks*, Montreal, July 31–August 4, IEEE Press, Washington, DC, pp. 1977–1982.

Bower, J.M. and Bolouri, H. (Eds.) (2001) *Computational Modelling of Genetic and Biochemical Networks*, MIT Press, Cambridge, MA.

Box, G. and Jenkins, G. (1970) *Time-Series Analysis: Forecasting and Control*, Holden-Day, San Francisco.

Box, G. and Tiao, G.C. (1973) *Bayesian Inference in Statistical Analysis*, Addison-Wesley, Reading, MA.

Bramer, M., Coenen, F., and Allen, T. (Eds.) (2005) Research and development in intelligent systems XII. In: *Proceedings of AI-2005, the Twenty-fifth SGAI International Conference on Innovative Techniques and Applications of Artificial Intelligence*, Springer, New York.

Brooks, M. (Ed.) (1999) *Quantum Computing and Communications*, Springer, New York.

Brown, C., Jacobs, G., Schreiber, M., Magnum, J., McNaughton, J., Cambray, M., Futschik, M., Major, L., Rackham, O., Tate, W., Stockwell, P., Thompson, C., and Kasabov, N. (2000a) Using bioinformatics to investigate post-trascriptional control of gene expression, *NZ Bio Science*, 7(4): 11–12.

Brown, C., Shreiber, M., Chapman, B., and Jacobs, G. (2000b) Information science and bioinformatics, In: N. Kasabov (Ed.) *Future Directions of Intelligent Systems and Information Sciences*, Physica Verlag (Springer Verlag), pp. 251–287.

Brown R.G. (1963) *Smoothing, Forecasting and Prediction of Discrete Time Series*, Prentice-Hall International, Upper Saddle River, NJ.

Brunelli, R. and Falavigna, D. (1995) Person identification using multiple cues, *IEEE Trans.Pattern Anal.Mach. Intell.* **17**: 955–966.

Bruske J. and Sommer, G. (1995) Dynamic cell structure learns perfectly topology preserving map. *Neural Comput.* **7**: 845–865.

Bruske, J., Ahrns, L., and Sommer, G. (1998) An integrated architecture for learning of reactive behaviours based on dynamic cell structures. *Robot. Auton. Syst.* **22**: 81–102.

Buckley, J.J. and Hayashi, Y. (1993) Are regular fuzzy neural nets universal approximators? In: *Proceedings of the International Journal Conference on Neural Networks (IJCNN)*, pp. 721–724.

Bukhardt, D. and Bonissone, P. (1992) Automated fuzzy knowledge base generation and tuning. In: *Proceedings of the First IEEE Fuzzy Systems Conference*, pp.179–188.

Bulsara, A. (1993) Bistability, noise, and information processing in sensory neurons. In: N. Kasabov (Ed.) *Artificial Neural Networks and Expert Systems*, IEEE Computer Society Press, Los Alamitos, CA, pp. 11–14.

Bunke, H. and Kandel, A. (2000) *Neuro-Fuzzy Pattern Recognition*, World Scientific, Singapore.

Bunke, H. and Kandel, A. (Eds.) (2000) *Neuro-Fuzzy Pattern Recognition*, Series in *Machine Perception Artificial Intelligence*, Vol. 41, World Scientific, Singapore.

Carpenter, G. and Grossberg, S. (1991) *Pattern Recognition by Self-Organizing Neural Networks*. MIT Press, Cambridge, MA.

Carpenter, G.A. and Grossberg, S. (1987) ART 2: Self-organization of stable category recognition codes for analogue input patterns, *Appl. Optics* **26**(23): 4919–4930, 1.

Carpenter, G.A. and Grossberg, S. (1990) ART3: Hierarchical search using chemical transmitters in self-organising pattern-recognition architectures. *Neural Netw.* 3(2): 129–152.

Carpenter, G.A., Grossberg, S., Markuzon, N., Reynolds, J.H., and Rosen, D.B. (1991) FuzzyARTMAP: A neural network architecture for incremental supervised learning of analogue multidimensional maps. *IEEE Trans. Neural Netw.*3(5): 698–713.

Chakraborty, S., Pal, K., and Pal, N.R. (2002) A neuro-fuzzy framework for inferencing. *Neural Netw.* **15**: 247–261.

Chan S. and Kasabov, N. (2004) Efficient global clustering using the greedy elimination method, *Electron. Lett.* **40**(25).

Chan, S.H. and Kasabov, N. (2005) Fast neural network ensemble learning via negative-correlation data correction, *IEEE Trans. Neural Netw.***16**(6): 1707–1710.

Chan, S.H., Collins, L., and Kasabov, N. (2005) Bayesian inference of sparse gene network, In: *Proceedings of the Sixth International Workshop on Information Processing in Cells and Tissues*, St William's College, York, UK, August 30–September 1, 2005, pp. 333–347.

Chan, S.Z., Kasabov, N., and Collins, L. (2005) A hybrid genetic algorithm and expectation maximization method for global gene trajectory clustering, *J. Bioinf. Comput. Biol.* 3(5): 1227–1242.

Chan, Z. and Kasabov, N. (2004) Evolutionary computation for online and off-line parameter tuning of evolving fuzzy neural networks, *Int. J. Comput. Intell. Appl.* **4**(3, Sept.): 309–319.

Chan, Z. and Kasabov, N. (2005) A preliminary study on negative correlation learning via correlation-corrected data (NCCD), *Neural Process. Lett.* **21**(3): 207–214.

Chan, Z., Collins, L., and Kasabov, N. (2006a) An efficient greedy K-means algorithm for global gene trajectory clustering, *Expert Sys. Appl..Int. J.* Special issue on intelligent bioinformatics systems, pp. 137–141.

Chan, Z., Kasabov, N., and Collins, L. (2006b) A two-stage methodology for gene regulatory network extraction from time-course gene expression data, *Expert Syst. App. Int. J.*, Special issue on intelligent bioinformatics systems, pp. 59–63.

Chan, Z.S. and Kasabov, N. (2004) Gene trajectory clustering with a hybrid genetic algorithm and expectation maximization method. In: *Proceedings of the International Joint Conference on Neural Networks, IJCNN 2004*, Budapest, 16–30 June, IEEE Press, Washington, DC.

Chaudhuri, D., Murthy, C., and Chaudhuri, B. (1992) A modified metric to compute distance, *Patt. Recogn.* **7**: 6687–677.

Chauvin, L. (1989) A backpropagation algorithm with optimal use of hidden units. *Adv. Neuro Inf. Process. Syst.* **1**: 519–526.

Chen, Y.Q., Thomas, D.W., and Nixon, M.S. (1994) Generating-shrinking algorithm for learning arbitrary classification. *Neural Netw.* **7**: 1477–1489.

Chen, W. and Smith, C. (1977) Adaptive coding of monochrome and colour images, *IEEE Trans. Commun.* **COM-25**(11): 1285–1292.

Cherkassky, V. and Mulier, F. (1998) *Learning from Data*, Series in Adaptive and Learning Systems for Signal Processing, Communications and Control, Haykin, S. (Series Ed.), Wiley Interscience, New York.

Chiu, S.L. (1994) Fuzzy model identification based on cluster estimation, *J. Intell. Fuzzy Syst.* **2**: 267–278,

Cho, R.J., Campbell, M.J., Winzeler, E.A., Steinmetz, L., Conway, A., Wodicka, L., Wolfsberg, T.G., Gabrielian, A.E., Landsman, D., Lockhart, D.J., and Davis, R.W. (1998) A genome-wide transcriptional analysis of the mitotic cell cycle, *Molec. Cell* **2**: 65–73.

Choi, B. and Bluff, K. (1995) Genetic optimisation of control parameters of neural networks. In: *Proceedings of the International Conference on Artificial Neural Networks and Expert Systems (ANNES* 1995), Dunedin, New Zealand, IEEE Computer Society Press, Washington, DC, pp. 174–177.

Chomsky, N. (1995) *The Minimalist Program.* MIT Press, Cambridge, MA.

Churchland, P. and Sejnowski, T. (1992) *The Computational Brain.* MIT Press, Cambridge, MA.

Clark, A. (1989) *Micro-Cognition: Philosophy, Cognitive Science, and Parallel Distributed Processing.* MIT Press, Cambridge, MA.

Cloete, I. and Zurada, J. (Eds.) (2000) *Knowledge-Based Neurocomputing*, MIT Press, Cambridge, MA.

Cockcroft D.W. and Gault, M.H. (1976) Prediction of creatinine clearance from serum creatinine,*Nephron* **16**:3–41.

Cole, R. et al. (1995) The challenge of spoken language systems: Research directions for the nineties. *IEEE Trans. Speech Audio Process.* 3(1): 1–21.

Collado-Vides, J., Magasanik, B., and Smith, T.F. (1996) *Integrative Approaches to Molecular Biology*, MIT Press, Cambridge, MA.

Comon, P. (1994) Independent component analysis, a new concept? *Signal Process.* **3**(36): 287–314.
Connor, C.E. (2005) Friends and grandmothers. *Nature* **435**(June): 23.
Coyle, D. and McGinnity, T.M. (2006), Enhancing autonomy and computational efficiency of the self-organizing fuzzy neural network for a brain-computer interface. In: *Proceedings of the IEEE International Conference on Fuzzy Systems*, Vancouver, July 16–21, IEEE Press, Washington, DC, pp. 10485–10492.
Crick, F. (1970) *Central dogma of molecular biology. Nature.* **227**: 561–563.
Culicover, P. (1999) *Syntactic Nuts: Hard Cases, Syntactic Theory, and Language Acquisition.* Oxford University Press, Oxford.
Cybenko, G. (1989) Approximation by superpositions of a sigmoidal function, *Math. Control Signals Syst.* **2**: 303–314.
D'haeseleer, P., Liang, S., and Somogyi, R. (2000) Genetic network inference; From co-expression clustering to reverse engineering. *Bioinformatics* **16**(8): 707–726.
Dasgupta, D. and Michalewicz, Z. (1997) *Evolutionary Algorithms in Engineering Applications*, Springer-Verlag, Berlin.
Darwin, C. (1859) *The Origin of Species by Means of Natural Selection.* London: John Murray.
de Bollivier, M., Gallinari, P., and Thiria, S. (1990) Cooperation of neural nets for robust classification, Universite de Paris-Sud, Report 547, *Informatiques.*
Deacon, T. (1988) Human brain evolution: Evolution of language circuits. In H. Jerison and I. Jerison (Eds.), *NATO ASI Series Intelligence and Evolutionary Biology*, Springer Verlag, Berlin.
Deacon, T. (1998) *The Symbolic Species. The Co-Evolution of Language and the Human Brain.* Penguin, New York.
Dean, T., Allen, J., and Aloimonos, Y. (1995) *Artificial Intelligence, Theory and Practice*, Benjamin/Cummings, Menlo Park, CA.
de-Boer, E. and de Jongh, H.R. (1978) On cochlea encoding : Potentialities and limitations of the reverse-correlation technique, *J. .Acoust. Soc. Am.* **63**(1): 115–135.
Delorme, A. and Thorpe, S. (2001) Face identification using one spike per neuron: Resistance to image degradation, *Neural Netw.* **14**: 795–803.
Delorme, A., Gautrais, J., van Rullen, R., and Thorpe, S. (1999) SpikeNet: A simulator for modeling large networks of integrate and fire neurons, *Neurocomputing* **26–27**:. 989–996.
Delorme, A., Perrinet, L., and Thorpe, S.J. (2001) Networks of integrate-and-fire neurons using Rank Order Coding B: Spike timing dependent plasticity and emergence of orientation selectivity. *Neurocomputing* **38–48**.
Deng, D. and Kasabov, N. (2000) ESOM: An algorithm to evolve self-organizing maps from online data streams. In: *Proceedings of IJCNN'2000*, vol. VI, Como, Italy, pp. 3–8.
Deng, D. and Kasabov, N. (2003) Online pattern analysis by evolving self-organising maps, *Neurocomputing* **51**(April): 87–103.
DeRisi, J.L., Iyer, V.R., and Brown, P.O. (1997) Exploring the metabolic and genetic control of gene expression on a genomic scale, *Science* **275**: 680–686.
Destexhe, A. (1998) Spike-and-wave oscillations based on the properties of $GABA_B$ receptors. *J. Neurosci.* **18**: 9099–9111.
Destexhe, A., Contreras, D., and Steriade, M. (1999) Spatiotemporal analysis of local field potentials and unit discharges in cat cerebral cortex during natural wake and sleep states. *J. Neurosci.* **19**: 4595–4608
Dimitrov, D.S., Sidorov, I.A., and Kasabov, N.K. (2006) *Computational Biology.* In: M. Rieth and W. Sommers (Eds.), *Handbook of Theoretical and Computational Nanotechnology*, vol. 6, American Scientific, Singapore, Chapter 1, 2–41.
Dingle, A., Andreae, J., and Jones, R. (1993) The chaotic self-organizing map. In: N. Kasabov (Ed.), *Artificial Neural Networks and Expert Systems*, IEEE Computer Society Press, Los Alamitos, CA, pp. 15–18.
Dowling, J. (2006) To compute or not to compute? *Nature* **439**(Feb.): 23.
Doya, K. (1999) What are the computations of the cerebellum, the basal ganglia, and the cerebral cortex. *Neural Netw.* **12**: 961–974.
Dubois, D. and Prade, H. (1980) *Fuzzy Sets and Systems: Theory and Applications*, Academic Press, New York.
Dubois D. and Prade, H. (1988) *Possibility Theory. An Approach to Computerised Processing of Uncertainty*, Plenum Press, New York and London.
Duch, W., Adamczak, R., and Grabczewski, K. (1998) Extraction of logical rules from neural networks, *Neural Proc. Lett.* **7**: 211–219.
Duda, R. and Hart, P. (1973) *Pattern Classification and Scene Analysis.* John Wiley and Sons, New York.
Duell, P., Fermin, I., and Yao X., (2006) Speciation techniques in evolved ensembles with negative correlation learning. In: *Proceedings of the IEEE Congress on Evolutionary Computation*, Vancouver, July 16–21, IEEE Press, Washington, DC, pp. 11086–11090.

Durand G., Kovalchuk, Y., and Konnerth, A. (1996) Long-term potentiation and functional synapse induction in developing hippocampus. *Nature* **381**(5): 71–75.

Edelman, G. (1992) *Neuronal Darwinism: The Theory of Neuronal Group Selection.* Basic, New York.

Elman, J. (1990) Finding structure in time, *Cogn. Sci.* **14**: 179–211.

Elman, J., Bates, E., Johnson, M., Karmiloff-Smith, A., Parisi, D., and Plunkett, K. (1997) *Rethinking Innateness (A Connectionist Perspective of Development).* MIT Press, Cambridge, MA.

Erdi, P. (2007) Complex explained. Springer, Berlin.

Eriksson, P.S., Perfilieva, E., Bjork-Eriksson, T., Alborn, A.M., Norborg, C., Peterson, D.A., and Gag, F.H. (1998) Neurogenesis in the adult human hippocampus. *Nature Med.* **4**: 1313–1317.

Eriksson, J.L. and Villa, A.E.P. (2006) Artificial neural networks simulation of learning of auditory equivalence classes for vowels. In: *Proceedings of IEEE International Joint Conference on Neural Networks*, Vancouver, July 16–21, IEEE Press, Washington, DC, pp. 1253–1260.

Ezhov, A. and Ventura, D. (2000) Quantum neural networks. In: N. Kasabov (Ed.), *Future Directions for Intelligent Systems and Information Sciences*, Springer Verlag, New York.

Fahlman, C. and Lebiere, C. (1990) The cascade-correlation learning architecture. In: Turetzky, D. (Ed.), *Advances in Neural Information Processing Systems*, Vol.2, Morgan Kaufmann, San Francisco, pp. 524–532.

Farmer, J. and Sidorowich (1987) Predicting chaotic time series, *Phys. Rev. Lett.* **59**: 845.

Feigenbaum, M. (1989) *Artificial Intelligence, A Knowledge-Based Approach*, PWS-Kent, Boston.

Feldkamp, L.A. et al. (1992) Architecture and training of a hybrid neural-fuzzy system. In: *Proceedings of the Second International Conference on Fuzzy Logic & Neural Networks*, Iizuka, Japan, pp. 131–134.

Feldman, J.A. (1989) Structured neural networks in nature and in computer science. In: R. Eckmiller and C.v.d. Malsburg (Eds.),*Neural Computers*, Springer-Verlag, Berlin.

Feynman, R.P., Leighton, R.B., and Sands, M. (1965) *The Feynman Lectures on Physics*, Addison-Wesley, Reading. MA.

Filev, D. (1991) Identification of fuzzy relational models. In: *Proceedings of the IV IFSA Congress*, Brussels, pp. 82–85.

Fisher, D.H. (1989) Knowledge acquisition via incremental conceptual clustering, *Mach. Learn.* **2**: 139–172.

Fisher, R. (1936) The use of multiple measurements in taxonomic problems. *Ann. Eugenics* **7**.

Fodor, J. and Pylyshyn, Z. (1988) Connectionism and cognitive architecture: A critical analysis. *Cognition* **28**: 3–71.

Fogel, D. (2002) *Blondie 24. Playing at the Edge of AI*, Morgan Kaufmann, San Diego.

Fogel, D., Fogel, L., and Porto, V. (1990) Evolving neural networks, *Biol. Cybern.* **63**: 487–493.

Fogel, D.B. (2000) *Evolutionary Computation: Toward a New Philosophy of Machine Intelligence*, 2d ed., IEEE Press, Piscataway, NJ.

Freeman J.A.S. and Saad, D. (1997) Online learning in radial basis function networks. *Neural Comput.***9**: 1601–1622.

Freeman, W. (1987) Simulation of Chaotic EEG Patterns with a dynamic model of the olfactory system, *Biol. Cybern.* **56**: 139–150.

Freeman, W. (1991) The physiology of perception, *Sci. Amer.* **2**: 34–41.

Freeman, W. (2000) Neurodynamics, Springer, London

Freeman, W. (2001) *How Brains Make Up Their Minds*, Columbia University Press, NY.

Freeman, W. and Skarda, C. (1985) Spatial EEG patterns, nonlinear dynamics and perception: The neo-Sherringtonian view, *Brian Res. Rev.* **10**: 147–175.

Friend, T. (2000) Genome projects complete sequence, *USA Today*, June 23.

Fritzke, B. (1995) A growing neural gas network learns topologies. *Adv. Neural Inf. Process. Syst.* 7 MIT Press, Ca, MA.

Fu, L. (1989) Integration of neural heuristics into knowledge-based inference. *Connect. Sci.* **1**(3).

Fu, L. (1999) An expert network for DNA sequence analysis. *IEEE Intell. Syst. Appl.* **14**(1): 65–71.

Fukuda, T., Komata, Y., and Arakawa, T. (1997) Recurrent neural networks with self-adaptive GAs for biped locomotion robot. In: *Proceedings of the International Conference on Neural Networks ICNN'97*, IEEE Press, Washington, DC.

Fukushima, K. (1987) Neural network model for selective attention in visual pattern recognition and associative recall, *Appl. Optics* **26**(23): 4985–4992.

Fukushima, K. (1997) Active vision: Neural network models. In: S. Amari and N. Kasabov (Eds.), *Brain Like Computing and Intelligent Information Systems*, Springer Verlag, New York.

Fukushima, K., Miyake, S., and Ito, T. (1983) Neocognitron: A neural network model for a mechanism of visual pattern recognition, *IEEE Trans. Syst. Man Cybern* **SMC-13**: 826–834.

Funahashi, K. (1989) On the approximate realization of continuous mappings by neural networks, *Neural Netw.* **2**: 183–192.

Furlanello, C., Giuliani, D., and Trentin, E. (1995) Connectionist speaker normalisation with generalised resource allocation network. In: D. Toretzky, G. Tesauro, and T. Lean (Eds.), *Advances in NIPS7*, MIT Press, Cambridge, MA, pp. 1704-1707.

Furuhashi, T., Hasegawa, T., Horikawa, S., and Uchikawa, Y. (1993) An adaptive fuzzy controller using fuzzy neural networks. In: *Proceedings of Fifth IFSA World Congress*, pp. 769–772.

Furuhashi, T., Nakaoka, K., and Uchikawa, Y. (1994) A new approach to genetic based machine learning and an efficient finding of fuzzy rules. In: *Proceedings of the WWW'94 Workshop*, University of Nagoya, Japan, pp. 114–122.

Futschik, M. and Kasabov, N. (2002) Fuzzy clustering in gene expression data analysis. In: *Proceedings of the World Congress of Computational Intelligence WCCI'2002*, Hawaii, May, IEEE Press, Washington, DC.

Futschik, M., Jeffs, A., Pattison, S., Kasabov, N., Sullivan, M., Merrie, A., and Reeve, A. (2002) Gene expression profiling of metastatic and non-metastatic colorectal cancer cell-lines, *Genome Lett.* **1**(1): 1–9.

Futschik, M., Reeve, A., and Kasabov, N. (2003a) Evolving connectionist systems for knowledge discovery from gene expression data of cancer tissue, *Artif. Intell. Med.* **28**:165–189.

Futschik, M., Sullivan, M., Reeve, A., and Kasabov, N. (2003b) Prediction of clinical behaviour and treatment of cancers, *Appl. Bioinf.* **3**: 553–558.

Fuzzy Logic Toolbox User's Guide (2002) MathWorks Inc., 3 Apple Hill Drive, Natick, MA, Ver.2.

Gallant, S. (1993) *Neural Network Learning and Expert Systems*, MIT Press, Bradford, Cambridge, MA.

Gallinari, P., Thinia, S., and Fogelman-Soulie, F. (1988) Multilayer perceptrons and data analysis. In: *Proceedings of IEEE International Conference on Neural Networks*, 24–27 July, USA, vol 2, pp. I391–I399.

Gates, G.F. (1985) Creatinine clearance estimation from serum creatinine values: An analysis of three mathematical models of glomerular function, *Am. J. Kidney Diseases* **5**: 199–205.

Gaussier, P. and Zrehen, S. (1994) A topological neural map for online learning: Emergence of obstacle avoidance in a mobile robot, *Animals Animats* **3**: 282–290.

Gerstner, W. and Kistler, W.M. (2002) *Spiking Neuron Models*, Cambridge University Press, Cambridge, UK.

Gervautz, M. and Purgathofer, W. (1990) A simple method for colour quantization: Octree quantization, In: Glassner (Ed.) *Graphics Gems*, Academic Press, New York.

Gevrey, M., Dimopoulos, I., and Lek, S. (2003) Review and comparison of methods to study the contribution of variables in artificial neural network models, *Ecol. Model.* **160**: 249–264.

Gevrey, M., Worner, S., Kasabov, N., Pitta J., and Giraudel, J.-L. (2006) Estimating risk of events using SOM models: A case study on invasive species establishment, *Ecol. Model*, 197, pp. 361–372.

Ghobakhlou, A. and Kasabov, N. (2004) A methodology and a system for adaptive integrated speech and image learning and recognition, *Int. J. Comput. Syst. Signals* **5**(2).

Ghobakhlou, A., Watts, M., and Kasabov, N. (2003) Adaptive speech recognition with evolving connectionist systems, *Inf. Sci.* **156**: 71–83.

Ghobakhlou, A., Zhang D., and Kasabov, N. (2004) *An Evolving Neural Network Model for Person Verification Combining Speech and Image*, LNCS, vol. 3316, Springer, New York, pp. 381–386.

Giacometti, A., Amy, B., and Grumbach, A. (1992) Theory and experiments in connectionist AI: A tightly coupled hybrid system. In: I. Aleksander and J. Taylor (Eds), *Artificial Neural Networks*, 2, Elsevier Science Publishers B.V., pp. 707–710.

Gibson, M.A. and Mjolsness, E. (2001) Modelling the activity of single genes. In: J.M. Bower and H. Bolouri (Eds.), *Computational Modelling of Genetic and Biochemical Networks*, MIT Press, Cambridge, MA, pp. 3–48.

Giles, L., Lawrence, S., Bollacker, K., and Glover, E. (2000) Online computing: The present and the future. In: P.P. Wang (Ed.), *Proceedings of the Joint Conference on Information Sciences - JCIS*, Atlantic City, p. 843.

Glarkson, T., Goarse, D., and Taylor, J. (1992) From wetware to hardware: Reverse engineering using probabilistic RAM's, *J. Intell. Syst.* **2**: 11–30.

Glassberg, B.R. and Moore, B.C. (1990) Derivation of auditory filter shapes from notched noise data, *Hearing Res.* **47**: 103–108.

Gleick, J. (1987) *Chaos: Making a New Science*, Viking Press, New York

Goh, L. and Kasabov, N. (2005) An integrated feature selection and classification method to select minimum number of variables on the case study of gene expression data, *J. Bioinf. Comput. Biol.* **3**(5): 1107–1136.

Goldberg, D.E. (1989) *Genetic Algorithms in Search, Optimisation and Machine Learning*, Addison-Wesley, Reading, MA.

Golub, T.R. et al. (1999) Molecular classification of cancer: Class discovery and class prediction by gene expression monitoring, *Science* **286**: 531–537.

Gottgtroy, P., Kasabov, N., and Macdonell, S. (2006) Evolving ontologies for intelligent decision support. In: E. Sanches (Ed.), *Fuzzy Logic and the Semantic Web*, Elsevier, New York, Chapter 21, pp. 415–439.

Gray, M.S., Movellan, J.R., and Sejnowski, T.J. (1997) Dynamic features for visual speech reading: A systematic comparison. In: M.C. Mozer, M.I. Jordan, and T. Petsche (Eds.), *Advances in Neural Information Proceeding Systems*, vol. 9, Morgan-Kaufmann, San Francisco, pp.751–757.

Greenwood, D. (1961) Critical bandwidth and the frequency coordinates of the basilar membrane, *J. Acoust. Soc. Am.* **33**: 1344–1356.

Greenwood, D. (1990) A cochlear frequency-position function for several species – 29 years later, *J. Acoust. Soc. Am.* **87**(6): 2592–2605.

Grossberg, S. (1969) On learning and energy - Entropy dependence in recurrent and nonrecurrent signed networks, *J. Statist. Phys.* **1**: .319–350.

Grossberg, S. (1982) *Studies of Mind and Brain.* Reidel, Boston.

Grossberg, S. (1988) Nonlinear neural networks: principles, mechanisms and architectures. *Neural Netw.* **1**: 17–61.

Grossberg, S. and Merrill, J.W.L. (1996) The hippocampus and cerebellum in adaptively timed learning, recognition and movement. *J. Cognitive Neurosci.* **8**: 257–277.

Grover, L.K. (1996) A fast quantum mechanical algorithm for database search. In: *STOC '96: Proceedings of the Twenty-Eighth Annual ACM Symposium on Theory of Computing*, New York, ACM Press, New York, pp. 212–219.

Guisan, A. and Zimmermann, N.E. (2000) Predictive habitat distribution models in ecology, *Ecol. Model.* **135**: 147–186.

Gupta, M. (1992) Fuzzy logic and neural networks. In: *Proceedings of the Second International Conference on Fuzzy Logic & Neural Networks*, Iizuka, Japan, July, 1992, pp. 157–160.

Gupta, M.M. and Rao, D.H. (1994) On the principles of fuzzy neural networks, *Fuzzy Sets Syst.* **61**(1): 1–18.

Hagan, M.T., Debut, H.B., and Beale, M. (1996) *Neural Network Design*,. PWS, Boston.

Hall, L., Bensaid, A.M., Clarke, L.P., Velthuizen, R.P., Silbiger, M.S., and Bezdek, J.C. (1992) A comparison of neural network and fuzzy clustering techniques in segmenting magnetic resonance images of the brain. *IEEE Trans. Neural Netw.* **3**: 672–682.

Hall, P. and Martin. R. (1998) Incremental eigenanalysis for classification. In: *British Machine Vision Conference*, Vol. 1, pp. 286–295.

Hamker, F.H. (2001) Life-long learning cell structures – Continuously learning without catastrophic interference, *Neural Netw*, Vol. 14, No. 4–5, 551–573.

Hamker F.H. and Gross H.-M. (1998) A lifelong learning approach for incremental neural networks. In: *Proceedings of the Fourteenth European Meeting on Cybernetics and Systems Research (EMSCR'98)*, Vienna, pp. 599–604.

Han, J. and Kauber, M. (2000) *Data Mining – Concepts and Techniques.* Morgan Kaufmann, San Francisco.

Han, K.-H. and Kim, J.-H. (2002) Quantum-inspired evolutionary algorithm for a class of combinatorial optimization, *IEEE Trans.Evol. Comput*, Vol. 6, No. 6, pp. 580–593.

Harris, C. (1990) Connectionism and cognitive linguistics, *Connection Sci.* **2**(1&2): 7–33.

Hartigan, J. (1975) *Clustering Algorithms.* John Wiley and Sons, New York.

Hartmann, W.M. (1998) *Signals, Sound, and Sensation*, Springer Verlag, New York.

Hassibi, B. and Stork, D. (1992) Second order derivatives for network pruning optimal brain surgeon. *Adv. Neural Inf. Process. Syst.* **5**: 164–171, Morgan Kaufmann.

Hassoun, M. (1995) *Fundamentals of Artificial Neural Networks*, MIT Press, Cambridge, MA.

Hauptmann, W. and Heesche, K. (1995) A neural net topology for bidirectional fuzzy-neuro transformation. In: *Proceedings of the FUZZ-IEEE/IFES*, Yokohama, Japan, pp. 1511–1518.

Havukkala, I., Benuskova, L., Pang, S., Jain, V., Kroon, R., and Kasabov, N. (2006) Image and fractal information processing for large-scale chemoinformatics, genomics analyses and pattern discovery. In: *Proceedings of the International Conference on Pattern Recognition in Bioinformatics, PRIB06.*

Havukkala, I., Pang, S., Jain, V., and Kasabov, N. (2005) Novel method for classifying microRNAs by Gabor filter features from 2D structure bitmap images. *J. Theor. Comput. Nanosci.* **2**(4): 506–513.

Hayashi, Y. (1991) A neural expert system with automated extraction of fuzzy if-then rules and its application to medical diagnosis. In: R.P. Lippman, J.E. Moody, and D.S. Touretzky (Eds.), *Advances in Neural Information Processing Systems* 3 Morgan Kaufman, San Mateo, CA, pp. 578–584.

Haykin, S. (1994) *Neural Networks – A Comprehensive Foundation*, Prentice-Hall, Upper Saddle River, NJ.

Haykin, S. (1999) *Neural Networks - A Comprehensive Foundation*, Prentice-Hall, Upper Saddle River, NJ.

Hebb, D. (1949) *The Organization of Behaviour*, Wiley, New York.

Hecht-Nielsen, R. (1987) Counter-propagation networks. In: *IEEE First International Conference on Neural Networks*, San Diego, vol.2, pp. 19–31.

Heckbert, P. (1982) Colour image quantization for frame buffer display, *Comput. Graph. (SIGGRAPH)* **16**: 297–307.

Hendler, J. and Dickens, L. (1991) Integrating neural network and expert reasoning: An example. In: L. Steels and B. Smith (Eds.), *Proceedings of AISB Conference*, Springer Verlag, New York, pp. 109–116.

Herik, J. and Postma, E. (2000) Discovering the visual signature of painters. In: N. Kasabov (Ed.), *Future Directions for Intelligent Systems and Information Sciences*, Physica Verlag (Springer Verlag), Berlin, pp. 130–147.

Hertz, J., Krogh, A., and Palmer, R. (1991) *Introduction to the Theory of Neural Computation*, Addison-Wesley, Reading, MA .

Heskes, T.M. and Kappen, B. (1993) Online learning processes in artificial neural networks. In: *Mathematic Foundations of Neural Networks*, Elsevier, Amsterdam, pp. 199–233.

Hey, T. (1999) Quantum computing: An introduction, *Comput. Control Eng. J.* **10**:(3, June): 105–112.

Hinton, G. (1987) Connectionist learning procedures, Computer Science Department, Carnegie-Mellon University, Pittsburgh, PA.

Hinton, J. (Ed.) (1990) Connectionist symbol processing, Special Issue of *Artif. Intell.* **46**.

Hirota, K. (1984) *Image Pattern Recognition.* McGraw-Hill, Tokyo.

Hodgkin, A.L. and Huxley, A.F. (1952) A quantitative description of membrane current and its application to conduction and excitation in nerve,*J. Physiol*, **117**: 500–544.

Hofstadter, D. (1979) *Godel, Escher, Bach: An Eternal Golden Braid.* Basic, New York.

Hogg, T. and Portnov, D. (2000) Quantum optimization, *Inf. Sci.* **128**: 181–197.

Holland, J. (1992) *Adaptation in Natural and Artificial Systems.* MIT Press, Cambridge, MA.

Holland, J.H. (1998) *Emergence.* Oxford University Press, Oxford, UK.

Hopfield, J. (1982) Neural networks and physical systems with emergent collective computational abilities, *Proc. Nat. Acad. Sci. USA* **79**: 2554–2558.

Hopfield, J. and Tank, D. (1985) Neural computation of decisions in optimization problems, Biol. Cybern. **52**: 141–152.

Hopfield, J., Feindtein, D., and Palmer, R. (1983) Unlearning has a stabilizing effect in collective memories, *Nature*, **304**: 158–159.

Hopkin, D. and Moss, B. (1976). *Automata.* Macmillan, New York.

Hoppensteadt, F. (1986) *An Introduction to the Mathematics of Neurons.* Cambridge University Press, Cambridge, UK.

Hoppensteadt, F. (1989) Intermittent chaos, self-organisation and learning from synchronous synaptic activity in model neuron networks, *Proc. Nat. Acad. Sci. USA* **86**: 2991–2995.

Hornik, K. (1991) Approximation capabilities of multilayer feedforward networks, *Neural Netw.* **4**: 251–257.

Hornik, K., Stinchcombe, M., and White, H. (1989) Multilayer feedforward networks are universal approximators, *Neural Netw.* **2**(5): 359–366.

Howell, W.N. (2006) Genetic specification of recurrent neural networks: Initial thoughts. In: *Proceedings of the IEEE International Joint Conference on Neural Networks*, Vancouver, July 16–21, IEEE Press, Washington, DC, pp. 9370–9379.

Huang, L., Song, Q., and Kasabov, N. (2005) Evolving connectionist systems based role allocation of robots for soccer playing. In: *Joint 2005 International Symposium on Intelligent Control & 13th Mediterranean Conference on Control and Automation (2005 ISIC-MED)*, June 27–29, Limassol, Cyprus.

Hubel, D.H. and Wiesel, T.N. (1962) Receptive fields, binocular interaction and functional architecture in the cat's visual cortex, *J. Physiol* **160**: 106–154.

Hull, J.H., Hak, L.J., Koch, G.G., Wargin, W.A., Chi, S.L., and Mattocks, A.M. (1981) Influence of range of renal function and liver disease on predictability of creatinine clearance. *Clin. Pharmacol. Ther.* **29**:516–521.

Hunter, L. (1994) Artificial intelligence and molecular biology. *Canad. Artif. Intell.* **35**(Autumn).

Hyvarinen, A., Karhunen, J., and Oja, E. (2001) *Independent Component Analysis*, John Wiley & Sons, New York.

Ishikawa, M. (1992) A new approach to problem solving by modular structured networks. In: *Proceedings of the Second International Conference on Fuzzy Logic and Neural Networks*, Iizuka, Japan, pp. 855–858.

Ishikawa, M. (1996) Structural learning with forgetting, *Neural Netw.* **9**: 501–521.

Israel, S. and Kasabov, N. (1996) Improved learning strategies for multimodular fuzzy neural network systems: A case study on image classification. *Austral. J. Intell. Inf. Process. Syst.* **3**(2): 61–69.

Izhikevich, E.M. (2003) Simple model of spiking neurons, *IEEE Trans. Neural Netw.* **14**: 1569.

Jang, J.-S., Han, K.-H., and Kim, J.-H. (2003) *Quantum-Inspired Evolutionary Algorithm-Based Face Verification*, LNCS, Springer, New York, pp. 2147–2156.

Jang, R. (1993) ANFIS: Adaptive network based fuzzy inference system. *IEEE Trans. Syst. Man Cybern.* **23**(3): 665–685.

Jordan, M. and Jacobs, R. (1994) Hierarchical mixtures of experts and the EM algorithm, *Neural Comput.* **6**: 181–214.

Joseph, S.R.H. (1998) Theories of adaptive neural growth, PhD Thesis, University of Edinburgh.

Jusczyk, P. (1997) *The Discovery of Spoken Language*, MIT Press, Cambridge, MA.

Kandel, E.R., Schwartz, J.H., and Jessell, T.M. (2000) *Principles of Neural Science*, fourth edn., McGraw-Hill, New York,.

Karayiannis, N.B. and Mi, G.W. (1997) Growing radial basis neural networks: merging supervised and unsupervised learning with network growth techniques. *IEEE Trans. Neural Netw.* **8**: 1492–1506.

Kasabov, N. (1996) *Foundations of Neural Networks, Fuzzy Systems and Knowledge Engineering.* MIT Press, Cambridge, MA.

Kasabov N. (1998a) ECOS: A framework for evolving connectionist systems and the ECO learning paradigm. In: S. Usui and T. Omori (Eds.) *Proceedings of ICONIP'98*, IOS Press, Kitakyushu, Japan, pp. 1222–1235.

Kasabov N. (1998b) Evolving fuzzy neural networks - Algorithms, applications and biological motivation. In: T. Yamakawaand and G. Matsumoto (Eds.), *Methodologies for the Conception, Design and Application of Soft Computing*, World Scientific, Singapore, pp. 271–274.

Kasabov, N. (1999) Evolving connectionist systems and applications for adaptive speech recognition. In: *Proceedings of IJCNN'99*, Washington DC, July, IEEE Press, Washington, DC.

Kasabov, N. (Ed.) (2000a) *Future Directions for Intelligent Systems and Information Sciences*, Physica-Verlag (Springer Verlag). Heidelberg.

Kasabov, N. (2000b) Evolving and evolutionary connectionist systems for online learning and knowledge engineering. In: P.Sincak, and J. Vascak (Eds.), *Quo Vadis Computational Intelligence? New Trends and Approaches in Computational Intelligence*, Physica-Verlag, Heidelberg, pp. 361–369.

Kasabov N (2001a) Evolving fuzzy neural networks for online supervised/unsupervised, knowledge-based learning. *IEEE Trans. SMC-B Cybern.* **31**(6): 902–918.

Kasabov, N. (2001b) Adaptive learning system and method, Patent, PCT WO 01/78003 A1.

Kasabov, N. (2001c) Ensembles of EFuNNs: An architecture for a multi-module classifier. In: *Proceedings of the International Conference on Fuzzy Systems*, Australia, vol. 3, 1573–1576.

Kasabov, N. (2002) *Evolving Connectionist Systems: Methods and Applications in Bioinformatics, Brain Study and Intelligent Machines.* Springer Verlag, London

Kasabov, N. (2003) Evolving connectionist-based decision support systems. In: X. Yu and J. Kacprzyk (Eds.), *Applied Decision Support with Soft Computing*, Series: Studies in Fuzziness and Soft Computing, vol. 124, Springer, New York.

Kasabov, N. (2004) Knowledge based neural networks for gene expression data analysis, modelling and profile discovery, *Drug Discovery Today: BIOSILICO* **2**(6, Nov.): 253–261.

Kasabov, N. (2007a) Brain-, gene-, and quantum-inspired computational intelligence: challenges and opportunities, in: W. Duch and J. Mandzink (Eds.), *Challenges in Computational Intelligence*, Springer, New York.

Kasabov, N. (2006) Adaptation and interaction in dynamical systems: Modelling and rule discovery through evolving connectionist systems, *Appl. Soft Comput.* **6**(3): 307–322.

Kasabov, N. (2007) Global, local and personalised: Modelling and profile discovery in bioinformatics, *Patt. Recogn. Lett.* 28, pp. 673–685.

Kasabov, N. and Benuskova, L. (2004) Computational neurogenetics, *Int. J. Theor. Comput. Nanosci.* **1**(1): 47–61.

Kasabov, N. and Dimitrov, D. (2004) Discovering gene regulatory networks from gene expression data with the use of evolving connectionist systems. In: L. Wang and J. Rajapakse (eds) *Neural Information Processing*, vol. 152, Springer Verlag, New York.

Kasabov N. and Iliev, G. (2000) A methodology and a system for adaptive speech recognition in a noisy environment based on adaptive noise cancellation and evolving fuzzy neural networks. In: H. Bunke and A. Kandel (Eds.), *Neuro-Fuzzy Pattern Recognition*, World Scientific, Singapore, pp. 179–203.

Kasabov, N. and Kozma, R. (Eds.) (1999) *Neuro-Fuzzy Techniques for Intelligent Information Systems*, Physica-Verlag (Springer Verlag), Heidelberg.

Kasabov, N. and Pang, S. (2004) Transductive support vector machines and applications in bioinformatics for promoter recognition, *Neural Inf. Process. Lett. Rev.* **3**(2): 31–38.

Kasabov, N. and Song, Q. (2002) DENFIS: Dynamic, evolving neural–fuzzy inference systems and its application for time-series prediction. *IEEE Trans. Fuzzy Syst.* **10**: 144–154.

Kasabov, N., Bakardjian, H., Zhang, D., Song, Q., Cichocki, A., and van Leeuwen, C. (2007) Evolving connectionist systems for adaptive learning, classification and transition rule discovery from EEG data: A case study using auditory and visual stimuli, *Int. J. Neural Systems,* in print.

Kasabov, N., Benuskova, L., and Wysoski, S.G. (2005a) Computational neurogenetic modeling: integration of spiking neural networks, gene networks, and signal processing techniques. In: W. Duch et al. (Eds.), *ICANN 2005,* LNCS 3697, Springer-Verlag, Berlin, pp. 509–514.

Kasabov, N., Benuskova, L., and Wysoski, S. (2005b) Biologically plausible computational neurogenetic models: Modeling the interaction between genes, neurons and neural networks, *J. Comput. Theor. Nanosci.* 2(4, December): 569–573(5).

Kasabov, N., Chan, S.H., Jain, V., Sidirov, I., and Dimitrov S.D. (2004) *Gene Regulatory Network Discovery from Time-Series Gene Expression Data – A Computational Intelligence Approach* , LNCS, vol. 3316, Springer-Verlag, New York, pp. 1344–1353.

Kasabov, N., Gottgtroy, P., Benuskova, L., Jain, V., Wysoski, S., and Josef, F. (2006b) Brain-gene ontology. In: *Proceedings of HIS/NCEI06,* Auckland, NZ, IEEE Computer Society Press, Washington, DC.

Kasabov, N., Israel, S., and Woodford, B. (2000a) Methodology and evolving connectionist architecture for image pattern recognition. In: Pal, Ghosh and Kundu (Eds.), *Soft Computing And Image Processing,* Physica-Verlag, Heidelberg.

Kasabov, N., Israel, S., and Woodford, B.J. (2000b) Hybrid evolving connectionist systems for image classification, *J. Adv. Comput. Intell.* **4**(1): 57–65.

Kasabov, N., Kim, J.S., Watts, M., and Gray, A. (1997) FuNN/2 - A fuzzy neural network architecture for adaptive learning and knowledge acquisition, *Inf. Sci. Appl.* **101**(3–4): 155–175.

Kasabov, N., Postma, E., and van den Herik, J. (2000c) AVIS: A connectionist-based framework for integrated auditory and visual information processing, *Inf. Sci.* **123**: 127–148.

Kasabov, N., Sidorov, I.A., and Dimitrov, D.S. (2005c) Computational intelligence, bioinformatics and computational biology: A brief overview of methods, problems and perspectives, *J. Comput. .Theor. Nanosci.* 2(4): 473–491.

Kasabov, N., Song Q., and Nishikawa I. (2003) Evolutionary computation for dynamic parameter optimisation of evolving connectionist systems for online prediction of time series with changing dynamics. In: *Proceedings of IJCNN'2003,* Portland, Oregon, July, vol.1, pp. 438–443.

Kater, S.B., Mattson, N.P., Cohan, C., and Connor, J. (1988) Calcium regulation of the neuronal cone growth, *Trends Neurosci.* **11**: 315–321.

Kecman, V. (2001) *Learning and Soft Computing (Support Vector Machines, Neural Networks, and Fuzzy Systems),* MIT Press, Cambridge, MA.

Kennedy, J. and Eberhart, R. (1995) Particle swarm optimization. In: *Proceedings of IEEE International Conference on Neural Networks,* Australia, vol. IV, pp. 1942–1948.

Kennedy, J. and Eberhart, R. (2001) *Swarm Intelligence.* Academic Press, San Diego.

Kennedy, J. and Eberhart, R.C. (1997) A discrete binary version of the particle swarm algorithm. In: *IEEE Conference on Computational Cybernetics and Simulation, Systems, Man, and Cybernetics,* vol. 5, pp. 4104–4108.

Khan et al. (2001) Classification and diagnostic prediction of cancers using gene expression profiling and artificial neural networks, *Nature Med.* **7**: 673–679.

Kidera, T., Ozawa, S., and Abe, S. (2006), An incremental learning algorithm of ensemble classifier systems. In: *Proceedings of the IEEE International Joint Conference on Neural Networks,* Vancouver, July 16–21, IEEE Press, Washington, DC, pp. 6453–6459.

Kim, J., Mowat, A., Poole, P., and Kasabov, N. (2000) Linear and nonlinear pattern recognition models for classification of fruit from visible-near infrared spectra, *Chemometrics Intell. Lab. Syst.* **51**: 201–216.

Kim, J.S. and Kasabov, N. (1999) HyFIS: Adaptive neuro-fuzzy systems and their application to nonlinear dynamical systems, *Neural Netw.* **12**(9): 1301–1319.

Kim, K., Relkin, N., Min-lee, K., and Hirsch, J. (1997) Distinct cortical areas associated with native and second languages, *Nature,* 388–392.

Kitamura, T. (Ed.) (2001) *What Should Be Computed to Understand and Model Brain Functions?* World Scientific, Singapore.

Kleppe, I.C. and Robinson, H.P.C. (1999) Determining the activation time course of synaptic AMPA receptors from openings of colocalized NMDA receptors. *Biophys. J.* **77**: 1418–1427.

Koch, C. and Hepp, K. (2006) Quantum mechanics in the brain, *Nature* **440**(30, March).

Koczy, L. and Zorat, A. (1997) Fuzzy systems and approximation, *Fuzzy Sets Syst.* **85**: 203–222.

Kohonen, T. (1977) *Associative Memory. A System-Theoretical Approach*, Springer-Verlag, Berlin.

Kohonen T. (1982) Self-organised formation of topology correct feature maps. *Biol. Cybern.* **43**: 59–69.

Kohonen, T. (1990) The self-organizing map, *Proc. IEEE* **78**(N-9): 1464–1497.

Kohonen, T. (1993) Physiological interpretation of the self-organising map algorithm, *Neural Netw.***6**.

Kohonen T. (1997) *Self-Organizing Maps*, 2d ed., Springer Verlag, New York.

Kolmogorov, A. (1957) On the representations of continuous functions of many variables by superpositions of continuous functions of one variable and addition, *Dokladi Academii Nauk USSR*, **114**(5): 953–956 (in Russian).

Koprinska, I. and Carrato, S. (1998) Video segmentation of MPEG compressed data. In: *Proceedings of ICECS'98*, Lisboa, Portugal, pp. 243–246.

Koprinska, I. and Kasabov, N. (2000) Evolving fuzzy neural network for camera operation recognition, In: *Proceedings of the International Conference on Pattern Recognition*, September 3–7, ICPR, Barcelona vol. II, pp. 523–526.

Kosko, B. (1992) Fuzzy systems as universal approximators In: *Proceedings of IEEE Fuzzy System Conference*, San Diego, pp. 1153–1161.

Kouda, N., Matsui, N., Nishimura, H., and Peper, F. (2005) Qubit neural network and its learning efficiency, *Neural Comput. Appl.* **14**: 114–121.

Koza, J. (1992) *Genetic Programming*, MIT Press, Cambridge, MA.

Kozma, R. and Kasabov, N. (1998) Chaos and fractal analysis of irregular time series embedded into connectionist structure. In: S. Amari and N. Kasabov (Eds.), *Brain-Like Computing And Intelligent Information Systems*, Springer Verlag, Singapore, pp. 213–237.

Kozma, R. and Kasabov, N. (1999) *Generic neuro-fuzzy-chaos methodologies and techniques for intelligent time-series analysis.* In: R. Ribeiro, R. Yager, H.J. Zimmermann, and J. Kacprzyk (Eds.), *Soft Computing in Financial Engineering.* Physica-Verlag, Heidelberg.

Kozma, R. and Kasabov, N. (1998) Rules of chaotic behaviour extracted from the fuzzy neural network FuNN. In: *Proceedings of the WCCI'98 FUZZ-IEEE International Conference on Fuzzy Systems*, Anchorage.

Krogh, A. and Hertz, J.A. (1992) A simple weight decay can improve generalization. *Adv. Neural Inf. Process. Syst.* **4**: 951–957.

Kuhl, P. (1994) Speech perception. In: F. Minifie (Ed.), *Introduction to Communication Sciences and Discorse*, Singular, San Diego, pp. 77–142.

Kurkova, V. (1992) Kolmogorov's theorem and multiplayer networks, *Neural Netw.* 501–506.

Lakoff, G. and Johnson, M. (1999). *Philosophy in the Flesh.* Basic, New York.

Lange, T. and Dyer, M. (1989) High-level inference in a connectionist network. *Connect. Sci.* **1**(2): 181–217.

Lawrence, S., Fong, S., and Giles, L. (1996) Natural language grammatical inference: A comparison of recurrent neural networks and machine learning methods. In S. Wermtner, E. Riloff, and G. Scheler (Eds.), *Symbolic, Connectionist and Statistical; Approaches to Learning for Natural Language Processing*, LNAI, Springer, New York, pp. 33–47.

Laws, M., Kilgour, R., and Kasabov, N. (2003) Modelling the emergence of bilingual acoustic clusters: A preliminary case study, *Inf. Sci.* **156**: 85–107.

Le Cun, Y., Denker, J.S., and Solla, S.A. (1990) Brain damage. In: D. Touretzky (Ed.), *Advances in Neural Information Proessing. Systems*, Morgan Kaufmann, San Francisco, No. 2, 598–605.

Lee, C., Gauvin, J., Pierracini, R., and Rabiner, L. (1993) Sub-word based large vocabulary speech recognition, *AT&T Tech. J.* (Sept/Oct): 25–36.

Lehner, W.G. (1988) Symbolic/subsymbolic sentence analysis: exploiting the best of two worlds. In J. Barnden and J. Pollack (Eds.), *Advances in Connectionist and Neural Computation Theory.* Ablex, Norwood, NJ.

Lek, S. and Guégan, J.F. (1999) Artificial neural networks as a tool in ecological modelling, an introduction, *Ecol. Model.* **120**: 65–73.

Lek, S., Delacoste, M., Baran, P., Dimopoulos, I., Lauga, J., and Aulagnier, S. (1996) Application of neural networks to modelling non linear relationship in ecology. *Ecol. Model.* **90**: 39–52.

Levey, A.S., Bosch, J.P., Lewis, J.B., Greene, T., Rogers, N., and Roth, D. (1999) For the modification of diet in renal disease study group, a more accurate method to estimate glomerular filtration rate from serum creatinine: A new prediction equation, *Annals Internal Med.* **130**: 461– 470.

Li, F., Xie, C., Zheng, D., and Zheng, B. (2006) Feedback quantum neuron for multiuser detection. In: *Proceedings of the IEEE International Joint Conference on Neural Networks*, Vancouver, July 16–21, IEEE Press, Washington, DC, pp. 5274–5278.

Liberman, A., Cooper, F., Shankweiler, D., and Studdert-Kennedy, M. (1967) Perception of the speech code. *Psychol. Rev.* **74**: 431–461.

Lim, C.P. and Harrison, F.P. (1997) An incremental adaptive network for unsupervised learning and probability estimation. *Neural Netw.* **10**: 925–939

Lin, C.T. and Lee, C.S.G. (1996) *Neuro Fuzzy Systems*, Prentice-Hall, Upper Saddle River, NJ.

Lippman, R. (1987) An introduction to computing with neural nets, *IEEE ASSP Mag.* (April):.4–21.

Lippman, R. (1989) Review of neural networks for speech recognition, *Neural Comput.* 1–38.

Liu, J., Xu, W., and Sun, J. (2005) Quantum-behaved particle swarm optimization with mutation operator. In: *Seventeenth IEEE International Conference on Tools with Artificial Intelligence (ICTAI'05).*

Liu, Y. and Yao, X. (1999) Simultaneous training of negatively correlated neural networks in an ensemble, *IEEE Trans. SMC B* **29**(6): 716–725.

Liu, Y., Yao, X., and Higuchi, T. (2000) Evolutionary ensembles with negative correlation learning, *IEEE Trans. EC* **4**(4): 380–387.

Lloyd, S.P. (1982) Least squares quantization in PCM. *IEEE Trans. Inf. Theor.* **28**(2): 129–137.

Lodish, H., Berk, A., Zipursky, S.L., Matsudaira, P., Baltimore, D., and Darnell, J. (2000) *Molecular Cell Biology.* 4th edn. W.H. Freeman, New York.

Luettin J., Thacker, N.A., and Beet, S.W. (1996) Active shape models for visual speech feature extraction. In: D.G. Storck and M.E. Heeneeke (Eds.), *Speechreading by Humans and Machines*, Springer, Berlin, pp. 383–390.

Lukashin, A.V. and Borodovski, M. (1998) GeneMark.hmm: New solutions for gene finding, *Nucleic Acids Res.* **26**: 1107–1115.

Maass, W. (1996) *Networks of Spiking Neurons: The Third Generation of Neural Network Models*, Australian National University, Canberra.

Maass, W. (1998) Computing with spiking neurons. In: W. Maass and C.M. Bishop (Eds.), *Pulsed Neural Networks*, MIT Press, Cambridge, MA, Chapter 2, pp. 55–81.

Machado, R.J. and da Rocha, A.F. (1992) Evolutive fuzzy neural networks. In: *Proceedings of the First IEEE Conference on Fuzzy Systems*, pp. 493–499.

Macilwain, C. et al. (2000) World leaders keep phrase on human genome landmark, *Nature*, **405**: 983–985.

Maclagan, M. (1982) On acoustic study of New Zealand vowels. *NZ Speech Therap. J.*37(1): 20–26.

MacQueen, J. (1967) Some methods for classification and analysis of multivariate observations. In: L.M. LeCam and J. Neyman (Eds.), *Proceedings of the Fifth Berkeley Symposium of Mathematical Statistics and Probability*, vol. I, University of California Press, San Francisco, pp. 281–297.

Mamdani, E. (1977) Application on fuzzy logic to approximate reasoning using linguistic synthesis, *IEEE Trans. Comput.* **C-26**(12): 1182–1191.

Mandziuk, J. and Shastri, L. (1998) Incremental class learning approach and its application to hand-written digit recognition. In: *Proceedings of the Fifth International Conference on Neuro-Information Processing*, Kitakyushu, Japan, October 21–23.

Manel, S., Dias, J.M., and Ormerod, J.S. (1999) Comparing discriminant analysis, neural networks and logistic regression for predicting species distributions: A case study with a Himalayan river bird, *Ecol. Model.* **120**: 337–347.

Marcus, G. (2004) *The Birth of the Mind: How a Tiny Number of Genes Creates the Complexity of the Human Mind.* Basic, New York.

Marshall, M.R., Song, Q., Ma, T.M., MacDonell, S., and Kasabov, N. (2005) Evolving connectionist system versus algebraic formulae for prediction of renal function from serum creatinine, *Kidney Int.* **67**: 1944–1954.

Martinez, T.M. and Schulten, K.J. (1991) A 'neural gas' network learns topologies. In: *Artificial Neural Networks*, Amsterdam, vol. I, 397–402.

Martinez, T.M., Berkovich, S.G., and Schulten, K..J. (1993) Neural gas network for vector quantization and its applications to time-series prediction, *IEEE Trans. Neural Netw.* **4**: 558–569.

Massaro, D. and Cohen, M. (1983) Integration of visual and auditory information in speech perception. *J. Exper.Psychol.: Hum. Percep. Perform.* **9**: 753–771.

MATLAB Tutorial Book (2000) *Mathworks.*

Matsumoto, G. (2000) Elucidation of the principle of creating the brain and its expression, *RIKEN BSI News*, No.8, May 2000, RIKEN, Japan, 2–3.

Matsumoto, G., Ichikawa, M., and Shigematsu, Y. (1996) Brain computing. In: *Methodologies for Conception, Design, and Application of Intelligent Systems*, World Scientific, Singapore, pp. 15–24.

Mattia, M. and del Giudice, P. (2000) Efficient event-driven simulation of large networks of spiking neurons and dynamical synapses. *Neural Comput.* **12**(10): 2305–2329.

McCauley, J.L. (1994) *Chaos, Dynamics and Fractals, An Algorithmic Approach to Deterministic Chaos.* Cambridge University Press, Cambridge, UK.

McClelland, J. and Rumelhart, D. et al. (1986) *Parallel Distributed Processing*, vol. II, MIT Press, Cambridge, MA.

McClelland, J., McNaughton, B., and O'Reilly, R. (1995) Why are there complementary learning systems in the hippocampus and neocortex: insights from the success and failure of connectionist models of learning and memory. *Psychol. Rev.* **102**: 419–457.

McClelland, J.L., Rumelhart, D.E., and Hinton, G.E. (1986) A general framework for PDP. In: D.E. Rumelhart, J.L. McClelland, and PDP Research Group, *Parallel Distributed Processing: Explorations in the Microstructure of Cognition*, vol. 1: *Foundations*, MIT Press, Cambridge, MA.

McCulloch, W.S. and Pitts, W. (1943) A logical calculus of the ideas immanent in nervous activity, *Bull. Math. Biophys.* **5**: 115–133.

McMillan, C., Mozer, M., and Smolensky, P. (1991) Learning explicit rules in a neural network. In *Proceedings of the International Conference on Neural Networks*, vol. 2, IEEE, New York, pp. 83–88.

Medsker, L. (1994) *Design and Development of Hybrid Neural Network and Expert Systems.* IEEE, Washington, DC, pp.1470–1474.

Mel, B.W. (1998) SEEMORE: Combining colour, shape, and texture histrogramming in a neurally-inspired approach to visual object recognition, *Neural Comput.* **9**: 777–804.

Mendel, J.M. (2001) *Uncertain Rule-Based Fuzzy Logic Systems: Introduction and New Directions*, Prentice-Hall, Englewood Cliffs, NJ, pp. 197–203.

Metcalfe, A. (1994) *Statistics in Engineering – A Practical Approach*, Chapman & Hall, London.

Michaliewicz, Z. (1992) *Genetic Algorithms + Data Structures = Evolutionary Programs.* Springer-Verlag, Berlin.

Miesfeld, M. (1999) *Applied Molecular Genetics.* John Wiley and Sons, New York

Miller, D., Zurada, J., and Lilly, J.H. (1996) Pruning via dynamic adaptation of the forgetting rate in structural learning, *Proc. IEEE* ICNN'96, **1**: 448.

Minku, F.L. and Ludermir, T.B. (2005) Evolutionary strategies and genetic algorithms for dynamic parameter optimisation of evolving fuzzy neural networks. In: *Proceedings of IEEE Congress on Evolutionary Computation, (CEC)*, Edinburgh, September, vol. 3, pp. 1951–1958.

Minku, F.L. and Ludermir, T.B. (2006) EFuNNs ensembles construction using a clustering method and a coevolutionary genetic algorithm. In: *Proceedings of the IEEE Congress on Evolutionary Computation*, Vancouver, July 16–21, IEEE Press, Washington, DC, pp. 5548–5555.

Minsky, M.L. and Papert, S. (1969) *Perceptrons: An Introduction to Computational Geometry*, MIT Press, Cambridge, MA, 2nd edn. 1988.

Mitchell, M.T. (1997) *Machine Learning*, McGraw-Hill, New York.

Mitra, S. and Hayashi, Y. (2000) Neuro-fuzzy rule generation: Survey on soft computing framework, *IEEE Trans. Neural Netw.* **11**(3): 748–768.

Mohan, N. and Kasabov, N. (2005) Transductive modelling with GA parameter optimisation. In: *IJCNN 2005 Conference Proceedings*, vol. 2, IEEE Press, Washington, DC, pp. 839–844.

Moody, J. and Darken, C. (1988) Learning with localized receptive fields. In: D. Touretzky, G. Hinton, and T. Sejnowski (Eds.) *Proceedings of the 1988 Connectionist Models Summer School, Carnegie Mellon University*, Morgan Kaufmann, San Mateo, CA.

Moody, J. and Darken, C. (1989) Fast learning in networks of locally-tuned processing units. *Neural Comput.* **1**: 281–294.

Moore, A.W. (1990) Acquisition of dynamic control knowledge for a robotic manipulator. In: *Proceedings of the Seventh International Conference on Machine Learning*, pp.244–252, Austin, TX: Morgan Kaufman, San Francisco.

Moore, B.C. and Glassberg, B.R. (1983) Suggested formulae for calculating auditory filter bandwidths and excitation patterns, *J. Acoust. Soc. Am.* **74**(3): 750–753.

Moore, D.S. and McCabe, G.P. (1999) *Introduction to the Practice of Statistics*, W.H. Freeman, New York.

Morasso, P., Vercelli, G., and Zaccaria, R. (1992) Hybrid systems for robot planning. In: I. Aleksander and J. Taylor (Eds.), *Artificial Neural Networks* 2, Elsevier Science Publ. B.V., pp. 691–697.

Morgan, D. and Scofield, C. (1991) *Neural Networks and Speech Processing*, Kluwer Academic, Dordrecht.

Mount, S. (1982) A catalogue of splice junction sequences. *Nucleic Acids Res.* **10**(2): 459–472.

Mozer, M., Smolensky, P. (1989) A technique for trimming the fat from a network via relevance assessment. In: D. Touretzky (Ed.), *Advances in Neural Information Processing Systems*, vol. 2, Morgan Kaufmann, San Francisco, pp. 598–605.

Murata, N., Müller, K.-R., Ziehe, A., and Amari, S. (1997) Adaptive online learning in changing environments. In: *Proceedings of the Conference on Neural Information Processing Systems (NIPS 9)*, MIT Press, Cambridge, MA, pp. 599–604.

Murphy, P. and Aha, D. (1992) UCI Repository of Machine Learning Databases. Department of Information and Computer science, University of California, Irvine.

Narayanan, A. and Meneer, T. (2000) Quantum artificial neural network architectures and components, *Inf. Sci.*, 128(3–4), 231–255.

Neisser, U. (Ed.) (1987) Concepts and conceptual development, Cambridge University Press, Cambridge, UK.

Nelson, M. and Rinzel, J. (1995) The Hodgkin-Huxley model. In: J.M. Bower and Beeman (Eds.), *The Book of Genesis*, Springer, New York, Chapter 4, pp. 27–51.

Neural Network Toolbox User's Guide (2001) The Math Works, ver. 4.

Newell, A. and Simon, H.A. (1972) *Human Problem Solving*, Prentice-Hall, Englewood Cliffs, NJ.

Nguyen, M.N., Guo, J.F., and Shi, D. (2006) ESOFCMAC: Evolving self-organizing fuzzy cerebellar model articulation controller. In: *Proceedings of the IEEE International Joint Conference on Neural Networks*, Vancouver, July 16–21, IEEE Press, Washington, DC, pp. 7085–7090.

Nijholt, A. and Hulstijn, J. (Eds.) (1988) Formal semantics and pragmatics of dialogue. In: *Proceedings Twendial'98 (TWLT13)*, University of Twente, The Netherlands.

Nikolaev, N.Y. and Iba, H. (2006) *Adaptive Learning of Polynomial Networks*, Series in Genetic and Evolutionary Computation, D.E. Goldberg and J.R. Koza, (Series Eds.), Springer, New York.

Nolfi, S. and Floreano, D. (2000) *Evolutionary Robotics*, MIT Press, Cambridge, MA.

Oja, E. (1992) Principle components, minor components and linear neural networks, *Neural Netw.* **5**: 927–935.

Okabe, A., Boots, B., and Sugihara, K. (1992) *Spatial Tessellations - Concepts and Applications of Voronoi Diagrams*, John Wiley & Sons, New York.

Omlin, C. and Giles, C. (1994) *Constructing Deterministic Finite-State Automata in Sparse Recurrent Neural Networks*, IEEE, Washington, DC, pp. 1732–1737.

Owens, F.J. (1993) *Signal Processing of Speech*. Macmillan, New York.

Ozawa, S., Pang, S., and Kasabov, N. (2004a) *A Modified Incremental Principal Component Analysis for online Learning of Feature Space and Classifier*, LNAI, vol. 3157, Springer-Verlag, Berlin, pp. 231–240.

Ozawa, S., Pang, S., and Kasabov, N., (2004b) *One-Pass Incremental Membership Authentication by Face Classification*, LNCS, vol. 3072, D. Zhang and A. Jain (Eds.), Springer-Verlag, Berlin, pp. 155–161.

Ozawa, S., Pang, S., and Kasabov, N. (2006) An incremental principal component analysis for chunk data. In: *Proceedings of the IEEE International Conference on Fuzzy Systems*, Vancouver, July 16–21, IEEE Press, Washington, DC, pp. 10493–10500.

Ozawa, S., Toh, S.L., Abe, S., Pang, S., and Kasabov, N. (2005b) Incremental learning for online face recognition. In: *Proceedings of the IEEE International Joint Conference on Neural Networks*, Montreal, July 31–August 4, IEEE Press, Washington, DC, pp. 3174–3179.

Ozawa, S., Too, S., Abe, S., Pang, S., and Kasabov, N. (2005a) Incremental learning of feature space and classifier for online face recognition, *Neural Netw.* (August): 575–584.

Pal, N. (1999) Connectionist approaches for feature analysis. In: N. Kasabov and R. Kozma (Eds.) *Neuro-Fuzzy Techniques for Intelligent Information Systems*, Physica-Verlag (Springer-Verlag), Heidelberg, pp. 147–168.

Pal, N. and Bezdek, J.C. (1995) On cluster validity for the fuzzy c-means model, *IEEE Trans. Fuzzy Syst.*, Vol. 3, Issue 3, 370–379.

Pal, S., Ghosh, and Kundu (Eds.) (2000) *Soft Computing and Image Processing*, Physica-Verlag (Springer Verlag), Heidelberg.

Pang, S. and Kasabov, N. (2004) Inductive vs transductive inference, global vs local models: Svm, tsvm, and svmt for gene expression classification problems. In *Proceedings of the International Joint Conference on Neural Networks, IJCNN* 2004, Budapest, 16–30 June, IEEE Press, Washington, DC.

Pang, S. and Kasabov, N. (2006) Investigating LLE Eigenface on Pose and Face Identification, LNCS, vol. 3972, pp. 134–139.

Pang, S., Havukkala, I., and Kasabov N. (2006) *Two-Class SVM Trees (2-SVMT) for Biomarker Data Analysis*, LNCS, vol. 3973, Springer, New York, pp. 629–634.

Pang, S., Ozawa, S., and Kasabov, N. (2005) One-pass incremental membership authentication by face classification, *Int. J. Comput. Vis.*

Pang, S., Ozawa, S., and Kasabov, N. (2005a) *Chunk Incremental LDA Computing on Data Streams*, LNCS, vol. 3497, Springer, New York.

Pang, S., Ozawa, S., and Kasabov, N. (2005b) Incremental linear discriminant analysis for classification of data streams, *IEEE Trans. SMC-B*, **35**(5): 905–914.

Pao, Y.-H. (1989) *Adaptive Pattern Recognition and Neural Networks*, Addison-Wesley, Reading, MA.

Parisi, D. (1997) An artificial life approach to language. *Brain Lang.* **59**: 121–146.

Penrose, R. (1989) *The Emperor's New Mind*, Oxford University Press, Oxford, UK.
Penrose, R. (1994) *Shadows of the Mind. A Search for the Missing Science of Consciousness*, Oxford University Press, Oxford, UK.
Perkowski, M.A. (2005) Multiple-valued quantum circuits and research challenges for logic design and computational intelligence communities, *IEEE Comp. Intell. Soc. Mag.* (November).
Perlovski, L. (2006) Towards Physics of the Mind: Concepts, Emotions, Consciousness, and Symbols, Phy. Life Rev. 3(1), pp. 22–55.
Perou, M.P. et al. (2000) Molecular portraits of human breast tumours, *Nature* **406**: 747–752.
Persidis, A. (2000) Data mining in biotechnology, *Nature* **18**(2):237–238.
Petersen, S. (1990) Training neural networks to analyse biological sequences. *Trends Biotechnol.* **8**(11): 304–308.
Pevzner, P. (2001) *Computational Molecular Biology*, MIT Press, Cambridge, MA.
Picone, J. (1993) Signal modelling techniques in speech recognition. In: *Proceedings of IEEE* **81**(9, Sept): 1215–1247.
Pinker, S. (1994) *The Language Instinct: How the Mind Creates Language*. Penguin, London.
Pinker, S. and Prince, A. (1988) On language and connectionism: Analysis of a PDP model of language acquisition, *Cognition* **28**: 1–2, 73–193.
Platt, J. (1991) A resource allocating network for function interpolation. *Neural Comput.* **3**: 213–225.
Plotkyn, H.C. (1994) *The Nature of Knowledge*, Penguin, London.
Plunkett, K. (1996) Connectionist approaches to language acquisition. In: Fletcher and MacWhinney (Eds.), *The Handbook of Child Language*. Oxford: Blackwell, pp. 36–72.
Poggio, T. (1994) Regularization theory, radial basis functions and networks. In: *From Statistics to Neural Networks: Theory and Pattern Recognition Applications*. NATO ASI Series, No. 136, pp. 83–104.
Poggio, T. and Girosi, F. (1990) Regularisation algorithms for learning that are equivalent to multiplayer networks. *Science* **247**: 978–982.
Port, R. and van Gelder, T. (Eds.) (1995) *Mind as Motion (Explorations in the Dynamics of Cognition)*. MIT Press, Cambridge, MA.
Potter, M.A. and De Jong, K.A. (2000) Cooperative coevolution: An architecture for evolving co-adaptive sub-components, *Evol. Comput.* **8**(1): 1–29.
Pribram, K. (1993) Rethinking neural networks: Quantum fields and biological data. In: *Proceeding of the First Appalachian Conference on Behavioral Neurodynamics*. Lawrence Erlbaum, Hillsdate, NJ.
Protégé Ontology Software Environment, http://protege.stanford.edu/.
Purves, D. and Lichtman, J.W. (1985) *Principles of Neural Development*, Sinauer, Sunderland, MA.
Qian, N. and Sejnowski, T.J. (1988) Predicting the secondary structure of globular protein using neural network models, *J. Molec. Biol.* **202**: 065–084.
Quartz, S.R. and Sejnowski, T.J. (1997) The neural basis of cognitive development: a constructivist manifesto, *Behav. Brain Sci.* **20**(4): 537–596.
Quinlan, J. (1986) Induction of decision trees, *Mach. Learn.* 1: 1.
Quiroga, R.Q. (1998) Dataset #3: Tonic-clonic seizures: www.vis.caltech.edu/~rodri/data.htm .
Rabiner, L. (1989) A tutorial on hidden Markov models and selected applications in speech recognition, *Proc. IEEE* **77**(2): 257–285.
Rabiner, L. and Juang, B. (1993) *Fundamentals of Speech Recognition*. Prentice-Hall, Upper Saddle River, NJ.
Rajapakse, J., Kruggel, F., Zysset, S., and von Cramon, D.Y. (1998) Neuronal and hemodynamic events from fMRI time series, *J. Advan. Comput. Intell.* **2**(6): 185–194.
Ralescu, A. and Iwamoto, I. (1993) Recognition and reasoning about facial expressions using fuzzy logic. In: *Proceedings of RO-MAN' 93 Conference, Tokyo.*
Ramaswamy, S., et al. (2001) Multiclass cancer diagnosis using tumor gene expression signatures, *Proc. Nat. Acad. Sci. USA* **98**(26): 15149.
Ray, K. and Ghoshal. J. (1997) Neuro-fuzzy approach to pattern recognition, *Neural Netw.* **10**(1): 161–182.
Reed, R. (1993) Pruning algorithms - A survey, *IEEE Trans. Neural Netw.* **4**(5): 740–747.
Regier, T. (1996) *The Human Semantic Potential: Spatial Language and Constrained Connectionism*. MIT Press, Cambridge, MA.
Renals, S. and Rohwer, R. (1989) Phoneme classification experiments using radial basis functions. In: *Proceedings of the International Joint Conference on Neural Networks - IJCNN*, Washington, DC, June, pp. 461–467.
Resconi, G. and Jain, L.C. (2004) *Intelligent Agents*, Springer, New York.
Resconi, G. and van Der Wal, A.J. (2000) A data model for the morphogenetic neuron, *Int. J. Gen. Syst.* **29**(1): 141.

Resconi, G., Klir, G.J., and Pessa, E. (1999) Conceptual foundations of quantum mechanics the role of evidence theory, quantum sets and modal logic. *Int. J. Mod. Phys. C* **10**(1): 29–62.

Richardson, K. (1999) *The Making of Intelligence*, Phoenix, London.

RIKEN (2001) *BSI News*, 1–20, 2001, Japan.

Rizzi, L., Bazzana, F., Kasabov, N., Fedrizzi, M., and Erzegovesi, L. (2003). Simulation of ECB decisions and forecast of short term Euro rate with an adaptive fuzzy expert system. *Eur. J. Oper. Res.* **145**: 363–381.

Robins, A. (1996) Consolidation in neural networks and the sleeping brain, *Connection Sci.* **8**(2): 259–275.

Robinson, A.J. (1989) Dynamic error propagation networks, PhD Thesis, Cambridge University.

Rolls, E.T. and Treves, A. (1998) *Neural Networks and Brain Functions*, Oxford University Press, Oxford.

Rosch, E. and Lloyd, B.B. (Eds.) (1978) *Cognition and Categorization*. Lawrence Erlbaum, Hillsdale, NJ.

Rosenblatt, F. (1962) *Principles of Neurodynamics*, Spartan, New York.

Rosipal, R., Koska, M., and Farkas, I. (1997) Prediction of chaotic time-series with a resource-allocating RBF network, *Neural Process. Lett.* **10**: 26.

Ross, A. and Jain, A.K. (2003) Information fusion in biometrics, *Patt. Recogn. Lett.* **24**(13, September): 2115–2125.

Royer, M.H., and Yang, X.B. (1991) Application of high-resolution weather data to pest risk assessment, *OEPP/EPPO Bulletin*, **21**: 609–614.

Rumelhart, D.E. and McClelland, J.L. (1986) (Eds.) *Parallel and Distributed Processing: Exploration in the Microstructure of Cognition*, vol. 1, MIT Press, Cambridge, MA.

Rumelhart, D.E. et al. (1986) Learning internal representation by error propagation. In: D.E Rumelhart, J.L. McClelland, and the PDP Research Group, *Parallel Distributed Processing: Explorations in the Microstructure of Cognition*. vol. 1, Foundation, MIT Press, Cambridge, MA.

Rummery, G.A. and Niranjan, M. (1994) Online Q-learning using connectionist system. Cambridge University Engineering Department, CUED/F-INENG/TR, pp. 166.

Saad, D. (Ed.) (1999) *online Learning In Neural Networks*, Cambridge University Press, UK.

Salzberg, S.L. (1990) *Learning with Nested Generalized Exemplars*, Kluwer, Boston.

Sammon, J.W. (1968) *IEEE Trans. Comp.*, **18**(5), 401–409.

Sankar, A. and Mammone, R.J. (1993) Growing and pruning neural tree networks, *IEEE Trans. Comput.* **42**(3): 291–299.

Santos, J. and Duro, R.J. (1994) Evolutionary generation and training of recurrent artificial neural networks. In: *Proceedings of IEEE World Congress on Computational Intelligence*, vol. 2, pp. 759–763.

Schaal, S. and Atkeson, C. (1998) Constructive incremental learning from only local information. *Neural Comput.* **10**: 2047–2084.

Schena, M. (Ed.) (2000) *Microarray Biochip Technology*, Eaton, Natick, MA.

Schiffman, W., Joost, M., and Werner, R. (1993) Application of genetic algorithms to the construction of topologies for multilayer perceptrons. In: R.F. Albrecht, C.R. Reeves, and N.C. Steele (Eds.), *Artificial Neural Nets and Genetic Algorithms*, Spring-Verlag, Wien, New York.

Schneider, G. and Wrede, P. (1993) Development of artificial neural filters for pattern recognition in protein sequences. *J. Molec. Evol.* **36**: 586–595.

Segalowitz, S.J. (1983) *Language Functions and Brain Organization*, Academic Press, New York.

Segev, R. and Ben-Jacob, E. (1998) From neurons to brain: Adaptive self-wiring of neurons, TR, Faculty of Exact Sciences, Tel-Aviv University.

Seidenberg, M. (1997) Language acquisition and use: Learning and applying probabilistic constraints, *Science* **275**: 1599–603.

Serpen, G., Patwardhan, A., and Geib, J. (2001) Addressing the scaling problem of neural networks in static optimization, *Int. J. Neural Syst.* **11**(5): 477–487.

Shastri, L. (1988) A connectionist approach to knowledge representation and limited inference. *Cogn. Sci.* **12**: 331–392.

Shastri, L. (1999) A biological grounding of recruitment learning and vicinal algorithms, TR-99–009, International Computer Science Institute, Berkeley.

Shavlik, J.W. and Towell, G.G. (1989) An approach to combining explanation-based and neural learning algorithms. *Connect. Sci.* **1**(3).

Shi, Y. and Eberhart, R.C. (1998) A modified particle swarm optimizer. In: *Proceedings of the IEEE Congress on Evolutionary Computation*, pp. 69–73.

Shigematsu, Y., Okamoto, H., Ichikawa, K., and Matsumoto, G. (1999) Temporal event association and output-dependent learning: A proposed scheme of neural molecular connections, *J. Advan. Comput. Intell.* **3**(4): 234–244.

Shipp, M. et al (2002) Difuse large B-cell lynphoma outcome prediction by gene expression profiling and supervised machine learning, *Nat. Med.*, 8, 68–74.
Shor, P.W. (1997) Polynomial-time algorithms for prime factorization and discrete logarithms on a quantum computer, *SIAM J. Comput.* **26**: 1484–1509.
Si, J., Lin, S., and Vuong, M.A. (2000) Dynamic topology representing networks, *Neural Netw.* **13**: 617–627.
Sinclair, S. and Watson, C. (1995) The Development of the Otago Speech Database. In: N. Kasabov, and G. Coghill (Eds.), *Proceedings of ANNES '95*, IEEE Computer Society Press, Los Alamitos, CA.
Slaney, M. (1988) Lyon's cochlear model, Apple Computer Inc.
Smith, E.E. and Medin, D.L. (1981) *Categories and Concepts*, Harvard University Press, Cambridge, MA.
Snow, C. and Ferguson, C. (1977) Talking to Children: Language Input and Language Acquisition. Cambridge University Press, Cambridge, UK.
Soltic, S., Pang, S., Kasabov, N., Worner, S., and Peacock, L. (2004) *Dynamic Neuro-fuzzy Inference and Statistical Models for Risk Analysis of Pest Insect Establishment*, LNCS 3316, Springer, New York, pp. 971–976.
Somogyi, R., Fuhrman, S., and Wen, X. (2001) Genetic network inference in computational models and applications to large-scale gene expression data. In: J.M. Bower and H. Bolouri (Eds.), *Computational Modelling of Genetic and Biochemical Networks*, MIT Press, Cambridge, MA, pp. 120–157.
Song, Q. and Kasabov, N. (2004) TWRBF – *Transductive RBF Neural Network with Weighted Data Normalization*, LNCS, vol.3316, Springer Verlag, pp. 633–640.
Song, Q. and Kasabov, N. (2005) NFI: A neuro-fuzzy inference method for transductive reasoning, *IEEE Trans. Fuzzy Syst.* **13**(6): 799–808.
Song, Q. and Kasabov, N. (2006) TWNFI –Transductive weighted neuro-fuzzy inference system and applications for personalised modelling, *Neural Netw.*, vol. 19, No. 10, pp. 159–596.
Song, Q., Kasabov, N., Ma, T., and Marshall, M. (2006) Integrating regression formulas and kernel functions into locally adaptive knowledge-based neural networks: A case study on renal function evaluation, *Artif. Intell. Med.* **36**: 235–244.
Song, Q., Ma, T.M., and Kasabov, N. (2005) *Transductive Knowledge Based Fuzzy Inference System for Personalized Modeling.* LNAI 3614, Springer, New York, pp. 528–535.
Spector, L. (2004) *Automatic Quantum Computer Programming: A Genetic Programming Approach*, Kluwer Academic, Hingham, MA.
Spellman, P. et al. (1998) Comprehensive Identification of Cell Cycle-regulated Genes of the Yeast Saccharomyces cerevisiae by Microarray Hybridization *Mol. Biol. Cell*, **9**: 3273–3297.
Stephens, C., Olmedo, I., Vargas, J., and Waelbroack, H. (2000) Self adaptation in evolving systems, *Artif. Life* **4**(2): 183–201.
Stork, D. (1991) Sources of neural structure in speech and language processing, *Int. J. Neural Syst. Singapore* **2**(3): 159–167.
Stork, D. and Hennecke, M. (Eds.) (1996) *Speechreading by Humans and Machines*, Springer Verlag, New York.
Strain, T.J., McDaid, L.J., Maguire, L.P., and McGinnity, T.M. (2006) A supervised STDP based training algorithm with dynamic threshold neurons. In: *Proceedings of IEEE International Joint Conference on Neural Networks*, Vancouver, July 16–21, IEEE Press, Washington, DC, pp. 6441–6446.
Sugeno, M. (1985) An introductory survey of fuzzy control, *Inf. Sci.***36**: 59–83.
Sutherst, R.W. (1991) Predicting the survival of immigrant insect pests in new environments, *Crop. Prote.* **10**: 331–333.
Sutton, R.S. and Barto, A.G. (1998) *Reinforcement Learning (Adaptive Computation and Machine Learning).* MIT Press, Cambridge, MA.
Szu, H. and Hsu, C. (1999) Image processing by chaotic neural network fuzzy membership functions. In: N. Kasabov and R. Kozma (Eds.) *Neuro-Fuzzy Techniques for Intelligent Information Systems*, Physica Verlag, Springer Verlag, Berlin, pp. 207–225.
Takagi, H. (1990) Fusion technology of fuzzy theory and neural networks - Survey and future directions. In: *Proceedings of the First International Conference on Fuzzy Logic and Neural Networks*, Iizuka, Japan, July 20–24, pp.13–26.
Takagi, T. and Sugeno, M. (1985) Fuzzy identification of systems and its applications to modeling and control. *IEEE Trans. Syst. Man Cybern.* **15**: 116–132.
Tanaka, S. (1997) Topology of visual cortical maps, *FORMA* **12**: 101–106.
Tavazoie, S., Hughes, J.D., Campbell, M.J., Cho, R.J., and Church, G.M. (1999) Systematic determination of genetic network architecture, *Nature Genet.* **22**: 281–285.
Taylor, J. R. and Kasabov, N. (2000) Modelling the emergence of speech and language through evolving connectionist systems. In: N. Kasabov (Ed.) *Future Directions for Intelligent Information Systems and Information Sciences*, Springer Verlag, New York, pp. 102–126.

Taylor, J.G. (1998) Neural networks for consciousness. In: S. Amari and N. Kasabov (Eds., *Brain-Like Computing and Intelligent Information Systems*, Springer Verlag, New York.

Taylor, J.G. (1999) *The Race for Consciousness*, MIT Press, Cambridge, MA.

Taylor, J.G. (2005) *The Human Mind: A Practical Guide*, Welley, London.

Taylor, J.R. (1995) *Linguistic Categorization: Prototypes in Linguistic Theory*. 2nd Edition. Clarendon Press, Oxford, UK.

Taylor, J.R. (1999) *An Introduction to Cognitive Linguistics*. Clarendon Press Oxford, UK.

Thorpe, S., Gaustrais, J. (1998) Rank order coding, in: Bauer, J. (ed) Computational Neuroscience, Plenum Press, NY.

Tomita, M. (2001) Whole-cell simulation: A grand challenge of the 21st century, *Trends Biotechnol.* **19**(6): 205–210.

Touretzky, D. and Hinton, J. (1988) A distributed connectionist production system, *Cogn. Sci.* **12**: 423–466.

Touretzky, D.S. and Hinton, G.E. (1985) Symbols among the neurons: Details of a connectionist inference architecture. In: *Proc. IJCAI'85*, pp. 238–243.

Towell, G., Shavlik, J., and Noordewier, M. (1990) Refinement of approximate domain theories by knowledge-based neural networks, *Proceedings of the Eighth National Conference on Artificial Intelligence AAAI'90*, Morgan Kaufmann, 861–866.

Towell, G.G. and Shavlik, J.W. (1993) Extracting refined rules from knowledge-based neural networks. *Mach. Learn.* **13**(1): 71–101.

Towell, G.G. and Shavlik, J.W. (1994) Knowledge based artificial neural networks, *Artif. Intell.* **70**(4): 119–166.

Townsend, A. and Vieglais, D.A. (2001) Predicting species invasions using ecological niche modeling: new approaches from bioinformatics attack a pressing problem, *BioScience* 51(5): 363–371.

Trentin, E. (2001) Hybrid hidden Markov models and neural networks for speech recognition, PhD thesis, University of Florence and IRST-Trento, Italy.

Tresp, V., Hollatz, J., and Ahmad, S. (1993) Network structuring and training using rule-based knowledge. In: M. Kaufmann (Ed.), *Adv. Neural Inf. Process. Syst.* **5**.

Trugenberger, C.A. (2002) Quantum pattern recognition, *Quant. Inf. Process.* **1**: 471–493.

Tsai, X.-Y., Huang, H.-C., and Chuang, S.-J. (2005) Quantum NN vs. NN in signal recognition. In: *ICITA'05: Proceedings of the Third International Conference on Information Technology and Applications (ICITA'05)* vol. 2, Washington, DC, IEEE Computer Society, pp. 308–312.

Tsankova, D., Georgieva, V., and Kasabov, N. (2005) Artificial Immune Networks As A Paradigm For Classification And Profiling Of Gene Expression Data, *J. Comput. Theor. Nanosci.* **2**(4, Dec.): 543–550(8).

Tsukada, M, et al. (1996) Hippocampal LTP depends on spatial and temporal correlation of inputs. Neural networks, 9(8), pp. 1357–1365.

Tsypkin, Y.Z. (1973) *Foundation of the Theory of Learning Systems*, Academic Press, New York.

Uchino, E., Yamakawa, T. (1995) System modelling by a neo-fuzzy-neuron with applications to acoustic and chaotic systems, *Int. J. Artif. Intell. Tools* **4**: 73–91.

Valova, I., Gueorguieva, N., and Kosugi, Y. (2004) An oscillation-driven neural network for the simulation of an olfactory system, *Neural Comput. Appl.* **13**: 65–79.

Van Hulle, M.M. (1998) Kernel-based equiprobabilistic topographic map formation. *Neural Comput.* **10**(7): 1847–1871.

Van Owen, A. (1994) Activity-dependent neural network development, *Network Comput. Neural Syst.* **5**: 401–423.

Vapnik, V. (1998) *Statistical Learning Theory*. John Wiley & Sons, New York.

Venayagamoorthy, G.K and Singhal, G. (2005) Quantum-inspired evolutionary algorithms and binary particle swarm optimization for training MLP and SRN neural networks, *J. Theor. Comput. Nanosci.*, vol. 2, 561–568

Venayagamoorthy, G.K., Wang, X., Buragohain, M., and Gupta, A. (2004) Function approximations with multilayer perceptrons and simultaneous recurrent networks. In: *Conference on Neuro-Computing and Evolving Intelligence*, Auckland, New Zealand, pp. 28–29.

Ventura, D. (1999) Implementing competitive learning in a quantum system. In: *Proceedings of the International Joint Conference of Neural Networks*, IEEE Press, Washington, DC.

Ventura D. and Martinez, T. (2000) Quantum associative memory, *Inf. Sci. Inf. Comput. Sci.* **124**: 273–296.

Vesanto, J. (1997) Using SOM and local models in time-series prediction, Proc. Worrsh. SOM (WSOM'97) ESpoo, Finland, pp. 209–214.

Waibel, A. et al. (1997) Multimodal interfaces for multimedia information agents, PNC. ICASSP, Muivid, IEEE Press, pp. 1997–2004.

Wang, J. and Jabri, M. (1998) A computational model of the auditory pathway to the superior colliculus. In: S. Amari, and N. Kasabov (Eds.), *Brain-Like Computing and Intelligent Information Systems*, Springer, New York.

Wang, J., Weiderhold, G., Firschien, O., and Sha, X.W. (1996) Applying wavelets in image database retrieval, Technical Report, Stanford University, Stanford, CA.

Wang, L. and Fu, X. (2005) *Data Mining with Computational Intelligence*, Springer, New York.

Wang, L.X. (1994) *Adaptive Fuzzy System and Control: Design and Stability Analysis.* Prentice-Hall, Englewood Cliffs, NJ.

Warren, R.M. (1982) *Auditory Perception: A New Synthesis*, Pergamon Press, Elmsford, NY.

Watts, M. and Kasabov, N. (1998) Genetic algorithms for the design of fuzzy neural networks. In: S. Usui and T. Omori (Eds.) *Proceedings of ICONIP'98 - The Fifth International Conference on Neural Information Processing*, Kitakyushu, Japan, 21–23 October 1998, vol. 2, IOS Press, Singapore, pp. 793–796.

Watts, M. and Kasabov, N. (1999) Neuro-genetic tools and techniques. In: N. Kasabov and R. Kozma, (Eds.) *Neuro-Fuzzy Techniques for Intelligent Information Systems*, Physica Verlag Heidelberg, pp. 97–110.

Watts, M. and Kasabov, N. (2001) Dynamic optimisation of evolving connectionist systems by pseudo-evolution strategies. In: *Proceedings of the IEEE Congress of Evolutionary Computation (CEC)*, Seoul, May, 2001, vol. 2, pp. 1335–1342.

Watts, M. and Kasabov, N. (2002) Evolutionary computation for the optimisation of evolving connectionist systems. In: *Proceedings of WCCI'2002 (World Congress of Computational Intelligence)*, Hawaii, May, 2002, IEEE Press, Washington, DC.

Watts, M.J. (2006) Nominal-scale evolving connectionist systems. In: *Proceedings of the IEEE International Joint Conference on Neural Networks*, Vancouver, July 16–21, 2006, IEEE Press, Washington, DC, pp. 4057–4061.

Wearing, H. (1998) Pest identification through image analysis on apples, AgResearch Report, New Zealand.

Weaver, D.C., Workman, C.T., and Stormo, G.D. (1999) Modeling regulatory networks with weight matrices. In: *Proceedings of the Pacific Symposium on Biocomputing* - Hawaii, vol. 4, World Scientific, Singapore, pp. 112–123.

Weigend, A. and Gershefeld, N. (1993) *Time-Series Prediction: Forecasting the Future and Understanding the Past.* Addison-Wesley, Reading, MA.

Weigend, A. et al. (1990) Predicting the future: A connectionist approach, *Int. J. Neural Syst.* **1**: 193–209.

Weng, J., McClelland, J., et al. (2001), Autonomous mental development by Rosots and animals science, vol. 291, no.5504, pp. 599–600.

Werbos, P.J. (1990) Backpropagation through time: What it does and how to do it. *Proc. IEEE* **8**(10): 1550–1560.

Werbos, P.J. and Pang, X. (1996) Generalized maze navigation: SRN critic solve what feedforward or Hebbian nets cannot. In: *Proceedings of World Congress on Neural Networks*, San Diego, pp. 88–93.

Wermter, S. and Lehnert, W.G. (1989) A hybrid symbolic/ connectionist model for noun phrase understanding. *Connect. Sci.* **1**(3).

Wessels, L.F.A., vanSomeren, E.P., and Reinders, M.J.T. (2001) A comparison of genetic network models. In: *Proceedings of the Pacific Symposium on Biocomputing*, vol. 6, pp. 508–519.

West, M. and Harrison, P.J. (1989) *Bayesian Forecasting and Dynamic Models*, Springer-Verlag, New York.

West, M., Blanchette, C., Dressman, H., et al. (2001) Predicting the clinical status of human breast cancer by using gene expression profiles, *PNAS* **98**(20, Sept. 25): 11462–11467.

White, J. and Kauer, J.S. (1999) Odour recognition in an artificial nose by spatio-temporal processing using an olfactory neuronal network, Neurocomputing, **26–27**: 919–924.

Whitley, D. (1995) Genetic algorithms and neural networks. In: G. Winter, J. Periaux, M. Galan, and P. Cuesta (Eds.), *Genetic Algorithms Engineering and Computer Science*, Wiley, New York, pp. 191–201.

Whitley, D. and Bogart, C. (1990) The evolution of connectivity: Pruning neural networks using genetic algorithms. In: *Proceedings of the International Joint Conference on Neural Networks*, No.1. pp. 17–22.

Widrow, B. (2006) Memory and learning. In: *Proceedings of IJCNN, 2006*, Vancouver, July, IEEE Press, Washington, DC.

Widrow, B. and Hoff, M.E. (1960) Adaptive switching circuits, *IRE WESCON Convention Rec.* **4**: 96–104.

Williams, C.P. and Clearwater, S.H. (1998) *Explorations in Quantum Computing.* Springer-Verlag, Berlin.

Wiskott, L. (2005) How does our visual system achieve shift and size invariance? In: J.L.van Hemmen and T.J. Sejnowski (Eds.), *Problems in Systems Neuroscience*, Oxford University Press, Oxford, UK.

Woldrige, M. and Jennings, N. (1995) Intelligent agents: Theory and practice, *Knowl. Eng. Rev.* (10).

Wolfe, A. (1985) Determining Lyapunov exponents from a time series. *Physica D* **16**: 285–317.

Wolpert, D.H. and Macready, W.G. (1997) No free lunch theorems for optimization, *IEEE Trans. Evol. Comput.* **1**(1): 67–82.

Wolpert, L. et al. (1998) *Principles of Development*, Oxford University Press, Oxford, UK.

Wong, R.O.L. (1995) Use, disuse, and growth of the brain, *Proc. Nat. Acad. Sci. USA* **92**(6): 1797–1799.

Woodford, B., Kasabov, N., and Wearing, H. (1999) Fruit image analysis using wavelets, in emerging knowledge engineering and connectionist-based systems. In: N. Kasabov and K. Ko (Eds.), *Proceedings of the Iconip/Anziis/Annes'99 Workshop Future Directions For Intelligent Systems And Information Sciences*, Dunedin, 22–23 Nov., pp. 88–92.

Woolsey, C.M. (1982) *Cortical Sensory Organization, Multiple Auditory Areas*, Vol. 3, Humana Press, Totowa, NJ.

Worner, S.P. (1988) Ecoclimatic assessment of potential establishment of exotic pests, *J. Econ. Entomol.* **81**: 973–83.

Worner, S.P. (2002) Predicting the invasive potential of exotic insects. In: G. Halman (Ed.), *Invasive Arthropods and Agriculture: Problems and Solutions.* Science, New Hampshire.

Wu, C.H. and McLarty, J.W. (2000) *Neural Networks and Genome Informatics*, Elsevier, The Hague.

Wu, X. (1992) Colour quantization by dynamic programming and principal analysis, *ACM Trans. Graph.* **11**: 348–372.

Wysoski, S., Benuskova, L., and Kasabov, N. (2006) Online learning with structural adaptation in a network of spiking neurons for visual pattern recognition. In: *Proceedings of ICANN 2006*, LNCS, Springer, New York.

Xie, G. and Zhuang, Z. (2003) A quantum competitive learning algorithm, *Liangzi Dianzi Xuebao/Chinese J. Quantum Electron. (China)*, **20**: 42–46.

Yager, R.R. and Filev, D. (1994) Generation of fuzzy rules by mountain clustering, *J. Intell. Fuzzy Syst.* **2**: 209–219.

Yamakawa, T. and Tomoda, S. (1989) A fuzzy neuron and its application to pattern recognition. In: J. Bezdek, Ed., *Proceedings of the Third IFSA Congress*, pp. 1–9.

Yamakawa, T., Kusanagi, H., Uchino, E., and Miki, T. (1993) A new effective algorithm for neo fuzzy neuron model. In: *Proceedings of Fifth IFSA World Congress*, pp. 1017–1020.

Yamakawa, T., Uchino, E., Miki, T., and Kusanagi, H. (1992) A neo fuzzy neuron and its application to system identification and prediction of the system behaviour. In: *Proceedings of the Second International Conference on Fuzzy Logic & Neural Networks*, Iizuka, Japan, pp. 477–483.

Yamauchi, K. and Hayami, J. (2006) Sleep learning - An incremental learning system inspired by sleep behavior, In: *Proceedings of IEEE International Conference on Fuzzy Systems*, Vancouver, July 16–21, IEEE Press, Piscataway, NJ, pp. 6295–6302.

Yao, X. (1993) Evolutionary artificial neural networks, *Int. J. Neural Syst.* **4**(3): 203–222.

Yao, X. (1996) Promises and challenges of evolvable hardware. In: *Proceedings of the First International Conference on Evolvable Systems - From Biology to Hardware*, Tsukuba, Japan, 7–8 October.

Zacks, R. (2001) Biology in silico, *MIT Technol. Rev.* (March): 37.

Zadeh, L. (1965) Fuzzy sets. *Inf. Contr.* **8**: 338–353.

Zadeh, L.A. (1988) Fuzzy logic. *IEEE Comput.* **21**: 83–93.

Zanchettin, C. and Ludermir, T.B. (2004) Evolving fuzzy neural networks applied to odor recognition in an artificial nose. In: *Proceedings of IEEE International Joint Conference on Neural Networks*, Budapest, July 26–29, IEEE Press, Washington, DC.

Zhang, D., Ghobakhlou, A., and Kasabov, N. (2004) An adaptive model of person identification combining speech and image information. In: *International Conference on Control, Automation, Robotics and Vision*, Kumming, China.

Zhou, X. and Angelov, P. (2006) Real-time joint landmark recognition and classifier generation by an evolving fuzzy system. In: *Proceedings of the IEEE International Conference on Fuzzy Systems*, Vancouver, July 16–21, IEEE Press, Washington, DC, pp. 6314–6321.

Zigmond, M.J., Bloom, F.E., Landis, S.C., Roberts, J.L., and Squire, L.R. (1999) *Fundamental Neuroscience*, Academic Press, San Diego, Chapter 25.

ZISC Manual (2001) *Zero Instruction Set Computer*, Silicon Recognition, Inc., California.

Zurada, J. (1992) *Introduction to Artificial Neural Systems*, West Puse. Comp., Singapore.

Zwicker, E. (1961) Subdivision of the audible frequency range into critical bands, *J. Acoust. Soc. Am.* **33**: 248.

Extended Glossary

Adaptation. The process of structural and functional changes of a system in order to improve its performance in a changing environment.

Alan Turing's test for AI. Definition for AI introduced by the British mathematician and computer scientist Alan Turing. It states approximately that a machine system is considered to possess artificial intelligence (AI) if while communicating with a person behind a 'bar', the person cannot recognise whether it is a machine or a human behind the bar.

Apparent error (training error). The error calculated on the basis of the reaction of a neural network to the data used for its training. It is usually calculated as a mean square error.

Approximate reasoning. A process of inferring new facts and achieving conclusions with the use of inexact facts and uncertain rules.

ART. Adaptive resonance theory. A neural network invented and developed by Carpenter and Grossberg.

Artificial life. A modelling paradigm that assumes that many individuals are governed by the same or similar rules to grow, die, and communicate with each other. Ensembles of such individuals exhibit repetitive patters of behaviour.

Artificial neural network. Biologically inspired computational model which consists of processing elements (called neurons) and connections between them with coefficients (weights) bound to the connections, which constitute the neuronal structure. To the structure are also attached training and recall algorithms.

Atom. In chemistry and physics, an atom is the smallest possible particle of a chemical element that retains its chemical properties. Most atoms are composed of three types of massive subatomic particles which govern their external properties: electrons, which have a negative charge and are the least massive of the three; protons, which have a positive charge and are about 1836 times more massive than electrons; and neutrons, which have no charge and are about 1838 times more massive than electrons. Protons and neutrons are both nucleons and make up the dense, massive atomic nucleus. (Adapted from http://en.wikipedia.org/wiki/.)

Automatic speech recognition system (ASRS). A computer system which aims at providing enhanced access to machines via voice commands.

Backpropagation training algorithm. An algorithm for adjusting the connection weights in a neural network (NN) where the gradient descent rule is used for finding the optimal connection weights w_{ij} which minimise a global error E. A change of a weight Δw_{ij} at a cycle $(t+1)$ is in the direction of the negative gradient of the error E.

Bayesian probability. The following formula, which represents the conditional probability between two events C and A, is known as the Bayes Formula (Thomas Bayes, 18th century): $p(A|C) = p(C|A).p(A)/p(C)$. Using the Bayes formula involves difficulties, mainly concerning the evaluation of the prior probabilities $p(A)$, $p(C)$, $p(C|A)$. In practice (e.g. in statistical pattern recognition), the latter is assumed to be of a Gaussian type. The Bayes theorem assumes that if the condition C consists of condition elements $C1, C2, \ldots, Ck$ they are independent (which may not be the case in some applications).

Catastrophic forgetting. Phenomenon which represents the ability of a network to forget what it has learned from previous examples when they are no longer presented to it but other examples are presented instead.

Cellular automata. A set of regularly connected simple finite automata. The simple automata communicate and compute together when solving a single global task. Cellular automata may be able to grow, to shrink, and to reproduce thus providing a flexible environment for computation with arbitrary complexity. They are also called 'non-von Neumann' models because of their difference from the standard digital von Neumann computer organisation. (This is in spite of the fact that von Neumann was one of the originators of the mathematical theory of self-reproducing automata.)

Centre-of-gravity defuzzification method (COG). Method for defuzzification, e.g. transforming a membership function B of an output fuzzy variable in a fuzzy system into a crisp value y such that y is the geometrical centre of the area occupied by B. The following formula is used: $y = \Sigma \mu_B(v).v / \Sigma \mu_B(v)$.

Chaos. A complicated behaviour of a nonlinear dynamical system according to some underlying rules.

Chaotic attractor. An area or points from the phase space of a chaotic process where the process goes often through time, but without repeating the same trajectory.

Classification problem. A generic AI problem which arises when it is necessary to associate an object with some already existing groups, clusters, or classes of objects.

Clustering. Based on a measured distance between instances (objects, points, vectors) from the problem space, subareas in the problem space of closely grouped instances can be defined. These areas are called clusters. They are defined by their cluster centres and the membership of the data points to them. A centre c_i of a cluster C_i is defined as an instance of the mean of the distances to which from each instance in the cluster is minimum. Let us have a set X of p data items represented in an n-dimensional space. A clustering procedure results in defining k disjoint subsets (clusters), such that every data item (n-dimensional vector) belongs to only one cluster. A cluster membership function

M_i is defined for each of the clusters $C_1, C_2, \ldots, C_k$: $M_i : X\text{-}> \{0,1\}, M_i(x) = 1$, if $x \in Ci$, 0, otherwise, where x is a data instance (vector) from X. In fuzzy clustering one data vector may belong to several clusters to a certain degree of membership, all of the degrees summing up to 1.

Computational Intelligence (CI). This encompasses methods for information processing based on learning, reasoning, dealing with incomplete and uncertain data, and their numerous applications in almost all areas of science, engineering, and human activities. These methods include probabilistic methods, neural networks, rule-based and fuzzy systems, evolutionary computation, and hybrid systems. Many methods of CI are inspired by human intelligence and aim at modelling brain data and brain functions along with other biological data.

Conditional probabilities. The probability $p(A|C)$ defines the probability of the event A to occur, given that the event C has occurred. It is given by the formula: $p(A|C) = p(A \wedge C)/p(C)$.

Connectionist production system. A connectionist system that implements production rules of the form IF C THEN A, where C is a set of conditions and A is a set of actions.

Control. Process of acquiring information for the current state of an object and emitting control signals to it in order to keep the object in its possible and desired states.

Data analysis. Data analysis aims at answering important questions about the process under investigation. Some exemplar questions are: what are the statistical parameters of the data available for the process, e.g. mean, standard deviation, distribution. What is the nature of the process: random, chaotic, periodic, stable, etc.? How are the available data distributed in the problem space, e.g. clustered into groups, sparse, covering only patches of the problem space and therefore not enough to rely on fully when solving the problem, uniformly distributed? Are there missing data and how many? Is there a critical obstacle which could make the process of solving the problem by using data impossible? What other methods can be used either in addition to, or in substitution for methods based on data?

Data, information, and knowledge. *Data* are the numbers, the characters, and the quantities operated on by a computer. *Information* is the ordered, structured, and interpreted data. *Knowledge* is the theoretical or practical understanding of a subject, gained experience, or true and justified belief, the way we do things.

Decision support system. This is an intelligent system that analyses variants and suggests decisions, e.g. automated trading systems on the Internet; systems that grant loans through electronic submissions; medical decision support systems for cardiovascular event prediction.

Defuzzification. Process of calculating a single output numerical value for a fuzzy output variable on the basis of the inferred resulting membership function for this variable (see **Centre-of-gravity defuzzification**) .

Destructive learning. A learning technique in neural networks that destroys the initial neural network architecture, e.g. removes connections, for the purpose of better learning.

Diagnosis. Process of finding faults in a system.

Discrete Fourier Transform (DFT). DFT transforms a vector X of N numbers taken from a signal (or time-series data) from the time domain into a vector F of N numbers in the frequency domain, i.e. finds the energy of the signal for certain N frequencies.

Distance between data points. A way of measuring difference between data vectors. The distance between two data points in an n-dimensional geometrical space can be measured in several ways, e.g. Hamming, $D_{ab} = \Sigma|a_i - b_i|$; Euclidean distance, $E_{ab} = sqrt(\Sigma(a_i - b_i)^2/n)$.

Distributed representation. A way of encoding information in a neural network where a concept or a value for a variable is represented by a collective activation of a group of neurons.

DNA information. Each cell of a living organism contains a significant amount of genetic information stored in the DNA molecules that are located in the nucleus of the cell. DNA is built of four types of small molecules called bases, and denoted A, C, G, and T. It is expected that the complete human genome will have been determined and it will contain about three billion bases (Human Genome Program, USA,http://www.ornl.gov/hgmis/publicat/primer/intro.html).

Dynamic system. A system which evolves in a continuous or in a discrete time.

Elitism (in genetic algorithms (GA)). The single most fit member of each generation is copied unmodified into the next generation. The intention of this strategy is to reduce the chance of losing the best genotypes, as may happen in a stochastic process such as GA's.

Evolutionary computation (evolutionary algorithms). This is a computational paradigm that uses principles from biological evolution, such as genetic representation, mutation, survival of the fittest, population of individuals, or generations of populations.

Evolutionary programming. Evolutionary algorithms applied to automatic creation or optimisation of computer programs.

Evolutionary strategies. Evolutionary algorithms that represent a solution to a problem as a single chromosome and evaluate different mutations of this solution through a fitness function, until a satisfactory solution is found.

Evolving intelligent systems (EIS). The book covers methods that facilitate the design of intelligent systems characterised by adaptation and incremental evolving of knowledge. Such systems are also called evolving intelligent systems (EIS). The methods are mainly based on neural networks, but include many other techniques from the area of CI.

Expert system. A program which can provide expertise for solving problems in a defined application area in the way the experts do. Expert systems are knowledge-based systems that provide expertise, similar to that of experts in a restricted application area. An expert system consists of the following main blocks: knowledge base, database, inference engine, explanation module, user interface, and knowledge acquisition module.

Explanation in an intelligent system. This is a desirable property for many AI systems. It means tracing, in a contextually comprehensible way, the process of inferring the solution, and reporting it. Explanation is easier for the AI symbolic systems when sequential inference takes place. But it is difficult for parallel methods of inference and especially difficult for the massive parallel ones.

Fast Fourier transformation (FFT). A nonlinear transformation applied on (mainly speech) data to transform the signal taken within a small portion of time from the time scale domain into a vector in the frequency scale domain. It is a fast version of the discrete Fourier transformation.

Feedforward neural network. A neural network in which there are no connections back from the output to the input neurons.

Finite Automaton. A computational model represented by a set X of inputs, a set Y of outputs, a set Q of internal states, and two functions f_1 and $f_2 : f_1 : X \times Q \text{->} Q$, i.e. $(x, q(t)) \text{->} q(t+1)$; f_2: $X \times Q \text{->} Y$, i.e. $(x, q(t)) \text{->} y(t+1)$; where: $x \in X$, $q \in Q$, $y \in Y$; t and $(t+1)$ represent two consecutive time moments.

Fitness. See **Goodness.**

Forecasting. See **Prediction.**

Fractals. Objects which occupy fractions of a standard (integer number for dimensions) space called embedding space.

Fuzzification. Process of finding the membership degree $\mu A(x')$ to which input value x' for a fuzzy variable x, defined on a universe U, belongs to a fuzzy set A defined on the same universe.

Fuzzy ARTMAP. Extension of ART1 when input nodes represent not 'yes/no' features, but membership degrees, to which the input data belong, for example a set of features {sweet, fruity, smooth, sharp, sour}used to categorise different samples of wines based on their taste.

Fuzzy clustering. A procedure of clustering data into possibly overlapping clusters, such that each of the data examples may belong to each of the clusters to a certain degree. The procedure aims at finding the cluster centres $Vi(i = 1, 2, \ldots, c)$ and the cluster membership functions μ_i which define to what degree each of the n examples belongs to the ith cluster. The number of clusters c is either defined a priori (supervised type of clustering), or chosen by the clustering procedure (unsupervised type of clustering). The result of a clustering procedure can be represented as a fuzzy relation $\mu_{i,k}$, such that: **(i)** $\Sigma\mu_{i,k} = 1$, for each $k = 1, 2, \ldots, n$ (the total membership of an instance to all the clusters equals 1); **(ii)** $\Sigma\mu_{i,k} > 0$, for each $i = 1, 2, \ldots, c$ (there are no empty clusters).

Fuzzy control. Application of fuzzy logic to control problems. A fuzzy control system is a fuzzy system applied to solve a control problem.

Fuzzy expert system. An expert system to which methods of fuzzy logic are applied. Fuzzy expert systems use fuzzy data, fuzzy rules, and fuzzy inference in addition to the standard ones implemented in ordinary expert systems.

Fuzzy logic. A logic system that is based on fuzzy relations and fuzzy propositions, the latter being defined on the basis of fuzzy sets.

Fuzzy neural network. A neural network that can be interpreted as a fuzzy system.

Fuzzy propositions. Propositions which contain fuzzy variables with their fuzzy values. The truth value of a fuzzy proposition 'X is A' is given by the membership function μ_A.

Fuzzy relations. Fuzzy relations link two fuzzy sets in a predefined manner. Fuzzy relations make it possible to represent ambiguous relationships such as 'the grades of the third and second year classes are similar', or 'team A performed slightly better than team B', or 'the more you eat fat, the higher the risk of heart attack'.

Generalisation. Process of matching new, unknown input data to the problem knowledge in order to obtain the best possible solution, or close to it.

Genetic algorithms (GA). Algorithms for solving complex combinatorial and organisational problems with many variants, by employing analogy with nature's evolution. There are three general steps a genetic algorithm cycles through: generate a population (cross-over); select the best individuals; mutate, if necessary; repeat the same.

Goodness functions (also fitness function). A function that can be used to measure the appropriateness of a prospective decision when solving a problem.

Hebbian learning law. Generic learning principle which states that a synapse, connecting two neurons i and j, increases its strength w_{ij} if repeatedly the two neurons i and j are simultaneously activated by input stimuli.

Homophones. Words with different spellings and meanings but sound the same, for example 'to, too, two' or 'hear, here'.

Hopfield network. Fully connected feedback network which is an autoassociative memory. It is named after its inventor John Hopfield (1982).

Image filtering. A transformation of an original image through a set of operations that alter the original pixel intensities of the image by applying a two-dimensional array of numbers, which is known as a kernel. This kernel is then passed over the image using a mathematical process called convolution.

Independent component analysis. Given a dataset (or a signal) which is a mixture of unknown independent components, the goal is to separate these components.

Inference in an AI system. The process of matching current data from the domain space to the existing knowledge and inferring new facts until a solution in the solution space is reached.

Information. Collection of structured data. In its broad meaning it includes knowledge as well as simple meaningful data.

Information entropy. Let us have a random variable X that can take N random values $x1, x2, \ldots, xN$. The probability of each value xi to occur is pi and the variable X can be in exactly one of these states, therefore $\sum_{i=1,\ldots,N} pi = 1$. The question is, 'What is the uncertainty associated with X?' This question only has a precise answer if it is specified who asked the question and how much this person knows about the variable X. It depends on both expectations and the reality. If we associate a measure of uncertainty $h(xi)$ to each random value xi which means how uncertain the observer is about this value occurring, then the total uncertainty $H(X)$, called entropy, measures our lack of knowledge, the seeming disorder in the space of the variable X: $H(X) = \sum_{i=1,\ldots,N} pi \cdot h(xi)$.

Information retrieval. The process of retrieving relevant information from a database.

Information science. This is the area of science that develops methods and systems for information and knowledge processing regardless of the domain specificity of this information. Information science incorporates the following subject areas: data collection and data communication (sensors and networking); information storage and retrieval (database systems); methods for information processing (information theory); creating computer programs and information systems (software engineering and system development); acquiring, representing, and processing knowledge (knowledge-based systems); and creating intelligent systems and machines (artificial intelligence).

Initialisation. The process of setting the connection weights in a neural network to some initial values before starting the training algorithm.

Instinct for information. A speculative term introduced in Chapter 7 that expresses human constant striving for information and knowledge, their active search for information in any environment in which they live.

Intelligent system (IS). An information system that manifests features of intelligence, such as learning, generalisation, reasoning, adaptation, or knowledge discovery, and applies these to complex tasks such as decision making, adaptive control, pattern recognition, speech, image and multimodal information processing, etc.

Interaction (human–computer). Communication between a computer system and the environment, or the user on the other hand, in order to solve a given problem.

Ion. An ion is an atom or group of atoms with a net electric charge. A negatively charged ion, which has more electrons in its electron shell than it has protons in its nucleus, is known as an anion, for it is attracted to anodes; a positively charged ion, which has fewer electrons than protons, is known as a cation as it is attracted to cathodes (adapted from http://en.wikipedia.org/wiki/).

Knowledge. Concise presentation of previous experience, the essence of things, the way we do things, the know-how.

Knowledge engineering. The area of science and engineering that deals with knowledge representation in machines, knowledge elucidation, and knowledge discovery through computation.

Knowledge-based neural networks (KBNN). These are prestructured neural networks to allow for data and knowledge manipulation, including learning from data, rule insertion, rule extraction, adaptation, and reasoning. KBNN have been developed either as a combination of symbolic AI systems and NN, or as a combination of fuzzy logic systems and NN, or as other hybrid systems. Rule insertion and rule extraction operations are typical operations for a KBNN to accommodate existing knowledge along with data, and to produce an explanation of what the system has learned.

Kohonen self-organising map (SOM). A self-organised map neural network for unsupervised learning invented by Professor Teuvo Kohonen and developed by him and other researchers.

Laws of inference in fuzzy logic. The way fuzzy propositions are used to make inferences over new facts. The following are the two most used laws illustrated on two fuzzy propositions A and B: (a) generalised modus ponens: A->B, and A′ ∴ B′, where B′ = A′o (A ->B); (b) generalised modus tolens (law of the contrapositive): A ->B, and B′, ∴ A′, where A′ = (A->B) o B′.

Learning. Process of obtaining new knowledge.

Learning vector quantisation algorithm (LVQ). A supervised learning algorithm, which is an extension of the Kohonen self-organised network learning algorithm.

Linear transformation. Transformation $f(x)$ of a raw data vector x such that f is a linear function of x; for example: $f(x) = 2x + 1$.

Linguistic variable. A variable that takes fuzzy values.

Local representation in a neural network. A way of encoding information in a neural network in which every neuron represents one concept or one variable.

Logic systems. An abstract system that consists of four parts: an alphabet, a set of basic symbols from which more complex sentences (constructions) are made; syntax, a set of rules or operators for constructing sentences (expressions) or alternatively more complex structures from the alphabet elements. These structures are syntactically correct 'sentences'; semantics, to define the meaning of the constructions in the logic system; and laws of inference, a set of rules or laws for constructing semantically equivalent but syntactically different sentences. This set of laws is also called a set of inference rules.

Logistic function. The function described by the formula: $a = 1/(1+e^{-u})$, where e is a constant, the base of natural logarithms (e, sometimes written as exp, is actually the limit of the n-square of (1 + 1/n) when n approaches infinity). In a more general form, the logistic function can be written as $a = 1/(1+e^{-c.u})$, where c is a constant. The reason why the logistic function has been used as a neuronal activation function is that many algorithms for performing learning in neural networks use the derivative of the activation function, and the logistic function has a simple derivative; i.e. $\partial g/\partial u = a\ (1-a)$.

Machine learning. Computer methods for accumulating, changing, and updating knowledge in a computer system.

Mackey–Glass chaotic time series. A benchmark time series generated from the following delay differential equation: $dx(t)/dt = [0.2 \times (t-D)]/[1+x^{10}(t-D)] - 0.1x(t)$, where D is a delay, for $D > 17$ the functions show chaotic behaviour.

Main dogma in genetics. A hypothesis that cells perform the following cycle of transformations: DNA ->RNA ->proteins.

Mel-scale filter bank transformations. The process of filtering a signal through a set of frequency bands represented by triangular filter functions similar to the functions used by the human inner ear.

Membership function. A generalised characteristic function which defines the degree to which an object from a universe belongs to a fuzzy concept, such as 'small'.

Memory capacity of a neural network. Maximum number m of the patterns which can be learned properly in a network.

Methods for feature extraction. Methods used for transforming raw data from the original space into another space, a space of features.

Modular system. A system which consists of several modules linked together for solving a given problem.

Molecule. In general, a molecule is the smallest particle of a pure chemical substance that still retains its composition and chemical properties. In chemistry and molecular sciences, a molecule is a sufficiently stable, electrically neutral entity composed of two or more atoms.

Monitoring. Process of interpretation of continuous input information, and recommending intervention if appropriate.

Moving averages. A moving average of a time series is calculated by using the formula: $MA_t = (\Sigma S_{t-i})/n$, for $I = 1, 2, \ldots, n$, where n is the number of the data points, s_{t-i} is the value of the series at a time moment $(t-i)$, and MA_t is the moving average of time moment t. Moving averages are often used as input features in an information system in addition to, or in substitution for, the real values of a time series.

Multilayer perceptron network (MLP). A neural network (NN) that consists of an input layer, at least one intermediate or 'hidden' layer, and one output layer, the neurons from each layer being fully connected (in some particular applications, partially connected) to the neurons from the next layer.

Mutation. A random change in the value of a gene (either in a living organism or in an evolutionary algorithm).

Neural networks (NN). SeeArtificial neural networks.

Noise. A small random ingredient that is added to the general function which describes the underlying behaviour of a process.

Nonlinear dynamical system. A system the next state of which on the time scale can be expressed by a nonlinear function from its previous time states.

Nonlinear transformation. Transformation f of a raw data vector $\boldsymbol{x}$ where f is a nonlinear function of $\boldsymbol{x}$; for example $f(x) = 1/(1+e^{-x.c})$, where c is a constant.

Normalisation. Transforming data from its original scale into another, predefined scale, e.g. [0, 1].

Normalisation – linear. Normalisation with the use of the following formula (for the case of a targeted scale of [0,1]): $v_{norm} = (v - x_{\min}) / (x_{\max} - x_{\min})$, where v is a current value of the variable x; $x_{\min}$ is the minimum value for this variable, and $x_{\max}$ is the maximum value for that variable x in the dataset.

Nyquist sampling frequency. Half of the sampling frequency of a signal; it is the highest frequency in the signal preserved through the sampling (e.g. *Sfreq.*= 22,050 Hz; Nfreq = 10,025 Hz).

Optimisation. Finding optimal values for parameters of an object or a system which minimise an objective (cost) function.

Overfitting. Phenomenon which indicates that a neural network has approximated, or learned, a set of data examples too closely, which may contain noise in them, so that the network cannot generalise well on new examples.

Pattern matching. The process of matching a feature vector to already existing ones and finding the best match.

Phase space of a chaotic process. The feature space where the process is traced over time.

Phonemes. Linguistic abstract elements which define the smallest speech patterns that have linguistic representation in a language.

Photon. In physics, the photon is the quantum of the electromagnetic field, for instance light. The term photon was coined by Gilbert Lewis in 1926. The photon can be perceived as a wave or a particle, depending on how it is measured. The photon is one of the elementary particles. Its interactions with electrons and atomic nuclei account for a great many of the features of matter, such as the existence and stability of atoms, molecules, and solids (adapted from http://en.wikipedia.org/wiki/).

Planning. An important generic AI-problem which is about generating a sequence of actions in order to achieve a given goal when a description of the current situation is available.

Power set of a fuzzy set A. Set of all fuzzy subsets of a fuzzy set A.

Prediction. Generating information for the possible future development of a process from data about its past and its present development.

Principal component analysis (PCA). Finding a smaller number of m components $Y = (y1, y2, \ldots, ym)$ (aggregated variables) that can represent the goal function $F(x1, x2, \ldots, xn)$ of n variables, $n > m$ to a desired degree of accuracy Θ; i.e. $F = M.Y + \Theta$, where M is a matrix that has to be found through the PCA.

Probability automata. Probability (or stochastic) automata are finite automata the transitions of which are defined as probabilities.

Probability theory. The theory is based on the following three axioms. Axiom 1. $O <= p(E) <= 1$. The axiom defines the probability $p(E)$ of an event E as a real number in the closed interval [0,1]. A probability $p(E) = 1$ indicates a certain event, and $p(E) = 0$ indicates an impossible event. Axiom 2. $\Sigma p(E_i) = 1, E_1 \cup E_2 \cup \ldots \cup E_k = U$, U, problem space (universum); Axiom 3. $p(E_1 \vee E_2) = p(E_1) + p(E_2)$, where E_1 and E_2 are mutually exclusive events. This axiom indicates that if the events E_1 and E_2 cannot occur simultaneously, the probability of one or the other happening is the sum of their probabilities.

Production system. A computer system consisting of three main parts: (a) a list of facts, considered a working memory (the facts being called 'working memory elements'); (b) a set of production rules, considered the production memory; and (c) an inference engine which is a reasoning procedure, the control mechanism.

Productions. Transformation rules which are applied for obtaining one sequence of characters from another.

Propositional logic. A logic system that can be dated back to Aristotle (384–322 B.C.). There are three types of symbols in propositional logic: propositional symbols (the alphabet), connective symbols, and symbols denoting the meaning of the sentences. There are rules in propositional logic to construct syntactically correct sentences (called well-formed formulas) and rules to evaluate the semantics of the sentences. A proposition represents a statement about the world, for example, 'The temperature is over 120.' The semantic meaning of a propositional symbol is expressed by two possible semantic symbols: true and false. Statements or propositions can be only 'true' or 'untrue' (false), nothing in between.

Pruning in artificial neural networks. This is a technique that is based on gradual removing from the network the weak connections (which have weights around 0) and the neurons which are connected by them during the training procedure.

Recall process. The process of using a trained neural network when new data are fed and results are calculated.

Recurrent fuzzy rule. A fuzzy rule which uses in its antecedent part one or more previous time-moment values of the output fuzzy variable.

Recurrent networks. Neural networks with feedback connections from neurons in one layer to neurons in a previous layer.

Reinforcement learning (or also reward-penalty learning) A neural network training method that is based on presenting input vector $\boldsymbol{x}$ and looking at the output vector calculated by the network; if it is evaluated as 'good', then a 'reward' is given to the network in the sense that the existing connection weights get increased, otherwise the network is 'punished': the connection weights, being considered 'not appropriately set', decrease.

Representation (in information science). A process of transforming existing problem knowledge to some of the known knowledge engineering schemes in order to process it in a computer program through the application of knowledge engineering methods.

Roulette wheel selection (in genetic algorithms (GA)). A selection strategy according to which each individual is assigned a slot in an imaginary roulette wheel, with the size of the slot dependent upon the fitness of the individual.

Therefore, the more fit the individual, the higher the chance is of being selected to breed.

Sampling. A process of selecting a subset of the data available. Sampling can be applied on continuous time-series data; for example speech data is sampled at a frequency of 22 KHz say, or on static data, a subset of the dataset is taken for processing purposes.

Sensitivity to initial conditions. A characteristic of a chaotic process which practically means that a slight difference in the initial values of some parameters that characterise the chaotic process will result in quite different trends in its future development.

Signal processing. Transforming a signal taken within a small portion of time into an n−dimensional vector, where n is the number of features used.

Sources of information. There are many sources of information in the world today. The 'macroworld' of information contains many different types of information, e.g. health and medical information, business, financial and economic information, geographic information, information about the universe, etc. The 'microworld' of information includes information about the human brain and the nervous system, genetic and molecular information, and quantum information.

Spatial-temporal artificial neural networks. Artificial neural networks that represent patterns of activities which have some spatial distribution and appear at certain times.

Stability/plasticity dilemma. Ability of a system to preserve the balance between retaining previously learned patterns and learning new patterns.

Statistical analysis methods. Methods used for discovering the repetitiveness in data based on probability estimation.

Supervised learning in ANN. A process of approximating a set of 'labelled' data; i.e. each data item (which is a data point in the input–output problem space) contains values for attributes (features), independent variables, labelled by the desired value(s) for the dependent variables. Supervised learning can be viewed as approximating a mapping between a domain and a solution space of a problem: $X \text{->} Y$, when samples (examples) of (input vector–output vector) pairs (x, y) are known; $x \in X, y \in Y, x = (x_1, x_2, \ldots, x_n), y = (y_1, y_2, \ldots, y_m)$.

Supervised training algorithm. Training of a neural network when the training examples comprise input vectors **x** and the desired output vectors **y**; training is performed until the neural network 'learns' to associate each input vector **x** to its corresponding and desired output vector **y**.

Test error. An error which is calculated when, after having trained a network with a set of training data, another set (test, validation, cross-validation), for which the results are also known, is applied for a recall procedure.

Time alignment. A process where a sequence of vectors recognised over time is aligned to represent a meaningful linguistic unit (phoneme, word).

Time-series prediction. Prediction of time series events.

Tournament selection (in genetic algorithms GA). A selection strategy when two individuals are selected using roulette wheel selection and their fitness is compared, the individual with the highest fitness being inserted into the breeding population.

Training error. See **Apparent error.**

Training of a neural network. A procedure of presenting training examples to a neural network and changing the network's connection weights according to a certain learning law.

Tree. Directed graph in which one of the nodes, called a *root*, has no incoming arcs, but from which each node in the tree can be reached by exactly one path.

Type-2 fuzzy inference system. A fuzzy rule-based system that uses type-2 fuzzy rules.

Type-2 fuzzy membership function. A fuzzy membership function to which elements belong with a membership degree that is represented not by a single number, but by an interval of min–max membership degrees

Type-2 fuzzy set. A fuzzy set to which elements belong with a membership degree that is represented not by a single number but by an interval of min–max membership degrees.

Universal function approximator (for NN). A theorem was proved by Hornik et al. (1989), Cybenko (1989), and Funahashi (1989) that a MLP with one hidden layer can approximate any continuous function to any desired accuracy, subject to a sufficient number of hidden nodes. The theorem proves the existence of such a MLP. As a corollary, any Boolean function of n Boolean variables can be approximated by a MLP. An easy proof can be shown by using 2^n hidden nodes, but the optimum number for these nodes is difficult to obtain.

Unsupervised learning algorithm. A learning procedure when only input vectors x are supplied to a neural network; the network learns some internal characteristics, e.g. clusters, of the whole set of all the input vectors presented to it.

Validation. Process of testing how good the solutions produced by a system are. The solutions are usually compared with the results obtained either by experts, or by other systems.

Validation error. See **Test error.**

Vector quantisation. A process of representing an n-dimensional problem space as an m-dimensional one, where $m < n$, in a way that preserves the similarity between the data examples (points in each of these two spaces).

Vigilance. Parameter in the ART network which controls the degree of mismatch between the new patterns and the learned (stored) patterns which the system can tolerate.

Wavelet transformation. A nonlinear transformation that can be used to represent slight changes of the signal within the chosen window from the time scale (for the FFT it was assumed that the signal does not change or at least does not change significantly within a window). Here, within the window, several transformations are taken from the raw signal by applying wavelet basis functions of the form $W_{a,b}(x) = f(ax - b)$, where f is a nonlinear function, a is a scaling parameter, and b is a shifting parameter (varies between 0 and a). Thus, instead of one transformation, several transformations are applied by using wavelet basis functions $Wa, 0, \ldots, Wa, 1, \ldots, Wa, 2, \ldots, Wa, a$. An example of such a set of functions is $f(x) = \cos(\pi x)$, for $-0.5 <= x <= 0.5$, and $f(x) = 0$, otherwise. Wavelet transformations preserve time variations of a signal within a certain time interval.

Index